A-Level

Geography

Exam Board: Edexcel

From lava flows to NGOs, Edexcel A-Level Geography
takes you on one heck of an intellectual journey.

But don't worry — with this fantastic book, you'll be 100% prepared for anything
the exams can throw at you. It's packed with concise notes for every topic,
incredibly helpful case studies and plenty of practice exam questions.

So buckle up, keep your arms and legs inside the vehicle at all times, and get
ready to discover everything you need to know for A-Level Geography.

How to access your free Online Edition

This book includes a free Online Edition to read on your PC, Mac or tablet.
You'll just need to go to **cgpbooks.co.uk/extras** and enter this code:

3377 2825 9867 3564

By the way, this code only works for one person. If somebody else has used
this book before you, they might have already claimed the Online Edition.

A-Level revision? It has to be CGP!

Contents

Topic 5 — The Water Cycle and Water Insecurity

Topic 6 — The Carbon Cycle and Energy Security

Topic 7 — Superpowers

Topic 8: Option 8A — Health, Human Rights and Intervention

Topic 8: Option 8B — Migration, Identity and Sovereignty

Independent Investigation & Fieldwork

Exam Skills

Exam Practice

Published by CGP

Based on the classic CGP style created by Richard Parsons.

Written by:
Katharine Howell and Chloë Searl

Editors:
Claire Boulter, Chris Lindle, Georgina Paxman and Adam Worster

Reviewers:
Mercy Dennis-Smith, Paul Logue and Claire Priddle

Contributors:
Paddy Gannon and Brenda Turnbull

ISBN: 978 1 78908 569 3

With thanks to Tom Carney, Becca Clifford and Amy Turner for the proofreading.
With thanks to Alice Dent and Laura Jakubowski for the copyright research.

Clipart from Corel®
Printed by Elanders Ltd, Newcastle upon Tyne.

Specification Map

Here's some guidance on how the A-Level Geography specification is structured. It's super useful stuff, so read on...

You're **either** studying **AS or A-Level**

The course you are studying is either the:

- Edexcel Level 3 Advanced Subsidiary GCE in Geography (if you're studying the **AS** course). It has the course code **8GE0**.
- Edexcel Level 3 Advanced GCE in Geography (if you're studying the **A-Level** course). It has the course code **9GE0**.

There are **eight** different topics to study

AS-Level students will study **four** of these topics, and A-Level students will study all **eight**. The topics are:

Topic	AS-Level	A-Level
1 — Tectonic Processes and Hazards	✓	✓
2 — **EITHER** 2A: Glaciated Landscapes and Change **OR** 2B: Coastal Landscapes and Change	✓	✓
3 — Globalisation	✓	✓
4 — **EITHER** 4A: Regenerating Places **OR** 4B: Diverse Places	✓	✓
5 — The Water Cycle and Water Insecurity		✓
6 — The Carbon Cycle and Energy Security		✓
7 — Superpowers		✓
8 — **EITHER** 8A: Health, Human Rights and Intervention **OR** 8B: Migration, Identity and Sovereignty		✓

Case studies support what you've learnt

Case studies are an important part of Geography. It's essential not just to know about how Geography works in **theory**, but how it works in a particular **place**. The following case studies are recommended and all feature in this book:

Case study	Page
Topic 1 — Tectonic Processes and Hazards	
2004 Indian Ocean tsunami	26
2010 eruption of Eyjafjallajökull, Iceland	26
2011 Tohoku tsunami, Japan	26
A multiple-hazard zone — the Philippines	27
Topic 2A — Glaciated Landscapes and Change	
Glacial mass balance — Greenland ice sheet	37
The tundra environment of Northern Canada	53
Threats to Alpine valleys	59
Changes in the Himalayan glaciers	60
Managing Yosemite Valley — protective strategies and sustainable management	62
Topic 2B — Coastal Landscapes and Change	
Portland Bill to Selsey Bill on the south coast of England	66
Development of coastal landscapes — Glamorgan Heritage Coast	68
Human activities in the Nile Delta	74
Flood risk in Bangladesh	81
Coastal management at Happisburgh, England	87
Coastal management in Chittagong, Bangladesh	87

Case study	Page
Topic 3 — Globalisation	
China's 1978 'Open Door Policy' welcomed FDI	91
North Korea remains largely 'switched off' from globalisation	92
Global shift of manufacturing — China	94
Outsourcing of services — India	95
Growth of megacities — Mumbai	98
International migration — Russian oligarchs to London	98
International migration — India to Qatar	99
Asia's changing diet	101
2016 Paralympic Games in Rio de Janeiro	101
Social relations — Papua New Guinea	102
The rise of nationalism in Europe	106
Transboundary water conflicts in southeast Asia	107
Censorship in China	107
Migration policies in Japan	108
The First Nations in Canada	108
Transition towns — Totnes, Devon	110
Local authority recycling	110

Specification Map

I prefer <u>stair</u>case studies.

Specification Map

Even more case studies...

I need to find brain space for all that? Guess I'll lose some family birthdays...

Remember — the examples in these tables are to guide you in all the areas where you need to know examples and case studies. It's not compulsory that you learn these exact case studies, nor are they the minimum you should study. You should be using contemporary examples to illustrate key concepts and ideas throughout your A-Level course.

Exam Structure

It's never too early to start thinking about the exams. Before we delve into all that Geography you've got to learn, here's a lovely page telling you all about how your A-level Geography course will be assessed.

You'll have to do **three exams** and some **fieldwork**

If you're doing AS Geography, the structure of your assessment will be different — ask your teacher for more information.

Paper 1

1) It's **2 hours 15 minutes** long and there are **105 marks** available (worth 30% of your A-Level grade).
2) It tests five physical geography topics, but you only have to answer questions on **FOUR** of them.
3) There are **three sections** in the paper — **A**, **B** and **C**:

Exams — who needs them, don't you know who my father is?

Section A

You have to answer the **multiple-part question** on Topic 1: Tectonic Processes and Hazards. The question is worth **16 marks**.

Section B

You have to answer **ONE multiple-part question** from a choice of:

- Topic 2A: Glaciated Landscapes and Change **OR**
- Topic 2B: Coastal Landscapes and Change.

Each question is worth **40 marks**.

Section C

You have to answer the **multiple-part question** that covers **both**:

- Topic 5: The Water Cycle and Water Insecurity **AND**
- Topic 6: The Carbon Cycle and Energy Security.

The question is worth **49 marks**.

Paper 2

1) It's **2 hours 15 minutes** long and there are **105 marks** available (worth 30% of your A-Level grade).
2) It tests six human geography topics, but you only have to answer questions on **FOUR** of them.
3) There are **three sections** in the paper — **A**, **B** and **C**:

Section A

You have to answer **two multiple-part questions** that cover both Topic 3: Globalisation and Topic 7: Superpowers. Each question is worth 16 marks, giving a total of **32 marks**.

Section B

You have to answer **ONE multiple-part question** from a choice of:

- Topic 4A: Regenerating Places **OR**
- Topic 4B: Diverse Places.

Each question is worth **35 marks**.

Section C

You have to answer **ONE multiple-part question** from a choice of:

- Topic 8A: Health, Human Rights and Intervention **OR**
- Topic 8B: Migration, Identity and Sovereignty.

Each question is worth **38 marks**.

Paper 3

- It's **2 hours 15 minutes** long and there are **70 marks** available (worth 20% of your A-Level grade).
- This exam is **synoptic** and draws on different parts of the course. This includes Topics 1, 3, 5, 6 and 7. You'll be provided with a **resource booklet** which the questions will refer to.
- There are six questions in the exam paper. Some of those questions may have multiple parts.

The Non-Examination Assessment (NEA)

The final **20%** of your A-Level grade will come from your independent investigation (worth **70 marks**).
You have to complete a **written report** for your investigation, which will be marked by your teacher — there is **no exam**.
See p.288-289 for more about completing your independent investigation.

Plate Tectonics Theory

The ground beneath your feet is moving all the time, albeit very, very slowly. Plate tectonics theory explains this movement.

The internal structure of the Earth has **different layers**

Our knowledge about the structure of the Earth is based **more on theory** than on direct evidence. These theories are based on evidence from **seismic activity** at the Earth's surface and the study of **magma**.

1) At the centre of the Earth is the **core**. Scientists believe that the large amount of heat at the core is produced by the **radioactive decay** of elements such as uranium. The core is split into:

- The inner core — a solid ball containing iron and nickel.
- The outer core — a semi-molten layer thought to contain iron and nickel.

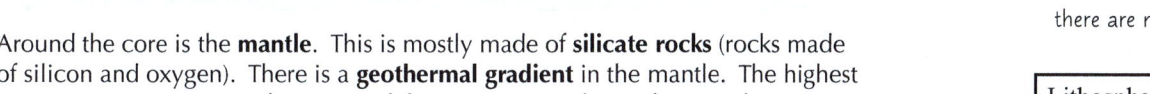

Don't be fooled — under Simon's hard exterior there are many layers.

2) Around the core is the **mantle**. This is mostly made of **silicate rocks** (rocks made of silicon and oxygen). There is a **geothermal gradient** in the mantle. The highest temperatures are nearest the core and the temperature drops closer to the crust.

- The lower mantle, nearest the core, is quite rigid. The rocks found here have a degree of plasticity though, which means that they can reshape under stress.
- The upper mantle is known as the asthenosphere. This is semi-molten, so has the ability to flow. As it is cooler than the lower mantle, the rocks here tend to be weaker and are more likely to break apart under stress. This is why the focus of earthquakes is found here.

3) The very top of the mantle and the **crust** are known together as the **lithosphere**. The crust is split into different sections called tectonic plates which move over the asthenosphere. There are two types of crust which make up the different tectonic plates:

- Continental crust is thicker (30 to 70 km thick) but has a lower density. It is mostly made up of granite.
- Oceanic crust is thinner (6 to 10 km thick) but more dense than continental crust. It is mostly made up of basalt.

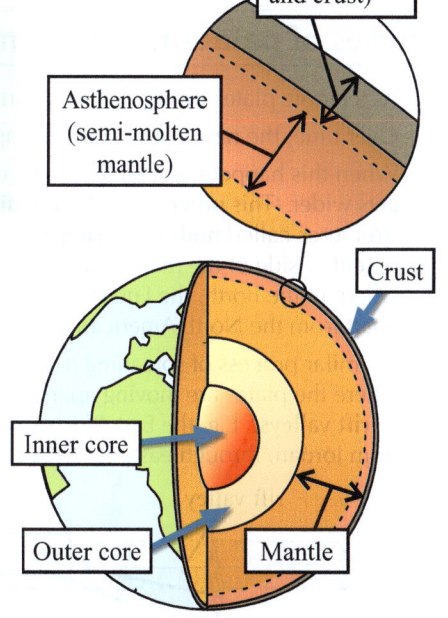

Lithosphere (rigid mantle and crust)

Asthenosphere (semi-molten mantle)

Crust

Inner core

Outer core

Mantle

Tectonic plates are in constant **motion**

- The seven major **tectonic plates** (African, Antarctic, Eurasian, Indo-Australian, North American, Pacific and South American), along with many other minor ones, fit together to cover the Earth's surface and are constantly moving.
- Geologists believe that around 300 million years ago the continents were in different positions from where they are today. They were believed to be held in one landmass (or 'super-continent') known as **Pangaea**. Since then, the continents have been **drifting apart and back together** as the tectonic plates holding them move around. This is known as **continental drift**.

Until recently, scientists thought that **convection currents** were the main cause of plate movement

1) The Earth's mantle is hottest close to the **core** (due to **radioactive decay**), so the lower parts of the **asthenosphere** heat up, become less dense and slowly rise.

2) As they move towards the top of the asthenosphere, they cool down, become denser and sink. These circular movements of **semi-molten** rock are called **convection currents**.

3) Convection currents create **drag** on the base of the tectonic plates, causing them to move.

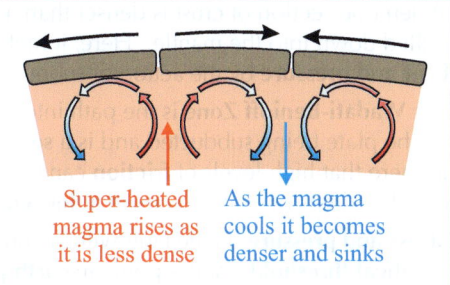

Super-heated magma rises as it is less dense

As the magma cools it becomes denser and sinks

Plate Tectonics Theory

Slab pull is now thought to be the main cause of plate movement

1) The **rising limb** of a **convection current** will make the **oceanic crust** above it **hotter** and **less dense**.

2) The part of the crust **furthest** from the **rising limb** will be much **cooler** and **more dense**. This means it will be more likely to be **forced underneath** another plate.

3) The sinking of the plate at one of its edges **pulls** the rest of the plate in that direction. This is known as **slab pull**.

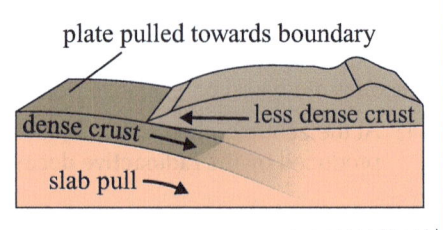

It is thought slab pull is **stronger** than any movement caused by convection currents, making it the **main driver** of movement. The plates with **larger subduction zones** (the point where one tectonic plate is **pulled under** another) tend to be the ones that move the **fastest**.

An example of this is oceanic crust moving beneath continental crust at a destructive plate boundary — see p.7.

Studies of palaeomagnetism suggest that the seafloor is spreading

1) As tectonic plates **diverge** (move apart), magma rises up to fill the gap created, then cools to **form new crust**.

2) Over time, the **new crust** is dragged apart and **even more new crust forms** between it.

3) When this happens at a plate margin under the sea, the seafloor gets wider. This process is called **seafloor spreading**. It creates structures called **mid-ocean ridges** — ridges of higher terrain on either side of the margin. E.g. the Mid-Atlantic Ridge, where in the north, the Eurasian and African plates are moving apart from the North American and the South American plates.

4) A similar process of spreading occurs at land margins where the plates are moving apart. Here they are known as **rift valleys**. E.g. the East African Rift System which stretches from Jordan, through eastern Africa to Mozambique.

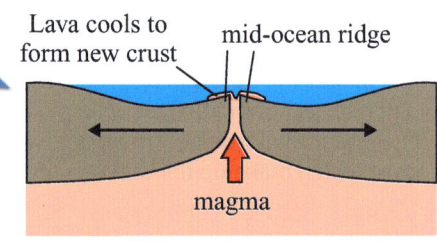

5) This process means that the **oldest rock** is normally found on the edge of the continents and the **youngest rock** is found at the mid-ocean ridges.

6) **Magma** is made up of **iron-rich** rocks and these become **polarised** when they reach the **surface**. This means it **aligns** with the **Earth's poles** through **magnetism**, which is created by the Earth's **core**.

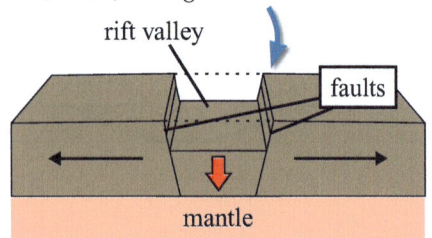

7) The Earth's **magnetic field** changes **polarity** roughly every 200 000 years — the last time this happened was 780 000 years ago. When the **polarity changes** the alignment of the iron-rich rocks in the magma will **change direction**. By studying the crust **either side** of a **divergent** boundary, scientists can see when it was formed and how quickly the plates are moving apart.

palaeomagnetism
the study of the Earth's ancient magnetic fields. It can be used to date different parts of the crust at a divergent plate boundary

Subduction is the process of one plate moving under another

Have a look at the slab pull diagram at the top of the page.

- When one section of crust is denser than another, it is pulled down into the mantle. Here, it melts under extreme **heat and pressure** by the action of subduction.

- The **Wadati-Benioff Zone** is the path into the mantle taken by the plate being subducted and is a **seismically** active area. It's here that high levels of **friction** can build up (sometimes over hundreds of years) to create a **'locked' fault**.

- **Stress and pressure** on the fault will rise until it reaches a **critical threshold**. At this point an **earthquake** will occur and the pressure will be released.

seismicity
a measure of the number and frequency of earthquakes in a particular area

A 'locked' fault is one that is not moving.

Carole blamed her faults on the pressures of the exam system.

Plate Tectonics Theory

There are **four types** of **plate margin**

(1) **Destructive margin** — This occurs where two plates are moving towards each other (**converging**). What happens at these margins depends on the types of crust on the converging plates.

Oceanic - Continental

- At this margin, the **denser oceanic crust** is subducted beneath the **less dense continental crust**. This forms a deep sea trench, e.g. the Peru-Chile trench in the Pacific Ocean.
- **Fold mountains** also form where the plates meet. They are made up of **sediments** that have accumulated on the continental crust. These are then **folded upwards** along with the edge of the crust.
- The oceanic crust is heated by **friction** and contact with the **upper mantle**, which melts it into **magma**.
- The magma is less dense than the crust above it, so it rises up and **volcanoes** can form on the surface. This process can occur on either type of crust.
- The plates are not smooth. As one tries to move under the other, they can lock together, creating a **locked fault**. This causes **pressure** to build and when the pressure becomes too much, the plates will slip and cause an **earthquake**.

Composite volcanoes (which can be very explosive) form at this plate margin.

Oceanic - Oceanic

- At this margin, the two sections of oceanic crust will have roughly the same density, but the one which is slightly denser will be **subducted**.
- This can form a **deep sea trench**, e.g. the Mariana Trench on the eastern edge of the Philippine Sea.
- **Earthquakes** can also occur in the **subduction zone**, leading to **tsunamis**. It is possible for **volcanic eruptions** to take place underwater too. These can create **island arcs**. These are **clusters of islands** that sit in a curved line, e.g. the Mariana Islands.

Pagan Island, North Marianas, with a volcanic eruption

(2) **Collision margin** — This is where convergence between two continental crusts takes place.

Continental - Continental

- At a collision margin, **neither crust is subducted** as they both have the same density, which is less than the underlying asthenosphere.
- Instead, the two sections of crust **fold up against** each other, creating belts of **fold mountains** as part of an enormous **tectonic uplift**.
- An example of this is the Himalayas, which continue to rise by 10 mm every year because of the convergence of the Indo-Australian and Eurasian plates.

The Himalayas

Volcanoes don't form at this plate margin because there isn't any subduction and therefore any rising magma.

(3) **Constructive margin** — This is where two plates are moving apart (**diverging**).

Continental - Continental or Oceanic - Oceanic

- The **mantle** is under pressure from the plates above. When they move apart, the pressure is released at the margin.
- The release of pressure causes the mantle to melt, producing **magma**. The magma is less dense than the plate above, so it rises through the gap to produce new crust or it can erupt to form a **volcano**.
- The plates do not move apart in a uniform way — some parts move faster than others. This causes pressure to build up. When the pressure becomes too much, the plate cracks, making a **fault line** and causing an **earthquake**. Further earthquakes may occur along the fault line once it has been created.
- Constructive margins create two different landforms — **ocean ridges** and **rift valleys** (see p.6).

lava cools and slides downslope

magma

Plate Tectonics Theory

④ **Conservative margin** — This occurs where two plates are trying to move past each other at what is known as a **transform plate boundary**.

Continental - Continental or Oceanic - Oceanic

- Crust is neither made nor destroyed at these margins and the overall movement is horizontal rather than into or out of the mantle.
- The two plates may be moving in opposite directions to each other or they may be moving in the same direction but at different speeds. For example, at the San Andreas Fault, the Pacific plate is moving northwards at 80 mm/year compared to the North American plate which is moving northwards at 25 mm/year.
- This movement creates a shearing action which leads to increased friction between the two plates. As the plates are not smooth, when they try to move past each other, they can lock together. The pressure can build up and an earthquake is felt when the plates jolt free of each other.
- There are no volcanoes formed at this margin because there is no magma rising. Instead, you can commonly see fault lines or cracks at the ground surface along conservative margins. These may become enlarged by the action of erosion and weathering.

The fault line at the San Andreas Fault

Plate boundaries are found at different locations around the world

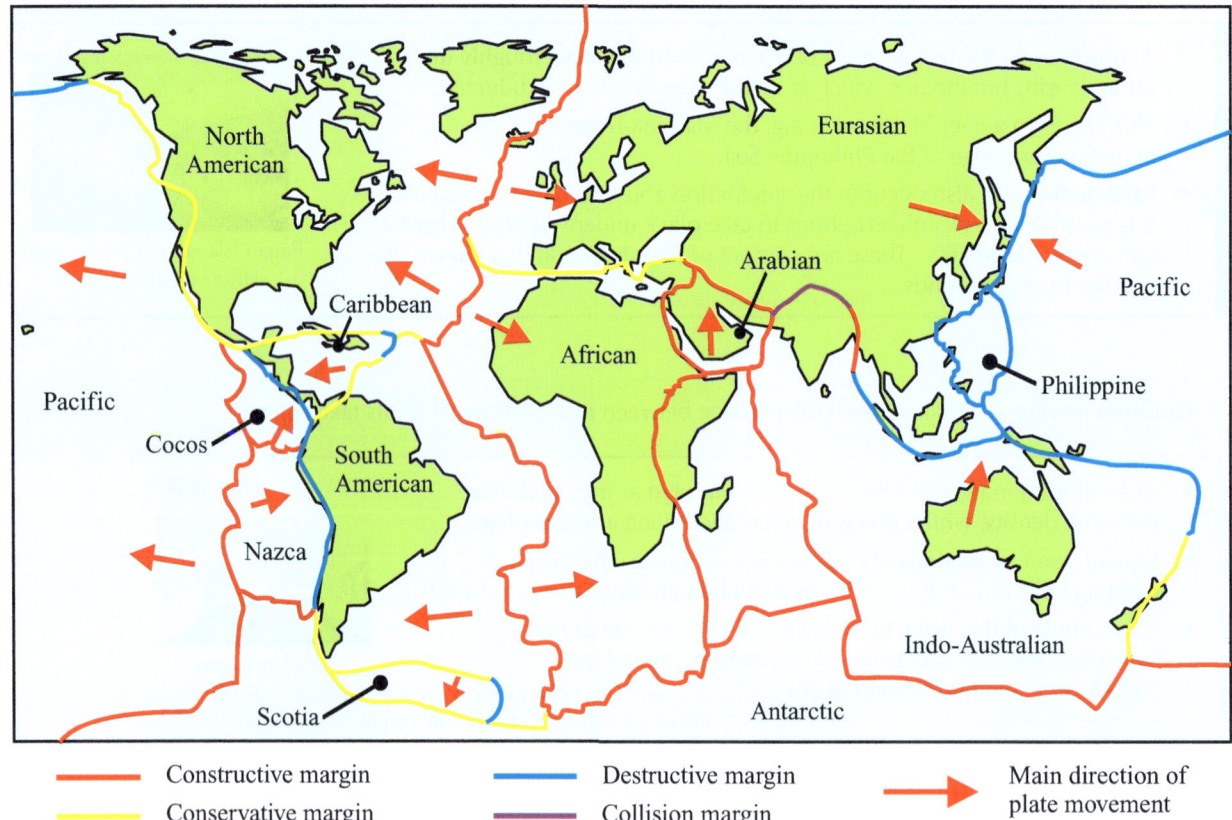

| Plate labels: North American, Eurasian, Arabian, Caribbean, African, Pacific, Philippine, Pacific, Cocos, South American, Nazca, Indo-Australian, Scotia, Antarctic |

— Constructive margin   — Destructive margin   → Main direction of plate movement
— Conservative margin   — Collision margin

Warm-Up Questions

Q1 Describe how tectonic plates move.

Q2 Explain how constructive plate boundaries can produce different landforms.

PRACTICE QUESTIONS

Exam Question — AS and A-Level

Q1 Explain how two features of a destructive plate margin are formed.   [4 marks]

It's important to know your faults...

Then you can work on them to become a better person. Little nugget of life advice there — you're welcome.

Earthquakes and Tsunamis

Earthquakes are seismic tectonic hazards and tsunamis are one of the other nasty hazards they cause. It's earth-shattering stuff.

Earthquakes and tsunamis mainly occur **on or near plate boundaries**

1) **Earthquakes** are caused by the build-up and sudden release of **tension** at all four types of plate margin (destructive, constructive, conservative and collision).

2) The majority of earthquakes occur around the edge of the **Pacific plate** and particularly around **Indonesia**, **Japan** and **The Philippines**.

3) Earthquakes can also occur in the middle of plates. These **intraplate earthquakes** are very rare compared to earthquakes produced at margins.

4) These may occur due to **old fault lines** moving into their resting positions as well as **new fault lines** developing as plates become stretched as they move, e.g. the East African Rift Valley.

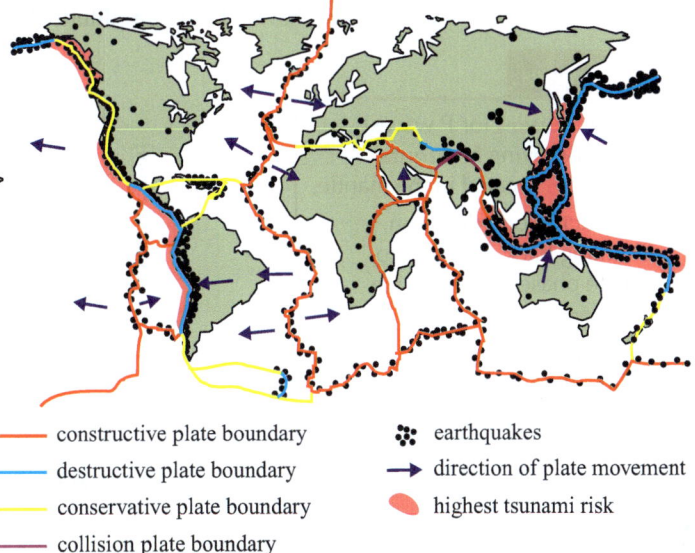

— constructive plate boundary
— destructive plate boundary
— conservative plate boundary
— collision plate boundary

⁕ earthquakes
→ direction of plate movement
🔴 highest tsunami risk

The **strength of earthquakes varies** due to physical processes

1) When plates jolt past each other at plate margins, they send out **seismic waves** (vibrations) along a **fault line** (a weak point in the crust). These seismic waves carry the energy produced by the initial movement.

2) The pressure that is released can cause **crustal fracturing** (where the deeper crust breaks apart) as well as **surface fracturing** and **buckling**, where the Earth's crust appears to fold up at unusual angles.

3) The shockwaves spread out from the **focus**. This is the place where the actual movement starts. The focus can be a single point or spread out along a large section of a fault line.

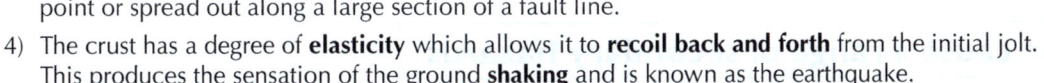

4) The crust has a degree of **elasticity** which allows it to **recoil back and forth** from the initial jolt. This produces the sensation of the ground **shaking** and is known as the earthquake.

5) The **epicentre** is the point on the Earth's surface where the earthquake is first felt. It's immediately above the focus.

6) The strength of an earthquake (its **magnitude**) is affected by **two main factors**:

① The type of plate margin

- The highest magnitude earthquakes occur at **destructive** plate margins, where one plate is forced beneath another creating a **subduction zone**. The subduction of a plate causes massive pressure to build up and a huge earthquake occurs when it is released.
- Earthquakes at **constructive** margins are common, but tend to be of a lower magnitude. At constructive margins, earthquakes also occur more frequently.

Earthquakes of lower magnitude can still have devastating impacts.

② The depth of the focus

- An earthquake's **focus** can be close to the Earth's surface or deep below it.
- **Deep-focus earthquakes** (those with a focus more than 300 km deep) tend to be of a higher magnitude than shallow focus earthquakes. Deep-focus earthquakes generally do less damage than **shallow focus earthquakes**. This is because shockwaves generated deeper in the Earth have to travel further to reach the surface, which reduces their power.

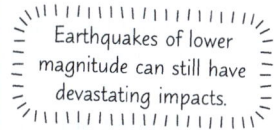

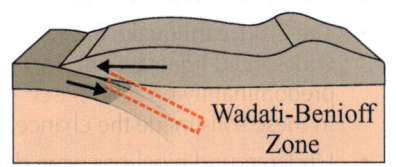

- The position of the **Wadati-Benioff Zone** maps the position of deep-focus earthquakes above the subduction zone at destructive plate boundaries. The strongest earthquakes tend to come from foci in the Wadati-Benioff Zone.

The fact that tectonic plates move at different rates to each other **does not** have an impact on magnitude.

Earthquakes and Tsunamis

Earthquakes produce **different** types of **waves**

For example, earthquake waves can be **primary (P) waves**, **secondary (S) waves** and **love (L) waves**. The severity of an earthquake is linked to the **amplitude** of the waves it produces. S and L waves have a **larger amplitude** than P waves.

amplitude
the size of wave disturbance compared to a straight line

1 Primary waves

- **Primary waves** (or P waves) are fast-moving **horizontal vibrations** created in the mantle.
- They act in one direction, shunting back and forth to create phases of **expansion and compression**.
- They can **travel through** solids, liquids and gases.
- They **don't** tend to produce **much damage**.

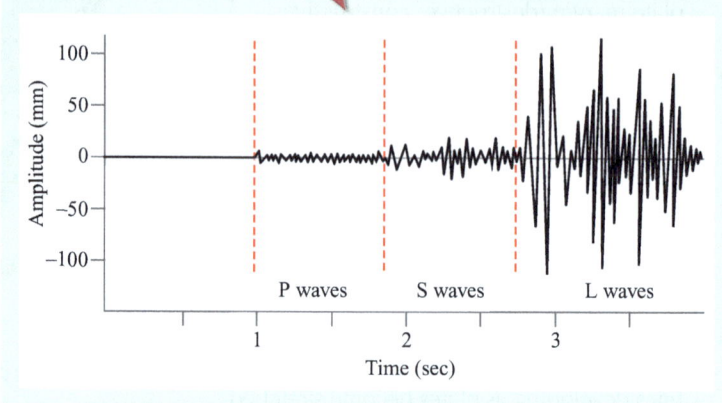

A seismograph reading showing the amplitude of the initial P waves, the S waves and the L waves for an earthquake.

2 Secondary waves

- **Secondary waves** (or S waves) are slower moving **vertical vibrations** that pass through the centre of the crust.
- They can **only** travel through **solids**.
- They cause **more** damage than **primary waves**, but are **not usually** as damaging as Love waves.

3 Love waves

- **Love waves** (or L waves) also move horizontally but at right angles to P waves.
- These are **surface waves** that follow from the S waves and so only move out from the epicentre on the surface of the crust.
- Love waves can **only** travel through solids.
- These waves tend to cause the **most damage**, such as fracturing the ground surface.

The closure of the beach toilets led to a lot of P waves.

Earthquakes can **cause** a **range** of **secondary hazards**

Secondary hazards happen as a **consequence** of the **primary hazard**. For example, the **ground shaking** in an **earthquake** (primary hazard) can **cause secondary hazards** such as **soil liquefaction**, **mass movement** (e.g. landslides) and **tsunamis**.

Soil liquefaction occurs in areas of loose, waterlogged ground

1) **Soil liquefaction** takes place when an earthquake's vibrations place **groundwater** under pressure
2) This pressure **forces** water through the **soil** particles, causing the ground surface to **lose** its **usual** structure and act like a **liquid**.
3) The soil becomes weaker and is very **easily deformed**. This means it is more likely to cause **building subsidence** and cause heavy objects such as cars to be partly swallowed by the ground.
4) Soil that is regularly saturated (such as that on a floodplain) is more vulnerable to liquefaction.

The 2011 earthquake in Christchurch, New Zealand

- The 2011 earthquake in **Christchurch**, New Zealand, witnessed widespread liquefaction. This was due to Christchurch being built predominantly on **sand-based soils**. The **small particle size** found in these soils made the chance of liquefaction much higher.
- Underground pipelines were damaged and vehicles became stuck in road surfaces. The liquefaction was so extreme in some areas that they have since been designated 'red zones' — areas where **no future rebuilding** should take place.

Damage caused by the 2011 Christchurch earthquake

Earthquakes and Tsunamis

Landslides happen when rocks and soil are dislodged

- **Landslides** occur when the shaking of the ground dislodges rocks and soil. These can move down slope very quickly.
- The shaking can also loosen ground material, making it easier for water to **infiltrate**.
  The weight of the extra water may trigger a landslide even after the ground shaking has stopped.
- They tend to be more common in areas where the slopes are relatively young in geological terms and have been deforested.

Damage caused by a landslide after an earthquake in Nepal in 2015

The 2015 earthquake in Nepal

Landslides played a significant role in the high number of deaths attributed to the **2015 Nepal earthquake**. This disaster took place in April, before the start of the Nepalese **rainy season**. The earthquake weakened the **rock and sediment structure** on large sections of the southern slopes of the **Himalayas** north of Gorkha, the earthquake's **epicentre**.

Landslides and **avalanches**, including one on **Mount Everest** which killed over 20 climbers, were frequent occurrences in the immediate aftermath. The continuous rainfall over the **monsoon season** (June to September) continued to create **slip planes** for **rockslides**. Large-scale **mass movements** continued months after the earthquake.

Tsunamis are a series of large waves

Tsunamis are secondary hazards that usually occur at **subduction zones**. They are a series of large waves caused by the **displacement** of large volumes of water.

1 Tsunamis can be triggered by **submarine earthquakes**. The earthquakes cause the **seafloor** to move quickly upwards or downwards, which displaces water (known as **water column displacement**).

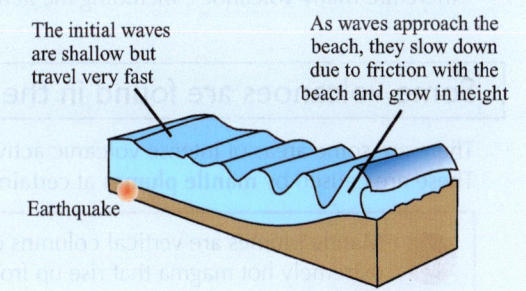

A tsunami wave hits the coastline of Japan in 2011

2 Waves radiate out in all directions from the epicentre of the earthquake. The greater the movement of the seafloor, the greater the volume of water displaced, and the bigger the wave.

3 Waves start very **small in height** but travel very **fast**. This means they pose little threat while out at sea. As they approach the coast the waves lose energy and slow down due to **friction with the rising seabed**. This causes the waves to grow in height as water is backed up.

4 This means a tsunami will usually be more powerful if the epicentre is **close** to the coastline, or is travelling through very **deep water**, as it has less time to lose this energy.

5 The **length of time** between tsunami **waves** can vary but is usually no more than a **few minutes**.

The initial waves are shallow but travel very fast

As waves approach the beach, they slow down due to friction with the beach and grow in height

Earthquake

Warm-Up Questions

Q1 Describe the differences between the three types of wave produced by an earthquake.

Q2 Explain how the position of the focus of an earthquake can affect its strength.

PRACTICE QUESTIONS

Exam Question — AS and A-Level

Q1 Using the map at the top of page 9, explain the global distribution of earthquake activity. [6 marks]

What did the ground say to the earthquake?

*...You crack me up. Not one of my best — but then again tectonic hazards are no laughing matter. Make sure you know all about the causes and effects of earthquakes and tsunamis so that you can **rock** the examiner's world (sorry, I couldn't help it).*

Volcanoes

Volcanoes — I could talk about them all day. That's probably why I'm such a favourite amongst my friends and relatives.

Volcanoes mainly occur **on or near plate boundaries**

- Most volcanic eruptions occur near **constructive** and **destructive** plate margins (see p.7).
- They **don't** occur at **conservative** or **collision** margins as no new crust is formed at these margins.

At constructive margins

- **Magma** rises to fill the space left by plates moving apart because it is **less dense** than the **surrounding** crustal **rock**.
- If the margin is underwater, these can form **ocean ridges** or lines of single volcanoes can form parallel to the plate boundary.
- These may eventually form **islands** as they grow and extend above the surface of the sea.
- If the margin is on land, **rift valleys** can be formed as the plates pull apart. Magma is also able to break through the surface here to create **volcanoes**.

Example: Heimaey — a volcanic island off the Icelandic mainland. It's formed by the **divergence** of the **North American** and **Eurasian** plates.

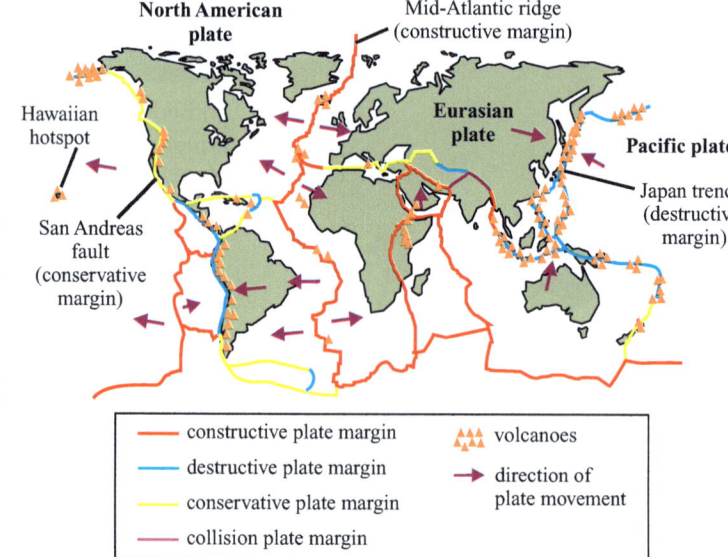

constructive plate margin
destructive plate margin
conservative plate margin
collision plate margin
volcanoes
direction of plate movement

At destructive margins

- At **subduction zones**, where one plate is pulled beneath another, the melting of the denser plate forms additional magma.
- This magma, which comes under pressure in the **upper asthenosphere**, then rises to the surface.
- Where there is a small **fault** or **weakness** in the crust, a volcano will be created.

Example: The subduction of the **Nazca** plate underneath the **South American** plate has formed the **Andes mountain range**, where there are many volcanoes, including the active **Nevado del Ruiz**.

The largest concentration of volcanoes is found around the edge of the Pacific plate, giving it its nickname, the Pacific Ring of Fire.

Italy sits on a point where the Eurasion and the African plates are colliding. This results in Italy having active volcanoes.

Some volcanoes are found in the **middle of plates**

There are some areas of intense volcanic activity that are not near any plate margins. These are caused by **mantle plumes** at certain points in the **mantle**.

1 Mantle plumes are vertical columns of extremely hot magma that rise up from the mantle. They are thought to be caused by **radioactive decay** in the asthenosphere. They melt and **weaken** the underneath of the **crust**, creating a body of **less dense magma**, which **forces** its way through to form volcanoes. These volcanoes are known as **hotspot** volcanoes.

2 Volcanoes form **above mantle plumes**. The mantle plume remains stationary over time, but the crust moves above it. This means that progressively newer volcanoes are formed in new locations as the crust moves.

3 A chain of volcanoes or a **volcanic island arc** may form. For example, the islands that make up Hawaii are an arc of progressively newer volcanoes.

Iyanu regretted mistaking the hotspot volcano for a holiday hotspot.

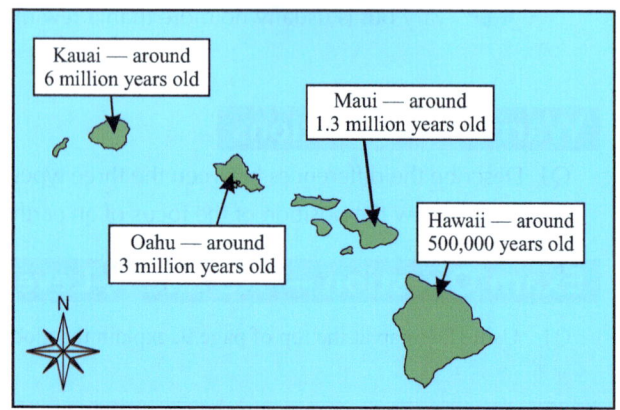

Kauai — around 6 million years old

Maui — around 1.3 million years old

Oahu — around 3 million years old

Hawaii — around 500,000 years old

The Hawaiian island arc is caused by mantle plumes.

Volcanoes

The **strength** of volcanic eruptions varies due to their **different lavas**

The **explosivity** (i.e. explosiveness) of a volcanic eruption is determined by:

1 How **easy gases** can **escape** from **lava** — if gas is trapped, pressure **builds** to cause a **more explosive** reaction.

2 The **viscosity** (thickness) of the **lava**. Lavas that contain a lot of **silica** or that are **cooler** tend to be **more explosive**.

Basaltic lava is formed at **constructive** plate margins

- Basaltic lava is very **hot** (1000 °C to 1200 °C) and has a **low silica content.**
- This means basaltic lava has a **low viscosity** (a runny texture) so it flows easily and quickly.
- Eruptions of basaltic lava are **frequent** and go on for a long time. Eruptions are usually less violent than other volcanoes.
- The shape of basaltic volcanoes tends to be quite **flat with gently sloping sides**. They're often called **shield volcanoes**.

Basaltic lava also forms at oceanic hotspots.

An example of a shield volcano is Mauna Kea in Hawaii.

Andesitic and **rhyolitic** lavas are formed at **destructive** plate margins

- These lavas are **cooler** (650 °C to 1000 °C) and have a **high silica content**.
- This means they are **more viscous** than basaltic lava so they flow less easily.
- As this lava erupts, it moves slowly down the sides of the volcano.
- As the lava moves, it cools down. This creates a **composite** (cone shaped) and more **dome-shaped** volcano.
- Andesitic and rhyolitic lavas usually **erupt intermittently** and the eruptions are **short-lived**.
- As the lava is viscous, it can form **blockages** in the **vent** of the volcano. This can cause pressure to build and a **violent eruption** can occur, which often ejects large volumes of **tephra** (rock fragments).

Rhyolitic lava also forms at continental hotspots.

Mount Toba in Indonesia erupted rhyolitic lava around 74 000 years ago. Lake Toba occupies the caldera of Mount Toba.

Volcanoes can generate a **variety of hazards**

Primary hazards are generated from the eruption itself

Lava flows

- **Lava** can flow from a volcanic **vent** down the side of the volcano.
- It can reach extremely **high temperatures** and takes a long time to cool down.
- The **speed** of the flow and the **distance** travelled depend on the **temperature** and **viscosity** of the lava as well as how **steep the sides** of the volcano are.
- **Low viscosity** (runny) lava can flow at up to 10 km/hour on a steep slope and may travel tens of kilometres. Most flows are relatively slow, so people have time to **evacuate** areas that will be affected. Lava flows **destroy** anything in their path, including buildings and vegetation, by burning, burying or knocking them down.

Pyroclastic flows

- A **pyroclastic flow** is a mixture of super-heated **gas**, **ash** and **volcanic rock** created in the crater of the volcano. It flows down the side of a volcano like a hot (700-800 °C) 'liquid'.
- It travels at **high speed** (often more than 80 km/hour) and flows a long way from the volcanic crater (up to 100 km away).
- The speed with which they travel means they can happen with relatively **little warning**.
- This means they can cause **widespread death and destruction**.

A pyroclastic flow is sometimes called a nuée ardente.

Pyroclastic flow from Merapi Volcano, Indonesia

Volcanoes

Volcanic gases

- Lava contains **gases** such as **water vapour**, **carbon dioxide** and **sulphur dioxide**, which are released into the atmosphere when a volcano erupts.
- Some of these gases can be harmful to human and animal health if they are inhaled. For example, sulphur dioxide is **poisonous** in high concentrations and can cause **breathing difficulties**.
- Volcanic gases which are **colourless and/or odourless** can be particularly dangerous as they are difficult to detect.

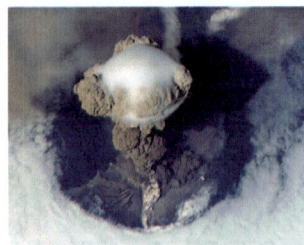

An ash cloud in Russia

Tephra

- **Tephra** is a term for all pieces of rock that are ejected into the air during an eruption, ranging from large pieces of rock (known as lapilli or volcanic bombs) measuring several metres across to **microscopic ash particles**.
- Material can travel thousands of kilometres from the volcanic crater due to **global atmospheric weather systems**. Heavier particles are deposited earlier than lighter ones, so material ends up being **sorted by mass**, with larger, heavier particles deposited **near** the volcano and smaller, lighter particles (e.g. ash) being deposited **further away**.
- **Large pieces** of tephra can damage buildings and kill or injure people, as well as **start fires** if it lands on flammable materials.
- **Fine tephra** can form a thick layer on vegetation as well as produce **atmospheric hazes**, which in turn can reduce **crop yields**. Road, rail and air **transport** can also be disrupted, and the weight of ash can cause buildings to **collapse**. Ash is also **harmful** to people if they breathe it in.

Secondary hazards are triggered by the primary hazards

Lahars

- **Lahars (mudflows)** occur when volcanic materials such as **ash** and **silt** mix with large amounts of water. This may come from **heavy rainfall**, **rivers** flowing on the slopes of volcanoes, or **ice** which has been melted by the eruption. Lahars can move very quickly (over 80 km/hour) and can travel for tens of kilometres.
- Lahars can **bury or destroy natural habitats**, **settlements** and **infrastructure** such as roads and bridges.

Example: In 1995, a lahar was produced when heavy rainfall mixed with ash from eruptions in the **Soufrière Hills** on the island of Montserrat, causing a **mass evacuation** of the island.

Jökulhlaups

- A **jökulhlaup** is a flood caused by the sudden release of **glacial meltwater** from a **subglacial lake** (one beneath a glacier's surface). This melting occurs due to the heat given off during a **volcanic eruption** below the glacier.
- The surface glacier will hold this water in place until the volume of water becomes so large that it can no longer be held back. The **sudden deluge** of water it releases (also known as a **glacial outburst**) may contain **ice fragments** and **moraine** (eroded rock debris).

Example: In 1996, the eruption of the **Grímsvötn volcano** in Iceland caused meltwater from the **Vatnajökull glacier** (which covered the summit) to pool in the volcano's caldera. This then created a jökulhlaup when the water level got too high.

> **caldera**
> a volcanic crater, often formed when a volcano collapses in on itself

Warm-Up Questions

Q1 Describe the distribution of different types of volcanoes around the world.

Q2 Explain what makes lahars and jökulhlaups secondary rather than primary hazards.

PRACTICE QUESTIONS

Exam Question — AS and A-Level

Q1 Explain why volcanic island arcs are formed in the middle of tectonic plates. [6 marks]

I like boiled eggs the consistency of basaltic lava — hot and runny...

Perfect for dipping. Whilst I'm enjoying my tea, check that you know all about volcanoes and the hazards they cause. Yum.

Natural Disasters

Hazards come in all shapes and sizes, and when they happen in populated areas they can be pretty nasty things. *I'm scared.*

There are key **differences** between **hazards** and **disasters**

Hazards

1) A **hazard** is something that is a **potential threat** to human life or property.

2) **Natural hazards** can be divided into three types:

1 **Geophysical hazards** are caused by processes in the lithosphere. These include tectonic hazards such as earthquakes, volcanic eruptions, landslides and tsunamis.

2 **Atmospheric** (or **hydrometeorological**) **hazards** are those caused by climatic processes. These include tropical cyclones, storms, droughts, extremes of hot or cold weather and wildfires.

3 **Hydrological hazards** are those caused by water movement. These include floods and tsunamis.

Disasters

1) A **disaster** is when a hazard occurs and has a **significant impact** on a **population**.

2) The **threshold** at which a disaster happens will be different for different countries — e.g. a hazardous event may be considered a disaster when there are:

- deaths of 10 or more people.
- 100 or more people affected in some way.
- at least US $1 million lost economically.

3) A **mega-disaster** is has a **regional** or **global impact**.

4) Different countries or communities will define a disaster differently, but generally they are said to have occurred when countries or communities can no longer cope with the challenges that the hazardous event has created.

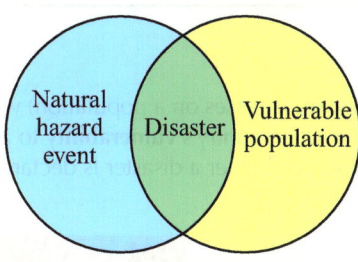

The Degg Model

- This model suggests that a disaster occurs where a natural hazard event meets a vulnerable population.

- The greater the natural hazard event, and the more vulnerable the people, the greater the disaster.

threshold
the intensity or magnitude level a natural hazard would have to surpass to be considered a disaster

Some **communities** are more **vulnerable** or **resilient** than others

1) Different communities experience the impacts of disasters in very different ways. This is in large part due to different places having different **levels of development**.

Everything is a disaster for Eve, especially when she doesn't get a second biscuit.

Resilience is the capacity for a community to cope with and recover from a natural disaster. This includes any systems in place to manage a natural hazard once it has occurred, and any pre-planning done to reduce the number of people at risk.

Vulnerability is the level to which a community or geographical area will be damaged or affected by the occurrence of a hazard. The vulnerability of a community relates to its level of hazard preparedness.

2) Some communities have a higher level of resilience due to their awareness of natural hazard risks. They may acknowledge **potential** hazards and **act** to **limit** the **damage** that a natural hazard event could cause.

3) Communities that have a **low resilience** and a **high vulnerability** will have a **lower disaster threshold**. This means that a hazard is more likely to be a disaster in that community.

High disaster threshold
(less likely to be a disaster)

Low disaster threshold
(more likely to be a disaster)

Resilience / *Vulnerability*

4) The vulnerability of a community can be determined by the factors below:

- location in relation to **topographical features**
- location in relation to the **site of the hazard**
- **isolation** from / **access** to other communities

- **age** and **gender** structure of their population
- level of **urbanisation** and **population density**
- degree of **poverty** / **wealth** of population
- level of **education** and **literacy rate**

5) This means that the vulnerability of a community to natural hazards can be determined before a natural hazard occurs.

Natural Disasters

The **Hazard Risk Equation** examines the relationship between hazards and disasters

- **Risk** is the **potential exposure** to a disaster.
- Risk is a complex idea based on a country's **vulnerability**, **resilience** and the **nature** of the hazard itself. The **unpredictable nature** of hazards means that there is always an element of risk, and so people always have a **degree of vulnerability**.

$$Risk = \frac{Hazard \times Vulnerability}{Capacity\ to\ Cope}$$

The Hazard Risk Equation

- The **Hazard Risk Equation** is a way of trying to quantify the different factors at play in determining the risk that a natural disaster poses. Countries are at greater risk if they are more vulnerable or if the intensity of the hazard itself is greater too.
- In a similar way to the Degg model, **increases** in the magnitude of the hazard will increase the risk of a disaster. Equally, if the population becomes **more vulnerable**, the risk and intensity also increases.
- Factors which increase the risk of a hazard:

  1) Size of the affected area
  2) Duration it is experienced for
  3) Level of intensity or magnitude
  4) Speed of onset (how quickly it happens)
  5) Time of day / season
  6) Frequency

- It's **difficult to stop** a hazard occurring — for example, we can't 'prevent' an earthquake. This means that governments tend to look at reducing a country's **vulnerability level** in order to better manage hazards.

The **Pressure and Release (PAR) model** maps the factors that create a disaster

- The **PAR model** explains a hazard's socio-economic context and the combination of conditions in which a disaster could emerge.
- It suggests that there are certain **root causes** of vulnerability that create **dynamic pressures** (stresses on a population) which in turn lead to **unsafe conditions** when a hazard occurs. All of these develop a picture of a country's **vulnerability** to a natural hazard. It also shows that the nature of the hazard itself has a large part to play in whether a disaster is declared.

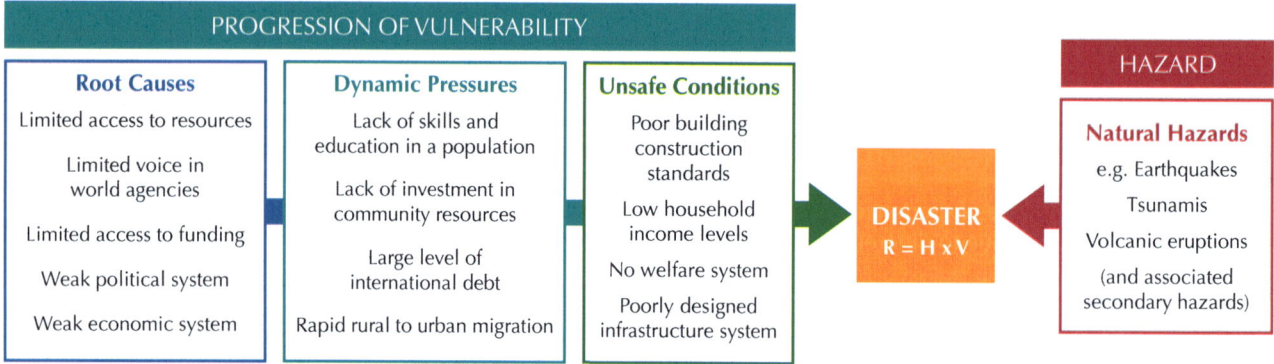

- The '**Pressure**' in the title of the model comes from the idea that there is increased **stress** from both causes of the disaster — the natural hazard itself — as well as layers of vulnerability. When either becomes too much, a disaster happens. In order to '**Release**' the pressure, either the nature of the hazard or the degree of vulnerability needs to be reduced.
- The PAR model highlights the importance of a country's **economic development** to its risk of experiencing a disaster. Many root causes for vulnerability are less likely to be a problem in developed countries.

Warm-Up Questions

Q1 Describe and explain the hazard risk equation.
Q2 Explain why some communities may be more vulnerable in natural hazard events than others.

Exam Question — AS and A-Level

Q1 Using the Pressure and Release model, explain the relationship between vulnerability and disasters. [4 marks]

Lack of revision could turn a hazardous exam into a disaster...

Lots of definitions to learn here, plus two models and an equation. Remember, a hazard isn't the same thing as a disaster.

Vulnerability to Tectonic Hazards

Improving a country's level of development is crucial for reducing its vulnerability to hazards.

Inequality, vulnerability and resilience are often linked

Sandy felt invulnerable in her new outfit.

Inequality is the unequal distribution of monetary and physical resources.

- A country's level of **vulnerability** and **resilience** to natural hazard events often reflects its levels of **economic**, **political**, **cultural** and **social development**.
- Generally, countries with a low level of development and a low score in the Human Development Index (see p.104) are more vulnerable and have an increased risk of experiencing a **disaster**.
- Where there are limited public funds, authorities are more likely to spend finances on **short-term**, urgent problems rather than long term **hazard mitigation strategies**. These are often very expensive and may not be needed for some time.
- Pre-existing issues within a developing country, such as **drought**, **external debt** or **civil unrest**, can increase the country's vulnerability when a natural hazard occurs.

Increased vulnerability and reduced resilience may be caused by inequalities in:

The **World Risk Index** measures the level of vulnerability and **exposure to hazards** in different countries. It's an index score based on four factors:

1) exposure to hazards
2) susceptibility to the impacts of hazards
3) the capacity to cope with the impacts of hazards
4) the capacity to adapt to the impacts of hazards

When compared with the income level, you can see a **correlation** between this and vulnerability.

Selected countries	World Risk Index (2021)	GDP per capita (US $ adjusted by PPP)
Switzerland	2.04	77 324
Luxembourg	2.53	134 754
Burundi	10.42	793
Niger	13.90	1 310

Education

- Populations with less access to **education** have a reduced chance of perceiving and understanding hazard risks.
- They may be unable to read **leaflets and posters** or lack access to **public information systems**, such as radio broadcasts, that advise them of safe practices. This means that when a natural hazard occurs, they are less likely to know how to react to the hazard and may make decisions which endanger their lives.
- Populations with low levels of education may view the ideas of **experts** and **scientists** with suspicion. When a natural hazard then occurs, they might ignore official advice that would make them less vulnerable.

Housing

- Populations living in **informal settlements** made from **low-grade materials** are at greater risk in a natural hazard event. These buildings may not follow **building codes**, so they are more likely to be unsafe. This means that homes and businesses are less likely to be able to withstand a hazard, putting lives and livelihoods at greater risk.
- After a natural disaster, populations from low-income groups do not always have the means to **rebuild**. People may still be living in substandard or temporary housing when the next natural hazard occurs.
- **Lower-income groups** are more likely to live in areas that have a higher level of **exposure** to hazards.

Health care

- Populations with poor access to **health care** are at greater risk of **illness**, **disease** and **death** after a natural hazard. This may be due to:
  1) a limited number of **healthcare facilities**
  2) fewer trained medical personnel
  3) a lack of medical materials and drugs
- Populations with poor healthcare systems may not have hazard **emergency response teams**.
- Populations from lower income groups are also more likely to be **undernourished** or suffering from lower levels of **immunity** to disease before a natural hazard occurs, so they have an increased vulnerability when that event happens.

Income

- Populations with a majority of people in **lower-income groups** are more likely to suffer in a disaster as they may lack the means to protect themselves. E.g. they may not be able to afford simple items (such as those you'd find in an **emergency kit**) that would keep them alive until help arrives.
- In the aftermath of a natural disaster, people from lower-income groups are unlikely to have **savings** that will allow them to **rebuild** their homes and lives. If work places are destroyed, people from lower-income groups are more likely to lose the ability to earn an income as they have no **means of employment**.
- The places of work of those in higher income brackets are more likely to have insurance measures in place that will protect the salaries of their employees.

Vulnerability to Tectonic Hazards

Governance influences a community's vulnerability and resilience to hazards

At both local and national levels, **governance** has a part to play in how communities are able to react to natural hazards.

- A key aspect of this governance is having **mitigation measures** (ways of being prepared for natural hazards) in place.

    E.g. widespread **education programs** for citizens and **operational systems** (such as flood defences) to deal with different eventualities.

- At a more basic level, governance needs to ensure that it is meeting the needs of its citizens and giving them the means to make themselves less vulnerable.

    E.g. a **healthy** and **literate population** is better equipped to cope with a disaster.

- Government agencies are also responsible for a degree of **environmental management** that can reduce the likelihood of **secondary hazards** occurring.

    E.g. programmes that **afforest bare slopes** can be put in place to reduce the likelihood of **landslides** during and after an earthquake.

- Some countries have **national disaster management agencies**. These work in the immediate aftermath of a disaster to try and reduce secondary impacts and to support people most in need.

    E.g. In 2022, **FEMA** (Federal Emergency Management Agency) in the USA had an annual budget of over US $29 billion, which it can use to **aid regions** once a state's governor has declared a **state of emergency**.

> **governance**
> *the systems of control that run a region or country*

FEMA set up emergency centres in response to Hurricane Harvey in 2017.

Local authorities are likely to have their own **natural disaster action plan**. This will be **location specific**, taking into account the particulars of a local population and the facilities available to it. If a natural disaster happens, it is local authorities who are best placed to coordinate emergency response teams and aid that may come in from other regions.

Geographical factors also influence a community's vulnerability and resilience

Population Density	Isolation and Accessibility	Degree of Urbanisation
• In some urban areas, infrastructure systems such as a **sewerage network** may come under pressure in normal conditions. A natural disaster may cause part of the system to shut down, decreasing its overall capacity to cope even further.	• Populations in **rural** or **isolated areas** are more vulnerable to natural disasters because it can be difficult for **rescue operations teams** to reach people in the immediate aftermath.	• Places with a high population density are more vulnerable to natural disasters as they are more difficult to **evacuate**, so it's more likely people will be left behind.
• Rapid levels of **urbanisation** mean that more people live in **informal settlements** which, are more vulnerable to hazards.	• Roads and bridges are more likely to be **poorly maintained** in these areas. This means these populations can suffer for longer without **aid** and **response teams**.	• In more densely populated areas, **emergency services** and **facilities**, such as hospitals and health clinics, are often **under pressure** due to the amount of people they need to cater for. This pressure increases after a natural disaster, as the need for medical facilities and emergency services is **likely to increase**.
• Urban areas have greater access to **food**, **shelter** and **medical interventions** that can aid a population's disaster recovery.	• Isolated areas may also suffer from a **lack of communication infrastructure** — e.g. it may be difficult to summon help if there is no **mobile phone service**.	
• Urban areas often have **larger** and **more vulnerable building structures** which are likely to be damaged by the natural disaster.		

Warm-Up Questions

Q1 Explain the link between education, economic development and vulnerability to hazards.

Q2 Describe the role that national governments have in reducing a community's vulnerability to hazards.

PRACTICE QUESTIONS

Exam Question — AS and A-Level

Q1 Assess the view that "location is the most important factor affecting a community's vulnerability to hazards". [12 marks]

My response to a disaster and these pages would be the same — HELP!

Eek, that was a lot of information to take in. Read over these pages again, then reward yourself with a nice cup of tea. Ahhh.

Hazard Profiles

To create a hazard profile, the magnitude needs to be measurable. Below are some examples of how this is done.

There are different ways of **measuring** the magnitude of hazards

Magnitude is the **size** and **intensity** of a natural hazard event. It is measured using a numerical score on a **scale** which is often based on scientific measurements from seismic instruments.

This means there are **inherent disadvantages** with using these scales:

- They can oversimplify the complexity and size of the hazard.
- Measuring a hazard based on its impact can be unfair, as a country's **economic development level** has a large role to play in this.

VEI	Volume of material ejected	Type of eruption	Size of Plume
0	< 10 000 m³	Hawaiian	< 100 m
2	> 1 000 000 m³	Strombolian	1-5 km
4	> 0.1 km³	Plinian	10-25 km

The Moment Magnitude Scale (MMS)

- The **MMS** (or **M_w**), measures the size of an earthquake in terms of the energy released.
- It's based on the total amount of **energy released** by an earthquake at the moment it occurs — this is the **seismic moment**. This is recorded at the **epicentre** by measuring the amount of shift detected at the **fault line** and the relative **resistance** of the affected part of the lithosphere.
- The scale is a **logarithmic scale** and has no upper limit. It has generally replaced the **Richter scale** because it is more accurate, especially for measuring the magnitude of large earthquakes.

The Volcanic Explosivity Index (VEI)

The **VEI** measures the magnitude of **volcanic eruptions** by grading volcanoes on a logarithmic scale from 0 to 8. This means that a VEI 4 volcanic eruption is ten times greater than one with a magnitude of VEI 3.
This composite index is based on:

- the amount of **material ejected**
- how **high** the plume reaches into the atmosphere
- the **duration** of the eruption
- a series of **qualitative** observations.

The Mercalli Intensity Scale

- This measures the impacts of an earthquake using **observations** of the event, e.g. reports and photographs. The scale is between 1 and 12.
- There are **disadvantages** to using the Mercalli scale:
  1) It is difficult to say that an entire earthquake episode is a certain point on the scale as one location may experience different impacts to another.
  2) It's **subjective** — witnesses may have **different opinions** about what happened during the tectonic event.

Intensity	Description
I	Only detectable by seismic instruments.
IV	Awakens people who are sleeping. Stationary vehicles are rocked. Standing crockery will rattle.
XII	Total damage. The ground surface moves like a wave. Large objects thrown into the air.

Hazard profiles aid the comparison of the **impacts** of tectonic hazards

Hazard profiles measure these **six physical characteristics**:

1 **Magnitude** — The higher the magnitude of a hazard the greater the chance of **fatalities** and **injuries**, and damage to **buildings** and **infrastructure**.

2 **Speed of onset** — The **quicker** the onset speed of a hazard, the **less time** people have to react it. This means they may be **less prepared** and will have less time to react to the hazard.

3 **Duration** — The longer a hazardous event lasts, the greater the delay there is in the **emergency response** and the **recovery**.

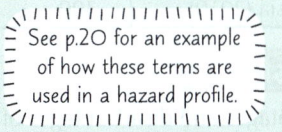
See p.20 for an example of how these terms are used in a hazard profile.

4 **Areal extent** — Typically, the larger the area affected by a hazard, the more people are at risk.

5 **Spatial predictability** — Being able to predict **when** and **where** a hazard will occur. If a hazard occurs on a well mapped and understood **plate boundary**, there is a higher level of spatial predictability and gives people time to plan and prepare for hazards.

6 **Frequency** — The more frequently a hazard occurs, the more systems may be in place to **manage its potential impact**. However, if the same hazard occurs frequently, it is **unlikely** a country will recover before the next hazard.

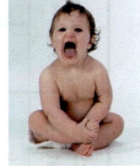
The magnitude, duration and frequency of Keith's screams made him a major hazard.

Hazard Profiles

Hazard profiles show the severity of tectonic hazards

Hazard profiles can be used to **compare** the same type of hazard taking place in different locations / different times. They can also be used to compare **different hazards** e.g. volcanoes and earthquakes.

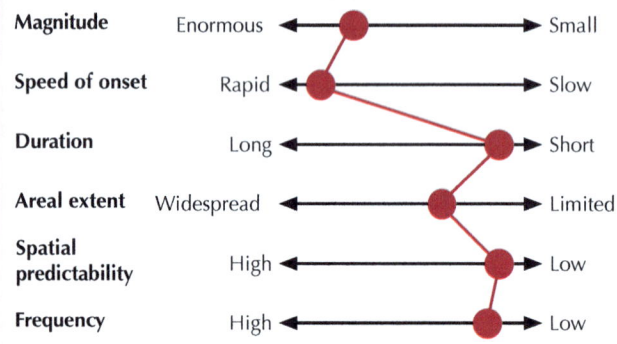

An example of a hazard profile for the 2008 Sichuan earthquake.

- The relationship between **magnitude** and **frequency** is one way in which populations can assess the degree of risk associated with a natural hazard. It helps those in hazard management **prepare** and **plan** for different eventualities.

- Hundreds of **low-magnitude earthquakes** happen around the world every day. Earthquakes of very high magnitude occur much less often. The number of earthquakes that occur globally varies from year to year. Seismic hazards do not seem to follow a clear pattern — their occurrence is largely **random**.

- The **magnitude** of volcanic events ranges from small, **slow lava** flows to **huge eruptions** of lava, ash and gas. The **frequency** of volcanic eruptions varies. Some **active** volcanoes erupt **once every 100 000 years** whereas others erupt every few months.

- Generally, less-frequent eruptions are larger in magnitude and more damaging. Some volcanoes erupt at **regular intervals**, whereas others may be **dormant** for hundreds or thousands of years, then erupt several times in quick succession.

- Hazard profiles can be thought of as useful **prediction tools** for potential impacts. For example, **characteristics** can link to both **direct impacts** and to other characteristics within the profile.

- However, each hazard is also a **unique event**. They are unlikely to follow any pattern of a previous, related hazard.

- Hazards are also **random in nature** and highly **unpredictable**. A hazard profile gives an **indication** of the types of impacts a population may feel, but it's often not useful for predicting **potential impacts** of a hazard.

	This table compares tectonic events that have taken place around the world.	Fatalities (a social impact)	Estimated total loss (US $mil) (an economic impact)	Magnitude	Speed of onset	Duration (of tectonic activity)	Areal extent	Spatial predictability	Frequency
Earthquakes	Christchurch, New Zealand (2011)	185	30 000	6.3 M_w	instant	seconds	220 km	medium	medium
	Sichuan, China (2008)	Over 70 000	130 000 -150 000	7.9 M_w	instant	seconds	1740 km	low	rare
Tsunamis (by epicentre)	Tohoku, Japan (2011)	Over 15 000	230 000 -300 000	9.0 M_w	2 minutes	2 days	18 256 km	high	rare
	Banda Aceh, Indonesia (2004)	Over 200 000	10 000	9.1 M_w	4 minutes	3 days	19 278 km	medium	very rare
Volcanic eruptions	Mount St Helens, USA (1980)	57	860	VEI 5	9 weeks	6 years	600 km^2	very low	every 100 yrs
	Mount Merapi, Indonesia (2010)	Over 300	700	VEI 4	1 month	20 months	~750 km^2	very low	every 10 yrs

Warm-Up Questions

Q1 Describe the advantages and disadvantages of using scales to measure magnitude and intensity.

Q2 Explain how a hazard profile may vary according to different types of tectonic hazard.

Exam Question — AS and A-Level

Q1 Assess the extent to which a hazard profile might indicate the potential impacts of a tectonic hazard. [12 marks]

Describe yourself in three words — spontaneous, exciting and a little fiery...

Ah. If a volcano had a dating profile. Just like dating profiles, hazard profiles provide all kinds of useful information to help you make comparisons between them. However, unlike dating profiles, they won't help you meet your soulmate. Probably...

Tectonic Hazards — Contrasting Impacts

There are lots of ways that tectonic hazards can impact a country. You need to know about how different types of tectonic events have affected countries at different levels of development.

The **severity** of a tectonic hazard's impacts depends on **many factors**

It is difficult to compare the impacts of one tectonic hazard with another as each occurs in its own **unique circumstances**:

- There tends to be **fewer fatalities** in **developed countries** compared to **developing ones**. This may be due to populations in developed countries being better **prepared** and **protected**.
- The **economic costs** of a hazard tend to be larger in developed countries. This may be due to the heightened cost of **rebuilding** what has been lost.
- **Volcanic eruptions** tend to have fewer impacts than **earthquakes** and **tsunamis**, largely because earthquakes happen without any warning.
- Hazards in developing countries often **displace** large numbers of people. Those left **homeless** often do not have the option of living in **temporary accommodation** provided by the state.
- The **time of day** and the **season** when the hazard occurs can have a large influence on the impacts it creates.

The **impacts** of earthquakes depend on a country's **development**

Developed Country — New Zealand

Date: 22nd February 2011, 12:51 pm   **Epicentre**: Christchurch   **Focus depth**: 5 km   **Magnitude**: 6.3 M_w

Plate movement: Conservative fault line created by movement of Pacific plate against Australian plate

- Seismologists believe the Christchurch earthquake was an **aftershock** of the September 2010 Canterbury earthquake. Unseen **structural damage** from the Canterbury earthquake may have contributed to the damage done by the aftershock.
- Seismologists think that **shockwaves** from the earthquake were amplified as they **rebounded** off the Port Hills to the south of the city.
- The region's **manufacturing zone** was located outside the major cities and so was relatively undamaged. This meant that business could return to usual quickly.
- Large areas of the city were built on **sandy deposits** with a **high-lying water table**, making them less stable.

The Christchurch earthquake happened in the middle of the day which meant there were a greater number of fatalities as more people were out and about.

Social	Economic	Environmental
• 185 fatalities • **Building infrastructure** undermined — 80% of water and sewerage system damaged • 10 000 homes demolished	• Over US $30 billion loss • **Cost of the NZ$ fell** amid market uncertainty • 60% of buildings in CBD damaged • Nights stayed by international visitors fell by 30%	Extensive areas of **soil liquefaction** (estimated upwelling of 200 000 tonnes of silt)

Emerging Economy — China

Date: 12th May 2008, 2:28 pm   **Epicentre**: Sichuan   **Focus depth**: 19 km   **Magnitude**: 7.9 M_w

Plate movement: Collision between the Indo-Australian and Eurasian plates

- Analysts believed that economic losses were going to be extremely high because of the many **businesses** and larger houses that have been built as China has progressed economically.
- The **level of damage** varied between neighbouring buildings, revealing evidence of poor building designs and companies not following building regulations adequately.

Social	Economic	Environmental
• Over 70 000 fatalities • At least 5 million made **homeless**	• US $130-150 billion worth of loss • **World price of oil fell** by $1.73 a barrel as it was feared that demand from China would drop • Large quantities of **livestock lost**	• Many large **landslides** occurred due to **destabilisation** • Over 2000 **dams** were **damaged** by the quake, including the **largest** dam in the area, Zipingpu Dam.

Tectonic Hazards — Contrasting Impacts

Developing Country — Nepal

Date: 25th April 2015, 11:56 am   **Epicentre**: Gorkha   **Focus depth**: 15 km   **Magnitude**: 7.8 M_w

Plate movement: Collision between the Indo-Australian and Eurasian plates

- Nepal's **mountainous landscape** made it more **susceptible to landslides** in the aftermath of the earthquake. **High deforestation rates** (for fuelwood and farming) made **steep mountain slopes** even more **vulnerable** to landslides.
- Residential buildings in Kathmandu are overwhelmingly **owner-built** (estimated 98%). Local authorities **didn't** check whether these buildings **met building permits** and people **weren't trained** to make their houses **earthquake-safe**.
- With the epicentre in a rural area, many villages were destroyed. There were increased fatalities as rescuers struggled to access the region due to the **mountainous terrain**.

Social	Economic	Environmental
• Around 9 000 • 4 million made homeless • Some **UNESCO world heritage sites** were destroyed in Kathmandu	• Loss of estimated US $5 billion • **Tourism numbers dropped** heavily in following months	• Aftershocks caused multiple **landslides**, raising the death toll • **Landslides** also blocked rivers • **Avalanches** occurred on Mt. Everest causing more deaths and injuries

The impacts of **tsunamis** also **vary**

Developed Country — Japan

Date: 11th March 2011, 2:46 pm   **Epicentre**: 130 km east of Sendai   **Focus depth**: 29 km   **Magnitude**: 9.0 M_w

Plate movement: Subduction of Pacific plate under Okhotsk plate

- Intricate patterns of **wave refraction** and **diffraction** (from islands) meant that tsunami waves heading south and east from the epicentre took nearly a day to reach other land masses, reducing their impact there.
- Japanese **lowlands** are found on the east coast, so waves were able to reach further inland.
- Waves rose up to 10 m in height at the coastline and travelled at around 800 km/hour in open water.

Social	Economic	Environmental
• Over 15 000 • 122 000 homes were **completely destroyed** • At least 4.4 million homes lost their **electricity supply**	• Estimated US $230-300 billion cost to the Japanese economy • Large scale **infrastructure damage** along the coastline (such as to **ports**)	• Damage to the Fukushima Daiichi **nuclear power plant** released dangerous levels of **radiation** (very long-term impact) • **Flooding** destroyed 21 000 hectares of **farmland**

Emerging Economy — Chile

Date: 27th February 2010, 3:34 am   **Epicentre**: 3 km off Pelluhue coast   **Focus depth**: 35 km   **Magnitude**: 8.8 M_w

Plate movement: Subduction of Nazca plate beneath South American plate

- A **tsunami warning system** was in place, but some say that the reactions were not quick enough to warn those living on the coast of the impending danger.
- The tsunami disproportionately killed **tourists** who were **camping** in **low-level coastal** areas. Local residents were **aware** that a tsunami was possible so **evacuated** to higher ground.

Social	Economic	Environmental
• Over 520 fatalities • **Shortages of food, water** and **fuel** led to widespread **looting** in Concepción	• Loss of US $30 billion • Talcahuano Port was badly damaged — **fishing** and **tourism** industries faced decline • **Prices of simple foodstuffs** tripled in some places	• Debris from collapsed buildings was dumped onto **fragile wetland** areas • **Pipe damage** meant that **unprocessed sewage** was **released** into the Biobio river

Tectonic Hazards — Contrasting Impacts

Developing Region — Indian Ocean

Date: 26th December 2004, 07:58 am   **Epicentre**: 250 km off the coast of Sumatra   **Focus depth**: 30 km
Magnitude: 9.1 M$_w$ **Plate movement**: Fault movement between Indo-Australian and Burma plate

- Though the epicentre was closest to Indonesia, fatalities resulting from the tsunami occurred across **14 countries** in total (Indonesia, Sri Lanka and Thailand were the worst affected).
- The coastlines around the Indian Ocean are particularly **low lying**, making **inland flooding** more likely.
- As the fault line lies in a **north-south orientation**, the greatest waves extended from the epicentre in an **east-west direction**. This meant that some low-lying countries (such as Bangladesh) received relatively few large waves, reducing their potential fatality rate.
- Due to the **high energy** of the waves and their **refractive tendencies**, the west coasts of India and Sri Lanka were also affected.
- **No tsunami warning systems** existed in the Indian Ocean prior to the tsunami.

Aerial photograph showing the aftermath of the 2004 tsunami on the Andaman and Nicobar Islands in the Indian Ocean.

Social	Economic	Environmental
• Over 200 000 fatalities • Over 1.5 million people made **homeless** around the Indian ocean coastline	• Economic cost totalled US $10 billion • **Jetties** and **fishing fleets** were washed away, causing the **fishing industry** to decline	• Damage to vulnerable **coral reefs** in the seas around the Maldives • Coastal **mangrove forests** were flattened • **Freshwater** supplies became **contaminated** with sea water

Volcanoes can cause huge losses too

The Mount St Helens volcano erupted on a Sunday. This reduced the number of casualties as there were fewer loggers in the forests surrounding the volcano.

Developed Country — USA

Date: 18th May 1980   **Volcano**: Mount St Helens   **Tephra types**: Pyroclastic flow, ash, gas, lava, volcanic bombs
Explosivity: VEI 5   **Plate movement**: Subduction of Juan de Fuca plate beneath the North American plate

- The eruption was triggered by a significant earthquake. This then created a **landslide** on the volcano slopes. The earthquake came after two months of **minor tremors** and **gas releases**.
- As the north face of the volcano started to **bulge**, an **exclusion zone** was set up around the volcano six weeks before the main eruption. However, due to **logging activities** within much of the area, the exclusion zone was **much smaller** than it should have been.
- A **large-scale media presence** made everyone in the area aware of the danger and gave advice about how to best prepare for the eruption in the days leading up to it.

Helen's name matched her explosive personality.

Social	Economic
• 57 fatalities • Poor **visibility** closed some highways around the volcano for over a week • **Commercial flights** were cancelled	• Estimated loss of US $860 million • **Harvests** of agricultural crops were **lost** from the ashfall • 9.4 million m³ of **destroyed timber** created widespread job losses in the timber industry

Environmental
- **Glacial meltwater** mixed with the ash to produce extensive **lahars**
- The whole fallout zone was reduced to an ash and mud-covered **wasteland**
- The **landslide** flowed into Spirit Lake and dammed the Toutle River
- 19 miles of **forest** were **levelled** from the **initial shockwave**

Aerial photograph showing the changed landscape around Mount St Helens.

Tectonic Hazards — Contrasting Impacts

Emerging Economy — Indonesia

Date: 25-26th October 2010   **Volcano**: Mount Merapi   **Tephra types**: Ash, lava, pyroclastic flow

Explosivity: VEI 4   **Plate movement**: Subduction of Indo-Australian plate beneath the Eurasian plate

- **Avalanches**, and increases in the size of the dome, were witnessed a month before the main eruption.
- During the eruption, **ash** fell heavily on **rice fields** and other **crops**, ruining the harvest and placing local farmers in a poor financial situation.
- Over 300 000 people were **evacuated**, including those from a **20 km exclusion zone** around the volcano, but many didn't want to leave their homes or land.

"My rice tastes a bit weird..." said Ash, "maybe it's organic."

Social
- Over 300 fatalities
- 400 000 people internally **displaced**
- Many people reported severe **respiratory problems**
- Large scale disruption to **airline activity**

Economic
- Loss of over US $700 million
- Heavy ash fall flattened crops
- The **price of basic foodstuffs** increased

Environmental
- Deposits of **ash** caused the **destruction of forests** found on the **slopes** in Mount Merapi National Park
- **Lava** flowed into the six main rivers on the volcano slopes, halting their flow, and created **new channels**

Developing Country — Democratic Republic of Congo

Date: 17th January 2002   **Volcano**: Mount Nyiragongo   **Tephra types**: Lava, gas

Explosivity: VEI 1   **Plate movement**: Divergent movement within fault line in the East African Rift Valley

- Years of **civil war** and violence in rural areas had caused many people to **flee to cities** such as Goma, increasing their vulnerability. Civil unrest also made some **relief efforts** more **difficult**.
- There were **warnings**, which meant around 350 000 people **fled** the **lava flow**.

Mount Nyiragongo releasing steam.

Social
- Over 100 fatalities and over 120 000 people made **homeless**
- Some deaths occurred due to **asphyxiation** and **respiratory problems**
- 300,000 people from the city of **Goma** were **evacuated** to **Rwanda**

Economic
- Lava flows destroyed 10-15% of the city of Goma on the bottom of the slope
- Large-scale **building collapses**
- Many **businesses** were **destroyed** in Goma, resulting in a 15% **increase** in **unemployment**

Environmental
- **Lava** flowed to Lake Kivu which created a new 120 m long **delta**
- **Trees** on the upper northern slope were **destroyed** by ash

Warm-Up Questions

Q1 Describe how the number of fatalities may be related to the type of hazard occurring.

Q2 Describe how a volcanic eruption might have an impact on the environment.

Exam Question — AS and A-Level

Q1 Assess the extent to which the impacts of a tectonic hazard are shaped by a country's level of economic development.

[12 marks]

Well, this section was certainly impactful...

For your exam, you need to know examples of volcanic eruptions, earthquakes and tsunamis in countries at different levels of development. You might have learnt different ones in class, and it's fine to use them — but make sure you learn all the details.

Tectonic Disaster Trends

Since 1960, there have been some interesting trends in tectonic disasters.

Trends in disasters have changed since 1960

1) Some hazards are not entirely natural. **Human activity** can affect the frequency and intensity of **atmospheric hazards**. E.g. **tropical storms** and **droughts** may occur more frequently due to the **enhanced greenhouse effect** and **climate change**.

2) The actions of humans on the natural environment, e.g. **deforestation** and **land degradation**, can exacerbate the impacts of tectonic disasters and, in some cases, make them occur more frequently.

3) In general though, tectonic disasters have not seen a big change in **frequency** or **magnitude**.

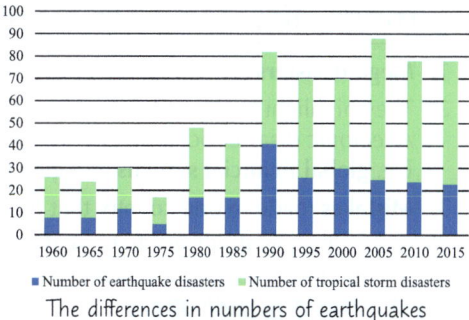

■ Number of earthquake disasters ■ Number of tropical storm disasters

The differences in numbers of earthquakes and tropical storm disasters over time.

The growth of global media gives the impression that disasters occur more regularly

- Tectonic disasters are more frequently in the **news**. The rapid increase in the scale of **global media communications** has made it easier to report on disasters. The media often focuses on natural disasters that have a **fast speed of onset**, as these may seem to be more dramatic. Natural disasters that affect **developed nations** tend to receive **more global media coverage** than disasters affecting developing nations.

- More disasters may be reported due to improvements in the ways they are **monitored** and **recorded**. Monitoring systems have become more **sensitive** and **integrated**, generally reducing the need for humans to manually take observations.

The reliability of tectonic disaster data varies for several reasons. E.g. data can be difficult to collect and quantify, especially in remote areas.

Make sure you remember the difference between a hazard (something that's a potential threat to human life or property) and a disaster (when a hazard seriously affects humans).

There are ways to explain the trends in the intensity of tectonic disasters

 **The number of people who are affected by tectonic disasters is increasing over time.**
- The number of people living in locations considered at **risk** of tectonic disasters is **increasing**.
- This means that as people live in more **urbanised** and **dense living conditions**, they become more vulnerable to the effects of tectonic disasters. The number of people who find themselves homeless or in need of long-term support following a disaster has increased

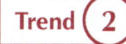 **The impact of tectonic disasters on the population has generally fallen.**
- The number of **fatalities** due to tectonic disasters has **fallen**. This is largely due to populations being **better prepared** for tectonic disasters. E.g. countries are more likely to invest in buildings designed to withstand earthquakes, and new buildings may have to conform to certain **building codes** and **regulations**. However, **developing countries** may **struggle** to invest to the **same extent** as **developed countries**.
- While a **development gap** exists between developed and developing nations, general wealth is rising and **absolute poverty is reducing**.
- Improvements in **communication technology** mean that warnings of, and the response to, tectonic disasters is often quicker and better coordinated e.g. **satellite communications**.

 **The economic impact of tectonic disasters is increasing over time.**
- A rise in relative wealth has led to an increase in losses. A developing country usually requires less money to **rebuild** to its previous state, compared to a developed country with more complex infrastructure.
- As more countries have growing numbers of **middle-class** citizens, **property**, **infrastructure** and **services** become more **expensive** to replace if lost in a tectonic disaster.

Warm-Up Questions

Q1 Outline the differences between trends of atmospheric and tectonic hazards since 1960.

Q2 Explain why the reporting of tectonic hazards may have increased.

Exam Question — AS and A-Level

Q1 Explain why the number of tectonic disasters has increased since 1960.   [4 marks]

Have you seen the latest trend? — it's pretty groundbreaking...

Now that you're learning about tectonic hazards, you're bound to start hearing about them non-stop.*

* If you don't believe me, look up the Baader-Meinhof phenomenon — it's not in your exam, but it's pretty fascinating stuff.

Tectonic Mega-Disasters

Mega-disasters are extremely rare, though three of significance happened in the first 15 years of the twenty-first century.

A **mega-disaster** is one that has a regional or global **significance**

1) **Mega-disasters** are sometimes known as **high-impact**, **low-probability** (HILP) events.

2) They highlight how much **interdependence** there is between different regions of the globe.
A tectonic disaster occurring in one country can have a knock-on effect in many others in direct and indirect ways.
E.g. **global financial systems** and **global supply chains** inherently link countries of every development level.

3) **Mistakes** made in one country that experiences a mega-disaster can **prompt others** to act to **prevent** a similar event.

4) Many disasters **don't reach** the **scale** of a **mega-disaster** because there are greater levels of **preparedness** in countries and **actions taken** in the aftermath may be **quicker** and **more efficient**.

2004 Indian Ocean tsunami

See p.23 for a hazard profile of the Indian Ocean tsunami.

- Several countries around the Indian Ocean received **tsunami waves** due to the earthquake off the Indonesian coast. 14 of these countries suffered fatalities. Additionally, **tourists** from all over the world were killed in the disaster.

- Many of the affected countries suffered losses in their **fishing** and **tourism** industries. The whole region suffered as visitors were scared of returning to resorts.

- Coastline **ecosystems** around the Indian Ocean that are environmentally sensitive (e.g. **mangrove forests**) were uprooted by excess sea water and debris. This affected the wider marine ecosystem and organisms at all **trophic levels**.

Mangrove forests, Sri Lanka

2010 eruption of Eyjafjallajökull, Iceland

- Countries across Europe had to cancel **commercial flights** due to the **dust and ash cloud** that was produced by the eruption. The aviation industry estimated that this cost them over €2.2 billion.

- The 110 million m³ of tephra prevented planes from being able to operate safely, leaving **tourists stranded** away from home.

- **Perishable goods**, such as vegetables and cut flowers from Kenya, could not be transported to the UK and other European countries through **air freight**. This led to temporary **unemployment** in Kenya and the **loss of income** to the industry.

- Other industries, such as **car assembly plants** in Asia, had to stop production when delivery of key **components** by air freight from Europe was affected.

Eruption of Eyjafjallajökull and its ash plume

See p.22 for a hazard profile of the 2011 tsunami in Japan.

2011 Tohoku tsunami, Japan

- The tsunami waves had a significant impact, but the wide-reaching economic impacts were also very significant.

- Disruption to **Japanese factories** and ports had an impact in the **global trade** of a range of **component parts** for consumer electronic goods, as well as higher order goods such as **cars**.

- The vulnerability of **nuclear power stations** to natural hazards was highlighted by this tsunami. This led to other countries questioning how safe their nuclear power plants were. E.g. following the closure of the Fukushima power plants, Germany decided to **decommission** all of its nuclear power stations by 2022.

- In the longer term, the tsunami increased **carbon emissions** from Japan as the country moved more to **fossil fuel based energy production**. The price of **natural gas** rose as demand from Japan entered the market.

- **Debris** from **collapsed buildings** and **infrastructure** was swept out to sea and affecting marine ecosystems.

Warm-Up Questions

Q1 Define the term 'mega-disaster'.

Q2 Explain why the eruption of Eyjafjallajökull could be considered a mega-disaster.

PRACTICE QUESTIONS

Exam Question — AS and A-Level

Q1 Assess the view that the world is likely to see an increasing number of mega-disasters in this century. [12 marks]

Mega-disaster 2 — Tectonic Terror
Unfortunately it's not the title of a block-buster hit... In reality, mega-disasters have large-scale, devastating impacts.

Multiple-Hazard Zones

Some locations are at risk of several hazard types. The Philippines is one example of a multiple-hazard zone.

Multiple-hazard zones are sometimes known as 'disaster hotspots'

Populations living in **multiple-hazard zones** are extremely vulnerable to **natural disasters**. Some parts of the world are extremely prone to hazards:

- Japan, Indonesia and the Philippines are all considered to be multiple-hazard zones. They combine **active tectonic areas** with **hydrometeorological hazards** such as **cyclones**. In some cases, there are also **mountain belts** which can be prone to **mass movements** such as **landslides**.

- Other multiple-hazard zones include those affected by **El Niño** and **La Niña**, where variations in **annual climate systems** create unusual temperature ranges in the sea and the atmosphere, as well as periods of extreme rainfall or drought.

The aftermath of Typhoon Haiyan (2013) in the Philippines

Vulnerability also comes from a **lack of recovery time**. In areas where multiple hazards can happen within one year, a country may have limited opportunity or funding to **rebuild and protect** itself from further hazards.

Disasters can occur when **multiple hazards combine**

CASE STUDY

1) The Philippines is a group of over 7000 islands in South East Asia. The area is vulnerable to a variety of hazards with **social**, **economic** and **environmental** impacts.

2) This **vulnerability** is increased by its **rapidly growing population** (which mainly occupies coastal areas) and relatively **high levels of poverty**.

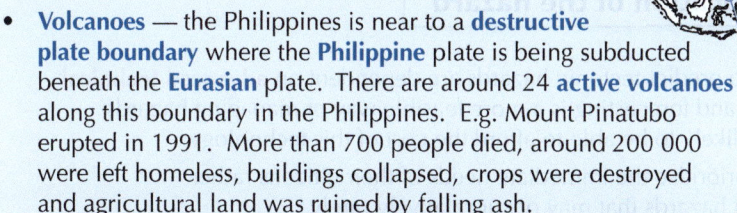

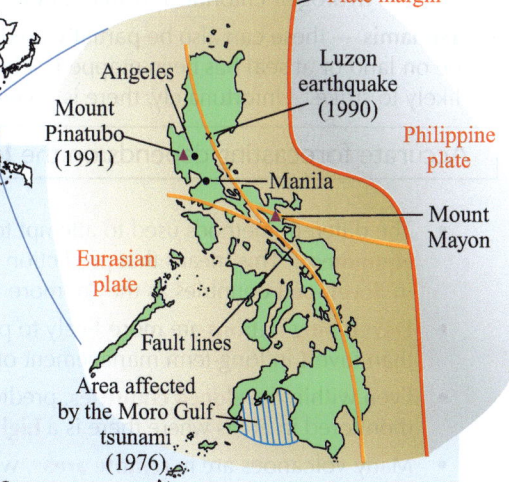

- **Volcanoes** — the Philippines is near to a **destructive plate boundary** where the **Philippine** plate is being subducted beneath the **Eurasian** plate. There are around 24 **active volcanoes** along this boundary in the Philippines. E.g. Mount Pinatubo erupted in 1991. More than 700 people died, around 200 000 were left homeless, buildings collapsed, crops were destroyed and agricultural land was ruined by falling ash.

- **Seismic hazards** — there are also **earthquakes** along the plate boundary and at **fault lines** where the plate has cracked under pressure. E.g. an earthquake of magnitude 7.8 M_w occurred on Luzon island in 1990, killing over 1500 people. Earthquakes in the surrounding oceans can cause **tsunamis**. E.g. in 1976 an earthquake of magnitude 7.9 M_w caused a tsunami that hit the coastline around the Moro Gulf. Thousands were killed and several cities were devastated.

- **Tropical storms** — the Philippines receives around twenty **tropical storms** every year. They develop in the Pacific Ocean and move westwards over the islands. E.g. Typhoon Haiyan swept across the city of Tacloban and the surrounding area in 2013. **High winds** and **torrential rain** destroyed homes and caused **flooding**, **landslides** and the **loss of power** and **water**. Around 8000 people died, and the total cost of damage was approximately US $13 million. **Atmospheric hazards** can increase the intensity of a **tectonic hazard** — e.g. during the 1991 Mount Pinatubo eruption, the Philippines was also struck by Typhoon Yunya. **Lahars** (see p.14) were created by the combination of volcanic ash and heavy rainfall from the storm.

Warm-Up Questions

PRACTICE QUESTIONS

Q1 Describe what makes a location a 'multiple-hazard zone'.

Q2 Explain why multiple-hazard zones are extremely vulnerable to the effects of natural disasters.

Exam Question — AS and A-Level

Q1 Explain what makes the Philippines a multiple-hazard zone. [4 marks]

They think it's all over — nah...

*Volcanoes, earthquakes, tsunamis **and** storms — the poor Philippines gets a pretty tough time of it. Make sure you learn the details of the multiple hazards in the Philippines if you want top marks. Then, turn the page for how hazards are managed...*

Managing Tectonic Hazards — Theory

Tectonic hazards aren't just scary when they happen — they can also be a nightmare to predict.
So, when it comes to managing them, things can get pretty complicated.

Scientists aim to **predict** and **forecast** tectonic hazards

Some **warning signs** before a volcanic eruption can help **predict** when they will occur. For example, **gases** may be released and **temperatures** around the volcano increase. A **forecast** will provide a **percentage chance** of the eruption happening in that location.

- **Prediction** involves stating **where and when** the next tectonic hazard will occur.
- **Forecasting** is giving a measure of the **likelihood** of a tectonic hazard happening.

Precursors (such as **foreshocks** before larger earthquakes) can be useful when trying to predict the larger and more significant hazard. There is no set pattern between precursors and the actual hazard, so there is a danger that false predictions might be made.

- **Earthquakes** — there are no proven methods for predicting earthquakes. Based on years of previous seismic data, scientists can use forecasting to identify **high-risk areas**. Much of this is based on **seismic gap theory**. This is the idea that areas along a **plate boundary** that have not experienced an earthquake for some time become more likely to experience one in the future.
- **Volcanoes** — volcanic eruptions can be partially predicted. Volcanoes give off **geophysical signs**, which indicate changes beneath the surface of the Earth. This helps scientists to predict that an eruption may be about to take place. Scientists also use a range of **volcanic monitoring equipment** to measure these changes. E.g. **tiltmeters** show changes in the gradient of the volcano's slope and **seismometers** record small earthquakes that may indicate movements in the **magma chamber**. **Gas meters** detect **gas emissions** (such as hydrogen chloride and sulphur dioxide) which might indicate **ruptures** just below the surface.
- **Tsunamis** — these can also be partially predicted. Once an earthquake has taken place and its location (which can be on land or at sea) has been mapped, scientists can use **modelling** to predict when and where tsunami waves are likely to strike. Unfortunately, there is no way of knowing where and when the initial earthquake will take place.

Accurate forecasting depends on the **location of the hazard**

- The different methods used to attempt to predict tectonic hazards are dependent on **advanced technical equipment**. This means that prediction and forecasting is a more feasible way of managing hazards in **developed** countries as they're more likely to be able to afford the cost of this technology.
- **Developing nations** are more likely to prioritise the immediate needs of their citizens rather than invest in long-term management of hazards that may occur relatively rarely.
- Even within **developed countries**, prediction equipment is more likely to be used and more intensely monitored in areas where there is a **high population density** — **rural areas** are more likely to be monitored less.
- Many volcanoes are in **remote areas**, which makes access for **monitoring difficult**.

Each stage of the **Hazard Management Cycle** is important

- The **Hazard Management Cycle** shows how **authorities** can think about the hazard in its entirety to best manage it. The responsibility for hazard management rests with many **key players** working together: governments, international NGOs, businesses, emergency planners and local community groups should all be able to work together.
- There are **four stages** that authorities go through in managing hazards. It is represented as a **cycle** because hazard events keep happening and efforts to prepare for them or mitigate their effects are ongoing. The lessons learnt from one disaster inform the management of the next.

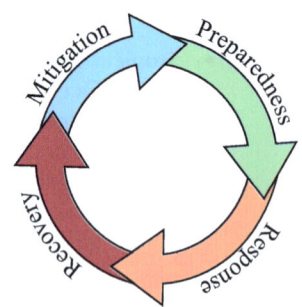

1 Preparedness — planning how to respond to a hazard

An authority can make sure there are **warning systems** in place and **educate** people about how to **evacuate** safely. **Evacuation holding zones** can be established and there may be the possibility of **stockpiling food and medical supplies** should an emergency take place. Some countries may also run practice **drills** so that citizens and those in the **emergency services** have experience of how to react to a hazard.

2 Response — how key players react when a disaster occurs

People may be evacuated and there might be **emergency services** rescuing people who are trapped underneath collapsed buildings. Those with serious injuries will be **prioritised**, and **access** to outside regions through roads, bridges and airstrips needs to be made available.

Andy's hazard management cycle was going badly — he kept falling off.

Managing Tectonic Hazards — Theory

(3) Recovery — getting the affected area back to normal

Houses might need to be **repaired** or **rebuilt**, and services such as **medical care** and **electricity** will need to be restored. Injured people will be cared for. **Businesses** will reopen alongside public services such as **schools**. **Infrastructure** such as roads, rail lines and ports will be re-established to allow for greater regional and international connections.

(4) Mitigation — minimising the impacts of future disasters

A **hard-engineering defence** (see p.30) could be built or **land-use zoning** could be introduced. Authorities may introduce **building codes** to ensure homes, businesses and **infrastructure** meet higher standards than they did previously. Mitigation can happen before a hazard occurs or afterwards when the area is recovering.

- **Different countries** spend different amounts of time in each stage — this depends on the hazard's **magnitude** and **intensity**.
- Stages like mitigation are never really reached in **multiple-hazard zones** as the country is in a constant state of response and recovery.

> Developing countries are more likely to focus on recovery as they may not be able to fund preparation or mitigation.

The Park Model shows how people respond to hazard events

- The **Park Model** (also known as a **Disaster Response Curve**) shows how responses progress during a hazard event.
- This model can help planners predict what resources will be needed at each stage and prepare for **future hazard events**. E.g. the reconstruction phase of the model shows that conditions can be improved after a hazard occurs, such as by designing **hazard resistant buildings** or installing **warning systems**. These then help to mitigate the impacts of future disasters.

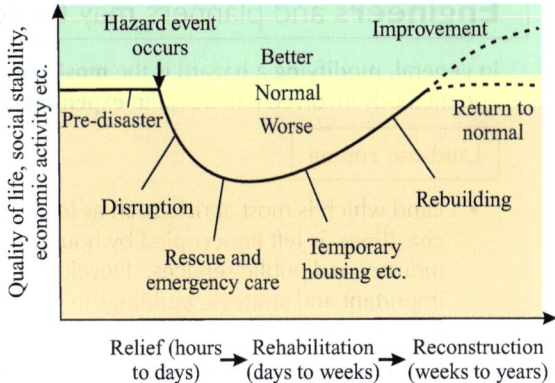

(1) Pre-disaster — before the hazard event, the situation is normal.

(2) Disruption — At the start of the disruption, there may be an advanced **warning** to allow for procedures such as **evacuations** to take place. During and directly after the hazard event, there may be the **destruction of property** and **loss of life**. People's **quality of life declines** along with their access to economic activities. The operation of services deteriorates.

(3) Relief — in the aftermath of the hazard event, rescue efforts focus on saving people and preventing further damage.

(4) Rehabilitation — once the immediate impacts are under control, people start to resolve longer-term problems such as providing **temporary shelter** and **aid** for those affected.

(5) Reconstruction — **rebuilding** permanent houses and infrastructure. This results in one of **two** outcomes:

1) If buildings etc. are built to the **same standard** as before, the area returns to normal and its vulnerability to hazard events remains the same.

2) If buildings etc. are built to a **higher standard** than before, the area improves and hazard **vulnerability** decreases. There will be eventual **economic recovery** and a return to focus on mitigation and prevention strategies.

- The shape of the curve will be different for different types of hazard. Those where the **speed of onset** is fast are likely to see more initial **disruption**. Hazards with a larger **magnitude** will possibly have an extended **recovery** time.
- **Developed countries** will likely have a shorter recovery time and greater opportunity to 'build back better'. Equally, if a country has an advanced warning system, the disruption is likely to be far less. **Developing countries** may spend **longer** in the **relief** and **rehabilitation** stages as they **may not** have had the available **funds** to adequately **prepare** for hazards, which may make them need more **immediate aid**.

Warm-Up Questions

Q1 Explain the rehabilitation stage of the Park Model.
Q2 How can volcanic eruptions be predicted?

Exam Question — AS and A-Level

Q1 Assess the importance of the different stages of the Hazard Management Cycle.   [12 marks]

Disaster Response Curves are a walk in the Park (Model)...

Sorry, that joke was pretty bad. If anything, the Park Model looks like a weird skateboard ramp. It's not quite as fun though.

Managing Tectonic Hazards

Unless you've been living in a cave, you'll probably be aware that seismic hazards have some pretty serious impacts. My response to this is to live as far away from the hazards as possible — funnily enough, in a cave.

There are **different approaches** to **hazard management**

- Hazard management may involve trying to **prevent** a hazard from occurring or **reducing** a hazard's **magnitude**. This isn't possible for some hazards, such as earthquakes, but for others, it may be possible by undertaking activities such as **building flood defences**.
- **Risk sharing** involves sharing the **costs** of reducing a hazard, the **benefits** of preventing it or the **costs** of not preventing it. E.g. people buy **insurance** to help them repair their property after a disaster. Most people will not be affected by a particular event, so they will not claim on their insurance. This means lots of people **contribute** so the cost is **shared**.
- People might try to **mitigate** the impacts of a hazard. This could be by **prediction** — working out where a hazard is likely to happen and then working out how to respond to it (e.g. **evacuations**). It could also be by **adaption**, such as by building earthquake-resistant features into homes and offices.
- **Governments** may **coordinate responses** to a hazard to manage it effectively. Management of a hazard may also involve the coordination of other countries' governments or organisations.
- Some people believe that hazards simply cannot be avoided, so they must be **accepted**. In this case, **no management** is applied.

Engineers and planners may try to **modify** a hazard

In general, **modifying** a hazard is the **most preferable option** available because the hazard hasn't impacted anyone yet. The methods involved can be quite **expensive** and so are often only available to **developed countries**.

Land-use zoning

- Land which is most at risk, such as **low-lying coastlines**, is left **unoccupied** by housing, industry and public services. Developers build important and **strategic buildings** in places that they perceive to be the safest areas.
- E.g. authorities can prevent the land around **volcano fallout zones** from being developed, or prevent construction work in areas where **soil liquefaction** may occur.
- Land-use zoning can be very **cost effective**, but it is **difficult to enforce**, particularly when building on high-risk land is attractive to businesses.

Earthquake-resistant buildings

- There have been large advances in the design of **earthquake-resistant buildings**:
  1) More **flexible materials** and **deep foundations** can absorb an earthquake's energy.
  2) **Cross bracing** and **counterweights** can reduce the amount of damage an earthquake can do as they **strengthen** the **structure** of the buildings.
  3) In tsunami-prone areas, buildings are sometimes raised up with **open foundation**, reducing the amount of damage to the buildings.
- While these initiatives work well on new buildings, it can be extremely difficult to **retrofit** older buildings with these modifications.

Hard-engineering defences

E.g. **embankments** or **offshore breakwaters** might be built to reduce the flood risk from tsunamis. These can give local people a strong **sense of security**.

Soft-engineering strategies

E.g. the **planting of mangrove forests** can protect coasts from natural hazards such as tsunamis and storms, as well as reducing coastal erosion.

Base isolation systems reduce the movement of buildings in an earthquake.

The construction of an 18 m high embankment wall to protect Hamaoka Nuclear Power Station, Japan from tsunamis.

Lava and lahar flow prevention

- In volcanic regions, **lava diversion channels** can be used to guide lava away from the most populous areas. Lava might also be **sprayed with seawater** to cool and solidify it before it potentially hits a populated area. Unfortunately, this is only useful for some types of volcanoes (those with basaltic lava (see p.13).
- **Lakes** can be **drained** to reduce the risk of **lahars**.

Managing Tectonic Hazards

There are ways of modifying the **vulnerability** of a community

Monitoring systems

- Authorities can install **monitoring systems** to predict when an eruption might occur, e.g. **tiltmeters** that measure how the angle of a volcano slope changes and **gas detectors** to sense minor emissions.
- **Remote sensing** via **satellite** and **submarine detectors** can monitor earthquakes and ocean waves after a tsunami and enable authorities to produce **tsunami warning maps**.
- Monitoring systems can be expensive but generally allow enough time for **evacuation procedures** to be carried out. However, a monitoring system is only worthwhile if people are **prepared** to act on the data it produces. Plans should be in place to evacuate people if there is an eruption and there should be a way of **restricting people** from entering the area around the volcano.

Warning systems

Warning systems rely on a country's ability to **communicate information** to a population rapidly. **Radio** and **social media** are increasingly used, as well as **television interruptions** across multiple networks.

Students practise an earthquake drill in Nepal.

Community preparedness

- Authorities should brief people on where their nearest **emergency shelter** is located and advise them to make an **emergency hazard kit** which can be taken if they need to evacuate quickly.
- Evacuations can be very **expensive** procedures, especially for a **large population**. This means the decision to evacuate must be based on **strong evidence**. If authorities ask people to evacuate too frequently, they are less likely to take these requests seriously.

Education

- **Evacuation routes** should be well **signposted** and people should be aware of simple ways to protect themselves when a hazard strikes. This might be done through **practice drills**.
- Children are educated in **schools** about safe and unsafe places during an earthquake. Children are also taught how to perform emergency **first aid**.

Losses can be **modified** in the **aftermath** of a hazard

Modifying the losses experienced during a hazard is the **least desirable** method, as the hazard has already had an impact.

Emergency aid

- Emergency aid may consist of **community search and rescue** teams or fire service units trained to deal with the immediate aftermath of a hazard.
- **Temporary infrastructure**, such as roads and generator-led electricity supplies, may also be set up, but emergency aid is usually more difficult to provide in isolated areas.
- It may be necessary to set up **field hospitals** as existing hospitals might be too badly damaged to use.
- In the immediate aftermath of the hazard, **local people** are the first to respond and so communities have an important role to play.

Overseas aid

- **Aid from overseas** may come at any stage of the aftermath.
- In the **short term**, this may consist of essential **medical supplies**, **food**, **water** and **fuel**, as well as personnel to work alongside local authorities.
- In the **long term**, this may be **financial aid** to redevelop and rebuild areas that have been destroyed. Financial aid may come from governments or from non-governmental organisations (NGOs) who redistribute **donations** made by citizens of other countries.

Insurance

It may be possible to **insure** businesses and homes against the effects of tectonic hazards. If insurance is in place, it can help people to **financially recover**. Unfortunately, insurance does little for those who can't **afford** it.

Warm-Up Questions

Q1 Describe what can be done to modify the vulnerability of a community.

Q2 Explain why some countries may not be able to modify a hazard before it has happened.

Exam Question — AS and A-Level

Q1 Assess whether modifying the event or modifying the losses is more effective in managing a hazard. [12 marks]

I'm struggling to *engineer* a joke out of these pages...

*Luckily, I **managed** — and you've managed to make it to the end of this extremely hazardous topic. Only 8 more to go... *gulp**

Climate Change and Glaciation

Climate change isn't a new thing. Planet Earth and the conditions on it are constantly changing.

Earth's climate has **changed many times**

Throughout Earth's history, the **climate** has shifted between **two** dominant forms:

Earth is experiencing icehouse conditions, but is currently in an interglacial period (see below).

Greenhouse conditions

- The climate is **relatively warm**, causing glaciers and ice sheets to **retreat** or **disappear**.
- The Earth becomes free from **very large ice masses**.
- **Meltwater** enters the oceans **raising** the sea level.

Icehouse conditions

- The climate cools to such an extent that **precipitation** falls as snow and hail instead of rain.
- This creates new glaciers and increases the spread of ice sheets at the **poles**.
- Ice holds water in a **solid state** lowering the sea level.

There have been **multiple glacial** and **interglacial periods** in the **Pleistocene**

1) The **Quaternary period** has lasted from about **2.6 million years ago** to the present day. It includes the **Pleistocene epoch** (from 2.6 million years ago to 11 700 years ago) and the **Holocene epoch** (from 11 700 years ago to present day).

2) During the Pleistocene epoch, there were fluctuations in global temperatures — colder **glacial periods** (when glaciers advanced and sea levels fell) were interspersed with warmer **interglacial periods** (when ice retreated and sea levels rose).

3) **Glacial periods** typically last for 100 000 years and **interglacial periods** last for 10 000-15 000 years.

Although there wasn't continual ice coverage during the Pleistocene epoch, there was much more than in the Holocene epoch.

- The **Last Glacial Maximum** (when **ice sheets** were at their largest) that affected the UK was about 21 000 years ago.
- Polar ice sheets covered much of the UK. Most of southern Europe was **periglacial** (an area that goes through frequent cycles of thawing and freezing) (see p.50).
- The Earth is currently in an **interglacial period**.

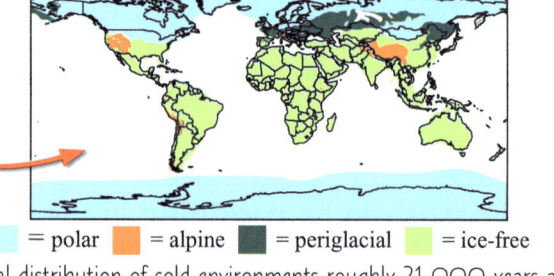

☐ = polar   ☐ = alpine   ☐ = periglacial   ☐ = ice-free

Global distribution of cold environments roughly 21 000 years ago

Long-term factors leading to climate change are outlined by the **Milankovitch cycles**

Milankovitch suggested that Earth's surface temperature varies due to **three cycles**:

Eccentricity Cycle
- Over time, the shape of the Earth's orbit changes from being circular to more elliptical. This changes the amount of solar radiation the Earth receives.
- The Earth is relatively further from the Sun when the orbit is more circular (meaning it receives less solar radiation), causing a glacial period to occur.

The shape of the orbit changes about every 100 000 years, which closely matches the glacial-interglacial cycle.

Obliquity Cycle
- The tilt of the Earth's axis changes between 22.1° and 24.5° — the current tilt is 23.4°.
- There is a smaller climatic difference between summer and winter when the tilt is 22.1°, meaning snow and ice is more likely to last the summer season, leading to a build up over time. At 24.5°, the temperature range between the seasons will be greater.

The axial-tilt cycle takes 41 000 years.

Axial Precession
- The Earth goes through long cycles of 'wobbles' on its axis, meaning different parts of the Earth go through phases of facing towards or away from the Sun.
- This causes the seasons to last different lengths of time. The longer the winter lasts, the more snow and ice remain on the Earth's surface.

These cycles last 26 000 years.

Milankovitch cycles are the **main driver** of **long-term** climate change. Some scientists believe that Milankovitch cycles (and the interaction between them) are a way of **kickstarting** feedbacks which lead to further climate change, e.g.:

- The **albedo effect** — where **land and sea ice** reflect **solar radiation** back into the **atmosphere**, creating cooler conditions which allow more ice to form (see p.34).
- During interglacial periods, **cloud cover** is greater due to increased **evaporation**. Clouds reflect solar radiation back into space, cooling the Earth's surface.

Climate Change and Glaciation

Solar output and volcanic eruptions are **short-term factors** leading to climate change

The climate changes on the previous page lasted hundreds of thousands of years, but some climate changes don't last nearly as long as others. **Two causes** of these relatively short-lived changes are **solar output** and **volcanic eruptions**.

Solar Output

- At certain times, **intense flares** of solar radiation have been known to leave the Sun's surface — these appear to coincide with high levels of **sunspot activity**. Sunspots are dark spots that are visible on the Sun's surface.

- The amount of **solar radiation** reaching the Earth's surface can affect the climate:

  1) The **Little Ice Age** (see below) corresponded to a time with no sunspot activity.
  2) The **Medieval Warm Period**, 900 to 1300, was connected to intense solar activity.

- Sunspot activity happens relatively regularly — approximately every 11 years.

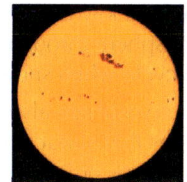

Sunspots visible on the Sun's surface

Intense magnetic activity in the Sun may cause sunspots but this phenomenon isn't fully understood.

Volcanic Eruptions

- **Volcanic eruptions** eject **ash** and **gases**, such as water vapour and carbon dioxide, into the atmosphere.

  1) Many gases released in volcanic eruptions are **greenhouse gases**, which can trap **infrared radiation** in the long term and warm the planet. However, the overall impact of volcanic eruptions is **global cooling**.
  2) **Ash plumes** reduce the amount of sunlight that the Earth's surface receives, causing lower global temperatures due to reduced **insolation**. As well as this, the ejection of **sulfur dioxide** gas reflects solar radiation back into space from the **upper atmosphere**, also leading to global cooling.

- The effects of an average eruption only last for a few years, so they affect the climate on a **relatively short timescale**.

The **Little Ice Age** is an example of **short-lived climate change**

- The Little Ice Age occurred between 1300 and 1850 across most of Europe and North America. The average temperature during this time was between 0.5 °C and 1 °C **cooler** than it has been since 1900.
- Scientists believe the cause was a combination of **sunspot** and **volcanic activity**.
- Standing and flowing water (such as the River Thames) **froze**, and sea ice in the Arctic became more plentiful. Glaciers in alpine regions advanced rapidly, and their subsequent movement down into mountain valleys caused the evacuation of many settlements. Crop harvests were much reduced because of the **shorter growing season** so food supplies became strained.

Another example is the **Loch Lomond Stadial**

- Around 12 700 years ago (during the Pleistocene), ice age conditions returned to some parts of the UK for about **1000 years**.
- This was caused by abrupt **atmospheric cooling** in the **northern hemisphere** which caused ice caps in mountainous regions to **grow quickly**.
- UK winter temperatures **fell** considerably during this period and **mountainous areas** exposed to wet weather conditions from the Atlantic were covered in ice. This included western Scotland, Snowdonia and the Lake District.

stadial
a time period in which temperatures fall and more ice forms on glaciers

Warm-Up Questions

Q1 Explain the difference between greenhouse conditions and icehouse conditions.
Q2 Describe the three different Milankovitch Cycles.

Exam Question — A-Level

Q1 Explain the natural causes of short-term climate change. [8 marks]

The Earth is like a unicyclist — both tend to wobble sometimes...

The Earth has been through lots of drastic climate change in its past, and will likely face drastic changes in the future. However, climate change can have big effects on how humans go about their business on this planet we call home.

Ice Cover on Earth

The cryosphere is basically all the frozen water on Earth, e.g. ice sheets, glaciers, the ice cube tray in my freezer.
OK maybe not that last one. It's pretty important for the planet though.

The **cryosphere** plays an important role in **global systems**

The cryosphere includes all the parts of Earth where **water** is **frozen**, e.g. glacial landscapes.
It stores roughly 69% of the world's freshwater but only covers around 15% of the Earth's surface.

- It plays a key role in the **hydrological cycle**. It removes water from the atmosphere through the **precipitation** of snow. It also releases water into the cycle when snow and glaciers **melt**, which contributes to **river systems**.

- The cryosphere also helps **regulate climate systems**. The **albedo effect** means that surface ice reflects solar radiation away from the Earth. This keeps polar regions relatively **cold**. Through indirect processes, it also keeps the rest of the world at a **moderate temperature**.

- Areas of **permafrost** (ground that is frozen) store carbon. This reduces the level of carbon dioxide in the atmosphere.

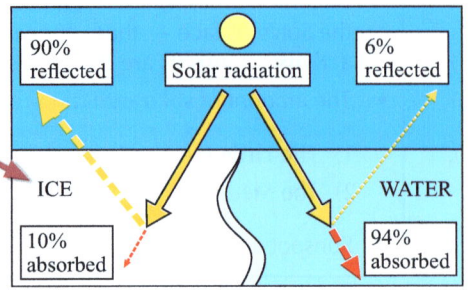

The albedo effect

There are **different** types of **ice masses** in the **cryosphere**

Ice sheets and ice caps cover the land beneath them. These are known as **unconstrained ice masses**. The sides of valleys usually create boundaries for ice fields, valley glaciers and cirque glaciers. These are known as **constrained ice masses**.

Ice sheets

- Ice sheets are **domes** of ice covering huge areas of land and sea. They can be up to several kilometres thick and are often more than 50 000 km^2 in area. There are only two ice sheets on Earth — the Antarctic and Greenland Ice Sheet.

- Where ice sheets extend from the land into the sea, they form a body of floating ice known as an **ice shelf**. Ice shelves can be vulnerable to accelerated **ablation**.

- **Topography** doesn't usually affect the **flow** of an ice sheet.

ablation
the loss of snow or ice from an ice mass

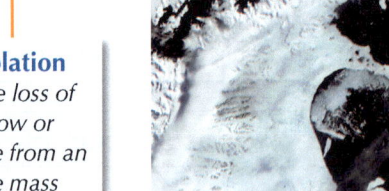

Collapse of the Larsen B ice shelf, Antarctica

Ice caps

- An ice cap (like the Vatnajökull ice cap in Iceland) is a smaller **dome-shaped** ice mass that covers large areas of land in polar regions or mountain ranges.

- The flow is usually **unaffected** by the **topography** below.

- An ice cap covers **less** than **50 000 km^2**.

Ice fields

- Ice fields are large series of **interconnected** valley glaciers. They form in **mountainous areas** where snow and ice can **build up** over time.

- Mountains can often be seen **peeking out** of an ice field, as their flow is determined by the **topography** of the area.

- An example of an ice field is **Patagonia**, South America.

Valley glaciers

- A valley glacier **fills valleys** (often old river valleys) and they may **flow** into the **sea**.

- Their **size** can **vary** between **several kilometres** to over **100 kilometres**.

- An example of a valley glacier is the 12 km long **Franz Josef Glacier** in **New Zealand**.

Stanley snuggled down between his ice sheets.

topography
the features of the landscape

Lower Curtis Glacier, Washington State, USA

Cirque glaciers

- A cirque glacier forms in **bowl-shaped** hollows high up in **mountains**, and can sometimes **flow** into valley glaciers.

- Cirque glaciers **vary** in **size** depending on how large the hollow is, but are usually around **1 km^2**.

- An example of a cirque glacier is the **Lower Curtis Glacier** in the **USA**.

Ice Cover on Earth

Ice sheets are found in **polar environments**

- The **Antarctic** and the **Arctic** are **cold deserts** — they receive little **precipitation** and experience **extreme temperatures** (often lower than –40 °C). This makes them perfect environments for **ice sheets**.

Polar Environments

1) The Arctic polar environment can be defined either by the **Arctic circle** (**66° N**) or by the **10 °C July isotherm** (areas north of this line have an average temperature **below** 10 °C in **July**, the **hottest** month).

2) The Antarctic polar environment includes land and sea at latitudes of **66° S** or greater. It is defined by the **10 °C January isotherm** (the hottest month in the southern hemisphere).

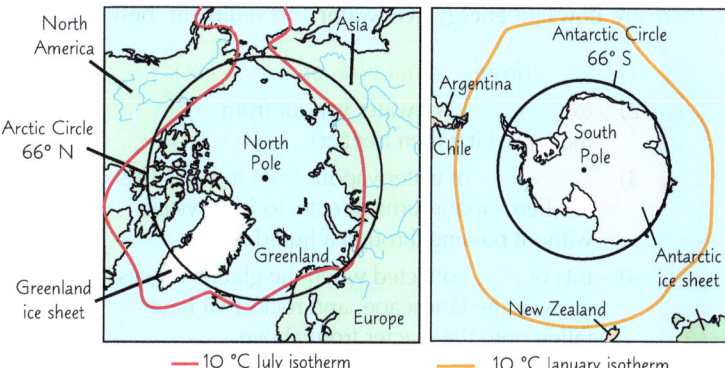

— 10 °C July isotherm — 10 °C January isotherm

- **Higher latitudes** are **colder** because they receive **less solar radiation**. This is because the Sun's rays hit the Earth's surface at a **less direct angle** than at the **Equator**.
- The solar radiation at the poles heats a larger area and so has a weakened effect on **surface temperatures**. This makes the **air temperatures** at the poles much colder.

85% of all current glacier ice is in the Antarctic.

1) During the **Last Glacial Maximum**, ice sheets covered much of northern Europe, Canada and Alaska and the area covered by Antarctica was also far larger.

2) The ice sheets in the northern hemisphere were larger than in the southern hemisphere. This may be because there is **more land** in the northern hemisphere than in the southern hemisphere (which has more ocean).

Temperate environments may have ice masses at **high altitudes**

- At high altitudes, it is **cold** enough for **precipitation** to fall as **snow** and **freeze** as ice. This is because as altitude **increases**, air **temperatures decline** by about **0.65 °C** for every **100 metres** gained.
- Temperatures normally **stay cold** enough for most of the accumulated snow and ice in glaciers to remain. However, some is **lost as meltwater**, which can cause **glacial movement** and, in extreme cases, the **retreat** of the glacier.
- These high altitudes contain **alpine environments**, such as the Alps, the Himalayas and the Andes.

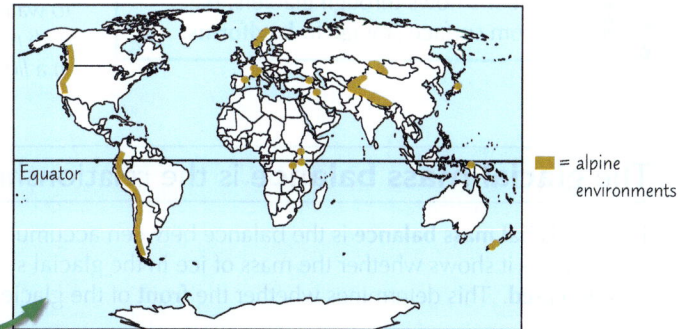

= alpine environments

Relict glacial landscapes show which areas once had ice cover

- A relict glacial landscape is one where there has been eroding and shaping of the **physical topography**. This shaping corresponds to the movement of glaciers during the **glacial periods** in the **Pleistocene epoch**.
- All but the southernmost areas of the UK (as well as much of northern Europe) had **ice cover** during the **Pleistocene epoch**. The Lake District and Snowdonia are relict landscapes as they contain **glacial features** formed in that era.

Warm-Up Questions

Q1 Name five types of ice masses.
Q2 Describe where glaciated landscapes are found.

Oo 'eck — we missed that one by a whisker...

Exam Question — AS and A-Level

Q1 Explain the role of the cryosphere in global systems. [6 marks]

Who stole my glacier? Must have been an iceberglar...

So, today, ice tends to be found at high latitudes and altitudes, but at times in the past, such as the Last Glacial Maximum, there was a heck of a lot more of it. Brrr... Geographers know which areas were covered by looking at the landscapes.

Glaciers

Since plenty of cold environments contain glaciers, you need to know a fair bit about them. I do like a nice glacier...

Glaciers are **systems**

The glacial system has **inputs**, **stores** and **outputs**.
There are **flows** (of **energy**, **ice**, **water** and **sediment**) between stores.

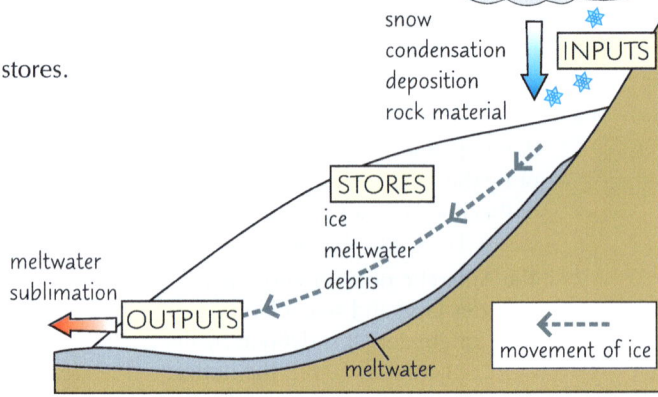

Inputs
1) **Snow** (from **precipitation** or **avalanches**).
2) **Condensation** of water vapour from the air (which then freezes).
3) **Deposition** of water vapour from the air. This is when vapour turns directly to ice crystals, without passing through a liquid stage.
4) Bits of **rock** collected when the glacier carves away at the landscape, and rocks that have fallen onto the glacier from above.

Stores
1) The **main** store is **ice** in the **glacier** itself.
2) **Meltwater** is stored **on** and **within** the glacier, e.g. in **supraglacial lakes** on top of the glacier.
3) **Rock** is also stored in or on glaciers, e.g. **debris** from freeze-thaw weathering (see p.51) falls onto the **surface** of glaciers.

Outputs
1) Ice can **melt** and **flow out** of the glacier as **meltwater**.
2) Surface snow can **melt** and **evaporate**.
3) Ice and snow can **sublimate** to water vapour.
4) Snow can be **blown away** by strong winds.
5) With glaciers that end at the **sea** or a **lake**, blocks of ice fall from the **front** (the snout) of the ice mass into water to create **icebergs**.

Flows
1) **Meltwater** flows through glaciers, e.g. from stores in supraglacial lakes to **channel storage** at the base of glaciers.
2) Debris flows through glaciers, e.g. from surface storage to **landforms**.

sublimation
the change of ice to water vapour gas with no time spent in a liquid state

Meltwater from Hintertux Glacier, Austria

The **glacial mass balance** is the relationship between a glacier's **inputs and outputs**

1) The **glacial mass balance** is the balance between accumulation and ablation over a year — it shows whether the mass of ice in the glacial system has **increased** or **decreased**. This determines whether the **front** of the glacier **advances** or **retreats**.

The mass balance is sometimes called the "glacial budget".

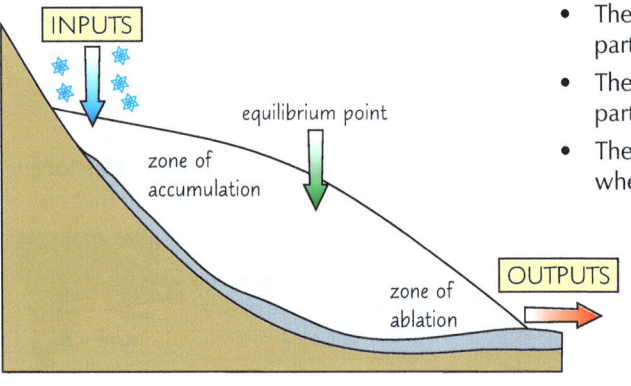

- There's more **accumulation** than ablation in the **upper** part of a glacier — known as the **zone of accumulation**.
- There's more **ablation** than accumulation in the **lower** part of the glacier — known as the **zone of ablation**.
- The glacier's **equilibrium point** is the place where accumulation and ablation are **equal**.

accumulation
the input of snow and ice into the glacial system

Most accumulation is snow.

Hannah and Betty had a very positive regime.

2) If there's **more accumulation** than ablation over a year, the glacier has a **positive regime** (or a positive mass balance). The glacier **grows and advances** (moves forward) in response to high accumulation in the upper zone.

3) If there's **less accumulation** than ablation over a year, the glacier has a **negative regime** (or a negative mass balance). The glacier **shrinks and retreats** (moves back) in response to low accumulation in the upper zone.

Glaciers

Glaciers can be in **dynamic equilibrium**

If there's the **same amount** of accumulation and ablation **over a year**, the glacier stays the **same size** and the **position** of the snout doesn't change — the glacier is in **dynamic equilibrium**.

Dynamic equilibrium is when a system has no overall change, despite short-term variations in inputs and outputs.

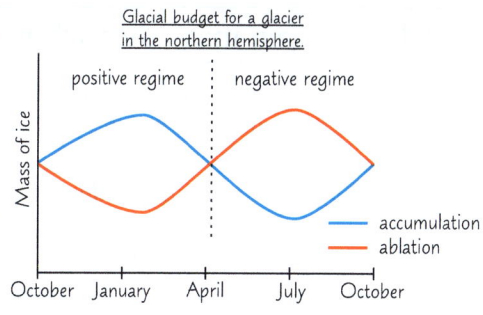

Glaciers can give information about **climate change**

When inputs or outputs change, there can be negative or positive feedbacks.

- Negative feedback mechanisms minimise the impact an input or output has by counteracting it. E.g. if the size of the ice input increases, a glacier's movement may speed up so that more meltwater and ice are output at its snout and the mass of the glacier remains constant.

- Positive feedback means that the response to a change makes the change even greater. E.g. ice has a high albedo (see p.34) so it reflects lots of the Sun's energy. If glaciers retreat, there is less ice — less of the Sun's energy is reflected and more is absorbed, so temperatures rise and glaciers retreat further.

Feedbacks are important because they can show how **change** is **impacting glaciers**. The Greenland Ice Sheet is in this **positive feedback loop**, and is therefore losing mass increasingly **rapidly**, as shown in the graph below.

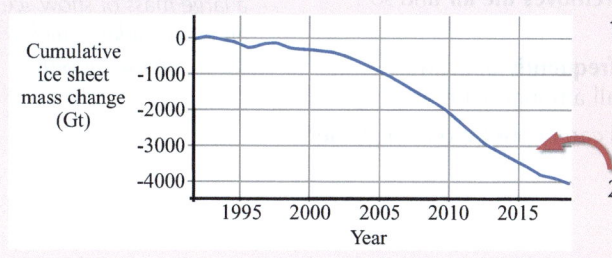

Cumulative change in mass balance for the Greenland Ice Sheet

1) Calculating the glacier mass balance can be a useful way of understanding the impact of climate change over a number of years.

2) Scientists believe the Greenland Ice Sheet is being affected by climate change and is closely monitored.

The Greenland Ice Sheet, Greenland

The glacial mass balance **changes over time**

1) As seen above, more ablation takes place during warmer times of the year but during colder months there tends to be more accumulation. Over the year this might **balance out** — the glacier advances in winter but retreats in summer, so overall the mass in the glacier **stays the same**.

2) However, there is **variation** in the amount of accumulation and ablation from year to year. For example, even though a glacier is in a state of **net retreat**, in some years there may be **advances** due to more accumulation or less ablation than usual.

3) Therefore, **longer-term records** are used to determine the '**health**' of the glacier.

4) Some glaciers are regarded as **benchmark glaciers** (such as Gulkana Glacier in Alaska) as they show annual changes that can **predict** changes to other glaciers.

5) Changes in **global temperature** over longer periods of time also affect the glacial mass balance. For example:

- Temperatures in the **Little Ice Age** were **colder** than periods before and after it.
- This meant that many glaciers **advanced** because they had a **positive regime** — the **Mer de Glace** in the French Alps advanced by over **1 km**.
- After the Little Ice Age, around 1850, global temperatures **increased** so glaciers since then have tended to have a **negative regime** and are **retreating**.

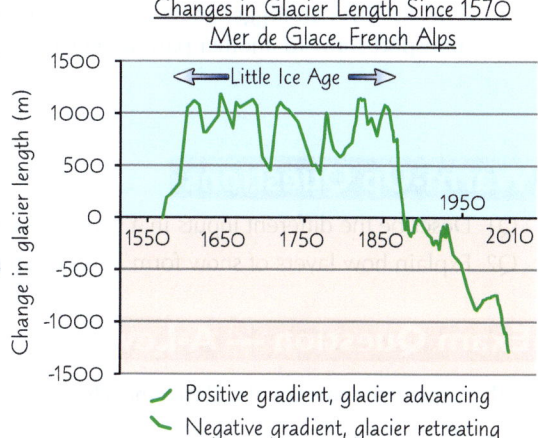

The Mer de Glace has retreated by nearly 2.4 km since 1850.

Topic 2: Option 2A — Glaciated Landscapes and Change

Glaciers

Glaciers gain mass through snowfall, avalanches and wind deposition

1) Glaciers initially form when snow accumulates over many centuries. It **compresses** under the cumulative weight to form **new ice**, which adds to the glacier as **ice mass**.

2) Each new layer increases the **pressure** on the one below it. This pressure causes **partial melting** of the lower layer **filling** in the **air gaps**. Over time, the snow turns to ice.

- **Snow** has a **low density** as it is about 90% air.
- Compression forms **granular snow** over time.
- **Firn** is compressed snow that has survived at least one melt season without becoming glacier ice. It can take **100-300 years** for firn to become glacier ice.
- Further compression creates **glacial ice** which is less than 20% air. It is **dense enough** to survive a year's ablation period before the next snow season arrives.

A magnified snowflake

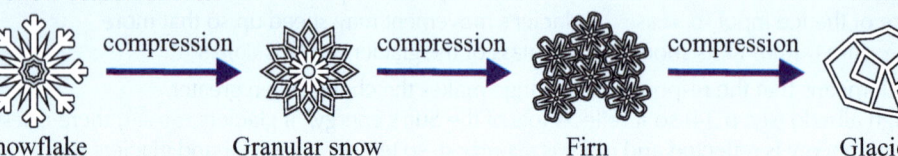

Snowflake	Granular snow	Firn	Glacier ice

compression → compression → compression →

3) It can take **hundreds of years** for glacial ice to fully form. The more **temperate** the area, the **quicker** ice is able to form. This is because, in temperate areas, snow can **melt relatively quickly** compared to polar regions. This **removes the air** and so when it refreezes, it does so as **denser ice**.

4) **Avalanches** also add snow to the system. They occur **infrequently** but can increase the amount of snow by a **large volume** in a small amount of time.

5) **Wind** blows powder snow from one area to another and can deposit it onto a glacier.

avalanche
a large mass of snow, ice and rock sliding quickly down a mountainside

Glaciers lose mass through ablation

There's more than one way for a glacier to lose mass:

- When ice **melts** on a glacier, it creates an **outflow** of **meltwater**.
- Ice and snow can **sublimate** (see p.36) into **water vapour** from the surface of the ice.
- **Calving** is the process by which ice **falls away** from the glacier at its snout. Wind and water erosion can create a **crack** in the glacier and the snout becomes **unstable**. This process creates **icebergs**.
- Snow can be **blown away** from the glacier by **strong winds**. This can trigger an **avalanche** which may transport snow to another part of the landscape.

Ice calving at the Perito Moreno Glacier, Argentina

Warm-Up Questions

Q1 Describe the different inputs in a glacial system.
Q2 Explain how layers of snow form ice over time.

PRACTICE QUESTIONS

Exam Question — A-Level

Q1 Explain the concept of a glacial mass balance.   [8 marks]

Shouldn't have left the cage outside — now I've got a glacial budgie...

It takes a pretty long time for a glacier to form. Snow has to get more and more squished until it is dense enough to not melt when summer comes. Don't forget to learn all the terms on the last few pages, like ablation, accumulation and sublimation.

Glacial Movement

They mightn't look like they're doing anything at all, but glaciers are constantly moving — some quicker than others.

Glaciers can be **polar** or **temperate**

Another way to classify glaciers is by the **temperature of their bases**.

Where are those
blasted elves?

Polar glaciers

- **Polar glaciers** are found in higher latitudes, such as in the Arctic and the Antarctic. They are also known as **cold-based glaciers**.
- Their bases are usually at a temperature **well below the ice's melting point**, meaning that the ice freezes to the **bedrock**.
- Typically, polar glaciers **melt less** at the surface than temperate glaciers.
- Polar glaciers tend to move **very slowly** (around 1 to 2 cm a day).

Temperate glaciers

- **Temperate glaciers** are found in **lower latitudes**, such as the **Himalayas**. They are also known as **warm-based glaciers**.
- Their bases are **warmer** than the melting point of ice, which means they produce **more meltwater**.
- The ice at the bottom of the glacier **melts**, and the meltwater acts as a **lubricant**, making it easier for the glacier to move downhill. Ice at the **surface** also melts if the temperature reaches 0 °C. This meltwater moves **down** through the glacier, lubricating it **even more**.
- As the glacier moves **downslope**, the **friction** it generates produces more **heat**, which further melts the glacier.
- This means the glacier moves **relatively quickly** (around 3 m a day).

Glaciers move **downhill** under their own weight

Gravity causes glaciers to move downhill. The **force** this exerts downslope is known as **shear stress**. There are three main processes that cause glaciers to move downhill:

Basal slip

Meltwater **underneath** a glacier allows it to **slide** over the ground. This is known as **basal slip** and it is the main way that **temperate glaciers** move. There are a number of processes which occur because of **basal slip**:

- **Rotational slip** — Glaciers move in an **arc shape** when they move out of a **hollow**. The increasing weight of the ice exerts **pressure** on the base of the glacier. If this pressure is large enough, the **melting point** of the ice **decreases**, and the ice melts more easily. This melting point is known as **PMP (pressure melting point)**.

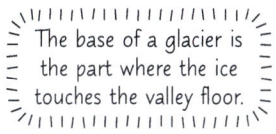

higher pressure

lower pressure

higher pressure

Rotational slip

more meltwater

- **Regelation creep** — Where there are bits of **rock** protruding from the valley floor, there is more **melting around them** because the rocks themselves exert **pressure** on the ice. The **larger** the piece of rock, the **greater** the pressure it exerts. This means the ice melts more easily there, even at temperatures **lower** than 0 °C. Meltwater then **refreezes** past the rock obstruction where there is **lower pressure**. This makes the flow **faster** upglacier and around the obstruction and **slower** downglacier — this is known as **regelation** creep.

Internal deformation

- Glaciers can also move in ways that **do not** involve sliding.
- **Internal deformation** is where the ice **bends** and **warps** to flow downhill like a **liquid**.
- This occurs due to ice crystals **shifting** and **deforming** as they try to move past each other. This process is known as **intergranular flow**.
- Internal deformation is the main way **cold-based glaciers move** as they are too cold for their bases to reach PMP (see above).

Glacier moving downhill, Norway

Topic 2: Option 2A — Glaciated Landscapes and Change

Glacial Movement

Extensional and compressional flow

The glacier can also move through the creation of a series of **minor faults**.

- At the **head** of a glacier, the valley is **steep** so there is a **strong gravitational force** pulling the ice downwards. This makes the ice move **quickly**. When ice moves quickly, there is more **tension** (pulling apart forces). This can cause the ice to **fracture** into **thick layers**, which then **slip downwards**. This is known as **extensional flow**.

- **Lower down**, within the glacier, the ice is moving more **slowly** because the valley is **less steep**. The faster ice from the head of the glacier **pushes** down on the slower ice and **compresses** it. The high pressure causes the ice to **fracture** into **layers** and the layers **slip forward**. This is known as **compressional flow**.

- Together, these are known as **laminar flows** — individual layers of the glacier moving **independently** of one another.

Extensional flow
glacier bends and warps here
Compressional flow
ice slips
ice slips

The **rate** of glacial movement can **vary**

Faster moving ice tends to:

- be found on **steeper** slopes.
- be at the **base** of a glacier.
- be part of a **larger glacier**.
- be **temperate** (warm-based).
- be found moving over **impermeable bedrock**.
- be found in **narrow valleys**.

Glacier moving along the mountain valley, Iceland

1) Slope

- If a glacier is on a **steep** slope, its **weight** will have more of an impact on its **movement**. This is because the **force** of the glacier's weight exerted downslope is **stronger** than the **friction** holding it in place.
- This means that glaciers on **steep valley slopes** will move far **faster** than those on flat valley floors.

2) Temperature

- The degree to which a glacier moves is dependent on its **melting point**.
- The **surface** of the glacier will have a melting point of **0 °C**. At **greater depths** within the glacier, the melting point will be **lower**. This is due to the **increased pressure** from the ice above.
- This means the ice **within** the glacier will flow at a **faster rate** than that at the surface.
- In polar environments, temperatures are **very low** and glaciers are more likely to be **cold-based**. This means they are more likely to **freeze** to the **bedrock** and **move slowly** due to the **absence** of meltwater.

3) Size

- The larger the glacier, the greater the **pressure** on the base of the glacier and the easier the ice there will **melt**.
- This will create a **faster** moving glacier.

4) Altitude

- Glaciers at higher altitudes are more likely to experience greater levels of **precipitation** which lead to higher rates of **accumulation** and a **positive mass balance**. This means the glacier will **advance** and move more quickly. However, at higher altitudes the glacier is more likely to **freeze** to the bed because of lower temperatures, which will slow down the glacier.
- Temperatures are warmer at lower altitudes. This means that glaciers here are more likely to undergo **melting** at the base of the glacier, which can increase **basal slip** (see p.39) and cause the rate of glacier movement to increase.
- **Temperate glaciers** are more likely to be found at **lower altitudes** or at **sea level**. The snouts of some glaciers at these lower altitudes can move much **faster** — at up to 10 m a day. This may be due to a **glacial surge**. This is where the **snout** moves very **quickly** due to **instability** within the end of the glacier.

Glacial Movement

⑤ Lithology

- A **permeable bedrock** will **soak up** some of the meltwater at the **base** of the glacier, reducing the glacier's ability to move. Glaciers on top of **impermeable rock** will move **faster** because the increased meltwater **reduces friction**.
- **Softer** rocks, like **clay** or **chalk**, allow glaciers to move **faster** because the glacier is **less likely** to be stopped by **obstructions**.

lithology
the physical characteristics of rocks

permeable
allowing liquids or gases to pass through

A permeable bedrock means drooling is no problem.

⑥ Topography

- **Narrowing** of valleys can cause a glacier to **thicken** and move **faster**.
- The **widening** of valleys causes glaciers to **spread out more** and **thin**, making them move more **slowly**.

Variations in mass balance also affect the rate of glacial movement

1) If the amount of **accumulation** and **ablation** is the **same**, a glacier is in a state of **equilibrium** and **won't move**.

2) If accumulation is **higher** than ablation or **vice versa**, the mass balance of a glacier **changes**, which affects the **speed** of the glacier.

- A glacier with an **overall negative mass balance** loses ice faster than it gains ice. These glaciers move more **slowly** because they are **shrinking** and exerting **less downward pressure**.
- Glaciers with an **overall positive mass balance** move at a **faster** pace because as they gain more mass each year, they can exert **more pressure**.

A change in a glacier's mass balance does not immediately affect its speed and movement.

Mendenhall Glacier in Juneau, Alaska

Positive and negative feedbacks affect glacial movement

Positive feedback loop

1) If the glacier has a large volume of **meltwater**, it will move relatively **quickly** due to **basal slip**.

2) This will generate greater levels of **heat** through **friction**.

3) This heat will further increase the amount of **meltwater** and therefore the **speed** the glacier moves.

Negative feedback loop

1) A **fast-moving glacier** will **become thinner** as the amount of **meltwater increases**.

2) This thinning will eventually **slow** the glacier down, as **less pressure** will be exerted on the base of the glacier. This means less **meltwater** is produced.

Warm-Up Questions

Q1 Explain how a glacier moves through regelation creep.

Q2 Describe the differences between a polar glacier and a temperate glacier.

PRACTICE QUESTIONS

Exam Question — AS and A-Level

Q1 Explain how the pressure melting point (PMP) of a glacier affects its rate of movement. [6 marks]

Glaciers move r e a l l y s l o w l y — bit like me on a Sunday morning...

So if you're going to place a bet on a glacier race, you want to pick the biggest one with the warmest bottom. Of course, if the race valley is narrow with impermeable bedrock, and at a low altitude, then the glacier will perform even faster. Ready, Set, Go...

Glacial Erosion — Processes and Landforms

Glaciers are moving all the time — they don't stop to think about the erosion they might be causing...

Glaciers shape the landscape through **erosion**

1) Glacial erosion is the **breakdown and removal** of rock material by the actions of **ice, rock and meltwater**.

2) Glacial erosion is a **powerful force**. The sheer **weight** of the glacier compared to water means that a glacier has **stronger erosive power** than a river.

3) **Warm-based glaciers** tend to have **stronger erosive power** than cold-based glaciers as they are able to move **faster**.

4) The amount and rate of erosion **increases** if:

- the glacier is **thick**
- there is **a lot** of **debris**
- the **debris** consists of **resistant rock**
- the **bedrock** is made of **less resistant rock**

Entrainment is when **rock debris** is **picked up** and **merged** into a glacier

Debris moves along with the glacier and causes **abrasion** on surrounding surfaces.

- **Supraglacial entrainment** is rock debris that is carried on the **surface** of the glacier.
- **Subglacial entrainment** is rock debris held on the **base** of the glacier. As the debris itself erodes under the weight and pressure of the ice above, it may **loosen from the glacier** and be transported by the **meltwater**.
- The **finest form** of debris is known as '**rock flour**' — this is debris which is less than 0.1 mm in diameter. Rock flour **polishes and smooths** the bedrock.

Glaciers **erode** in **different** ways

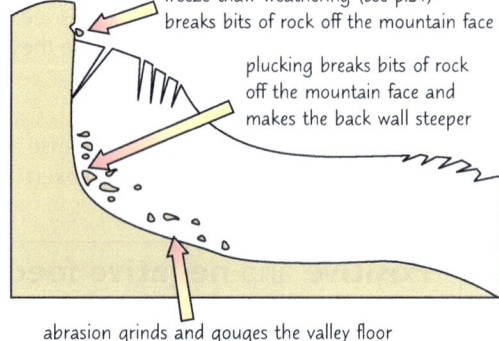

freeze-thaw weathering (see p.51) breaks bits of rock off the mountain face

plucking breaks bits of rock off the mountain face and makes the back wall steeper

abrasion grinds and gouges the valley floor

Plucking

- **Plucking** (or **quarrying**) occurs when the bedrock is jointed or cracked by **fracture** and **traction** (the crushing of rock by ice).
- Ice in contact with rock surfaces can **thaw** slightly then **refreeze** around rocks protruding from the valley sides and floor.
- When the glacier moves forward, it **plucks** these rocks away, leaving behind a **highly jagged** rock face.

Abrasion

- **Abrasion** occurs when **debris** carried by a glacier **scrapes** the valley walls and floor.
- The debris behaves like **sandpaper** as it rubs against the bedrock.
- The scraping of one rock against another can leave **striations** in the rockface.
- The striations record the **direction** in which the glacier moved.

Glacial striations

Crushing

- **Crushing** is where the sheer weight of the ice exerts **pressure** on the bedrock below.
- This **shatters** and **chips** the rock into large **fragments**.
- **Fractures** in the rock can **deepen** or **expand** when the **weight** of the glacier **varies**. This is called **dilation**.

Basal melting

- **Basal melting** (meltwater erosion) is where the **meltwater** underneath a glacier erodes like a river.
- This meltwater can cause **abrasion**, erosion by **hydraulic action** (the hammer-like impact of the meltwater being squeezed and forced through the bedrock), **attrition** (where rocks held within the meltwater erode **each other**) and **corrosion** (where the meltwater dissolves the rock).
- This meltwater is under a high **pressure** and can create **channels** and **hollows** beneath the glacier.

Glacial Erosion — Processes and Landforms

Freeze-thaw weathering and mass movement can erode rock too

Subaerial processes (happening in the open air) can also affect the erosion of rock in glaciated areas.

- **Freeze-thaw weathering** (also known as **frost shattering**) can create **cracks** in rocks. These processes cause fragments (called scree) to **separate** from the main bedrock.
- The **mass movement of rocks** e.g. landslides, rock falls and **avalanches** can erode rock faces in glacial areas too.

Turn to p.51 for more on freeze-thaw weathering.

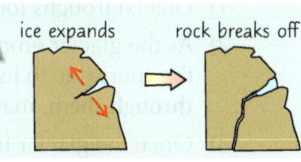
ice expands   rock breaks off

Glacial erosion leads to the formation of particular landforms

Glaciers normally form on one side of a mountain peak. This is usually the side that gets the **least sun** and the **coldest winds**. This is where there will be the **most accumulation** and **least ablation**. As the glaciers grow and start to move down the mountainside, they begin to create **particular landforms**.

Corries

Corries are also known as cwms or cirques.

For a **corrie** to form, there needs to be a **hollow** in the slope.

1) **Snow** collects in this hollow and turns to **ice**.
2) **Basal slip**, **abrasion** and **plucking** deepen the hollow into a corrie (a bowl-shaped hollow). The amount of **abrasion** that takes place **depends** on the amount of **rock material** that is **eroded**, e.g. by plucking.
3) When the ice in the hollow is thick enough, it flows **over the lip** and **downhill** as a glacier.
4) **Freeze-thaw** weathering and **plucking** steepen the back wall of the corrie. This creates a large, **arm-chair shaped** depression.
5) Once the glacier has melted, a **corrie lake** or tarn remains. There will be a lip of **deposited scree** at its downglacial edge.

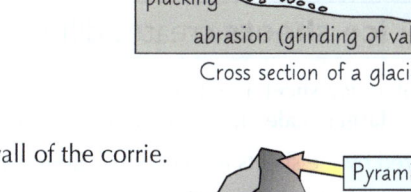
freeze-thaw weathering on exposed rock
Corrie lip
plucking
abrasion (grinding of valley floor)
Cross section of a glacier forming a corrie

Arête

An **arête** is a narrow **steep-sided ridge**.

1) Arêtes form when two glaciers flow in **parallel valleys**.
2) The glaciers **erode** the sides of the valley, **sharpening** the mountain ridge in between them.

Pyramidal peak

A **pyramidal peak** is a pointed mountain peak with at least **three sides**.

1) It forms where **three or more corries** form back to back.
2) Their **steep back walls** make the mountain peak.

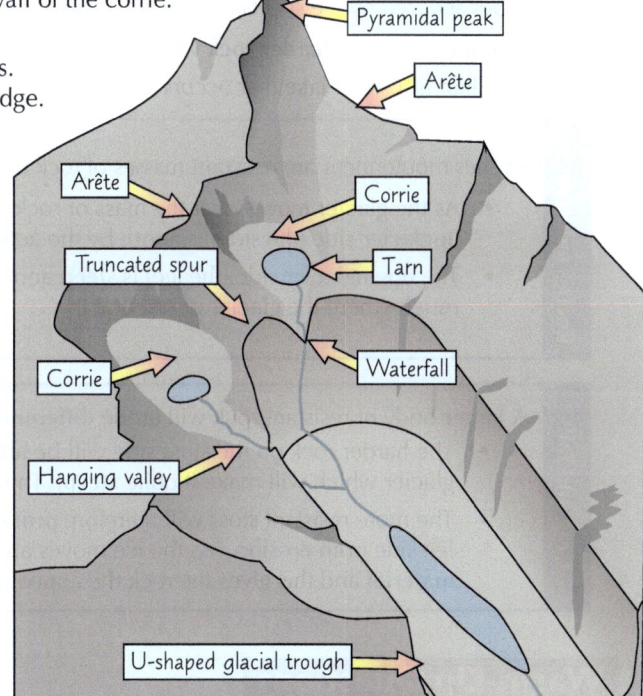
Pyramidal peak
Arête
Arête
Corrie
Truncated spur
Tarn
Corrie
Waterfall
Hanging valley
U-shaped glacial trough

Example: Ama Dablam, Himalayas

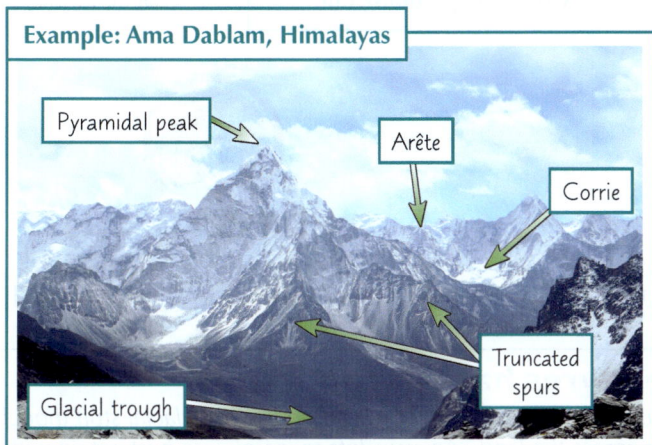
Pyramidal peak
Arête
Corrie
Truncated spurs
Glacial trough

Striding Edge, an arête in the Lake District

See the next page for glacial troughs, truncated spurs and hanging valleys.

Glacial Erosion — Processes and Landforms

Glacial troughs

~Glacial troughs are also known as U-shaped valleys.~

Glacial troughs are **steep-sided** valleys with **flat bottoms**.

1) Glacial troughs form through the **erosion** of V-shaped river valleys by glaciers.

2) As the glacier erodes through the valley, it **doesn't wind** around the spurs due to its **lack of flexibility**. Instead, it erodes **straight through them**, making the valley deeper and wider.

3) Once the glacier has **melted**, a long and narrow lake can form on the valley floor, known as a **ribbon lake**. This is where a band of **softer rock** was **eroded more** than the surrounding **hard rock**.

A U-shaped valley with a ribbon lake, Buttermere, Lake District

4) **Truncated spurs** form when the main glacier **chops off** (or truncates) the ridges of land (spurs) that stick out into the main valley as it moves past. The glacier erodes the ends of the spurs through **plucking** and **abrasion**.

Hanging valleys

~A tributary glacier is a smaller glacier that flows into the main glacier.~

Hanging valleys are valleys formed by **tributary glaciers**.

1) Tributary glaciers erode the valley floor **much less deeply** than the main glacier because they are **smaller**.

> Once the ice has melted, it's common to see waterfalls flowing over hanging valleys.

2) This means that when the tributary glaciers melt, the valleys remain at a **higher** level than the glacial trough formed by the main glacier.

Bridal Veil Falls flows over a hanging valley into Yosemite Valley, USA.

Scouring by ice sheets creates different types of landforms

The dimensions of an **ice sheet** aren't constrained by a valley, so they erode on a **larger scale** and their features are **less isolated**.

Landscapes scoured by ice sheets are known as **knock and lochan landscapes**:

- A **knock** is a hill of **harder rock** which has **resisted** glacial erosion.
- A **lochan** is a small **lake** that occurs where there are **weaker** patches of rock that have been **eroded**.

Roches moutonnées

Roches moutonnées are **resistant** masses of rock on the **valley floor**.

- As the glacier moves over the mass of rock, it makes the **upglacier** side (the stoss) **smooth** by the action of **abrasion**.
- The **downglacier** side (the lee) is **steep** and **rough** where the glacier **plucked** at it.

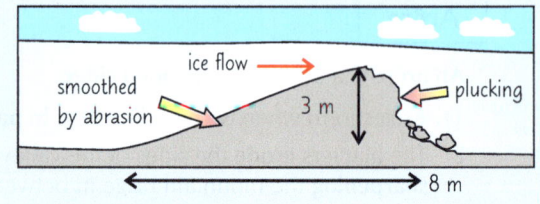

smoothed by abrasion

ice flow

3 m

plucking

8 m

Crag and Tail

A **larger** body of resistant rock will erode **differently** to a roche moutonnée.

- The harder rock on the stoss side will be able to **resist** the abrasiveness of the glacier which will make it quite **steep**. This will cause the glacier to **slow** down.
- The more resistant stoss will therefore **protect** the softer (less resistant) rock on the lee side from **erosion**. As the ice moves around the resistant stoss, it will **deposit material** and this gives the rock the appearance of having a **long, sloping tail**.

> A crag and tail is an outcrop of resistant rock that could be the remains of a volcanic plug. It has a 'tail' of deposited moraine on the lee side.

Warm-Up Questions

Q1 Name two ways that glaciers can be eroded by rocks.

Q2 Explain how a glacial trough forms.

PRACTICE QUESTIONS

Exam Question — AS and A-Level

Q1 Assess how important different erosional processes are in the shaping of a corrie. [12 marks]

What do glacial troughs and punctured tyres have in common?

They both have flat bottoms. Well, if that joke didn't cheer you up, I don't know what will? I do like this glacial stuff though. It's pretty mind-blowing to imagine these giant ice slugs sliding across the land. Anyway, lots to learn — so best get on with it.

Glacial Deposition — Processes and Landforms

More pages about the landscapes that glaciers leave behind them. Mountain climbers and geography teachers get very excited about the beauty of glacial landscapes. Whether or not they move you, you still need to learn about them.

Debris can be **transported** in different parts of the glacier

 Supraglacial material is carried on top of the glacier's **surface**.

 Englacial material is carried **within** the glacier.

 Subglacial material is moved along at the **base** of the glacier.

Glacial deposits in the Pyrenees

Glaciers **deposit** their load as they **move** and **melt**

- Glacial deposition can happen at the **edges** or the **base** of a glacier.
- The mixture of rock material deposited by the glacier is known as **till**. Till is **unsorted**. At any point, a glacier can drop till made up of debris of any size. Till is also usually **angular**.
- Glaciers deposit material through a number of processes:

> Till includes everything from large boulders down to pebbles and clay.

Ablation

- Debris held within the glacier's structure is abandoned as the glacier **thaws**.
- **Meltwater** provides a means of **moving** the abandoned debris.

Flow

Flow is the process by which high levels of **meltwater** cause the material deposits to **slide** to different locations.

Braided river carrying glacial water in Iceland. The meltwater deposits sediment, which the river splits to flow around.

Lodgement

- **Lodgement** happens when **subglacial material** becomes **lodged** against uneven bedrock.
- The moving glacier cannot overcome the **friction** that holds this material in place, so the material remains behind while the glacier moves on.
- This form of deposition is more common when the glacier is very **slow moving**.

Deformation

- **Deformation** occurs where the bedrock is particularly soft.
- The **pressure** of the moving glacier on this soft bedrock means it picks up the **whole layer of bedrock** and repositions it in a **deformed shape**, such as a **fold**.
- This type of deposition is **relatively rare**.

Sue shut her eyes in disbelief when confronted by yet another amazing set of deposition features.

Different factors affect the **rate** of glacial deposition

- The **speed** of the glacier affects the amount of deposition occurring. As a glacier moves into a new valley, it may be forced to **change shape**, leading to it **slowing down**. The reduction in the glacier's **energy** means it won't be able to carry as much rock material.
- The glacier deposits some of its rock material if it exceeds its **carrying capacity**. If a glacier comes across a rougher area, the **increased friction** may stop it moving. If this happens, **gravitational pressure** builds up. This causes the glacier to **melt** a little and the meltwater reduces the friction at the base and sides. The glacier is now **smaller** and its carrying capacity is **reduced**, so it deposits some material.
- If the glacier has a **negative regime**, it will have greater rates of ablation and will be **losing mass**. This means the glacier **won't be able** to carry as much material, so it will **deposit debris**. In addition, if a glacier moves over **rougher bedrock**, more material will catch and be **held back** as the glacier moves on (lodgement). This will create higher levels of deposition.

Topic 2: Option 2A — Glaciated Landscapes and Change

Glacial Deposition — Processes and Landforms

Moraines and drumlins form in areas previously covered by ice

Ice contact depositional features are those made by **eroded and transported sediment**. Deposition can occur straight onto the ground, or into meltwater channels.

- The deposited till creates landforms called **moraines**.
- Unlike sediment deposited by a river, till is **unsorted material**. This means that **different-sized rocks** of varying levels of **roughness** make up a moraine feature.
- Areas of moraine tend to be **linear and elongated**. They occur from the top of mountain slopes to the base which reflects the **continuous melting** that takes place **around** the glacier as it moves downslope.

Moraine on the approach to K2, Pakistan

There are different types of moraine formations

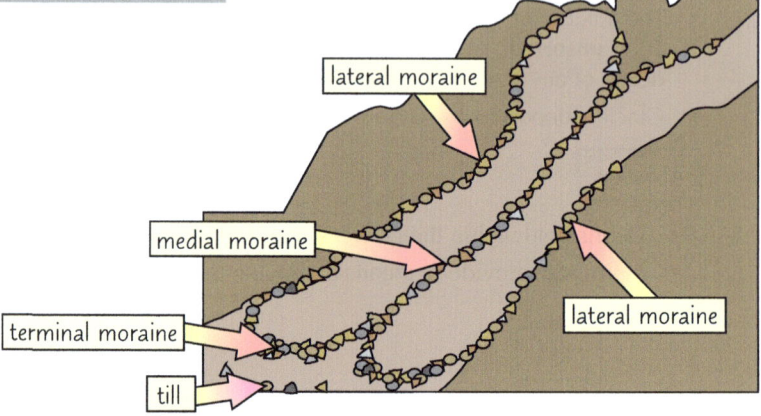

lateral moraine

medial moraine

terminal moraine

till

lateral moraine

Lateral moraine

- Lateral moraine is deposited where the **sides** of the glacier once were, so they often occur along the sides of valleys.
- Much of this moraine will have come from **freeze-thaw weathering** (see p.51) acting on the valley sides.
- In a relict glacial landscape, lateral moraine appears as a **tall ridge** that runs the length of the valley.

© Tom Bean / Alamy Stock Photo

Medial moraine

- Medial moraine is deposited in the **centre** of the valley where two glaciers converge.
- Medial moraine is in effect **two lateral moraines** joined together, made from the weathered material of the valley sides. It's found in lines **parallel** to the ice flow.

Terminal moraine

- Terminal moraine builds up at the **end** of the glacier in semi-circular **hillocks of till**.
- It is highly likely that **streams of meltwater** will have cut this semi-circle into a **series of hills** set in a line.
- These hills will be **steeper** on the **upglacier side** as the glacier would have deposited **more material** there once the hill had started to form.
- In a relict glacial landscape, terminal moraine indicates the **lowest extent** of the snout of the glacier.

© Benjamin R. Jordan / Shutterstock.com

Recessional moraine

- Recessional moraines are often found in lines behind a **terminal moraine**.
- A terminal moraine shows the **furthest extent** of a glacier, whereas a recessional moraine shows the **movement** of the glacier's snout during **shorter** periods of **change** when the melting of the glacier slowed down.
- As recessional moraines accumulate over shorter periods of time, they **aren't** usually as **tall** as terminal moraines.

Glacial Deposition — Processes and Landforms

Drumlins are half egg-shaped hills of till

- Drumlins can be up to 1500 m long and 100 m high.
- Drumlins have a **parallel orientation** to the direction of glacial flow.
- They occur in large clusters known as **swarms**.
- The **upglacier** (stoss) end is wide and tall and the **downglacier** (lee) end is narrow and low.

A drumlin faces the opposite way to a roche moutonnée (see p.44)

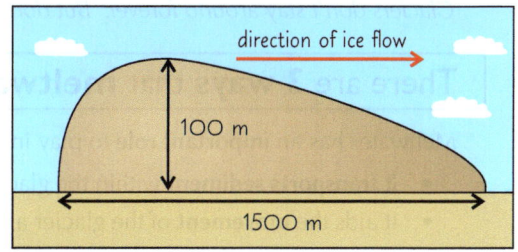

direction of ice flow
100 m
1500 m

Clew Bay, Ireland

- It is **unclear** how drumlins form, and there are a number of ideas, **all** of which **probably** occur.
- One theory is that **till** may have got **stuck** around a rock or small hill sticking out into the glacier. When the ice **readvanced** over it, it **streamlined** the original mound of dropped till.

This is an example of equifinity — when multiple processes lead to the same outcome.

- There are a large number of drumlins under the water level in **Clew Bay, Ireland**. There are also drumlins in the **Ribble Valley** in Lancashire.

Different types of depositional features are found in lowland areas

Lodgement till
- Moving glaciers spread and press **lodgement till** into the **valley floor** beneath the ice.
- Lodgement is more likely to produce and contain deposits of **slightly rounded material**, of varying sizes (from boulders to rock flour).

Ablation till
- **Ablation till is highly angular debris** that a glacier drops as it **melts**.
- Mostly, this till is deposited close to the glacier's **snout** because this is where most ablation happens.

Till plain
- A **till plain** is formed when a large section of ice **detaches** itself from the main ice sheet and **melts in place**. They typically form in **lowland areas**.
- The till that is being transported then **deposits** where the ice melts. The result is a **large expanse** of gently rolling **hills** of till.
- **Example**: Northern Ohio, USA, is a large till plain.

The landscape may appear largely flat.

Some depositional features are found in both lowland and upland areas

- **Erratics** are **rock boulders** found in both lowland and upland areas.
- A glacier or ice sheet picked them up, transported them, then dropped them in an area with a completely **different geology** (giving the erratic its name).

Example: in the Yorkshire Dales at Norber, loose black Silurian rocks sit on top of white Carboniferous limestone.

Loose, black Silurian rock

Limestone

Warm-Up Questions

Q1 Explain the factors that can affect the rate of glacial deposition.
Q2 Describe the defining features of a drumlin.

PRACTICE QUESTIONS

Exam Question — AS and A-Level

Q1 Explain how different types of moraines are formed. [4 marks]

Rock flour — makes fab rock buns...

It gets a bit confusing because moraine can be a landform or sediment. Then there is till, which is also a jumble of rocks dropped by glaciers. Then there's different moraine and till formations... And as for those erratics, they're just a bit random.

Fluvioglacial Processes and Landforms

Glaciers don't stay around forever. But don't worry — even when they're melting they still manage to change the landscape.

There are **3 ways** that **meltwater moves** through a glacial system

Meltwater has an **important role** to play in glacial processes:

- It **transports sediment** within the glacial system.
- It aids the **movement** of the glacier as a unit.
- It facilitates **erosive processes** such as plucking and abrasion.

fluvioglacial
erosion or deposition caused by meltwater from glaciers

Meltwater moves through a glacial system in **3 main ways**:

(1) **Supraglacial** flows of meltwater travel on a **glacier's surface** — mostly in the **zone of ablation** (see p.36). Supraglacial flows may **meander** and **filter** through the glacier through **crevasses** (deep fractures in the ice).

(2) **Englacial** flows of meltwater travel through the **centre** of the glacier, usually in **tunnels**. Meltwater may also move through **moulins** (tunnels that run **vertically** through the ice).

(3) **Subglacial** flows of meltwater flow through channels **underneath** the glacier. The meltwater flows under **pressure** above the bedrock until it exits through the **snout** of the glacier and onto the **proglacial area**.

proglacial
area in front of ice

Fluvioglacial deposits come from glacial meltwater

- STRATIFICATION — fluvioglacial deposition creates a stratified landscape (where the sediment is found in **clear layers**). This is because there are **seasonal variations** in **sediment accumulation**. Glacial landscapes don't have clearly stratified layers.
- SORTING and GRADING — unlike sorted glacial deposits, fluvioglacial deposits are **sorted and graded** by **size**. The larger materials are found **closest** to the glacier's snout whereas the finer sediment is transported **furthest away** from the glacier, as it requires **less energy** for it to be carried.
- IMBRICATION — this is where **rock deposits** are laid down in **distinctive directions**. The **longest axis** of a deposit will run **parallel** to the direction of flowing **meltwater**. Large rocks will dip upstream and overlap each other, so deposits tend to appear like rows of **toppled dominoes**.

Contact with the melting ice creates **eskers**, **kames** and **kame terraces**

- **Eskers** are **long, winding ridges** of sand and gravel that run in the **same direction** as the glacier.
- They're deposited by meltwater streams flowing in **subglacial tunnels**.
- When the glacier **retreats** and the stream dries up, the load remains as an esker made of **sorted** and **stratified** materials.
- Eskers show where **glacial tunnels** used to be and can be several hundreds of kilometres in length and up to 50 m high.

- **Kames** are **stratified mounds** of sorted sand and gravel found on the valley floor near to the **snout** of the glacier.
- **Supraglacial** meltwater streams collect in **depressions** in the ice and lose **velocity**. This reduces their **carrying capacity**, so **debris** is deposited in layers. When the ice **melts**, the debris is **dumped** onto the valley floor.

- **Kame terraces** are piles of deposits left against the **valley walls**.
- They are formed by meltwater streams that run **between** the glacier and the valley sides.
- They look like lateral moraine, but they are in **sorted layers**.

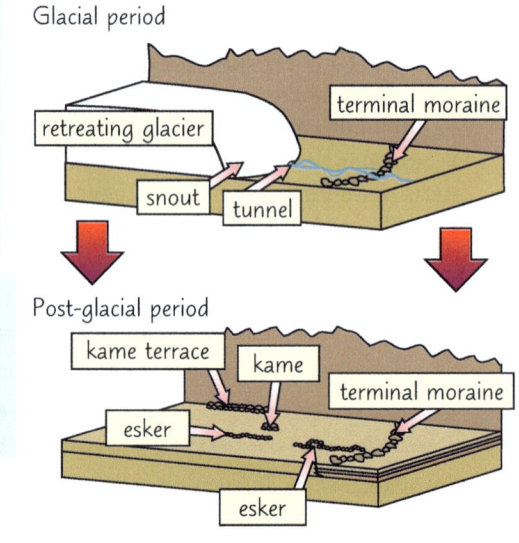

Meltwater streams deposit their heaviest loads first, so kames and kame terraces have gravel at the bottom and finer sediment at the top.

Fluvioglacial Processes and Landforms

Meltwater flow **beyond the edge** of the ice extent creates **proglacial landforms**

Meltwater channels

- Meltwater streams can form **troughs** in the landscape called **meltwater channels** (also known as **overflow channels**). They usually **follow** the path of pre-existing **river channels**.
- As meltwater streams can have a lot of **erosive power**, the meltwater channels they produce can be very **wide** and **deep**.
- After the glacier has **retreated**, shallow streams run through the deep meltwater channels left behind. If the meltwater has a high sediment load, **braided streams** (channels that interweave with each other) may form. This is because some of the sediment is deposited as **islands**, altering the course of the meltwater.

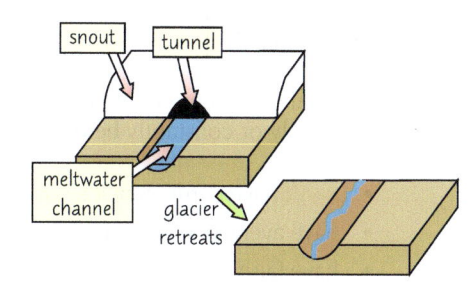

Sandurs

- A **sandur** (also known as an **outwash plain**) is a layer of gravel, sand and clay that forms in **front** of where the snout of the melting glacier used to be. Meltwater flows out of the glacier, and carries the sediment with it.
- Sediments on sandurs are **sorted** into layers. **Gravel** gets dropped **first** because it's **heavier** than sand and clay, so it forms the **bottom layer** of the outwash plain. **Clay** is dropped **last** and gets carried furthest away from the snout because it's the lightest sediment — it forms the **top layer** of the sandur.

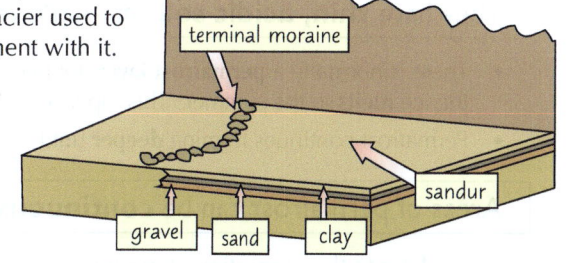

Kettle holes

- The retreating glacier can leave behind **individual blocks** of ice on a sandur.
- Meltwater will continue to wash sediment **past** the block, which can **surround** the ice or even **bury** it.
- As these blocks of ice **melt**, they will reveal individual **depressions** and any overlying **sediment** will **collapse** into the depression. These are known as **kettle holes**.
- Lakes known as **kettle lakes** often form in these kettle holes.

Peter's Pool, a kettle lake in New Zealand

Proglacial lakes

- **Proglacial lakes** can form in front of glaciers when terminal moraine **dams** the flow of meltwater streams.
- As meltwater streams flow into a proglacial lake they **slow** down and **deposit** their sediment on the ice (this is called a delta).
- Each layer of sediment is known as a **varve**. The thickness of the varves is directly proportional to the amount of meltwater (and sediment) coming from the glacier.

Warm-Up Questions

Q1 Explain the main differences between glacial deposits and fluvioglacial deposits.

Q2 Describe how eskers are formed.

Exam Question — AS and A-Level

Q1 Explain how meltwater can lead to the formation of different fluvioglacial features. [6 marks]

If you're proglacial, this section must be a dream come true...

Well, this is just typical of glaciers. Not content with ripping bits of rock out of mountains and scattering them all over landscapes, glaciers then have to go and melt and wash all kinds of bits of rock all over the place, giving you more to learn.

Periglacial Processes and Landforms

Periglacial areas aren't covered in ice. There's usually ice in the soil though — I knew it'd be there somewhere...

Periglacial landscapes have special characteristics

They're cold

Periglacial environments are places where the temperature is frequently or constantly **below freezing**. They:

- **don't** have permanent ice cover.
- have average temperatures ranging from **1 °C to –15 °C**.
- have fairly **low precipitation levels** (380 mm or less).
- have **clearly defined seasons** — brief, mild summers and long, cold winters.

Slumps caused by the melting of permafrost on Herschel Island, Yukon, Canada

permafrost
ground that remains frozen for two or more years

They have thin, acidic soil that isn't very fertile

- There is normally a **permafrost layer**, topped with a relatively **thin** layer of soil, in which the ice **melts** in the summer. This top layer is known as the **active layer**.
- Permafrost continues forming **deeper** until the influence of **geothermal heat** stops it. This may be as deep as 1 km.

Areas of permafrost can be continuous, discontinuous, sporadic or isolated

This relates to the proportion of ground area that is frozen:

- **Continuous permafrost** is where **at least 90%** of the ground area is frozen. To form continuous permafrost, the mean annual temperature needs to be **below –5 °C**.
- **Discontinuous permafrost** is a more fragmented layer where **50% to 90%** of the ground area is frozen. For discontinuous permafrost to form, the mean annual temperature needs to be **below 0 °C for at least two years**.
- **Sporadic permafrost** is where **10% to 50%** of the ground area is frozen. The permafrost will be in **patches** because it is reliant on specific, **microclimatic conditions**.
- **Isolated permafrost** is when **less than 10%** of the area is frozen.

They have vegetation that is slow-growing and short

- **Harsh climatic conditions** hinder plant growth.
- **Grasses, mosses** and **lichens** are the most common plants.
- There may be some **small, dwarf trees** that can grow in more sheltered areas.

The tundra biome (see p.57) is part of the periglacial environment.

Caribou on Alaskan tundra

Periglacial landscapes are found at high latitudes

- Periglacial environments are particularly prevalent at high latitudes, e.g. countries on the edge of the **Arctic Circle**.
- They're also found at **high altitudes**, such as in the Himalayan plateaus. In mountain ranges, they exist around **ice masses**.
- Periglacial environments at lower latitudes tend to be in the **centre of continents** due to the warming influence of the sea at the coast.

> **The distribution of periglacial landscapes has changed**
>
> - During the **Last Glacial Maximum**, the periglacial extent was as far south as the **Mediterranean** and **northern Japan** (see map on p.32).
> - Today, **20%-25%** of the Earth's land surface is permafrost compared to **33%** during the **last Ice Age**.

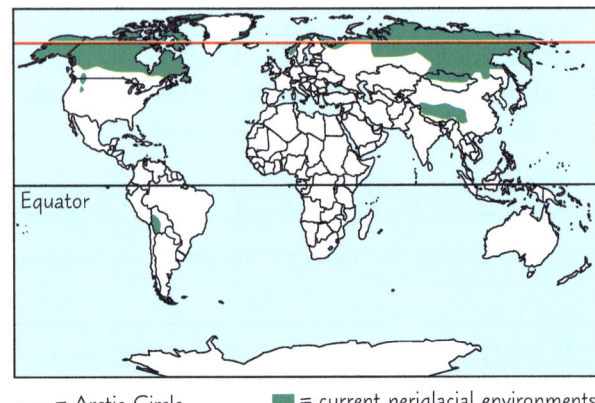

Equator

— = Arctic Circle   ■ = current periglacial environments

Periglacial Processes and Landforms

Periglacial processes include **mass movement** and **erosion**

Solifluction

Solifluction can occur in areas of permafrost as the ground starts to **thaw**.

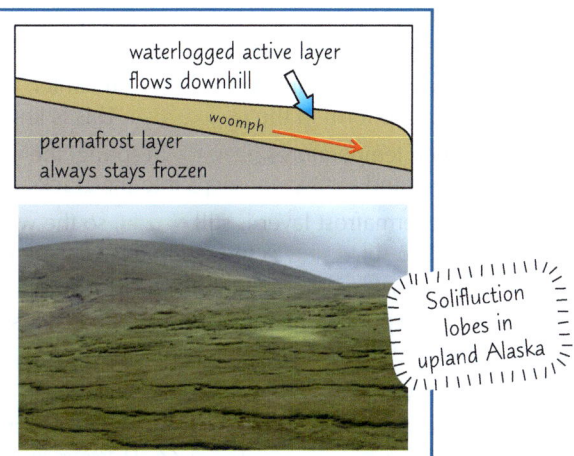

waterlogged active layer flows downhill

woomph

permafrost layer always stays frozen

Solifluction lobes in upland Alaska

- The layer of permafrost is **impermeable** (water cannot flow through it). If the temperature rises above 0 °C in the summer, the **active layer melts** but the meltwater **cannot** drain away.

- The active layer becomes **waterlogged** and **heavy** and **flows** relatively easily. This flow is known as **solifluction** and it can occur wherever there is a **gradient**.

- **Solifluction** produces **lobe formations**. This happens when one section of soil moves **faster** than the soil around it (often because it is on **steeper** ground). The faster flowing section of soil flows **further** and forms a **tongue shape**.

Freeze-thaw weathering

Regular **freeze-thaw** weathering can also lead to mass movement on **slopes**.

- Water can pool in **cracks** in the rock. The **volume** of this water **expands** by 9% when it freezes. The ice then **thaws** when temperatures rise. Continual cycles of freezing and thawing of this water **weaken** the rock and shatter it (**frost shattering**).

ice expands   rock breaks off

- When water in the soil **expands** due to freezing, it causes soil particles to move **upwards** at **right angles** to the slope.

- When the ground **thaws**, the soil particles move **vertically downwards**, meaning they end up **further down** the slope. This is known as **frost heave**.

Nivation

Nivation is especially common in periglacial areas because temperatures fluctuate around 0 °C.

Nivation makes **hollows** in a **sloped** landscape deeper by **freezing** and **thawing**.

- When **snow** collects in a hollow, **freeze-thaw weathering** can cause erosion. **Frost-shattering** breaks bits off the rock at the base of the hollow. When the snow **melts**, the meltwater carries the broken bits of rock away.

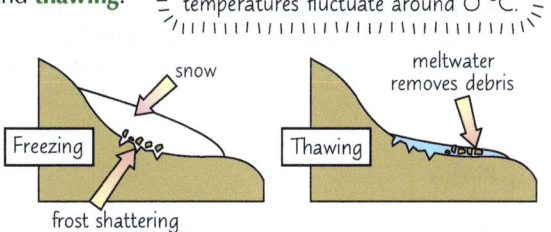

snow   meltwater removes debris

Freezing   Thawing

frost shattering

- Slopes **collapse** because they're **waterlogged** and they've been **eroded** — the material is **washed away** by the meltwater.

- Eventually the hollow becomes **deeper** and **wider**. The processes that cause this are collectively called nivation. Nivation can also lead to the formation of a **corrie** (see p.43).

Wind erosion (aeolian)

- **Strong winds** are likely in periglacial landscapes due to there being few tall trees and buildings to act as **windbreaks**.

- Wind erosion can **break up loose till** and **weaken** features of rocky landscapes.

- **Fine material** can be **carried** and **deposited** in glacial areas such as **outwash plains**.

Wind shaped rocks in Evolene, Switzerland

Meltwater erosion

This happens mostly in summer when there are greater rates of **ablation** (see p.34) and **meltwater discharge** is far higher.

Periglacial Processes and Landforms

Periglacial processes create **distinctive landforms**

Ice wedges

- Ice wedges develop in **permafrost soil**.
- When temperatures are **very low** in winter, the ground **contracts**, causing **cracks** to form in the **permafrost**. This is known as **frost contraction**.
- When temperatures **increase** in spring, the **active layer thaws** and **meltwater** seeps into the cracks.
- The permafrost layer is still **frozen**, so the water **freezes** in the cracks. The **ice-filled cracks** formed in this way are known as **ice wedges**.
- Frost contraction in following years can **reopen** these cracks in the **same place**, splitting the ice wedge. More water seeps in and freezes, **widening** the ice wedge. Over time, this creates a **polygon shape** on the land surface.

Ice wedge polygons, Spitzbergen, Svalbard

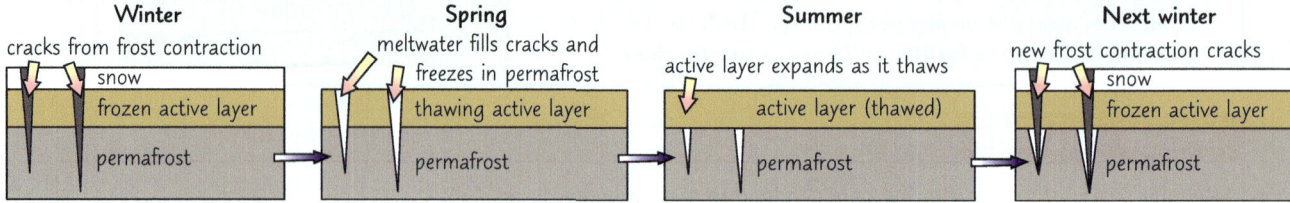

Patterned ground

- Patterned ground is where **stones** are arranged on the permafrost surface in **circles**, **polygons** or **stripes**.
- Patterned ground can form in two ways — by **frost heave** or by **frost contraction**.
- **Frost heave** happens when water underneath stones **freezes** and **expands**. This forces the stones **upwards**. Once the stones reach the **surface**, they roll down to the edges of the **mounds** that have formed, forming **circles** around them. **Polygons** form when the mounds are close together. If the mounds are on a slope, the stones roll downhill and form **lines**.
- **Frost contraction** causes the ground to **crack** in polygon shapes. The cracks fill with **stones**, forming **raised polygon patterns** on the surface.

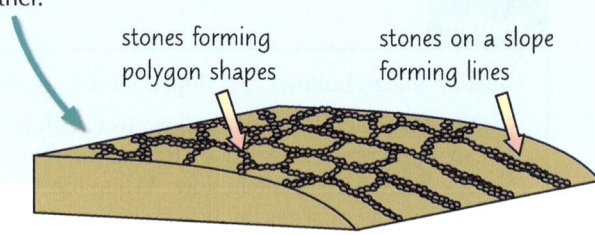

stones forming polygon shapes

stones on a slope forming lines

Pingos

- A pingo is a **conical hill** with a **core** of **ice**. Pingos can be as large as 90 m high and 500 m wide.
- There are **two types** of pingo — **open-system** and **closed-system**.
- **Open-system pingos** form where there's **discontinuous** permafrost. **Groundwater** is forced **up** through the **gaps between** areas of permafrost (from unfrozen layers lower down). The water **collects** together and **freezes**, forming a **core** of ice that **pushes** the ground above it **upwards** to create a **dome**.
- **Closed-system pingos** form in areas of **continuous** permafrost where there's a **lake** at the surface. The lake **insulates** the ground, so the area beneath it remains **unfrozen**. When the lake **dries up**, the ground is no longer insulated and the permafrost **advances** around the area of unfrozen ground. This causes water to **collect** in the centre of the unfrozen ground. The water eventually **freezes** and creates a **core** of ice that **pushes** the ground above it **upwards**.
- If the ice core **melts**, it can create a **crater** at the top of the pingo which can be **filled** with **meltwater**.

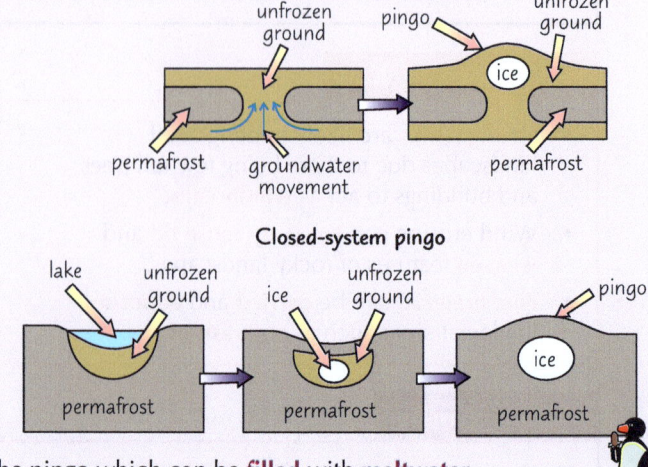

Open-system pingo

unfrozen ground   pingo   unfrozen ground

ice

permafrost   groundwater movement   permafrost

Closed-system pingo

lake   unfrozen ground   ice   unfrozen ground   pingo

ice

permafrost   permafrost   permafrost

Periglacial Processes and Landforms

Loess fields

- **Aeolian** (wind related) **processes** also shape periglacial landscapes.
- A **loess field** is a large area full of **wind-blown sediment**. This sediment is often **fine material** that has been **deposited** on extensive glacial **outwash plains**.
- Strong winds blow the fine, eroded silt material (loess) across a **plateau** that is **unobstructed** by large trees.
- The loess **improves** the soil, which may make it possible to set up **arable** farms in warmer areas.

The tundra environment of Northern Canada

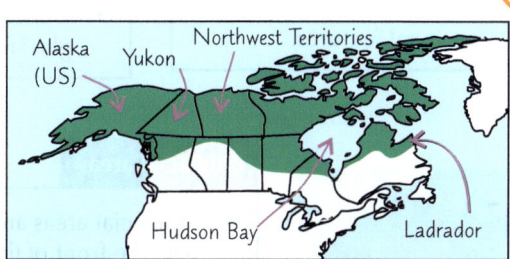

= Permafrost extent in Canada and Alaska

- **Permafrost** extends across 40-50% of Canada. The tundra environment stretches from the Yukon territory in the west of Canada to Hudson Bay and Labrador in the east.
- The landscape has many **periglacial features**. In northern Canada, the combination of **lowland** and **mountains areas** aid the formation of these features.
- **Patterned ground** can be found throughout Northern Canada, in particular in the **Mackenzie River Delta**.
- Just over 1300 **pingos** are located in the Northwest Territories. The tallest of these, Ibyuk, reaches a height of 49 m and is estimated to be over 1000 years old.
- In the northernmost parts of Canada, the growing season may be less than two months due to the **climate**. This means that mosses, grasses, sedges and other **low-lying plants** dominate the landscape.
- There is a range of **fauna**, such as musk oxen and Arctic hares that are adapted to survive in these conditions.

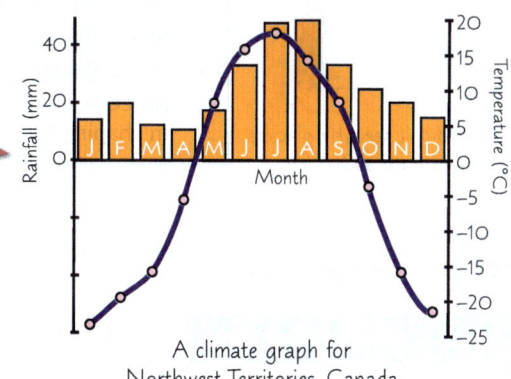

A climate graph for
Northwest Territories, Canada

An aerial view of a pingo landscape in
Tuktoyaktuk, Northwest Territories, Canada

- **Human actions** and **climate change** threaten the different areas of permafrost.
- **Rising temperatures** could significantly alter tundra environments. A warmer climate could cause **carbon** stored in permafrost layers to be released as **carbon dioxide**, causing **temperatures to rise** further.
- Though humans inhabit areas throughout the tundra environment, there is also a network of **National Wildlife Areas** to aid the **protection** and **management** of the tundra.

Warm-Up Questions

Q1 Describe how the global distribution of periglacial landscapes has changed over time.

Q2 Explain the process of freeze-thaw weathering.

Exam Question — AS and A-Level

Q1 Explain how ice wedges are formed. [4 marks]

I always thought Pingo was one of the Beatles...

The trouble with this lot is that there are so many different processes going on, and all of them are to do with water freezing and then thawing. But you can sleep soundly now you know that patterned ground is caused by frost activity and not aliens.

Reconstructing Glacial Dynamics

The landforms that glaciers leave behind aren't just pretty — they also tell us about the glaciers that used to be there.

Different **landforms** are associated with **different** parts of an **ice mass**

Different landforms develop **under**, **next to** and **in front of** an **ice mass**, which creates distinctive landscapes.

Subglacial areas

- Subglacial areas are those that are **underneath** the **ice**.
- These landscapes show evidence of the ice through their high numbers of **erosional features**.

Marginal areas

- Marginal areas are the **edges** of the glaciers or ice sheets.
- These landscapes show evidence of the ice through their **depositional features**.
- Deposition occurs due to both **glacial** and **fluvioglacial processes**.

Glacial deposition in the Yorkshire Dales

Tommy didn't see why glaciers should have all the fun.

Proglacial areas

- Proglacial areas are those associated with the **front** of the ice mass.
- These landscapes show evidence of the ice through features created by **meltwater** and the actions of the **wind**.
- **Fluvioglacial depositional processes** have a strong influence in shaping these areas.

Periglacial areas

- Periglacial areas are **cold** areas found **near to glaciers** and they often contain **permafrost**.
- **Weathering** actions such as **freeze-thaw actions** are influencing the features found in these areas.
- The movement and melting of the ice itself **don't** have a strong influence.

Features vary in **scale**

Micro-scale are the smallest

- They are usually less than 1 m long.
- Examples include **striations** and some **erratics**.

Meso-scale are medium-sized

- They're between 1 m and 1 km long.
- Examples include **roches moutonnées**, **drumlins**, **eskers** and **kettle holes**.

Macro-scale are the largest

- They can **dominate** the landscape.
- Examples include **glacial troughs**, **corries** and **sandurs**.

Relict landscapes reveal the **extent of past ice cover**

- **Active glacial landscapes** are those that are **currently** experiencing glaciation.
- **Relict landscapes** are those that show evidence of a **past** glacial period by an **analysis** of the features that remain there.
- Reading the **provenance** (the history or origin) of a landscape is **not** always **straightforward**:

  1) If **more than one** glacial period has shaped the landscape, then reading it is more challenging.
  2) **Human activity** may also have altered the natural landscape and made it more difficult to read.

> Understanding past glacial events by analysing the remaining features is known as inversion modelling. It draws on Hutton's principle of uniformitarianism, which is the idea that the present informs us about our past.

- Different features occur in **upland areas** (those in locations at higher altitudes such as mountain slopes) to those in **lowland areas** (those in locations at lower altitudes such as valley floors).

  1) **Upland** areas are characterised by features that are mostly formed by **erosion**.
  2) **Lowland** areas are characterised by features that are mostly formed by **deposition**.

Pyramidal peak, Dolomites, Italy

Reconstructing Glacial Dynamics

Glacial features give information about **glacier motion**

Some features show the **direction** the glacier moved...

Glacier direction

Erratics

Identifying the **rock type** of an erratic can help determine where it has come from, and therefore the direction the glacier was moving.

Drumlin, Clew Bay, Ireland

Drumlins

- The **orientation** of a drumlin tells you in which direction a glacier was moving.
- Drumlins run **parallel** to ice flow with the **taller** end on the **upglacier** side and the **shorter** end on the **downglacier** side.

Crag and tail landscapes

These are also **parallel** to ice flow with the **resistant rock** facing **upglacier** and the deposited **tail** of smaller sediment facing **downglacier**.

Though remember they don't if forming a pyramidal peak (see p.43).

Corries

- These usually orientate in the **same direction**.
- This indicates the **route** of a glacier down a mountain slope.

Castle Rock in Edinburgh is the resistant rock of a crag and tail.

...while others show the glacier's **specific path**, **extent** and **speed**

- **Lateral moraine** — this would have formed a ridge **alongside** the glacier, mirroring its path.
- **Recessional and terminal moraine** — these show the position of the glacial **snout** or the ice sheet's **furthest extent**.
- **Drumlins** — calculating the **elongation ratio** shows how fast the glacier moved. Dividing the length of the longest axis by the width of the widest point gives you the elongation ratio. The **greater** the elongation ratio, the **faster** the glacier was moving.

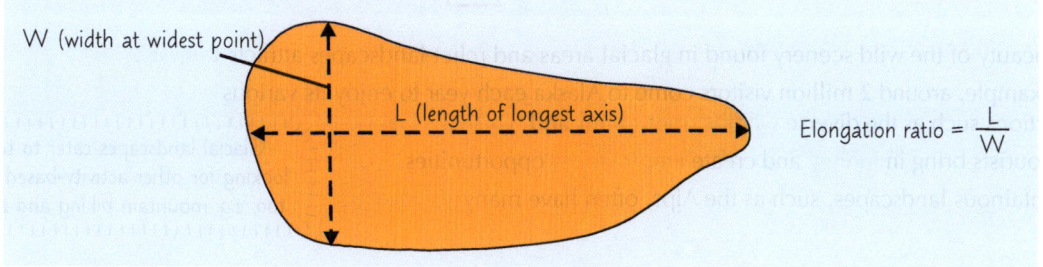

W (width at widest point)

L (length of longest axis)

Elongation ratio $= \dfrac{L}{W}$

- **Till fabric analysis** is a method of studying the **orientation** and **dip** of the longest length of deposited materials. Studying a **large sample** of till can reveal the **probable path** that a glacier took.

Warm-Up Questions

Q1 Explain the key differences between marginal and proglacial areas.
Q2 Describe the concept of inversion modelling.

PRACTICE QUESTIONS

Exam Question — A-Level

Q1 Explain how studying glacial landforms can reveal the past movement of a glacier. [8 marks]

Corries? I'm more of an Emmerdale fan myself...

There weren't many geographers around in the last ice age — the only way we know that most of Europe was covered in glaciers is because of the landforms left behind. Scientists also think that Mars's surface may have been carved by glaciers too. Wow.

Value of Glaciated Landscapes

Glaciated landscapes aren't just handy for giving you something to learn about in Geography, they also provide resources and habitats for wildlife. They're pretty important to humans for lots of reasons as you'll see...

Glacial and periglacial landscapes have **environmental** and **cultural** value...

1) Some landscapes can be sacred or **culturally significant** to people. For example:

View of Mount Everest and Nuptse with Buddhist prayer flags

- **Storytelling** amongst **Inuit** communities is often based around legends that involve a deep understanding of the North American tundra.
- The **Himalayas** are considered highly **sacred** to local people. The Nepalese have adorned some Himalayan mountains, such as Mount Everest, with Tibetan prayer flags.
- For hundreds of years, **Tibetan Buddhists** have hung these flags outside their houses.

> Buddhists believe that the prayers written on these flags will travel further due to the strong upland winds that blow through these environments.

2) Glaciers also have environmental value. For example, extensive **polar research stations** such as the British 'Halley VI' have been set up in **Antarctica**.

- These are used to study a variety of **climatic and oceanic processes**.
- Studying the climate of Antarctica and how it has changed over time has given scientists an insight into **past climate change events**.
- This allows them to **model** future **climate change**.

The British Antarctic Survey research station 'Halley VI'

3) Glacial landscapes can also be areas of **wilderness recreation**. Many people enjoy **hiking** and **climbing** in glacial landscapes. These wilderness areas often offer an escape from busy and stressful aspects of their life.

...as well as **economic** value

Mining
- Mining in glacial areas exploits **mineral resources**, such as gold, silver, iron ore, lead, zinc and copper.
- Mining has a considerable impact on local **economies**. For example, Alaska's mining industry was worth US $2.9 billion in 2018.

Forestry
- **Cold** weather and **high rainfall** tend to dominate relict glacial uplands making them well-suited to the growth of **coniferous trees**.
- Large businesses in both Canada and northern Europe manage glacial landscapes as **forestry plantations** to supply timber and paper.

Tourism
- The beauty of the wild scenery found in glacial areas and relict landscapes attracts **tourists**.
- For example, around 2 million visitors come to Alaska each year to enjoy its various attractions such as the diverse **wildlife**, **mountains** and **Northern Lights**.
- The tourists bring in **money** and create **employment** opportunities.
- Mountainous landscapes, such as the Alps, often have many **ski resorts**.

> Glacial landscapes cater to tourists looking for other activity-based holidays too, e.g. mountain biking and ziplining.

Herdwick sheep, Lake District

Farming
- Upland areas in relict glacial landscapes often suit **pastoral farming**.
- For example, in the Lake District National Park, farmers have a duty to **protect** the beauty in the landscape as well as use it as a source of income. In the National Park, 28% of the land has a **common land designation**. This means that the land is **communal** and all local farmers can use it to graze livestock.

Energy production
- Energy companies are using the increasing levels of **glacial meltwater** around the Arctic Circle for **hydroelectric power (HEP) generation**. Current HEP plants already generate more than 80 GW of electricity within the region. Both northern Canada and Russia each have five hydroelectric plants with the capacity to generate 2 GW each.
- **Micro-hydro systems** have been constructed in some **low-income countries** e.g. Darbang in Nepal, which have helped to power **industry** and **businesses** in rural areas. They typically generate up to **100 kWh** of electricity.
- The plants work well in valleys with **ribbon lakes** (see p.44) as they can easily be **dammed**.
- The **steep relief** and **low population** levels in some upland glacial areas makes HEP a viable and efficient choice.

Value of Glaciated Landscapes

The tundra provides unique **biodiversity**

1) Due to the harsh and often inhospitable **climate**, tundra regions have relatively **low levels** of biodiversity. The higher the **latitude** of the tundra, the **harsher** these conditions will be.

Tundra environments are very fragile and don't recover quickly from change.

2) Flora and fauna found here have **adapted** to this environment:

- Plants have **short growing seasons** due to the long, dark winters and relatively short summers. This means that plants tend to grow to a **limited height**. They've also adapted to do this to cope with **strong winds**.

The Arctic Willow tree only grows to an average height of 15 cm.

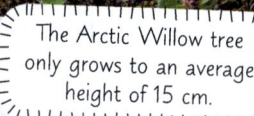

- The dominant flora is **low level plants**, such as grasses and mosses, which are an important **food source** for **herbivores** such as caribou, musk oxen and Arctic hares.

- There are **relatively few mammals** in the tundra compared to other ecosystems. However, some mammals are able to survive in the tundra. For example, **Arctic hares** are common throughout northern Russia and Canada. Larger animals such as **caribou** move through the tundra landscape with the seasons. They **migrate** north in the summer to make use of the grasses found there and migrate south as soon as the first snow arrives.

The biodiversity of the tundra in the summer can be vastly different to that of the winter.

- As the permafrost and ground ice starts to **melt** in spring, it can create **wet breeding grounds** for different fly and insect species. This attracts **birds**, which feed on them.

3) Despite there being limited biodiversity in these regions, little is known about many of the species. Their potential value in, for example, **future pharmaceutical development** is unknown.

Natural systems are maintained by glacial and periglacial landscapes

- The relationship between ice, meltwater and the glacial and periglacial landscape is a key part of the **water cycle**.

Irrigation channels carry glacial meltwater through farmland in Peru.

- The **cryosphere** represents a major store of water — around 2% of the world's water is ice, representing over 68% of the world's freshwater.

- A large amount of meltwater in Alpine areas replenishes **drinking water supplies**. In drier areas, such as in the Bolivian foothills of the Andes, farmers use meltwater from glaciers to **irrigate farmland**, provide **drinking water** and for **HEP**.

- Falling as snow, hail and rain, precipitation in glacial areas is more **varied** than in other regions. Each type of precipitation has a different influence. Snow and hail can accelerate gains in **ice stores** in the **zone of accumulation**. Warmer precipitation, like rainfall, can **accelerate ablation** and increase stores of **meltwater** both in and around a glacier.

- **Permafrost** areas are important **carbon stores**. They hold dead organic material in a state which means it will not decompose. Scientists estimate that there is around 1500 gigatonnes of carbon frozen into permafrost.

- Glacial and periglacial areas have long held stores of carbon in the form of **fossil fuels**. There is an estimated 2.9 trillion kWh of energy to come from the natural gas stored under the Canadian state of Yukon.

Billy recognises the value of glaciated landscapes.

River levels increase in spring and summer as temperatures rise, but can be almost zero in winter.

Warm-Up Questions

Q1 Describe the cultural value that glacial and relict glacial landscapes provide.

Q2 Explain why glacial landscapes play an important role in the water and carbon cycles.

Exam Question — AS and A-Level

Q1 Explain how the natural resources found in a glacial area may be used for economic gain. [6 marks]

What's the difference between reindeer and caribou? Caribou can't fly...

As you'll have seen from the past couple of pages, these landscapes have a multitude of uses. It's pretty handy the way permafrost locks away those greenhouse gases. The trouble is, the world is warming and the permafrost is melting. Eeek...

Threats to Glaciated Landscapes

We'd miss glaciated landscapes if they vanished. Trouble is, human actions can upset the balance... and the penguins.

Natural hazards pose a threat to glacial upland landscapes...

Avalanches

- Avalanches (mass movements of snow, ice and rocks) are occurring **more frequently** in glacial areas.
- In a changing climate, **snowfall** events are likely to be **heavier**. In winter, this means that there can be a sudden increase in a glacier's **zone of accumulation**. This is more vulnerable to movement because the fresh snow has not had time to firm up like the older layers of snow.

Powder avalanche in the Caucasus

- Basal meltwater (see p.39) can cause **slab avalanche** as it creates a weak layer at the bottom of the snowpack. The meltwater refreezes into a **more crystalline** and **less cohesive** structure, causing large sections of overlying snow and ice to fall at high speed away from the main structure.
- In **powder avalanches**, snow particles become suspended within the air as they flow downslope.
- **Dry avalanches** don't need much meltwater to aid their movement.
- **Wet avalanches** have a lot more meltwater built into their flow, making the snow **slushy** as it descends.

- An increased number of avalanches mean people are at **greater risk** in glacial areas. The risk has also increased as more people use upland areas for recreation.
- Some avalanches also create **lahars** (see p.14). Lahars can occur when a **volcanic eruption** triggers an **avalanche**. This can be where the moving snow combines with loose soil to create a **mud flow**.

> Between 2010 and 2019, 256 people lost their lives to avalanches in the USA compared to 53 between 1960 and 1969.

Glacial outburst floods

- A **glacial outburst flood** (or **jökulhlaup**) is the rapid movement of **meltwater** from beneath an ice mass (see p.14).
- **Geothermal activity** in the Earth's crust causes **basal heating**. This increase in temperature causes the glacier to **partially melt**, releasing **englacial meltwater** (see p.48).
- The sudden breakup of **terminal moraine walls** that dam meltwater, can also cause glacial outburst floods.

> As well as being a threat to human life, avalanches and outburst floods change glacial landscapes. They can move glacial deposits far from their original position and create new erosional features.

...and so can human activities

Tourism

- Increased demand for **tourism** in polar glacial areas has meant that **larger cruise ships** are visiting these areas. This can potentially increase **marine pollution** in sensitive areas.
- Tourists can also bring **noise** to otherwise quiet areas. In Antarctica, this has caused stress in some **penguin colonies** and has affected their ability to breed successfully. Tourism developments have **disrupted wildlife** and habitats, which has **reduced biodiversity** in an area that already has low biodiversity.

Fed up of the racket the tourists were making, Arnold relocated.

Construction

- The construction of **dams** and **reservoirs** can block the **migratory routes** of fish so they can't reach the **spawning grounds** where they breed. This reduces the total fish population and impacts other parts of the **food chain**.
- **Hydroelectric power plants** (HEP) can also disrupt the natural regime of a river, altering the normal **flow** and **discharge** of the river. This can then affect **sediment levels**, **deposition locations** and affect **aquatic life**.

> Fish, such as salmon, can travel long distances to spawn, so this has a widespread impact.

Urbanisation

- Where glacial areas contain fossil fuels or minerals, companies may build **small settlements** to house workers, e.g. Alaska's North Slope in areas around Prudhoe Bay. These settlements tend to **grow** over time as more shops and services are added.
- **Urbanisation** in glacial and periglacial areas creates small **urban heat islands** that can affect their surroundings. As temperatures increase over time, houses built on areas of permafrost may start to **subside** into the tundra. This is not only a problem for **residents**, but also affects the **habitats** that depend on the permafrost.

Threats to Glaciated Landscapes

Threats to the Alpine valleys

- As well as using the Alps for winter sports like skiing and snowboarding, tourists use the mountains in the **summer season** for activities such as mountain biking, bird watching and climbing.
- This increase in the number of people requires an increased amount of **resources**. Companies **transport** these resources into the region by **road**. Local authorities have **widened** the roads to cope with increased levels of traffic. This increased traffic means **air pollution** in the upland areas has increased.

Mountain biking in the Swiss Alps

Levelling a ski run in summer

- Patterns of **landslides** and **avalanches** have become more **variable** as **climate change** affects the times and places where meltwater flows. This can make the Alps more **dangerous** to people who visit and live there as **flood risks** increase.
- **Trees** have been removed to make way for larger resorts and ski runs, which many people believe spoils the **overall look** of the landscape. This can also **increase** the risk of landslides and avalanches as the ground becomes **less stable**.

Human activity can also degrade glacial landscapes

Litter

- **Increased numbers of visitors** to a glacial landscape can create increased levels of litter.
- Litter can **pollute water** in glacial landscapes, which can have a knock-on effect for both animals and humans.
- Litter can also **harm wildlife** and **damage habitats**, especially because **decomposition** rates in cold environments are **slow**.

Soil erosion and landslides

- **Increased numbers of hikers** increases the risk of people **straying** from defined areas. **Soil erosion** and **trampling** of **sensitive vegetation** off the main paths increases in the busiest visitor seasons.
- Such activity causes **soil compaction** which reduces **infiltration rates** (increasing the risk of **flooding**) and also makes it more difficult for **vegetation to re-establish**.
- Compacted soil lacks **vegetation roots** which hold the soil in place meaning that **landslides** are more common after periods of heavy rainfall.
- **Meltwater** washes away the **soil** that has been **exposed** by human activity and makes remaining soil more prone to erosion.

infiltration
the process by which water on the surface of the ground enters the soil

Signage in the Himalayas (top) and in the foothills of the Lake District (bottom) aims to keep visitors from straying off paths and damaging ecologically sensitive areas.

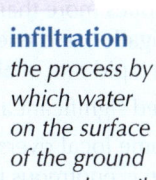

Introduction of alien species

- Increased visitors from elsewhere means **alien species** are more likely to be introduced into **ecologically sensitive areas**.
- Walkers may have **seeds** in their boots and clothing, while ships visiting polar regions may have **alien species**, e.g. Chilean mussels, stuck to their hulls.

Cruise ships visiting Antarctica may introduce alien species.

Deforestation

- Mature trees may also be removed (deforestation) to allow **access** to key areas.
- New **roads** and **carparks** can also increase the rate of **saturated overland flow** of meltwater, which can increase the risk of **flooding**.

Threats to Glaciated Landscapes

Glacial **mass balances** are affected by **climate change**

1) Globally, most glaciers have a **negative mass balance** (see p.36) and are **retreating** due to rising temperatures.

2) Melting glaciers and ice sheets, particularly those in Greenland and Antarctica, are causing **sea levels** to **rise**. In the future, this could cause **flooding** in low-lying coastal environments.

3) **Permafrost** is melting in places like Alaska, creating a **positive feedback loop**:

- This melting releases carbon trapped in the soil as methane and carbon dioxide.
- More atmospheric methane and carbon dioxide enhances the greenhouse effect.
- An enhanced greenhouse effect causes temperatures to rise.
- This leads to further melting of permafrost.

> **water table**
> *an underground boundary below which the rocks or soil are permanently saturated with water*

4) These changes caused by global warming may disrupt the hydrological cycle.

5) Sudden increases in **meltwater** lead to sudden increases in **river discharge**. This can cause **flooding** as soils can't cope with **increased** volumes of water so become **saturated** very quickly. If the **water table** is close to the surface, there are few places for the extra water to go.

6) In the longer term, meltwater supplies may **run out** as glaciers **retreat**:

- This may mean that **hydroelectric power plants** do not receive enough water to make their operations viable.
- It could result in a **reduction** in **aquatic life**, especially fish species.
- A decrease in **river flow rate** means that any pollution in the river becomes **more concentrated**, which decreases the **water quality**.
- Slower flowing rivers and streams will also have a **lower capacity** to carry sediment (i.e. a lower **sediment yield**). This causes the **deposition** of sediment within the channel, reducing its **size** and increasing the risk of **flooding**.

Changes in the Himalayan glaciers

CASE STUDY

- Advances in mountaineering equipment and increased media attention have encouraged **increasing** numbers of people to attempt to climb **Mount Everest** than ever before.

Over 800 people summited Everest in 2018, whereas only about 1300 between 1953 and 2000.

- This has an **impact** on the **local area**. Scientists believe the **snowline** is moving progressively **higher** up the mountains each year as glaciers **melt** due to changes in the **local microclimate**.

Around 75% of the Himalaya's glaciers are in retreat.

Glacial lakes at the snouts of glaciers in Bhutan, Himalayas

- There are over **600 dams** either in operation or construction in the Himalayas. These are set to generate power for India, Bhutan, Nepal and Pakistan. As levels of meltwater reduce, some of the reservoirs behind the dams may **dry up**. The reservoirs and the turbines in the dams may fill with **sediment deposits** making them inefficient and inoperative.

- **Meltwater from glaciers** also supplies more than 1 billion people in the Himalayan region with **water** for different uses including drinking water, irrigation and energy. This water supply is more useful if it feeds into the river system at a **steady rate**. With the rapid rate of **ablation**, river systems can **flood**, wasting some of this resource.

- **Deforestation** in Nepal has created significant areas of **exposed soil** in upland areas. This has lead to the following:

  1) Increased **sedimentation** in some local rivers. With 64% of Nepal's working population employed in **agriculture**, there is enormous pressure to use the land to **increase productivity**.

  2) An increased risk of **landslides** — this is concerning as many Nepalese live on the mountain slopes.

Warm-Up Questions

Q1 Explain the natural threats that affect glacial areas.

Q2 Describe how tourists can threaten a glacial area.

PRACTICE QUESTIONS

Exam Question — A-Level

Q1 Evaluate the idea that the greatest threat to glacial areas are the humans that live there.   [20 marks]

Glacial mass balances must be huge...

Although how they get the glacier onto them is beyond me. Anyway, no time for joking, you best get all these facts learnt.

Managing Glaciated Landscapes

Management of glacial areas is tricky — there's usually a bunch of 'stakeholders' who are all likely to have different priorities.

Managing **threats** to glacial areas involves **different** stakeholders

Local and regional governments

- Local and regional **governments** have a role in ensuring that **visitors** to a glacial area use it **responsibly**.
- They may also provide **funding** to volunteer groups that aim to **conserve** and **protect** these areas.
- Additionally, they have a duty to consider the **development** of glacial areas for **economic purposes**.

National governments

- **National governments** have the authority to decide whether to develop glacial areas or not. They can also designate an area a **National Park** or to provide an area with a **legislative protection status**.
- They also control funding for **larger conservation projects**.

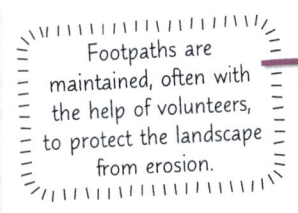

Footpaths are maintained, often with the help of volunteers, to protect the landscape from erosion.

The path to Catbells, Lake District

Conservationists

Kayaks created a flotilla in Elliott Bay, Seattle, in protest against Shell's plan to drill for oil in the Arctic.

- Conservation groups are likely to be **volunteers**, as well as **scientists** and **academics** that understand the wider importance of the glacial area.
- They may work in **partnership** with other stakeholders to provide manpower to **protect areas**, as well as advise local authorities on different **practices**.
- They may also be involved in **wider scale advocacy** and **protests** as well as **public awareness campaigns**.

Non-governmental organisations

- Non-government organisations aim to speak on behalf of the glacial environment and its **biodiversity**.
- They may also work with **indigenous cultures** to raise awareness of their need for protection and preservation.

Businesses

- Transnational corporations are often looking for opportunities to **expand**. However, this can have negative impacts on the **environment** in which they wish to work.
- Some businesses may actively work to **protect** the environment, or have **initiatives** and **policies** in place to **offset** any damage.
- Local businesses that are directly associated with the landscape can feel **conflicted** between providing **employment** for local people and looking after the **landscape** that creates the jobs.

A restaurant at a ski resort in the Austrian Alps

Global organisations

- Global organisations, such as the United Nations Environment Programme, work to monitor **wider scale conservation efforts**.
- They also police **international agreements** between countries.
- A large part of their work is **collecting data** to show the **impacts** conservation and protection have.
- Global organisations can **face challenges** such as **limited communication** between stakeholders, which can affect **decisions** being made.

Topic 2: Option 2A — Glaciated Landscapes and Change

Managing Glaciated Landscapes

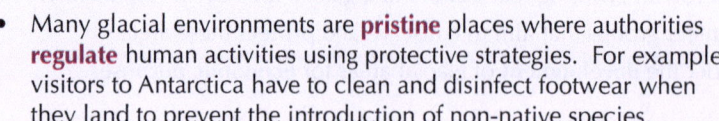

A **range of approaches** are used to manage threats

There is a **spectrum** of approaches which can be used to **manage** threats to **glacial environments**. Most of these fall between these **two strategies** and can be a source of **conflict**:

- **Total protection** of the environment where **no human activity** is permitted.
- **Total exploitation** of the environment where **no protection** occurs.

Protective conservation strategies

- Many glacial environments are **pristine** places where authorities **regulate** human activities using protective strategies. For example, visitors to Antarctica have to clean and disinfect footwear when they land to prevent the introduction of non-native species.
- Some glacial environments may have **protected status** that **prohibits** any unnatural change to the landscape such as the construction of **buildings** or **roads**. There may also be **limits** to the number of **visitors** who can enter some areas such as **wildlife reserves**.
- This form of management is most successful in areas that have **low human population densities**, such as those within the polar regions.
- The **exploitation** of resources is often **prohibited** in protected areas. Authorities can monitor the exploitation of glacial areas to assess its **impact**, and activities can be **strictly managed**.

> Protected areas, such as wildlife reserves, limit the amount of human activity that can take place within them. The ecological functions of a place are prioritised.

Sustainable management

- Sustainable management of glacial areas tries to **balance** the needs of a local population, both **economically** and **socially**, with the **protection** of the environment so that **future generations** can meet their needs.
- **National parks**, such as the Lake District, allow a degree of **economic development** (for example through farming or water management) but also have **conservation projects** taking place. No economic development is able to take place without first carrying out an **environmental impact assessment** to see if that activity is viable.
- This means there is often **conflict** between **visitors, local people** and **conservation officials** in these areas. It can be very difficult to **balance** the needs of the **local economy** with the needs of a **national park**.

Multiple economic use model

- A **multiple economic use model** means that the **local economy** has priority over the conservation of the area.
- This approach tends to be used where there are **limited opportunities** for economic development for a local population.
- It's usually used in areas that hold **mineral** or **fossil fuel** resources (which make them economically valuable).

Managing Yosemite Valley — protective strategies and sustainable management

CASE STUDY

- **Yosemite National Park** in California is a landscape shaped by **glaciers**. Two glaciers remain **active** in Yosemite Valley — the Lyell Glacier and the Maclure Glacier in the east of the park. Both are **retreating** and supply water to the Tuolumne River.
- The park is **vast** (over 3000 km²) and has the capacity to receive **huge numbers of visitors**. **Policing** those visitors is extremely difficult as there are **multiple access routes** to the park. It's therefore **impractical** to manage the park under a model of **complete protectionism**.
- Instead, Yosemite National Park is **open to visitors** (millions of people a year) and some tourism businesses, such as mule trekking excursions and holiday lodges. Other commercial activities, such as fracking for oil and natural gas are prohibited. However:

Yosemite Valley,
Yosemite National Park

  1) There's a concern that the high numbers of visitors reduces the feeling of **wilderness**.
  2) **Water quality** requires continuous **monitoring** and there are concerns over the fragmentation of wild meadows by the creation of informal **footpaths** through them.

- A **Visitor Use and Impact Monitoring Program** is used to identify areas that are **vulnerable**, e.g. by collecting **data** such as the percentage of bare soil in meadows. This means decisions can be made to **restrict visitors** to certain areas of the park, mostly by changing access routes away from the most vulnerable areas.
- There are some **restrictions on visitor use**, for example:

  1) To enter Yosemite National Park, there is a **visitor fee** which varies depending on the method of transport.
  2) Buses entering the park must be **under a certain size**, and groups of over 15 people **can't** stay overnight.
  3) In winter, only 2300 ski lift tickets are available per day to **restrict** the number of people on the slopes.

Managing Glaciated Landscapes

Legislative frameworks can be effective in protection and conservation

- There is growing demand for glacial landscapes to receive greater **legal protection** through **legislative frameworks**.
- Some countries have passed **laws** to prevent certain activities.

> For example, authorities have **forbidden development** and **limited access** in the North Slope area of Alaska that are designated **wilderness areas** (Arctic National Wildlife Refuge). In 2015, President Obama proposed extending the wilderness area to prevent further **oil exploration**, a move that generated considerable **debate** within the USA.

International agreements are sometimes needed

- International legal agreements are important when a sensitive glacial landscape crosses the **border** between two or more countries. The agreements put a **legislative framework** in place to try and meet the **needs** and **expectations** of all the countries involved. This is **challenging** when bordering countries have different levels of economic development, or when there is already **political tension** between them.
- There are a number of international agreements which are in place that also affect glaciated landscapes. For example, the **United Nations Convention on the Law of the Sea** (see p.280) is an international agreement that governs the world's **oceans** and the activity that takes place there.
- Some glacial environments are **internationally important** and there are **global management strategies** to protect them, such as **treaties**.

Ice camp of a polar research expedition, Antarctica

> For example, 55 different countries have agreed to the **Antarctic Treaty** (see p.281). It ensures that countries only use Antarctica for **peaceful purposes**, including **scientific research**. The treaty defines how countries may use the continent for **tourism** and for **fishing**. The Antarctic Treaty is considered to be **highly successful** and has effectively **settled disputes** over territorial claims.

Climate change makes management challenging

- Conservation and protection measures need to be **viable** within the wider context of a **changing climate**.
- This is known as the **context risk**. It requires decision makers to understand the **likely impact** of climate change globally in the future — any form of management needs to work in the context of these **predicted changes**.
- Given the global challenge climate change presents, countries need to use a **coordinated approach**.

  1) **International protocols and pledges** have tried to reduce carbon dioxide emissions, e.g. the Kyoto Protocol (1997), the Paris Agreement (2015) and COP26 in Glasgow (2021).
  2) Between 1987 and 2015 these protocols focused on **mitigation** — trying to **reduce** or **stop** climate change.
  3) The 2015 Paris Agreement shifted towards a focus on **adaptation** to **climate change**. The agreement highlighted how developed nations could **economically support** developing nations that were more **vulnerable** to the effects of climate change.

- Glacial areas are **already** feeling the effects of climate change.
- This means that **national governments** are thinking about how glacial areas can **adapt** to changes.
- Some countries have decided to implement their **climate change management strategies** at national, regional and local levels. This **multi-layered approach** may be more effective because decisions are more collaborative, and they're applied at each level of **governance**.

Warm-Up Questions

Q1 Describe how conservationists work to manage glacial landscapes.

Q2 Explain the term 'context risk' in relation to glacial landscape management.

Exam Question — A-Level

Q1 Evaluate whether glacial areas can be managed successfully. [20 marks]

Remember when glaciers were cool?

Climate change affects a lot of things, including the management of glacial landscapes. Planning for a changed climate is key.

The Coastal System

Coastal systems are the areas where the land meets the sea. And they're almost as exciting as they sound...

The **littoral zone** is a dynamic area of change

The **littoral zone** refers to the **coastal area** that is affected by **waves**. It includes:

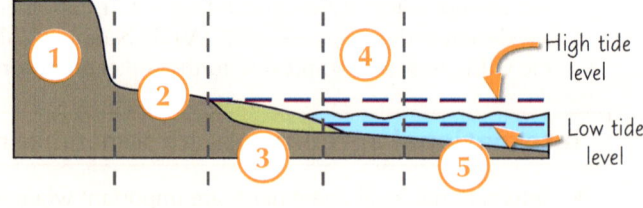

High tide level

Low tide level

1 The **coast** is the land affected by the sea. It may include coastal **settlements** and **farmland**.

2 The **backshore** is the area that sits above the **high tide level**. Coastal processes only affect the backshore when there's an **extreme weather event**, such as a storm or an extreme high tide.

3 The **foreshore** marks the zone between the **high** and **low tide marks**. This is where many **wave processes** take place.

4 The **nearshore** is the area just off the coastline beyond the **low tide** level. It consists of **shallow seawater**.

5 The **offshore** is the open sea where **waves** start to **break** and water is **deeper**.

The littoral zone is an area of **dynamic equilibrium**. This means that different parts of the coast (both **terrestrial** and **marine**) are constantly undergoing **short-term** changes. Coasts vary in their **rate of change**. This variety creates different classifications of coastline.

Changes can take place over seconds to millennia.

Long-term changes create coastlines that are:

- **emerging** or **submerging**, depending on whether **sea level** is **falling** or **rising** (see p.79-80).
- **rocky** or **estuarine**, depending on their **geological resistance** to weathering and erosion (see below).
- **concordant** or **discordant**, depending on the relative position of **geological strata** and **wave action** (see p.67).

Short-term changes create coastlines that are:

- classified by their **tidal range**.
- **retreating** (losing land) or **advancing** (gaining land), depending on whether **erosion** or **deposition** dominates the site.
- **high** or **low energy**, depending on the balance between erosion and deposition taking place there (see below).
- dominated by **land-based processes** (known as **primary coasts**) or by marine-based processes (**secondary coasts**).

tidal range
the difference in height between their high and low tide

Coasts can be classified according to their tidal range.

Type of coast	Tidal range
microtidal	less than 2 m
mesotidal	2-4 m
macrotidal	over 4 m

Coasts can be **high-** or **low-energy** environments

In the coastal system, **air** transfers energy as **wind** while **water** transfers energy as **waves**, **tides** and **currents**.

High-energy coasts:

- receive **large, powerful destructive waves** (see p.65).
- are created by **exposure to strong winds, long fetches** and **steeply shelving offshore zones**.
- often have higher rates of **erosion** than the rate of **deposition**, although the **geology** of some high energy coastlines make them more **resistant** to erosion.
- tend to be **rocky coasts** with **sandy coves** and **rocky landforms** such as **cliffs, stacks** and **arches**. There's a clear **distinction between land and sea**.
- have cliffs with **high erosion** levels that tend to be **steep** with **little vegetation**. Waves **undercut** the cliff and wash away debris. These cliffs have a **marine cliff profile**.
- have cliffs with high levels of **weathering** that tend to be **less steep**, with **weathered material** at their base. These cliffs have a **subaerial cliff profile**.

Low-energy coasts:

- receive **small, gentle constructive waves** (see p.65).
- are created by **gentle winds**, a **sheltered** location, **short fetches** and **gentle sloping offshore zones**. A **reef** or offshore **island** may protect the coast from the waves.
- often have a higher **rate of deposition** than erosion.
- characterised by **sandy beaches**, salt **marshes, estuaries** and **tidal mud flats**, but no **cliffs** next to the beach (there may be cliffs set back from the beach — see p.78).
- have land with a **gentle relief** and relatively **low elevation**. This makes it harder to identify the different parts of the littoral zone. High levels of **deposition** take place here due to a relative **stillness** in the water in both the **estuaries** and the **nearshore zone**. This **sediment** comes from both the **land** (through **river systems**) and from the **sea** (through **offshore movements**). Coastal **accretion** takes place (expansion of land into the sea).

The Coastal System

The shape of the beach is defined by different wave types

1) Wind blowing over the **surface of the sea** creates waves. The **friction** between the wind and the sea surface gives the water a **circular motion** as energy is transferred.

2) This creates **ripples** which will form into larger waves over time. The effect of a wave on the shore depends on its **height**. Wave height depends on the **wind speed**, the **fetch** of the wave and the **depth of the seabed**.

3) A high wind speed and a long fetch create **taller** and **more powerful waves**. As the waves approach the shore, they break. **Friction** with the **seabed** slows the bottom of the waves and makes their motion more **elliptical** (squashed and oval shaped). The **crest** (top) of the wave continues at the same speed, rises up and then collapses (the wave breaking). There are **two** types of wave:

fetch
the maximum distance the wind has blown over open water

swash
water washing up the beach

backwash
water washing back towards the sea

Destructive waves

- **Destructive waves** are **high** and **steep**.
- They have a more **circular cross profile**.
- They have a short wave length, so have a **high frequency** (10 to 14 waves a minute) and a high energy. The strong backwash scours and **removes** material from the beach.

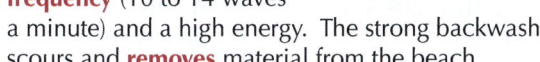

Constructive waves

- **Constructive waves** have a **low frequency** (around 6 to 8 waves a minute) and a low energy.
- They are **low** and have a **long wave length**, which gives them a more **elliptical cross profile**.
- The powerful swash carries material up the beach and **deposits** it at a higher point.

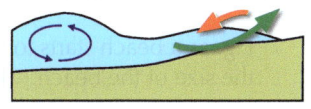

4) Waves can change over the long-term and short-term:

Short-term changes

1) A coastline may experience more destructive waves during a single **storm event**.

2) **Wind levels** can affect the characteristics of waves. A **higher** wind speed can create **taller waves**.

Long-term changes

1) **Seasonal changes** can affect waves:
- **Winter** beaches tend to be **steeper** in the back section due to destructive waves removing sediment and smoothing the beach profile in the lower section.
- **Summer** beaches tend to be **taller** in the back section where constructive waves create a large **berm** (see p.76).

2) In the long term, **climate change** is likely to create more 'storm-like' conditions in both summer and winter. This will give coastlines more **destructive waves** overall.

The height of the high and low tides will also determine where the waves strike the beach. This also affects its shape.

Sediment cells show how sediment moves around the coast

1) **Sediment cells** (or **littoral cells**) divide the coast into **separate sections**. These are lengths of coastline which act as **self-contained units** in the **movement** of sediment. Sediment cells often lie between two significant headlands.

2) There are **11 sediment cells** around the **English** and **Welsh** coastline.

3) Sediment doesn't move between the cells (except in very **extreme weather events**). Processes going on in one cell don't affect the movement of sediment in another cell.

4) Within each sediment cell are **inputs** (**sources** of sediment), **flows** (**transfers** of sediment) and **stores** (**sinks** of sediment):

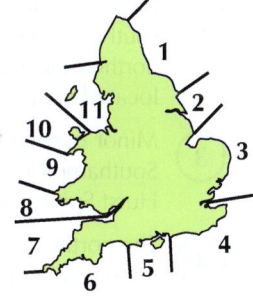

Sources	Transfers	Sinks
• Rivers carry **eroded sediment** from **inland** into the coastal system	• **Longshore drift** moves sediment along a shoreline	• Flocculation (binding together) of sediment in **salt marshes**
• **Erosion** and **weathering** of a **cliff** face	• **Offshore currents** move sediment inland from out at sea	• Formation of **spits** and **bars** by the loss of wave energy
• Accumulation of the **crushed shells of organisms**	• The **action of the wind** (aeolian processes) can move small material	• **Sand dunes** form by wind action
• Movement of material from **offshore deposits** (e.g. sandbanks) by waves, tides and currents	• **Tidal currents** will move material in and out of the coast twice daily	• **Beaches** receiving material from both waves and cliff faces
• Very small material eroded from **sand dunes**		• **Offshore bars** can build beneath the waves

Topic 2: Option 2B — Coastal Landscapes and Change

The Coastal System

The **sediment budget** can be positive or negative

The **difference** between the amount of sediment entering and leaving the system is known as the **sediment budget**.

- If more sediment enters an area than leaves it, the area has a **positive sediment budget**. This means beaches, sand dunes and spits will **grow**.
- If more sediment leaves an area than enters it, the area has a **negative sediment budget**. Beaches, sand dunes and spits are likely to **shrink**.

Sandra regretted blowing the whole budget on sediment.

Sediment cells are in **dynamic equilibrium**

Dynamic equilibrium is the way that a **balance** is maintained in a **system** where there are a number of **changing processes**. A change in one source or sink often causes **negative feedback** that try to **restore the balance** of the system. A negative feedback is when a change causes other changes that have the **opposite effect**.

E.g. as waves erode a beach, the cliffs behind it become exposed to erosion. The eroded sediment is then deposited on the beach. This causes the beach to increase in size again, which will lower the rate of erosion and bring the system back into balance.

Coastal systems also experience **positive feedback**. A positive feedback is when a change in the system causes other changes that have a **similar effect**.

E.g. as a beach starts to form, friction causes waves to slow down. This causes sediment deposits to build up and increase the size of the beach. The beach then has the capacity to slow down even larger waves and deposition continues.

An example of a **sediment cell** is from **Portland Bill** to **Selsey Bill**

Portland Bill to Selsey Bill on the south coast of England

The Portland Bill to Selsey Bill sediment cell has a **strong south-westerly prevailing wind** (wind that consistently blows from the same direction). **Erosive** processes dominate a large part of the coast on the western side of the cell.

1 **Eroded material** from the Isle of Purbeck moves into the west of Poole Bay and **accumulates** around the entrance to Poole Harbour.

2 At Poole Harbour, a **spit** has formed on its southern edge. Another spit has formed on the northern edge of the harbour. Further spits are located at Hengistbury Head and near Hurst Castle.

3 Minor **currents** emerging from the estuary at Southampton move into the Solent and prevent Hurst Spit from joining up to the Isle of Wight.

4 The northeast side of the Isle of Wight has **calmer waters**. These protect the inlets around Portsmouth and allow **accretion** (see p.79) to take place there.

CASE STUDY

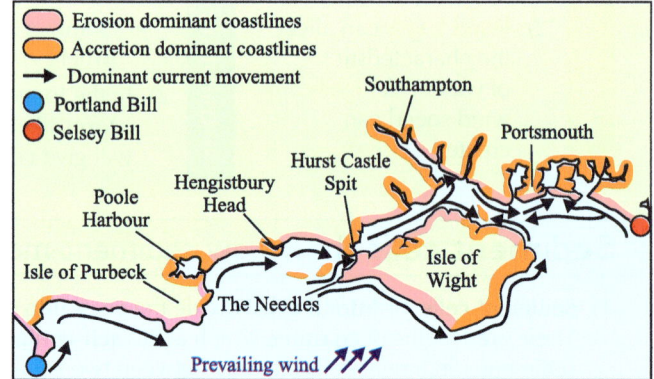

coastal accretion
the expansion of land into the sea

Warm-Up Questions

Q1 Describe the characteristics of the different parts of the littoral zone.

Q2 Name four different sources of sediment within a sediment cell.

PRACTICE QUESTIONS

Exam Question — AS and A-Level

Q1 Explain the characteristics of constructive waves. [6 marks]

Sediment cell — where delinquent sand ends up...

There are certainly a lot of definitions sprinkled across these pages. Make sure you learn them now — they'll stand you in good stead for the rest of the section. And remember, whether a beach grows or shrinks depends on the sediment budget.

Coastal Landscapes and Geology

Walk this way for some coastal landscapes... It may not sound very rock and roll, but trust me, you'll be hooked.

Coastal landscapes are made up of **discordant** coasts...

Headlands and Bays

1) A **discordant** coastline is where there are bands of **alternating hard** and **soft rock** at right angles to the shoreline. **Headlands** and **bays** can form on this type of coast.

- The **less resistant** rock is eroded **quicker**, forming a **bay**.
- The **more resistant** rock is eroded **less quickly** and sticks out as a **headland**.

2) Once the headland has formed it will erode at a **quicker rate**. This is partly due to **exposure** and partly due to **wave refraction** (the process by which wave direction curves, see p.72).

3) Headlands also experience **higher wave heights** compared to bays.

- At a headland, waves slow down. **Water** builds up behind the **wave crest**, increasing the height of the wave and giving it greater erosive power.
- This means headlands have **steep cliff faces**. Features like **arches** and **caves** can also develop (see p.72).
- The shape of the bay **dissipates the wave energy** which reduces their height. Over time, the difference between the headlands and the bays may become less marked as the headlands are eroded faster.

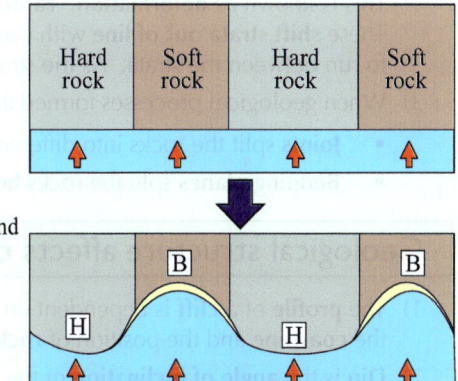

= waves
= beach
H = headland
B = bay

Cape of Good Hope, South Africa

...and **concordant** coasts

Coves

A **concordant** coastline is one where there are bands of rock running **parallel** to the shoreline. **Coves** can form on this type of coastline.

1) Once waves break through the **more resistant** outer rock (e.g. limestone), it **exposes** the **less resistant** rock (e.g. shale) behind it.

2) Waves erode the softer rock **at a faster rate** until they reach the next layer of harder rock.

3) This harder rock **slows** the growth of the cove further inland.

= waves
= beach
C = cove

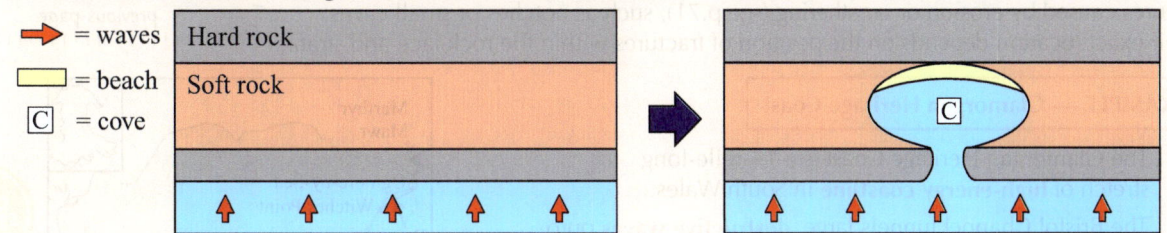

Dalmatian coastlines

In areas where **valleys** lie parallel to the coast, an increase in sea level can form a **Dalmatian coastline**. These valleys are mostly formed by **tectonic folding**. The folding creates **anticlines** (downward sloping strata) and **synclines** (upward sloping strata). Valleys become flooded as the sea level rises, leaving islands parallel to the coastline (see p.80).

Dalmatian coastline, Croatia

Haff coastlines

Haff coastlines are where deposits of sand (and other materials) run parallel to the coastline on top of offshore bars. **Lagoons** form in between the bars and the shoreline.

Topic 2: Option 2B — Coastal Landscapes and Change

Coastal Landscapes and Geology

Coastal morphology is also influenced by **geological structure**

1) Most coastlines consist of different rock types laid down in **strata** (layers). The positions of the strata create different **coastal morphology** and landforms. **Hard, resistant rocks**, such as **granite** or **basalt**, erode more slowly than **softer, less resistant rocks**, such as **clay**.

Some rock causes erosion.

2) Strata can **fold** (bend and crumple) and **dip** (angle towards or away from the sea) due to **tectonic activity**. This is known as **deformation**. **Faults** (cracks and weaknesses in the rock) can develop in the strata. These **shift strata out of line** with each other. Faults are **more susceptible to erosion** as they allow water to run between the strata. As the strata fold and dip, they experience different amounts of **pressure**.

3) When geological processes formed the strata, **intense heat and pressure** created **joints** and **bedding planes**.

- **Joints** split the rocks into different blocks **vertically**.
- **Bedding planes** split the rocks **horizontally**.

These splits can also come about as naturally wet rocks dry out over time.

Geological structure affects **cliff profiles**

1) The **profile** of a **cliff** is dependent on the **resistance** of the **rock**, the relative **energy** of the **coastline** and the position of **rock strata** in relation to the shoreline.

2) **Dip** is the **angle of inclination** of the rock strata from the **horizontal**. The dip is caused by **tectonic folding** and **faulting**. **Sedimentary** rocks are deposited **horizontally**, but then tectonic activity can move these layers forming **various profiles**:

1 If the strata dip **horizontally**, waves will erode **softer** rocks within them. This can create **notches** in the cliff face. If the wave-cut notch becomes **too large** to support the rocks above, it can cause a **rockfall**.

2 If the strata dip **away** from the sea (landward), there is **no undercutting** and the cliff will be very **stable**.

3 If the strata dip **towards** the sea (seaward), waves will **undercut** the more resistant rock. This can cause **mass movement**.

If this dipping is **gentle**, the large sections of rock may break away as a **wedge**.

Where this dipping is **steep**, large slabs of rock may instead **slide** down the cliff face.

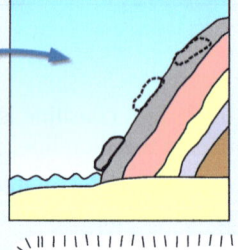

Cliff profiles affect the formation of the coastal landscapes on the previous page.

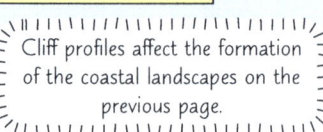

3) There might also be **micro-features** within the cliff profile. These are small-scale features caused by erosion or weathering (see p.71), such as notches or small caves. Their exact location depends on the position of fractures within the rock face and strata.

EXAMPLE — Glamorgan Heritage Coast

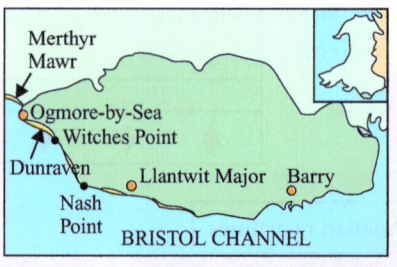

- The Glamorgan Heritage Coast is a 14-mile-long stretch of **high-energy coastline** in South Wales.
- The Bristol Channel funnels large, **destructive waves** onto the coast from the south-west. This has the effect of creating a highly-eroded **discordant** coastline which is prone to **cliff falls**.
- It's characterised by **high vertical cliffs** made of **sedimentary rocks**. These are largely **Carboniferous limestone**, **sandstone**, **shale** and **mudstone**.
- There are **headlands** at Witches Point and Nash Point and **sandy bays** near Dunraven and Merthyr Mawr.
- There are large areas of **wave-cut platforms** (see p.72) made from limestone, especially around Nash Point. These are the **original limestone strata** left behind when waves eroded the **layers of softer shale** in the cliff.
- The Southerndown Coast, which is part of the coastline, is a **Site of Special Scientific Interest**.

CASE STUDY

Headland at Nash Point

Coastal Landscapes and Geology

Coastal recession is dependent on **bedrock lithology**

Lithology refers to the **physical properties** of rock, such as its hardness and resistance. The pace of coastal recession depends on the type of rocks found along the coast. This means there can be **differential erosion** — where some parts of the coast erode more slowly than others

coastal recession
the movement of the coastline inland due to erosion

- **Igneous rocks** are **highly resistant** (they erode by 1 mm to 5 mm a year). They have a hard, **crystalline structure** which means they have few joints that wave action can break into. **Cooling magma** creates igneous rocks.

- **Metamorphic rocks** have **some resistance** (they erode by 1 mm to 10 cm a year). They have a **crystalline structure**, which makes them more resistant. However, the crystals are **aligned** in **one direction**, making them more prone to **folding** and **faulting**, which weakens the rock. **Intense heat and pressure** on sedimentary rocks forms metamorphic rocks.

- **Sedimentary rocks** have **limited resistance** (they erode by 2 cm to 6 cm a year). They can be **clastic** (made of fragments of other rocks), which means they may have many **fractures** that make them **vulnerable** to erosion. They can also be quite **porous**, which make them more **permeable**. Layers of the remains of dead animal and plant life create some types of sedimentary rocks.

Basalt is the most abundant igneous rock.

Slate is an example of a metamorphic rock.

Limestone is an example of a sedimentary rock.

Fingal's Cave on Staffa Island in western Scotland has basalt rock formations.

Coastlines can also consist of **unconsolidated material** such as **clay**. Unconsolidated materials are loose rock layers that aren't cemented in one large mass. Due to this, coastlines are **extremely vulnerable to erosion** with rates of up to 10 m a year.

Coastal recession is also dependent on **lithological structure**

It's rare for a cliff to consist of just one rock type. Most have **complex cliff profiles**. They may also contain unconsolidated layers of **glacial till** (see p.45) from previous ice ages as well as **fluvial deposits** from **ancient river systems**.

Rock type order in strata

Cliffs that have strata of less resistant rock at their base and more resistant rock at the top are more likely to experience **undercutting** and cliff collapse. This accelerates the rate of coastal recession. Cliffs that have strata in multiple **folds** are more likely to have **cracks** and **fractures** running through them. This will **reduce their resistance** to wave action.

Rock permeability

Impermeable rocks, such as **clays**, create greater amounts of **surface runoff**. This acts with its own **erosive force** on the face of the cliff. **Permeable** rocks allow water to pass through them. The pressure exerted by water held in permeable rocks (e.g. chalk) is known as **pore water pressure**. This pressure **weakens rock structure** and **enlarges any cracks and joints**. Permeable rocks also gain mass by taking in rainwater. This can lead to **slumping** (a type of landslide) when strata become too heavy. There may be **land slips** as the water creates **slip planes** on which rocks can move.

This is more common in cliffs made of unconsolidated materials.

Rock reactivity

Some rocks contain compounds that **chemically react** with seawater, this makes them more vulnerable to erosion. E.g. carbon dioxide in the atmosphere reacts with water to produce a **weak carbonic acid**, which breaks down the calcium carbonate in limestone. This is **chemical weathering** (see p.70).

Warm-Up Questions

Q1 Describe the main differences between concordant and discordant coastlines.

Q2 Explain how the dip of cliff strata affects their resistance to erosion.

PRACTICE QUESTIONS

Exam Question — AS and A-Level

Q1 Explain how the permeability of a cliff face may affect its rate of recession. [6 marks]

Bedrock lithology — yabba dabba doo...

It isn't that surprising that cliffs erode. They do tend to get bombarded with waves quite a lot. The whole cliff doesn't usually just wear away in a uniform fashion either — some bits wear away quicker than others, which can lead to land slips and slumping.

Coastal Weathering and Mass Movement

Coasts are affected by two types of processes — those caused directly by marine processes (erosion, transport and deposition), and subaerial processes, which aren't directly caused by the sea (weathering and mass movement). Confused? Read on...

Weathering produces sediment within the coastal system

Subaerial processes are the gradual **weathering** and breakdown of rock by agents such as ice, salt, plant roots and acids. Weathering **weakens cliffs** and makes them **more vulnerable to erosion** and mass movement.

Here are **three types** of weathering that affect coasts:

> Some coasts will be affected by one type of weathering more than another. This is due to their exposure to different weather conditions, as well as the dominant rock type.

1 Mechanical weathering

Mechanical weathering is the breakdown of rock material without change its chemical composition.

ice expands   rock breaks off

- **Freeze-thaw** weathering occurs in areas where **temperatures fluctuate above and below freezing**. Water enters fractures in the rock face, which **freezes** in the cracks and **expands** when the temperature drops **below 0 °C**. Repeated freeze-thaw action **weakens** the rocks and **dislodges** pieces more easily.
- **Saline water** causes **salt weathering**. It enters the **pores** and **cracks** in the rocks at **high tide** or sprays onto the cliff face as waves break. As the tide goes out, the rocks dry and **salt crystals** form as the water **evaporates**. These salt crystals **expand** and **exert pressure** on the rock. This pressure causes pieces of rock to fall off.
- Some rocks contain **clay**. When clay gets wet, it **expands** — when it dries it **contracts**. Cycles of **wetting and drying** weaken the rock and fragments can **break** off.

2 Chemical weathering

Chemical weathering is the breakdown of rock by changing its **chemical composition**.

- **Carbonation** is where **carbon dioxide** in the atmosphere dissolves in rainwater to produce a weak **carbonic acid**. This acid reacts with rocks that contain **calcium carbonate**, such as **Carboniferous limestone**. This gradually dissolves the rock. Some **seaweeds** also contain mild acids in their cells. If these break, the acid reacts with the rock face.
- **Oxidation** may also occur on coastlines — this happens when rocks containing iron are exposed to **oxygen** in air or water. The **iron** can **react** with oxygen to form iron oxide, which is quite **weak** — this makes the rock **crumble** more easily. Oxidation is why some rocks have a rusty red colour.

3 Biological weathering

Biological weathering is weathering through the actions of **plants** and **animals**.

- A rock face will have multiple **fractures** on its surface. **Organisms** may use these for their **habitats**. They **burrow** into the rock and break it apart which **undermines the rock structure**. Some marine animals **secrete chemicals** that dissolve the rock.
- **Seeds** may also fall into cracks in the rock. As these plants start to grow, their **roots** will widen the crack. Over time, this can break the rock apart.

Some coastlines experience a lot of mass movement

1) **Mass movement** is the shifting of material (such as rock, sand, soil and clay) downhill due to **gravity**. In coastal areas, it's most likely to occur when wave action **undercuts** cliffs. This creates an **unsupported overhang**, which is likely to collapse.

2) Types of mass movement include **landslides**, **blockfalls** and **rotational slumping**. **Unconsolidated rocks** are more susceptible to mass movement as there is **little friction** between particles to hold it together. **Weak lithological structures** (see p.69) are also vulnerable to mass movement.

3) **Heavy rain** can **saturate** unconsolidated rock. This reduces friction between parts of the rock face, meaning it's more likely to **collapse**. **Surface runoff** can erode fine particles such as sand and silt and transport them downslope.

Landslide

- A **landslide** is the movement of material in a **straight line** down a slope. Landslides can happen very quickly. They occur on steep slopes, often after a period of very **heavy rainfall** or after **storm conditions**. Rainfall acts as a lubricant, allowing the material to overcome the **friction** holding it in place. This creates a slip plane on which the large volumes of material slides.
- A **landslide scar** (an area of unvegetated rock face) is likely to be left behind above the material that has moved.

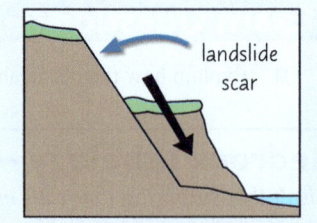
landslide scar

Coastal Weathering and Mass Movement

Rotational slumping
- **Rotational slumping** occurs when material slides down the slope at a **curved angle**. The material **retains its shape** and slides as one large mass, usually over a **long period**.
- Water in the cliff face can create a **slip plane** on which the material is able to move. This usually happens where **softer, permeable rocks** lie on top of **harder, impermeable** ones. The **vegetation layer** on top of the slump usually remains in place.

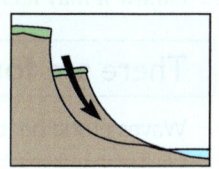

Blockfall
- A **blockfall** (or rockfall) is the movement of broken **blocks** of material downslope. This is more common on **steep** cliff faces with numerous **joints** and **bedding planes**.
- A **rock topple** can occur when the **cliff profile** has a steep **dip towards the sea**. This occurs when waves **undercut** the cliffs and large blocks of material become unstable and fall on top of each other. The material that collects at the base of the cliff is called **talus** or **regolith**.

Some coastal landforms are **created** by **mass movement**

A **rotational scar** forms after a **rotational slump** occurs. It's a **curved mark** behind the slump, left on the cliff face from where the material has moved. It's **devoid of vegetation**, making it look like a **scar** on the cliff face.

Rotational scar

A cliff may display a **terraced cliff profile** when **multiple rotational slumps** happen over different time periods. These have the appearance of **multiple clifftops** in small terraces down the cliff face.

Terraced cliff profile

A **talus scree slope** is a **steep mound of rockfall** at the base of a cliff. It **fans out** from the point of movement. During the rockfall, the material is **naturally sorted** — larger rocks fall to the bottom and smaller ones remain at the top. They then **protect the cliff** from further undercutting until the waves have transported the scree away.

Talus scree slopes

Weathering and **mass movement** influence **coastal recession**

Subaerial processes (weathering and mass movement) influence the **rate of coastal recession**.

> Subaerial processes can both increase and decrease the rate.

Weathering
- **Weathering** slowly weakens the **cliff face**. This makes it **more vulnerable** to the impact of wave action.
- Weathering also provides material (**scree**) for the waves to use as an **erosive tool** (e.g. through **abrasion**, see p.72). This means weathering can **increase** the rate of coastal recession.

Mass movement
- **Mass movement** can add **rock material** to beaches.
- **Strengthening beaches** by increasing their size means that waves encounter more **friction** as they come onto the shore. This slows their speed and force — a processes known as **wave dissipation**. This **reduces** their ability to **erode**.
- This means that mass movement, although a **form of coastal recession**, can **reduce** the rate of coastal recession by protecting the coastline.

> This is an example of a negative feedback.

Different types of **weathering** can make the rock **vulnerable** to other types of weathering. For example, **mechanical weathering** can **widen cracks** that animals or plants can use and exploit. **Cracks** and weaknesses in the rock caused by **chemical weathering** can then be vulnerable to mechanical weathering.

Warm-Up Questions

Q1 Describe the difference between a landslide and rotational slump.

Q2 Explain the formation of a rotational scar.

PRACTICE QUESTIONS

Exam Question — A-Level

Q1 Explain how different types of weathering work together to increase the rate of coastal recession. [8 marks]

Experiencing a mid-revision session slump? Have a biscuit...

It's pretty surprising how many ways there are for a cliff to collapse. Whatever the type of collapse, it's always caused by gravity. Of course, not all cliffs are created equal, and those made of unconsolidated material are more likely to fall apart.

Coastal Erosion — Processes and Landforms

I know it may feel like your brain is eroding away with all this info, but it'll mean you're on top form for the exam...

There are **four** main **erosive** processes

Waves erode beaches, rocks, cliffs and **sediment** held in the water itself, by four **main processes**.

(1) **Hydraulic action** — Waves **compress air** into cracks in the cliff face. The **pressure** exerted by the compressed air breaks off rock pieces.

(2) **Abrasion** — Waves transports **rock** and **sediment** which smashes and scrapes against rocks and cliffs. This breaks sections off the cliff face and **smooths** the surface.

(3) **Attrition** — Bits of rock in seawater **smash** against each other and break into **smaller**, more **rounded** pieces.

(4) **Corrosion** — Seawater gradually **dissolves** soluble rocks, such as limestone and chalk.

The Old Man of Hoy, Orkney — a stack formed by erosive processes.

The coastal landscape has **landforms** created by erosion

Cliffs are **vertical**, often **very steep**, rock faces that form as the sea **erodes** the land. Their shape is determined by **lithology** and **wave action**. Over time, cliffs **retreat** due to the action of **waves** and **weathering**.

1) Weathering and wave erosion (e.g. hydraulic action) cause a **notch** to form at the high water mark, usually where there's an exposed bedding plane or fracture.

2) Through further erosion, e.g. by abrasion, the **notch** gets bigger and eventually develops into a **cave**.

3) Rock above the cave becomes **unstable** with nothing to support it, and it **collapses**.

4) A new notch forms in the new cliff face and the process repeats, leaving behind a **wave-cut platform**.

Cliffs and platforms near Lannacombe Bay in South Devon

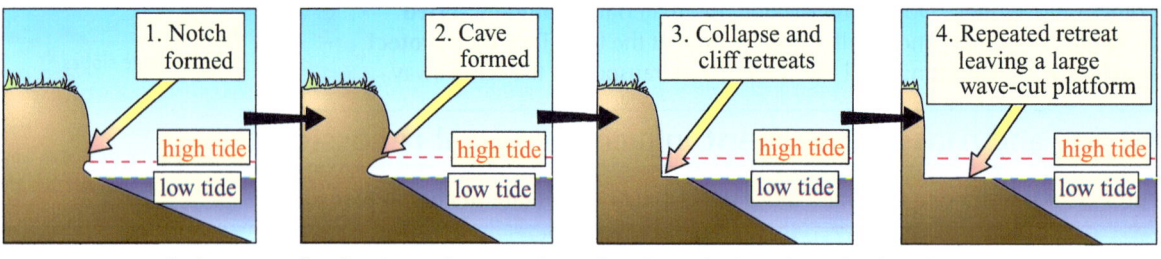

This is a negative feedback.

- Wave-cut platforms are flat, **horizontal outreaches of rock** on the beach or the foreshore.
- They show the **position** of the original cliff and can consist of **both more and less resistant** layers of rock.
- The **larger** the wave-cut platform becomes, the **less** the incoming waves erode the cliff face because they **slow down** due to **friction** with the platform. Once the platform **reduces in size**, the waves **continue their erosion** of the cliff.

Refraction can create **arches, blowholes** and **stacks**

As a wave approaches a headland, it enters **shallow** water and **slows down** due to **friction**. Away from the headland, the wave is still in **deep** water and is travelling **faster**. This makes it **refract (curve)** around the headland.

Wave **refraction** exposes both sides of a headland to erosion, which can create some landmarks:

- If **caves** form (following the erosion of weak areas of rock) on the opposite sides of a narrow headland, they may join up to form an **arch**.

- Sea spray may erode the top of the arch and create a **blowhole** (small opening through the rock) where there is a weakness in the **joints** of the rock.

- Further erosion of the arch and the blowhole will eventually make the top of the arch **collapse**, leaving behind a **stack**.

- Exposure makes the stack vulnerable to wind and rain, so weathering may **reduce** its size.

- Wave action may cut a **notch** into the stack at the high tide mark which can **undercut** it. Eventually the stack will **topple**, leaving behind a **stump** which is only visible at low tide.

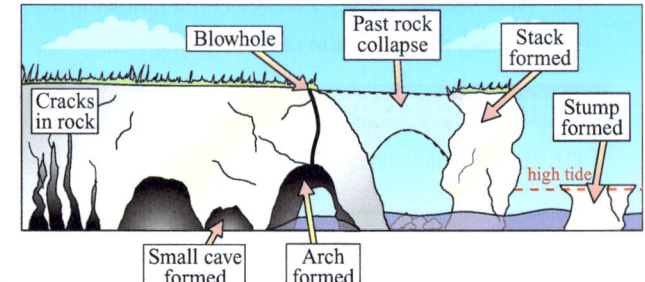

Topic 2: Option 2B — Coastal Landscapes and Change

Coastal Erosion — Processes and Landforms

Many factors influence rates of erosion...

Wave type
- **Destructive waves** have the most energy and a strong backwash, so are more effective at eroding (e.g. through hydraulic action and abrasion) as they hit the cliff face with **greater force** than constructive waves.
- Their greater volumes of **sea spray** also cause corrosion, sometimes high up the cliff.

Wave size

The size of the wave affects **erosion rates**. Larger waves have **more energy** than smaller waves, so they exert more power and will erode the coast **more quickly**.

Lithology
- **Soluble** rock as well as **softer, less resistant** rock is more vulnerable to **erosion**.
- Rocks that have **defined strata** or a large number of **fractures** erode more easily through hydraulic action as the waves easily penetrate into the cracks.

Destructive waves can generate large volumes of sea spray.

...and many factors influence rates of recession

Tides
- Waves will reach the coastline at high tide **twice** in a **twenty-four hour period**. These waves will reach **higher up** the cliff face than those at low tides. This means they have the potential to erode a **larger area** of rock. Some low tides may not even reach the cliff face, **reducing** the rate of coastal erosion.
- There are also extremes of both high and low tides known as **spring** and **neap** tides, respectively. Each happens twice a month. A spring tide creates greater gravitational pull on the waves and creates a higher high tide which moves waves higher up the beach and cliff. A neap tide occurs when the sun and the moon are at **right angles** to each other. This **weakens** the gravitational pull on the waves and the high tide is much lower.

Wind direction and fetch
- The fetch is the **maximum distance of sea** the wind has blown over in creating waves. Winds that blow over longer fetches create **stronger** waves. Although the size of the fetch doesn't change, the **dominant wind direction** does.
- Winds and waves that approach the UK from the **southwest** tend to be stronger, as in some locations the fetch is over 6500 km long (for example between Florida and Cornwall). This means that the rate of coastal recession on the south coast of the UK **increases** when winds blow from the south west.

A long fetch

Weather systems, seasons and storms
- Wind moves from areas of **high pressure** to areas of **low pressure**, so certain weather systems are likely to create stronger winds.
- **Low-pressure systems** are more common in winter due to large temperature differences between the equator and the poles. These systems create **stronger winds**, leading to more powerful waves. **High-pressure systems** are more common in summer. These mean **calmer wind conditions** and smaller, less powerful waves.
- A **depression** (a low-pressure weather system) can form out at sea and create **storm conditions**. A storm generates waves of great erosive power, which increase the rate of coastal recession and may lead to slumps and rock falls.
- In the long term, **climate change** may alter the types of weather that the coast experiences. It's likely coasts will witness an increase in the number and intensity of storms due to climate change.

Warm-Up Questions

Q1 Describe the four main ways that waves might erode a coastline.
Q2 Explain how changing tides result in different rates of coastal recession.

PRACTICE QUESTIONS

Exam Question — AS and A-Level

Q1 Explain how a wave-cut platform is formed. [6 marks]

There are plenty of blowholes in Wales...

And one in each whale too. Landforms like arches and stacks are pretty impressive, but they don't stick around forever. The erosive forces that caused them keep on going and eventually wear them away. It takes a while though, luckily.

Coastal Erosion and People

Oh, I do love a day at the seaside. Unfortunately, it's not all ice creams and sandcastles — human actions can impact coastal processes and cause problems. What's more, living on the coast comes with some pretty serious risks, too...

Human actions can influence coastal recession

Sediment cells behave **autonomously** — they can reach a state of **dynamic equilibrium** if left free from human intervention. The actions of humans can **disrupt** the sediment cell and increase and decrease rates of coastal recession. For example:

- Building a **dam** on a river can interfere with the movement of sediment through the river. There will also be **less sediment** for **longshore drift** (see p.76) to move within the cell.
- This will lead to **increased** rates of **erosion** further along the coastline as their is **less sediment** input into the system.

Dredging the harbour at Whitby.

- Building **coastal defences** such as **groynes** (see p.83) can retain sediment in a certain area by preventing its movement along the coast by **longshore drift**.
- This increases the size of the beach next to the groynes but starves beaches of sediment further along the coast. Beaches with no groynes would be **smaller** than they naturally should be, and coastal recession will occur at a **faster** rate than it would happen naturally.

- **Sand** and **shingle** are common building materials in the construction industry. They are obtained by **dredging** offshore bars, harbours and beaches. Dredging might also take place to **deepen** shipping lanes to allow for larger ships to dock at a port. Dredging also provides sand for **beach nourishment projects** (see p.84).
- **Removing sediment** from the cell permanently can **increase recession rates** — the removal starves the beaches of new sediment supplies that can **reduce** the effects of the incoming waves.

Human activities in the Nile Delta

This is a low-energy coastal environment.

CASE STUDY

The Nile Delta, in northern Egypt, is home to 50% of the country's population. The 22 000 km² area contains the final stretch of the River Nile before it empties into the Mediterranean Sea. Alexandria, one of the largest cities in Egypt, sits on the northwest edge of the delta and has a population of over 5.4 million people. The Nile Delta is considered a **densely populated coastal area**.

The Nile Delta

Saltwater intrusion is saltwater infiltrating the soil from below.

- Its coastline is under threat from **rising sea level** in the Mediterranean and high amounts of **erosion**.
- The rising sea level is making low-lying areas at the front of the delta more **saline**. **Saltwater intrusion** is occurring and floods leave salt deposits on the land.

- This makes it increasingly difficult to farm the delta as the **fertility** of land is decreasing. Freshwater lakes and lagoons are now too salty to use for **irrigation**. Some fish species are unable to survive in the water in the delta, which threatens the future of the **fishing industry**.

- The **transportation of sediment** along the Nile has reduced significantly since the building of the **Aswan Dam** in 1970. Before then, 9.5 million tonnes of sediment washed down the Nile and onto the delta every year. Now, **fluvial sediment** coming down the River Nile is not replacing the eroded material from the coast of the delta. The Aswan Dam **holds back** over 98% of the Nile's sediment in Lake Nassar.

- There's a combination of rising sea level and low levels of sediment deposition from the Nile. This means that the Nile Delta is **retreating** — up to 140 m a year in some places.

The Aswan Dam

Topic 2: Option 2B — Coastal Landscapes and Change

Coastal Erosion and People

Coastal recession can have **economic losses**...

Homes at Birling Gap, Sussex are slowly falling into the sea.

- Few people in the UK live or work on receding cliffs but low-lying coastlines can have **densely populated developments**. Those who live on receding coastlines face **economic challenges** and a lack of **economic assurance** for the future.

- Coastal settlements tend to suffer from **long-term economic problems**, e.g. limited **employment opportunities** due to lack investment from the government or private companies.

Land with different uses has different economic values

When coastal recession causes the **loss of land**, economists calculate the **actual economic loss** by examining the **land use**. Agricultural land has the **lowest value** at around £23 000 per hectare, while residential land has a much **higher value**. For example, land in the south of Purbeck, Dorset, has a value for residential purposes at £3.8 million per hectare. This means that agricultural land at the coast is often **less likely** to be **protected** and farmers will therefore lose farmland and income.

- It's unusual for **heavy industry** to be located on the coast. Essential services located near coastlines, such as power plants, tend not to be built in areas with **high levels of coastal recession**. Shutting down a large industrial plant and rebuilding it elsewhere is **costly**. Planning permission for new developments considers the likelihood of coastal recession making buildings lose value and becoming unoccupied in future years.

- Smaller businesses on receding coasts (e.g. hotels) may be **abandoned**, causing **job losses** for local people. This may impact other small businesses.

- **Selling** coastal property can be difficult. It's often impossible to buy **home insurance** for houses on receding coastlines. Residents sometimes have to **abandon** their homes. Homes on receding coastlines **lose their value** very quickly. This influences the value of other properties that are less at risk and can **devalue entire villages**.

- **Infrastructure** on receding coasts, such as roads and rail lines, can be costly to **reroute**. This can become an **economic burden** on a local authority (or national government).

Abandoned holiday camp at Brighstone, Isle of Wight

Small businesses often rely on each other to be successful. If one business is affected, this could impact multiple other businesses.

Although these costs are high, coastal recession is usually **predictable** and happens **slowly**. **Contingency funds** can be set aside for these costs and it may be possible to make **economic preparations** in advance of the coastline receding.

...and **social losses** too

Some areas have **a significant number of people** living in communities on receding coastlines. Coastal recession forces people to **relocate**, which can break up communities and cause **depopulation**.

- Those who stay may feel **isolated**, particularly if loss of infrastructure leaves them **disconnected** from other places.

- Remaining residents and business owners may suffer from **stress** due to **uncertainty** about the future.

- An areas's **amenity value** (the value the landscape and the people in an area add through their culture and wellbeing) can be lost. As coasts recede and areas become abandoned, they can lose their **character**. They may become **visually unattractive** as people stop caring for them.

- Local services may also close down as people move away. This can negatively affect those on low-incomes and people who are less mobile. It also may mean a **loss of livelihood** to people who would normally work in the area.

- Extremely rarely, cliff collapse and coastal recession can lead to **fatalities**, e.g. of people walking along unstable cliffs that **collapse**. Also, **falling debris** can injure people on beaches, particularly after a storm.

Warm-Up Questions

PRACTICE QUESTIONS

Q1 Describe the effect that dredging might have on a coastline.

Q2 Explain how human actions have increased the rate of coastal recession in the Nile Delta.

Exam Question — A-Level

Q1 Evaluate the idea that the economic losses of coastal recession are greater than the social losses. [20 marks]

The wise man built his house upon the rock...

But the wiser man built his house inland. Living next to the sea might be the dream of many, but it becomes less desirable if your house is likely to fall into the sea someday. As the coast recedes, many will move away, making communities smaller.

Coastal Deposition — Processes and Landforms

There's a lot of information to take in and learn on the next few pages, but trust the process and you'll be grand...

Different factors influence **sediment transportation**

The **energy** provided by waves, tides and currents **transports eroded material**.

Some cliff faces show clear high tide marks.

- **Wave actions** create and destroy most **landforms** in the area of land between maximum high tide and minimum low tide.

- Tides affect the **position** at which waves break on the beach — at high tide they break higher up the shore.

- The higher the **tidal range**, the more powerful **currents** will be. Some coastlines have tidal ranges up to 15 m or as low as 0.5 m.

- When **seawater** is funnelled into estuaries or other **narrow areas**, the **tidal currents** are particularly **powerful**. These currents can **transport** a lot of **sediment**.

- A **current** is the general **flow of water** in a **particular direction**. **Currents** move along the coast. Some currents are almost **permanently** in place, while others might last for just a few hours.

- **Rip currents** occur when water is **funnelled** between **obstacles** underwater, causing it to **move faster** than the surrounding water. This **carries material** out to sea and **moves sediment** from the beach and foreshore to the nearshore.

- Waves that approach the beach at an **acute angle** are likely to transport **sediment quickly** compared to waves that approach **perpendicularly** (at 90° to the shore).

There are **four** main **transportation processes**

Solution	Substances that can **dissolve** are carried along **in the** water. E.g. **limestone** is dissolved into water that's slightly **acidic**. 	**Suspension**	**Very fine** material, such as **silt** and **clay** particles, is whipped up by **turbulence** (**erratic swirling** of water) and carried along in the water column. **Most** eroded material is transported this way. 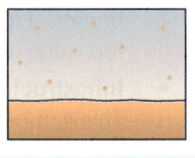
Saltation	**Larger particles**, such as **pebbles** or **gravel**, are **too heavy** to be carried in suspension. Instead, the **force** of the water causes them to **bounce** along the sea bed. 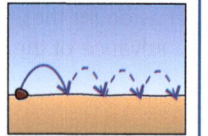	**Traction**	**Very large** particles, e.g. **boulders**, are **pushed** and **rolled** along the sea bed by the force of the water. 

The **prevailing** (most frequent) wind direction and the **dominant** (strongest) wind influence how waves travel. Since the direction of the wind **changes**, so too will the direction of the waves as they approach the shore.

Longshore drift transports sediment along the shore:

1) **Swash** carries sediment (e.g. shingle, pebbles) **up** the beach, **parallel** to the prevailing wind. **Backwash** carries sediment back **down** the beach, at **right angles** to the shoreline.

2) When there's an **angle** between the prevailing wind and the shoreline, a few rounds of swash and backwash move the sediment **along** the shoreline in a **zig-zag pattern**.

Beaches have distinctive features

1) **Beaches** form when constructive waves **deposit sediment** on the **shore**.

2) From the top of the **beach** to the **shoreline**, there is 'natural sorting' of beach sediment. Larger and more angular sediment is at the **top of the beach** and smaller, more rounded sediment is **closer to the shore**. This is because, while the **swash** deposits material of all sizes, the backwash only has enough **energy** to move the **smallest material**, leaving the larger material at the top.

3) **Berms** are ridges of sand and pebbles standing 1-2 m high. They occur at the high tide mark.

4) Below these, **runnels** (grooves) run in the sand parallel to the shore. They form when **backwash** drains back to the sea.

5) **Cusps** are crescent-shaped indentations created by **constructive waves** approaching the beach at an angle.

These features are more pronounced on a shingle beach compared to a sandy beach. Shingle beaches also have a steeper gradient than sandy beaches.

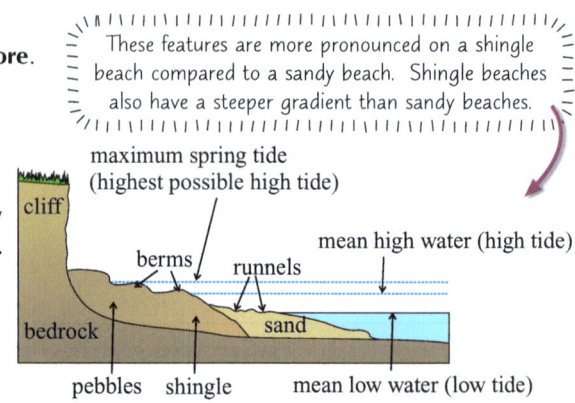

Coastal Deposition — Processes and Landforms

There are other **coastal landforms** created by **transportation** and **deposition**

Sediment can come from the **seabed**. It can also come from **river systems** and from the action of **subaerial processes** on the surrounding landscape, e.g. **weathering**. **Deposition** occurs when the water **loses** the ability to transport sediment. This may be due to **reduced water volume**, which lowers its **carrying capacity**. It may also be due to a reduction in **water velocity**. This means it has **less energy** to transport sediment. Waves will deposit large pieces of sediment **before** smaller pieces.

Spits

Spits tend to form where the coast suddenly **changes direction**, e.g. across river mouths.

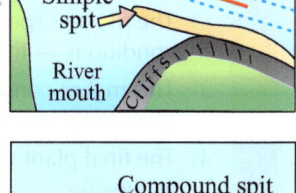

- **Longshore drift** (see p.76) continues to **deposit** material across the river mouth, leaving a bank of **sand** and **shingle** sticking out into the sea. A **straight** spit that grows out roughly **parallel** to the coast is called a **simple spit**.

- Occasional **changes** to the dominant wind and wave direction may lead to a spit having a **curved end** (the fancy name for this is a **recurved spit**).

- If the waves return to their **original direction**, longshore drift abandons the recurved spit and the spit continues to grow **straight**. A spit that has multiple recurved spits is known as a **compound spit** (see right).

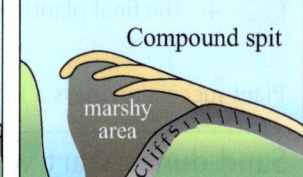

- The area **behind** the spit is **sheltered** from the waves and often develops into **mudflats** and **saltmarshes** (see p.78).

When **longshore drift** happens in **alternating directions**, a couple of landforms can occur:

- **CUSPATE FORELANDS** — These occur when **two spits** form in **opposite directions** and **merge** into a **low-lying triangular** shape. In some cases, they can stretch several kilometres out to sea.

- **DOUBLE SPIT** — These occur on **opposite sides** of a bay, both extending towards its middle. The **outflow** of river discharge **stops** the two spits from **joining**.

Bars and Tombolos

- Bars are formed when a **spit joins two headlands together**. This can occur across a **bay** or a **river mouth**.

- A **lagoon** forms **behind** the bar as seawater is trapped. Over time, if there is a **stream** flowing into this the water can become **less saline**.

- Bars can also form off the coast when material moves **towards** the shoreline (normally as the sea level rises). These may remain **partly submerged** by the sea — in this case they're called **offshore bars**.

- A bar that **connects** the shore to an **island** (often a stack) is called a **tombolo**. A tombolo can also be formed by **wave refraction** occurring around an **island**.

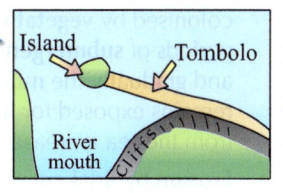

Paul and Mary left the bar to spend some time enjoying the lagoon.

Barrier beaches

- Barrier beaches are long, narrow islands of sand or gravel that run **parallel** to the shore and are detached from it. They tend to form in areas with a good supply of **sediment**, a **gentle slope** offshore, fairly **powerful** waves and a **small tidal range**.

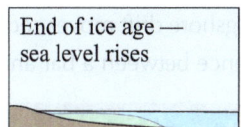

- It's not clear exactly how barrier beaches form, but scientists think that they probably formed after the **last ice age**, when ice melt caused rapid **sea level rise**. The rising waters **flooded** the land behind beaches and **transported** sand offshore, where it was **deposited** in shallow water, forming islands.

- Another theory is that the islands were originally **bars** attached to the coast and rising sea level breached them in places.

- A **lagoon** or **marsh** often forms behind the barrier island, where the coast is **sheltered** from wave action.

Coastal Deposition — Processes and Landforms

Vegetation **stabilises** deposited **material**

Some sites of deposition are **more resistant** to the forces of erosive waves due to vegetation.

- Plants colonise and then stabilise deposited material. The roots act as webs that help cement the sediment together.
- The upper branches or leaves of the plant buffer the wind which may otherwise blow away surface material.
- As parts of the plant die, they add organic matter to the sediment and soil can begin to form.

Succession

1) Succession is the process by which an ecosystem changes over time.
2) Smaller pioneer species colonise an area of bare sediment first. The pioneer species change the abiotic (e.g. water availability) conditions — they die and decompose, which forms a basic soil.
3) This makes conditions less hostile, which means new organisms can grow. Over time, more complex plant structures and trees start to grow.
4) The final plant species in the succession is known as the climatic climax community of plants.

Plant succession plays a key role in **coastal accretion**.

Plants that grow in sand dunes and salt marshes tend to be halophytic (salt tolerant) and xerophytic (drought tolerant).

Yellow dunes in Forvie National Nature Reserve, Aberdeenshire

Sand dunes start with **embryo dunes**

- Sand dunes form when **longshore drift** (see p.76) **deposits** sand, which is then transported up the beach by the wind. Sand trapped by **driftwood** or **berms** (see p.76) is **colonised** by **pioneer species**, such as couch grass.
- This vegetation **stabilises** the sand. It also encourages more sand to **accumulate** there, which forms **embryo dunes**. Further plants and grasses, such as marram grass in the **fore dunes**, grow in each stage of succession. **Yellow dunes** tend to be larger and have small amounts of soil in them. Over time, the oldest dunes occur further inland as newer **embryo dunes** form closer to the shore.

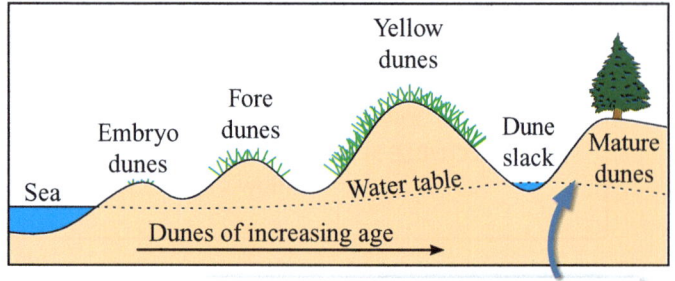

These mature dunes can reach heights of up to 10 m and are populated with trees such as birch.

Mudflats and **saltmarshes** form in **sheltered** places

Saltmarsh with water-filled channels

- Mudflats and saltmarshes form in **sheltered, low-energy environments** such as **river estuaries** or **behind spits**, by a process called **flocculation**.
- As silt and mud are **deposited** by a river or the tide, **mudflats** develop. The mudflats are colonised by **vegetation** (such as eelgrass) that can survive the **high salt levels** and long periods of **submergence** by the tide. The plants trap more mud and silt and gradually the mudflats grow. This creates an area of **saltmarsh** that remains exposed for longer and longer between tides. As the distance from the sea increases, salt levels fall and species diversity increases.
- **Erosion** by tidal currents or streams forms **channels** in the surface of mudflats and saltmarshes. Seawater can **flood** these areas at high tide, and they can be **dry** at low tide. They can also be **permanently** flooded.

flocculation
the settling and clumping together of sediment in water

Warm-Up Questions

Q1 Describe how longshore drift moves sediment along a coastline.

Q2 Explain the difference between a bar and a spit.

PRACTICE QUESTIONS

Exam Question — AS and A-Level

Q1 Explain how vegetation helps form mudflats.

[4 marks]

Mudflats — executive homes for lugworms...

Sand dunes are great for rolling down — just watch out for dog poo. Speaking of watching out for things, there are lots of different depositional landforms to learn on these pages, so make sure you know the difference between each one.

Sea Level Change

There's a fair bit to learn on these pages. Just try and keep your head above the water...

Sea level has changed due to **eustatic processes**...

Eustatic sea level change is caused by a change in the **volume of water** in the sea, or by a change in the **shape** of the **ocean basins**. The causes and effects are **global** and happen relatively **quickly**:

Changes in climate affect sea levels in different ways

- An **increase in temperature** causes the **melting of ice sheets**. This increases the **volume** of water in the **sea**, so **increases** sea level. Higher temperatures also cause water to **expand** because water particles have more energy and take up more space. This **raises** sea level further.
- A **decrease in temperature** causes more **precipitation** to fall as snow. This increases the **volume** of water stored in **glaciers** and decreases the volume of water in the sea. This **lowers** sea level.

Eustatic change is thought to be responsible for the sea level rise in Solva in Pembrokeshire, South Wales. This low-lying estuary is experiencing sea level increases of around 3 mm a year.

Tectonic movements of the Earth's crust can alter the shape of ocean basins

The **spreading** of the sea floor in the Atlantic slowly increases the volume of the basin, **lowering** sea level.

...and **isostatic processes**

Isostatic sea level change is caused by **vertical movements** of the land **relative** to the sea. A **depression** of land causes sea level to **rise** and an **uplift** of land causes sea level to **fall**. The effects occur at a **local** scale and can take a **long time** to happen.

- **POST-GLACIAL ADJUSTMENT** — The large volume of **ice** that built up in the last ice age caused some crust to **sink**. This is known as **isostatic subsidence**. The melting of these ice sheets can slowly **uplift** the land as the lost weight of ice **releases** the downward pressure. This is known as **isostatic rebound**. It can continue for thousands of years after the retreating glacier has gone, causing sea levels to slowly **fall**. Some areas of Scotland are still moving upwards at up to 1.5 mm a year.

There's definitely sea down there somewhere. Nigel, you go first.

- **SUBSIDENCE** — The **subsidence of land** due to shrinkage after **abstraction of groundwater** (withdrawing water from the ground) can cause the land to **compact**. This means the surface can become **flooded with seawater**.
- **ACCRETION** — The **accretion of sediment**, mostly in the mouths of major rivers and at the coast, or the **accumulation of ice** can also add **weight** to the Earth's crust. This makes it **sink** down and causes sea levels to **rise**.
- **TECTONIC** — Tectonic (crustal) processes, e.g. as one plate is **forced beneath** another at a plate margin. This can cause **uplift** of the **crust**, which can either **increase or decrease sea level**.

Isostatic uplift had caused problems for Ranj, Lois and Nigel.

Emergent coastlines have formed by **falling** sea level

When sea levels **fall** relative to the coast, new **coastlines** emerge from the sea, which creates different landforms.

Raised Beaches

Raised beaches form when the **fall** in sea level leaves beaches **above the high tide mark**. Over time, this beach sediment becomes **vegetated** and develops into **soil**. A fall in sea level also exposes **wave-cut platforms**. Leaving them raised above their former level. In the UK, raised beaches frequently occur along the west coast of Scotland and particularly on the **Isle of Arran**.

Raised beach on the Isle of Iona, Scotland

© Lynne Sutherland / Alamy Stock Photo

Fossil Cliffs

The sea no longer erodes the cliffs above raised beaches, meaning **vegetation** slowly **establishes and covers them**. These are known as **fossil cliffs**. It is common to see **wave-cut notches**, **caves**, **arches** and **stacks** within fossil cliffs. Over time, **weathering** changes the shape and size of these raised features.

Fossil cliff

Current beach and cliff

Raised beach and wave-cut platform

Sea Level Change

Submergent coastlines have formed by rising sea level

When sea level **rises** relative to the coast, the sea **submerges** the existing coastline and creates **new landforms**.

Rias

Rias are formed where **river valleys** are partially **submerged**, e.g. Milford Haven in South Wales is a ria. Rias have a **gentle** long- and cross-profile. They're **wide** and **deep** at their **mouth**, becoming **narrower** and **shallower** the further **inland** they reach.

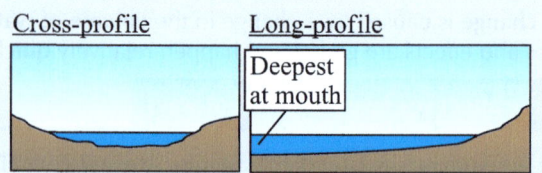

Cross-profile

Long-profile

Deepest at mouth

Fjords

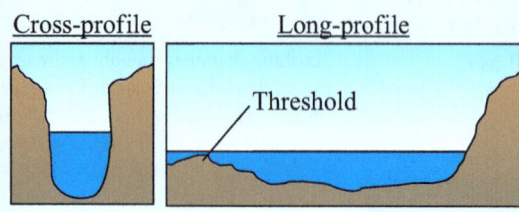

Cross-profile

Long-profile

Threshold

Fjords are a lot like rias, but are **submerged glacial valleys** rather than submerged river valleys. They're relatively **straight** and narrow, with very **steep sides**. They have a **shallow mouth** caused by a raised bit of ground (called the **threshold**) formed by deposition of material by the glacier. They're very **deep** further **inland**, e.g. Sognefjorden in Norway is over 1000 m deep in places.

Dalmatian coastlines

A Dalmation coastline is characterised by **numerous islands** lying **parallel** to the coast. These islands are the tops of a series of anticlines and synclines (see p.67) that become **isolated** when there is an **increase** in **sea level** and the **valleys** between them **flood**.

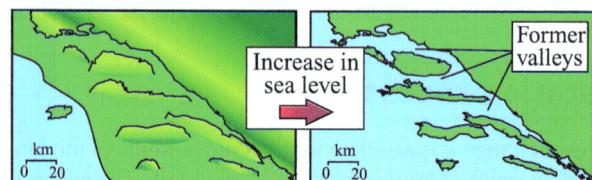

Increase in sea level

Former valleys

km 0 20

km 0 20

Sea level is now changing for **other reasons**

1) Over the **last century**, global temperatures have **increased rapidly** due the increase in **greenhouse gas emissions** from **human activity**. This is called **global warming**.

2) Increases in temperature are likely to cause **increases in sea level**, through the melting of the **cryosphere** (the part of the Earth covered by ice and snow). Warming has also caused the **thermal expansion** of water in oceans.

3) Global sea level is currently **rising** at almost 2 mm each year. This threatens to flood low-lying coastal areas and settlements. If **greenhouse gas emissions** remain high during the 21st century, scientists predict sea level will increase by 8 mm to 16 mm a year by 2100.

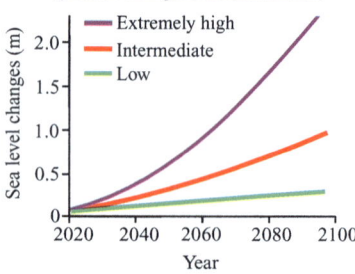

Predicted sea level rise with different greenhouse gas emission levels

— Extremely high
— Intermediate
— Low

Sea level changes (m)

2020 2040 2060 2080 2100

Year

Although the instruments used to record and model sea level changes have improved in accuracy, the multiple variables that can affect sea level rise can affect our ability to predict sea level rise.

4) **Tectonic processes** also play a role in sea level change. Although these changes are rare, they can cause significant changes in various different ways. Large **earthquakes** and **volcanic eruptions** can create changes in the shape of the land. Movements in **entire plate positions** can change the capacity of the ocean basin, so can raise or lower sea level.

Warm-Up Questions

Q1 Describe the difference between eustatic and isostatic sea level change.

Q2 Explain how a fossil cliff is formed.

PRACTICE QUESTIONS

Exam Question — AS and A-Level

Q1 Explain how climate change can cause sea level change. [6 marks]

What's black and white and sandy? A dalmatian coastline...

Hmm... not the best. Better stick to penguins rolling down hills. Now I've got your attention, don't forget that eustatic changes refer to a change in the level of the SEA itself and isostatic changes refer to a change in the LAND relative to the sea.

Coastal Flooding

Coasts are risky areas to live for a number of reasons. Keep on reading to learn why...

Some coasts are at **greater risk** than others

The risk of coastal flooding is **increasing**. **Extreme weather events** (e.g. storm surges) can cause **quick onset coastal flooding**, whereas **sea level rise** causes **slow onset coastal flooding**. Some coasts are more likely to flood than others due to their **physical structure** and **shape**. Coasts with high rates of coastal recession are **not always** areas with a high risk of coastal flooding.

- **Low elevation areas**, such as estuaries and deltas, are **more susceptible** to coastal flooding. Storm waves and a higher sea level can more easily flood these features. **Low-lying islands** are also at risk of **submergence**.

- Some **deltas** are subsiding due to increases in river flow washing their sediment out to sea. They're also sinking due to the **increased building of settlements** on the land. The amount of low-lying land making up deltas is **increasing**, making them **more susceptible** to flooding.

- The **removal of vegetation** on some coastlines also places them at a **higher risk** of coastal flooding. Coastal species such as mangroves can establish new land by encouraging **deposition** and **stabilisation**. This makes the area less susceptible to flooding particularly during storm surges as a result of tropical storms. Mangroves also **reduce the height and power** of oncoming waves which stops them from eroding the coastline.

If the sea level rises by 0.5 m, it will submerge most of the Maldives. The highest point of this group of coral atolls sits just 2.4 m above sea level.

Flood risk in Bangladesh

Three major rivers (the Brahmaputra, the Ganges and the Meghna) all flow through Bangladesh and enter the Indian Ocean through the Bay of Bengal. Bangladesh has many **physical features** that makes it more **vulnerable** to **coastal flooding**:

- Around 50% of the country is at an elevation of **10 m above sea level or less**. Approximately **20 million people** live in areas **less than 1 m** above sea level. A lot of this low-lying land is on the coast where there is a **large delta**.

- Tropical cyclones regularly cause **storm surges** on the coast and **monsoon rains** increase the amount of floodwater.

- Hundreds of **small islands** make up the Bay of Bengal area. **Unconsolidated material** makes up the majority of the islands. This material comes from the three main rivers as **sediment deposits**, which makes the islands highly **unstable** and **susceptible to flooding**. As sea level rises, more of these islands are at risk. Scientists estimate that 8000 km² of land in Bangladesh will **disappear** if sea level rises 0.3 m from its current level.

A farm that uses floating beds in the Pirojpur District of the delta

Some of these islands are more susceptible to flooding due to **human actions**. Fertile farmland is increasingly in **short supply** as the population grows, causing some farmers to clear areas of mangrove forest on the delta and drain it to **increase farmland**. The delta used to be a site of net deposition but some parts are now a site of **net erosion**. This, in combination with its low elevation, makes it more vulnerable to flooding from the **rising sea level**.

The islands change shape with the monsoon each year.

Extreme weather events can cause flooding

Typhoon Haiyan struck southeast Asia in 2013, causing a devastating storm surge (see p.27).

Storm surges are one cause of flooding during an extreme weather event.

1) A storm surge is a **short-term rise in sea level** caused by **high winds** pushing towards the coast, and by the **low pressure** of a storm. A drop in air pressure by 1 millibar causes a rise in sea level of 1 cm.

2) During a tropical cyclone, pressure can drop by around **100 millibars** (raising the sea level by a **metre**).

3) Air generally moves from areas of high pressure to areas of low pressure. The lower air pressure can cause **stronger winds** to move across the sea and a **storm surge** may occur.

> This will only happen in a very localised area.

- During a storm surge, waves move onshore with **greater force**. Water **builds up** behind the crest of the waves (a process known as **wave shoaling**), making them taller and more likely to flood the land as they break. If a storm surge occurs at the same time as **high tide**, the waves will reach an **even greater height** — further increasing the risk of flooding.

- Storm surges are also more dangerous if the shape of the land **funnels** the water into a smaller area, such as the mouth of a river. This raises the sea level and makes flooding more likely. The rise of the seabed or riverbed near the coast also causes **wave shoaling** (in a similar way to a **tsunami** — see p.11), increasing the flood risk.

Coastal Flooding

Future **climate change** is likely to change the **level of risk** of coastal flooding

- Climate change is unlikely to increase the number of tropical cyclones that strike coastlines, but the intensity (or magnitude) of those storms is likely to increase as sea temperatures rise. This increase in temperature may cause storms to grow larger and creates larger storm surges which increase the risk of coastal flooding.
- Increased global temperatures are also likely to increase sea level through the thermal expansion of water in ocean basins. Sea level will also rise due to an increased volume of seawater brought about by melting ice.
- As climate change warms the ocean, the strength of the winds will increase, which will produce stronger waves. These stronger waves will erode the coast more violently, which may break up or wash away embankments that protect low-lying areas, increasing their vulnerability to coastal flooding.

Although scientists believe sea levels **will rise**, they're not sure **how fast** it will happen, **which areas** of the world will be most affected and how much **more intense** tropical cyclones might be. Also, they don't know if there'll be any **negative feedback** when wind speeds get above a certain strength. Despite the uncertainty, scientists are fairly certain that coastal flooding will start to **affect more people** in the coming decades. Exactly what those effects will be depends on a variety of **less predictable human variables**.

Coastal flooding and storm surges have **economic** and **social** impacts

The impacts of coastal flooding can be felt **widely** and **long after** the initial event. Storm surges brought on by tropical cyclones can have a number of **economic** and **social** impacts in **developed** and **developing countries**:

Economic impacts of storm surges

- Salt water may inundate and **contaminate** bodies of freshwater, making water **unsuitable for farmland irrigation**. This is especially damaging to countries that rely on money from **agricultural exports**.
- Damage to **infrastructure** (such as ports) affects **trade** for both developed and developing countries. **Investment** in new and expensive port infrastructure that can withstand sea level rises is needed.
- Developing countries that are often affected by cyclones may not have enough funding or time to 'build back better' between a storm surge and the next cyclone.

Social impacts of storm surges

- Flood water can also contaminate **drinking water** sources. Standing inland water can **restrict movement** and leave communities **isolated**. Developed countries, where people rely on the **daily transport of food and fuels** into settlements, can be more vulnerable.
- People may be left homeless as flood water can **damage buildings and infrastructure**. Timber also expands when standing in water, meaning buildings may weaken or collapse.
- **Sediment** in floodwater may remain as **deposits** in built-up areas long after floodwaters have receded. Authorities may have to **unblock roads and drains** in order to make areas safer.

Environmental refugees are likely to increase

- As sea levels rise, more people will become **environmental refugees**. This may be because the land which they **live** on, or earn a **livelihood** from, becomes **submerged**, or because of **saltwater intrusion** (where saltwater infiltrates upwards). Saltwater can **contaminate soil** and **drinking water supplies**.
- Inhabitants of **Small Island Developing States (SIDS)** with low-lying coastlines, like Tuvalu, may be the first to suffer. Tuvalu has little high ground and relatively high population densities. Around **2400 Tuvaluans** have already been **forced to move** to **larger islands** due to **rising sea levels**, with some seeking **refuge** in other countries e.g. New Zealand.

Funafuti Atoll, Tuvalu

Warm-Up Questions

Q1 Describe how human activity has affected Bangladesh's vulnerability to coastal flooding.

Q2 How does air pressure affect sea level?

PRACTICE QUESTIONS

Exam Question — A-Level

Q1 Explain why some places are more vulnerable to coastal flooding than others. [8 marks]

A change is as good as a rest — or not...

If climate change is causing cyclones of increased intensity and rising sea levels, then change is definitely worse than a rest.

Coastal Management — Hard and Soft Engineering

Coastal management is a complex thing. Fixing up one coastal area can unintentionally mess up another area...

Hard engineering aims to change coastal processes

Hard engineering defences involve **built structures** that:

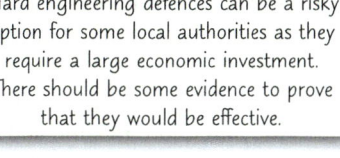

Hard engineering defences can be a risky option for some local authorities as they require a large economic investment. There should be some evidence to prove that they would be effective.

- tend to be **single solutions** that are built and put in place **once**.
- can have a detrimental impact on **local ecosystems**.
- can impact the **visual attractiveness** of a coast and may **reduce beach accessibility**.
- can work well in the **intended location** but may increase erosion on other coastlines. They can **interrupt** the **movement** of **sediment** from one part of the sediment cell to another and **unbalance the system**.

Defence	How it works	Cost	Impact on physical processes
Sea wall	A concrete wall **reflects** waves back out to sea, preventing **erosion** of the coast. It also acts as a **barrier** to prevent **flooding**. They can have different **designs** such as stepped or curved frontages. People can use the top of the wall for **access** and as a **promenade**.	Expensive to build and maintain	It creates a strong **backwash**, which can **erode** under the wall or **scour** away beach material. It can also be very **unsightly**.
Revetment	Revetments are **slanted structures** built at the foot of cliffs. They can be made from concrete, wood or rocks. Waves break against the revetments, which **absorb** the wave energy and prevent cliff erosion.	Expensive to build, but relatively cheap to maintain	They create a strong **backwash** which can cause further **erosion**. They tend to have a 20-50 year lifespan before they need **replacing**.
Offshore breakwaters	Breakwaters are usually concrete blocks or boulders deposited off the coast in **shallow waters**. They force waves to break **offshore**, reducing their energy and erosive power before they reach the shore. Over time, **deposition** may occur around the breakwater, increasing the effectiveness of the defence. The area between the breakwater and the beach can provide a **sheltered area** for swimming, water sports and harbour entrances.	Expensive	Storms can **damage** breakwaters easily so they may need **constant maintenance**. The deposition of sediment around the breakwater can **starve** another area of **sediment**.
Rip rap	Boulders piled up along the coast are known as rip rap (or rock armour). The irregular shaped boulders have a **large surface area**. This absorbs wave energy by increasing **friction** and so reducing **erosion**.	Fairly cheap	Can **shift** in storms. Can be **unattractive** as they are often not made from **local rocks**.
Groynes	Groynes are fences built at **right angles** to the coast. They can be made from wood, concrete or rock. They trap beach material transported by **longshore drift**. This creates **wider beaches**, which slow the waves (reducing their energy) and gives greater protection from **flooding** and **erosion**.	Quite cheap	Groynes **starve** down-drift beaches of material which means they can be more easily **eroded**. This is known as the **terminal groyne effect**. Some people consider groynes **unsightly**. They also make it more difficult for people to walk along the beach.

Rip rap.
Do not confuse with rickrack.

Topic 2: Option 2B — Coastal Landscapes and Change

Coastal Management — Hard and Soft Engineering

Soft engineering aims to work with coastal processes

- Soft engineering defences involve **working with natural processes** to reduce the effects of flooding and erosion, rather than trying to stop them. These options can have a more **natural 'look'** that fits in with the coastline.
- Soft engineering options might be more compatible with **managing sea level changes**. They are more **flexible** and can be adapted to suit new and different conditions.

Beach nourishment can be beneficial.

Defence	How it works	Cost	Impact on physical processes
Beach nourishment	Beaches have **sand and shingle** added to them, replacing material lost to **erosion** and **longshore drift**. The material often comes from **dredging** offshore areas. **Wide, natural-looking beaches** are formed, which reduce erosion of cliffs more than thin beaches.	Expensive	Nourishment needs to be **regular** and **continuous** as longshore drift will continue to remove material. Organisms, such as **corals** and **sponges**, can be **killed** by this defence method.
Dune stabilisation	Dunes are stabilised by **reducing the slope angle** and **planting vegetation**. Stakes and old tree trunks can stabilise the sand and prevent **wind erosion**. It also creates **wide beaches**, which reduces erosion and flooding.	Cheap	Needs **regular maintenance** as dunes will **shift position** naturally.
Cliff regrading	This changes the **shape** of the cliff so it a **gentler gradient** and is less susceptible to **erosion**. **Vegetation** can grow on the shallow cliff face which further stabilises the cliff slope and improves the local **ecosystem**.	Expensive	This can be hugely **disruptive** and may involve temporarily or permanently **rehoming** local people.
Cliff drainage	This is the **removal of water** from a cliff face by building **drainage channels** into the rock structure. This reduces the **pressure** within the cliff face and therefore the risk of **collapse**.	Fairly expensive	**Effectiveness** and **ease** of installation may depend on the **geology** of the **cliff face**.

Sustainable management of coasts can lead to local conflicts

Sustainable management involves **meeting the needs** of coastal communities **now**, without compromising the ability for **future** communities to **meet their needs**:

- Sea levels are likely to continue to **rise** but the rate and magnitude of this rise is **unpredictable**.
- The frequency and intensity of **storms** striking the coast is likely to **increase**, leading to increased **erosion** in some areas and increased **deposition** in others.

> Adaptation (as well as mitigation) is central to sustainable management of these threats because they are happening now.

Sustainable coastal management involves looking at the **wider coastal system** rather than individual beaches and towns. This includes managing the **physical coastline** and the **people** who live on, work at and visit the coast. It should also **minimise impacts** on coastal habitats and ecosystems. Management practices can lead to **conflict**:

- **Limiting access to resources** (e.g. stocks of fish or sand) may negatively affect **livelihoods** and **local economies**.
- **Protecting land** with a **high economic value** may leave residents of unprotected land feeling **abandoned**.
- Management strategies may try to **solve issues** that **aren't currently a problem** (such as storm surges at an increased height). This can seem like a **waste** of taxpayers' money.
- **Preventing access** to coastal areas (such as sand dunes) can upset people who wish to enjoy these spaces.

Warm-Up Questions

Q1 Explain how rip rap can reduce coastal erosion.

Q2 Explain why beach nourishment may not always be a suitable coastal management option.

(PRACTICE QUESTIONS)

Exam Question — AS and A-Level

Q1 Assess the view that soft engineering defences are a suitable form of coastal management. [12 marks]

Terminal groyne effect — sounds painful...

Hard engineering involves trying to change coastal processes, whereas soft engineering tries to work with natural coastal processes and tends to be cheaper (unless of course you're changing the slope of a cliff, which can be pretty pricey).

Integrated Coastal Zone Management

I know what you're thinking, not another acronym... But you've almost finished this section, so don't zone out just yet...

Integrated Coastal Zone Management uses a **holistic approach**

1) **Integrated Coastal Zone Management (ICZM)** is a strategy that considers **all elements** of the coastal system such as land, water, people and the economy. This means it's **holistic** — it looks at the **whole** rather than individual parts.

2) ICZM considers the **long-term management** of the coastline as well as its **immediate needs**. Previous strategies often focused on **local impacts**, where decisions were made independently. This means that there are many stretches of UK coastline where defences such as groynes have increased erosion in **neighbouring sediment cells**. ICZM schemes aim to protect the coastal zone in a **relatively natural state** while allowing people to use and develop it in different ways.

3) **Integrated** means:

- the **land** and **water** environments are **interdependent**.
- **different economic uses** of the coast (such as fishing and tourism) have **equal value**.
- **local**, **regional** and **national levels of authority** all have a say in the plan.

ICZM strategies manage coastlines in different parts of the world

In **developing** countries, ICZM focuses on **adaptation to coastal erosion and flooding** as it's cheaper to implement. In **developed** countries, ICZM focuses on **both adaptation and mitigation** as there's greater demand on authorities to control natural processes.

> ICZM is a dynamic strategy. Decisions are re-evaluated if the demands within the area change.

In the **UK**, the English and Welsh coastline is separated into 11 different littoral cells. The ICZM breaks the coastline into 22 smaller areas, each with its own **Shoreline Management Plan (SMP)** designed by the Environment Agency and local authorities.

There are **four main decisions** in coastal zone management

SMPs (see above) generally make **one of four decisions** on how to **manage** an area of **coastline**. There are **advantages** and **disadvantages** to using each:

No active intervention

- **No active intervention** means **building no coastal defences at all**, so local authorities have to deal with the **consequences of erosion and flooding** as they happen.
- Any defences already in place will **remain there** but will have **limited impact** in the long term as they're likely to erode away. This is the **cheapest** of the four decisions but it's **rarely viewed favourably** by those whose properties are most vulnerable.

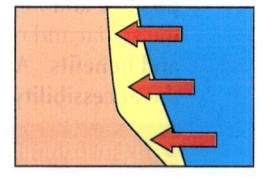

Strategic realignment

- **Strategic realignment** involves allowing the shoreline to **move naturally**.
- The process aims to cause the **least damage to important areas**. E.g. allowing low-quality farmland to **flood**, rather than a town that has far more **economic potential**.
- Although some people argue this decision is more **ecologically sustainable** than others, it's not suitable for use on much of the built-up UK coastline.

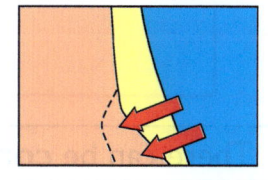

Hold the line

- **Hold the line** involves **maintaining the existing coastal defences**, meaning the **position of the shoreline stays the same**. These are often used on coastlines with **high value** land and often use a mixture of **hard and soft engineering defences**.
- This decision can involve spending a lot of **money** on defences that were **poorly designed initially**. It can also involve schemes that detract from the **attractiveness** of the beach, such as building **concrete sea walls**.

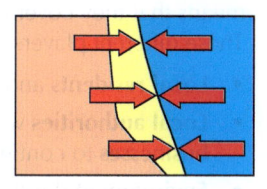

Advance the line

- **Advance the line** involves building new coastal defences **further out to sea than the existing line of defence**, e.g. by using **beach nourishment**. This increases the size of the beach or allows for **land reclamation**.
- These decisions are used on coastlines where there are likely to be **reasonable levels of deposition**. They can be **expensive** but can also make a coastline more **attractive**.

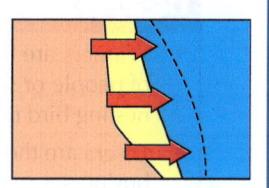

Integrated Coastal Zone Management

Deciding which strategy to adopt is a complex decision

The **Department for Environment, Food and Rural Affairs (DEFRA)** oversees coastal management decision-making in the UK. One of its aims is to protect homes, businesses and the environment from the **social**, **economic** and **environmental impacts** of **erosion** and **flooding**. It considers **many factors** when deciding which of the four strategies (see p.85) to use.

1) **Land value** — The more valuable the land, the more likely a SMP will try to **protect or manage** the area. The value of the land often relates to its **land use** (see p.75).

2) **Engineering feasibility of defences** — The **shape and structure of the land** affects the way coastal processes work within the sediment cell. This may make building some defences **impractical** or cause them to not reach their **full potential**. E.g. trying to **stabilise** very **mobile** sand dunes or **unstable cliffs**.

3) **Impact on coastal processes** — The strategy chosen may impact the **surrounding coastal areas** as well as the **defended area**. This will impact how **coastal processes work** within the **sediment cell**.

4) **Sensitivity of local environment** — Building a certain defence may destroy an area with **high ecological value**. Not defending an area may increase the vulnerability of **another habitat**.

5) **Cultural heritage of area** — Areas of coastline may be **culturally** or **historically** important. These may not have an economic value, but they may **represent cultural value** for local people and beyond.

6) **Local opinions** — Strong local feelings for and against a strategy as well as the viewpoints of local councils can **influence decisions**.

7) **Costs** — The **initial** and **ongoing maintenance** costs of any defences adopted must be considered.

Sometimes, authorities choose a management strategy because it gives **confidence** to local people and businesses. E.g. a **soft engineering option** may be best for a location, but a hard engineering option **shows** that there's something permanent in place. Increased local confidence can increase **consumer spending** and **outside investment**.

Cost-benefit analysis can help authorities to make decisions

1) All coastal settlements want to have coastal defences, but there's often **limited money** available to allow this. Priority might go to **large settlements** over isolated or small settlements.

2) Choosing defences for different settlements is based on **cost-benefit analysis (CBA)**. This weighs up the **positive and negative impacts** a particular decision will have. It tries to assign **comparable values** to the social and environmental impacts of a plan, so they can be weighed up against the **economic costs and benefits**. Analysts also consider the costs and benefits of **immaterial values**. These might include the **accessibility**, the **visual appeal of the structure** or the environmental value (e.g. biodiversity).

An Environmental Impact Assessment (EIA) can add detail to the analysis

An EIA looks at the impact of construction on:
- **natural processes** such as longshore drift.
- particular **species** (marine plants, fish, marine mammals).
- the balance in **ecosystems**.
- **environmental quality**, for example air and water pollution.

There can be conflict between different players

The holistic nature of ICZM means that SMPs are unlikely to benefit everyone. Their scale means that they cover a wide area, multiple habitats and many different groups of people. These different players have **different priorities** that can lead to **disagreement** and **conflict**.

- **Local residents** and **homeowners** want to see their property and their land protected.
- **Local authorities** want to ensure that settlements are operational. They want **local businesses** to continue trading and **transport infrastructure** in place to serve them.
- **Environmental pressure groups** want to **conserve habitats and species**.

The complexity of coastal management and the conflicts that exist between players occur in both developed and developing countries.

Decisions about coastal management produce perceived 'winners' and 'losers'

- **Winners** are those who stand to **gain** from the decision. They might be groups of people or species. E.g. a cliff face which shelters a particular species of nesting bird may get protection, increasing their numbers.
- **Losers** are those who stand to **miss out** from the decision. E.g. local businesses may have to close or relocate, leaving people unemployed.

Kittiwake on cliffs at Bempton, Yorkshire

Integrated Coastal Zone Management

Different strategies are suited to **different locations**

Coastal management at Happisburgh, England — developed country

1) Happisburgh is a small village in North Norfolk with around 1400 residents and 600 properties. **Sea defences** built in the 1950s have been destroyed due to **weathering** and **mass movement**.

2) In **1996**, a policy of **managed retreat** was adopted. It was predicted that this policy would mean that around **200 m** of coastline would be lost by 2105. This would result in large number of properties, a caravan park, 45 hectares of agricultural land and a number of heritage buildings (such as a listed church) at risk of **destruction** as the cliff continued to **erode**.

3) In **1998**, local residents set up the 'Coastal Concern Action Group' which **campaigned** for their voices to be heard when policies were being made. They raised **repeated objections** to the managed retreat policy. However, Happisburgh **didn't qualify** for new sea defences under the cost-benefit analysis done by the Ministry of Agriculture, Fisheries and Food. The **defences** were estimated to have cost **£15 million** — more than the value of the properties they would have been protecting.

4) The North Norfolk District Council was awarded around **£3 million** of funding to:
- purchase the otherwise **unsellable properties** most at risk, at 40% of their value.
- **relocate** the caravan site, carpark and public toilets as well as create **new access** to the beach. This would create a **buffer zone** between residents and the receding cliffs.
- build **emergency rock armour** at the base of the cliffs to **slow** the **rate of erosion**.

The coastline consists of boulder clay, which is vulnerable to coastal erosion.

Houses at risk of collapse in Happisburgh

In 2022, the UK government launched the Coastal Transition Accelerator Programme which aims to help coastal communities such as Happisburgh.

Coastal management in Chittagong, Bangladesh — developing country

1) Chittagong is a coastal city in southeastern Bangladesh. It is at risk of coastal flooding, both from **rising sea levels** and **storm surges from tropical cyclones**.

2) Since 2012, the 'Coastal Climate Resilient Infrastructure Project' has been trying to make the coastal areas of the city **more resilient to future changes in climate**. It has funding from the Asian Development Bank (ADB). This project has a **multi-pronged** approach which aims to help **people**, the **economy** and the **environment**.

3) Some of the schemes within the project include:
- building **coastal embankments** lined with **halophytic vegetation** to act as protecting walls.
- **raising the levels of roads** and making them more **hardwearing** to prevent storm damage.
- upgrade 25 **cyclone shelters** and improve the **access** to cyclone shelters.
- upgrade 37 **boat landing stages** so they are designed to cope with **rising floodwater**.
- **education programmes** covering how to react to cyclones and land management practices.
- improving **water supply pipelines** and creating **sewage and sanitation networks** with protection from floodwater.

4) There were **concerns** about the amount of vegetation that authorities cleared in order to create the embankments, roads, markets and cyclone shelters. However, planting young trees and shrubs **enhanced the look** of the new areas. **Local people** were employed in stewardship roles to oversee the construction phase. 346 people received **compensation** as they had to undertake **involuntary resettlement**, mostly from roadsides.

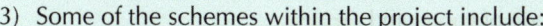

Cyclone shelters are also schools or community centres.

Halophytic vegetation grows in salty soil or water.

While funding came from the ADB, the projects were locally organised and run, which ensured greater local engagement with the objectives.

Warm-Up Questions

Q1 Describe the key differences between the four main coastal defence decisions.

Q2 Explain how a cost-benefit analysis might help authorities to make decisions.

Exam Question — A-Level

Q1 Evaluate the view that coasts are managed primarily for economic sustainability. [20 marks]

Hold the line vs no active intervention — wanna arm wrestle for it?

Deciding how to manage the coast is pretty tricky, and there'll more than likely be some losers. There's no point saving a pretty stretch of coast if a large town nearby is going to end up flooding as a result. That's why CBAs and EIAs are done.

Introduction to Globalisation

Globalisation affects how we communicate, how we trade and how we move around. Fairly important then, really...

Globalisation is the process of countries becoming more connected

- Globalisation has led to the world becoming more **interconnected** over time. This means people, industries and places in different countries interact like they belong to a **single community**. How different countries rely on each other is known as **interdependence**.

- Globalisation has been happening for a long time, but the **pace of change** has been **rapidly increasing** in the twentieth and twenty-first centuries.

There are **five different aspects** of globalisation:

1. **Economic** — e.g. the growth of companies into Trans-National Corporations (TNCs) (see p.93) that operate in more than one country.

2. **Cultural** — e.g. advances in communications technology, which allows for the easy flow of information and ideas.

3. **Political** — e.g. the development of international bodies such as the UN which works to reduce conflict across different countries.

4. **Social** — e.g. the growth of international migration.

5. **Environmental** — e.g. the development of treaties and international agreements seeking to protect the environment.

Globalisation is driven by flows of resources, people and information

1) Globalisation means that the **connections** people (and places) have with one another are **widening** — they stretch over a **larger spatial area** than ever before. The connections are also **deeper** — **more** parts of people's lives are **increasingly dependent** on the rest of the world.
   E.g. many **people** can buy products imported from all over the world in their **local supermarket**.

2) Connections are formed through **flows** of commodities, capital, information, tourists and migrants:

Commodities	• The volume of **raw materials** (e.g. food and fuel) as well as **manufactured goods** being traded around the world has increased rapidly since 1950.
	• This may be made easier by the increased numbers of international **trade deals** and improvements in the **technology** that enables international trade.
Capital	• Money is bought and sold globally through **currency exchanges** in banks every day. Banks also trade in **stocks and shares** internationally.
	• Money flows between countries through the trade of **goods** and **services**.
	• Individuals and companies from one country might invest money into an industry in another country — this is known as **Foreign Direct Investment (FDI)**.
	• Capital (wealth in the form of money or assets) is able to flow more easily and more quickly due to **online banking** and **cryptocurrency**.
	• Migrants often send money home to financially support family and friends (known as **remittance payments**).
Information	• Information (e.g. news of current events) can spread across the world very **quickly** and **easily**.
	• The rapid development and adoption of **email**, the **internet** and **social networking** means that large amounts of information can be exchanged quickly across the globe. This means people who live and work in **different countries** can communicate and work together more easily.
	• People can **learn** about different countries and cultures without leaving their country.
Tourists	• Increased international tourism has increased many people's **first-hand experience** of other countries.
	• Increases in **technology** (e.g. **jet planes**) and the **decreasing cost** of travel have allowed more people to travel to a wider range of countries.
	• Flows of tourists are often matched by flows of **foreign currency** — this generates wealth for the host country.
Migrants	• People **move** between countries **permanently** and **seasonally** (e.g. for employment). **International** migration **connects** people with other countries around the world. For example, the UK depends on over 70 000 **seasonal agricultural workers**, primarily from Romania and Bulgaria.
	• Migrants might be seeking better **economic opportunities** and greater stability. Migrants might be **refugees** — people seeking a **safer home** away from the threat of persecution, war, famine or extreme effects of climate change.
	• Flows of people are often **regulated** — e.g. governments use policies to encourage or discourage **migration**.

In 2019, tourists to the UK spent a total of £28.5 billion.

Introduction to Globalisation

Developments in transport have created a 'shrinking world'

1) A number of **technological innovations** in **transport** during the nineteenth and twentieth centuries have changed **how and where** trade takes place.

2) Improvements in technology and transport have enabled:
   - commodities to be traded in **greater volumes**
   - goods to be traded **between countries** who are further away from each other
   - **perishable exported goods** (such as food) to reach receiving countries more quickly.

 Steam railways — goods could be moved across a continent.

 Steam ships — goods could be traded overseas.

 Jet aircraft — goods could be moved quickly between continents and to places without a coastline.

 Containerisation — large volumes of goods could be transported all over the world efficiently in shipping containers.

The 'shrinking world' effect

1) All these improvements in transport technology have made the world seem **progressively smaller** over time. This is known as the **'shrinking world' effect**.

2) As people become **more connected** to places in the world that are **further away**, their perception of the world changes. Distant places 'feel' **closer** because it takes **less time** to get to them. This is known as **time-space compression**.

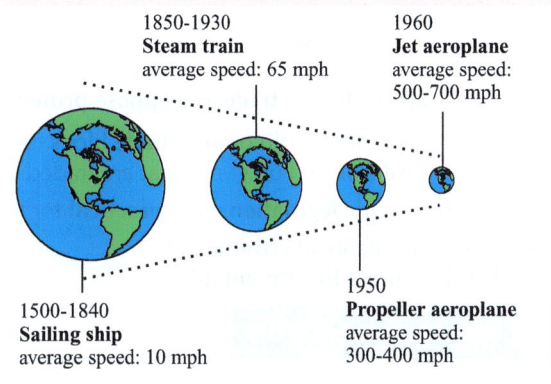

1850-1930
Steam train
average speed: 65 mph

1960
Jet aeroplane
average speed: 500-700 mph

1500-1840
Sailing ship
average speed: 10 mph

1950
Propeller aeroplane
average speed: 300-400 mph

Developments in ICT and communication have contributed to globalisation

Technological advances in ICT and communication systems allow people to connect with each other more **easily**, more **quickly** and **over greater distances**. This has contributed to **time-space compression**. The **decreasing cost of communication** has made it more accessible to more people, e.g. smartphones with mobile data are increasingly affordable. Improvements in communications technology include:

- **Telegraph cabling** (electrical cabling used to transmit Morse code messages) — allowed the first type of 'instant' communication overseas.
- **Mobile phones** — have become much more advanced and are used for calling, texting, e-mailing, instant messaging and much more.
- **Internet** — increased use (helped by **fibre-optic cabling**) allows people to find information and communicate with each other at the **touch of a button**.
- **Social media** — makes it easier to **share information** in a variety of formats (e.g. text, images, audio and video).
- **Electronic banking** — electronic money transfers and banking are increasingly used, and have become **more convenient** as people can access banking online through computers and smartphones.

"How strange!" said Jane, "This white box seems to be a type of computer."

Warm-Up Questions

Q1 What is globalisation?

Q2 Name the five flows that have created wider and deeper global connections.

PRACTICE QUESTIONS

Exam Question — AS and A-Level

Q1 Explain how developments in transport and trade have created a 'shrinking world'. [4 marks]

Around the world in (less than) eighty bullet points...

There's an almost unhealthy amount of jargon on these pages, but learning it now will give you a boost for the rest of the section.

Globalisation — Key Players

Various organisations affect how fast globalisation occurs and how it affects people's lives. Read on to find out more...

International organisations play an important role in globalisation

Some international organisations influence the **pace** and **nature** of globalisation. Here are **three** you need to know about:

International Monetary Fund (IMF)

The IMF promotes **financial cooperation** and **trade** between countries. One way it does this is by providing **loans** to member countries in exchange for **lifting trade restrictions**.

World Bank

Member countries pay a **subscription** to the bank, which can then **loan** money to **less developed** countries.

World Trade Organisation (WTO)

The WTO was set up to increase **trade** and help resolve **trade disputes** between member countries. It sets rules about how countries should trade with each other.

> These three organisations formed after the Second World War to try to build **stronger relationships** between countries. It was hoped that these relationships would strengthen the world economy, as well as prevent the outbreak of another world war.

These organisations **advocate free trade** and **oppose protectionism**:

- They encourage countries to join or form **trade blocs** (see next page). This reduces the **tariffs** placed on goods that are traded internationally.
- They work to make trade **legislation** more **practical** for trading nations.
- They also encourage national governments to accept **Foreign Direct Investment** (**FDI**).

free trade
unrestricted trade between countries

protectionism
when a country uses tariffs and quotas to limit trade in order to protect their own industries from foreign competition

Foreign Direct Investment (FDI)

1) Foreign direct investment is when a person, company or other group spends money in another country in order to generate a **profit**, e.g. by opening a new branch of their business or investing in local infrastructure.
2) FDI can also take the form of a **merger** between two companies operating in different countries. Or it can be an **acquisition**, where a TNC takes over the running of a company in a different country.

The work of international organisations can be **controversial**

Some people believe international organisations hold **too much power** over global flows of capital and goods and act in ways that are **unfair to some countries**.

- To receive a loan from the IMF and the World Bank, countries often have to abide by **strict rules and conditions** (e.g. Structural Adjustment Programmes, see p.241) which can be **difficult** for poorer or developing countries to follow.
- The **governance** of the WTO, IMF and the World Bank is predominantly found in **developed countries** — this means developed countries cast most votes in key decisions. Some decisions may serve the needs of the **richer nations** over the needs of the **poorer ones**.

National governments also make decisions that affect globalisation

1) Some countries are **protectionist**, while others decide to follow policies that promote **free-market liberalisation**. These policies **remove restrictions** placed on the trade of goods and capital between countries (promoting a **stronger connection** between them).
2) National governments may **privatise services** that are run by the state (such as a rail network) and allow **buyouts** from overseas companies. E.g. in 2013, a share of Manchester Airports Group was sold to IFM Investors, an Australian investment management company.
3) Governments can give **incentives to foreign companies** (e.g. **subsidies** or lower business rates) to encourage them to relocate their operations overseas. Equally, they may provide **grants** to encourage **international business start-ups**.
4) National governments can decide whether their country will be part of a **trade bloc**. A trade bloc is a group of countries that all agree to **remove tariffs** on goods traded between them (e.g. the EU and ASEAN — see next page).

> This is a form of **inward investment**.

Globalisation — Key Players

Trade blocs have a number of advantages for members

- Trade is encouraged within a trade bloc because member states receive goods at a **cheaper price**.
- There is a **larger market** for goods supplied by a member state — this may increase the volume of trade and allow producers to reduce production costs.
- Trade blocs can set up **trade barriers** with non-member states so that certain, or more vulnerable, industries within the trade bloc are **protected**.

- **Mergers** between smaller TNCs within the bloc can occur more easily, which can streamline their operations and make them more profitable.
- TNCs can **utilise the different strengths** of the countries in the trade bloc, e.g. a TNC might use cheaper labour from one country and development resources from another.
- There is greater **political security** between member states due to their mutual economic dependence.

The European Union (EU)

- The EU began as the **European Economic Community** in 1957 with 6 member states.
- In 2022, the EU had 27 members. **Free trade** exists between members and there are **common external tariffs** on goods imported into the bloc. The EU has its own **currency** (the euro) and a **European Parliament** to pass EU laws.
- The **Schengen Agreement** allows the free movement of European workers across its borders.

The Association of South East Asian Nations (ASEAN)

- ASEAN was established in 1967 and by 1999 it had 10 members. There is **free trade** between its nations — this has encouraged **local manufacturing** and **banking industries** to grow, making ASEAN countries more economically **competitive** with other countries.
- In 1995, member states agreed to not use any **nuclear weapons**, making their union more politically stable. In 2007, members signed the **ASEAN Charter** — this formalised the way that the member states would operate. ASEAN is one of the world's largest trade blocs.

Political and economic decisions accelerate globalisation in new regions

1) Governments in **emerging economies** have also developed **incentives** to encourage foreign businesses to **invest** there.
2) **Special Economic Zones (SEZs)** are areas of land, often on the coast, where special economic rules apply, such as **low tax rates** or **tax breaks** for foreign businesses and the **removal of import and export tariffs** on goods traded in and out of the zone.
3) SEZs often have **infrastructure networks**, such as road and port connections, already established to encourage TNCs to set up there.
4) Governments can invest in **improving transport networks in SEZs** to encourage trade. This is particularly important for **land-locked countries** or those with challenging topography.
5) Governments can also offer **subsidies** to support local and international businesses in the first few years while they become established.

China's 1978 'Open Door Policy' welcomed FDI — CASE STUDY

- In China, industry was under **state control** from 1949 until the late 1970s. China was on the **outskirts** of the **global economy** and it experienced famines and high levels of **poverty**.
- From 1978, radical **economic** and **political reforms** made China more competitive in the global economy. This was known as the **'Open Door Policy'** as it opened China to overseas investment.
- **Four SEZs** were established along the coast to encourage foreign companies to set up in China. **TNCs** like Apple and Dell™ outsourced jobs to China e.g. Shenzhen Special Economic Zone. China soon became known as the 'factory of the world', especially in **consumer electronics**.
- The **FDI** from these companies helped the Chinese economy **grow rapidly**. In 2019, China received US $187 billion in FDI and 27% of their GDP came from **manufacturing**.

Warm-Up Questions

Q1 What is a Special Economic Zone (SEZ)?
Q2 Briefly outline China's 'Open Door Policy'.

Exam Question — AS and A-Level

Q1 Explain the role of national governments in driving globalisation. [6 marks]

SEZs — Super Energetic Zebras?

There's a lot of acronyms in this section — the best way to keep track of them is to write a Nice Acronym List (NAL).

Extent of Globalisation

Very few places are unaffected by globalisation, though some are affected more than others.

The **effects** of globalisation are **not evenly felt**

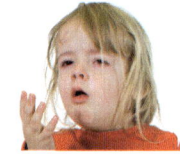
Suzie's cough index was off the chart.

1) Historically, some countries have found it **hard** to make connections. This could be for **geographical** reasons (e.g. distance from a coastline), or by **choice**. Countries that didn't form strong connections in the **past** are less likely to be well-connected in the **present**.

2) It is **difficult** to measure the extent to which a place is affected by globalisation. However, **indices** can be used to measure **certain elements** of globalisation:

1 The AT Kearney Global Cities Index (GCI)

A measure ranking **156 cities'** level of globalisation. It takes into account:
- their amount of **business activity**
- their level of **political engagement**
- the **cultural experience** offered to people
- the ways that **information** can flow through the city (e.g. through their internet connection)
- the number of people who create **global connections** (e.g. through being foreign-born).

New York, London, Paris, Tokyo and Beijing are the top five cities ranked by the GCI.

2 The KOF Globalisation Index

A measure of **195 countries'** level of globalisation.
- The index measures 42 variables covering indicators of **economic**, **social** and **political** globalisation.
- It then combines them to produce an **overall score** between 1 (low levels of globalisation) and 100 (high levels of globalisation).

Some countries are '**switched off**' from globalisation

Countries that are 'switched off' are relatively **untouched** by globalisation. This might be due to:
- **physical geography** — e.g. physical isolation due to mountain belts or deserts.
- **political decisions** — e.g. countries may favour nationalist policies.
- **lack of economic development** — e.g. poverty may stop a country being able to trade competitively.
- **environmental resilience** — e.g. countries may be highly vulnerable to climate change, which may limit their ability to trade cash crops.

North Korea remains largely 'switched off' from globalisation

CASE STUDY

1) Since the establishment of North Korea in 1948, the country has pursued an **isolationist, socialist** regime — the pursuit of isolationism means connections with other countries through trade and politics are very restricted.

2) The country's economy is based on an ideology of **self-sufficiency**, which concentrates the production of goods on the **immediate needs** of its citizens rather than on making trade connections with other countries.

3) North Korea has **limited physical communication** connections with other countries (such as internet cabling) in an attempt to **restrict** people's exposure to information and ideas from overseas.

4) However, North Korea has allowed some **external influences**:
- Since a period of famine and rapid economic **downturn** in the 1990s, North Korea has appealed for and accepted **international aid** from capitalist economies such as South Korea and the USA.
- The two Korean countries have competed together in a **unified team** in some sporting events, including ice hockey at the 2018 Winter Olympics.
- In 2018, the North Korean leader Kim Jong-Un met US President Donald Trump at a **summit** in Singapore, where they committed to a **partial denuclearisation** of North Korea and the lifting of some sanctions.

© Hoo me / Storms Media Group / Alamy Stock Photo

Warm-Up Questions

PRACTICE QUESTIONS

Q1 Briefly outline how indices can be used to measure globalisation.

Q2 Explain what is meant by a country being 'switched off' from globalisation.

Exam Question — AS and A-Level

Q1   Assess the significance of the different factors that might affect the extent of globalisation.   [12 marks]

Make sure you don't switch off during your revision...

So, it turns out that globalisation isn't as simple as it seemed... The world's getting more interconnected, but not everywhere is experiencing it to the same extent. You need to know about somewhere switched off from globalisation, so get learning.

Trans-National Corporations (TNCs)

TNCs play a central role in global systems of trade, connecting countries economically, socially and culturally.

TNCs are companies that operate in two or more countries

- Trans-National Corporations (TNCs) are companies that produce or sell products and services and are located in **two or more** countries. E.g. Sony® manufacture electronic products in China and Japan, and sell many of them in Europe and USA.
- TNCs play an **important role** in the global economy — around 80% of global trade is linked to TNCs.

Globalisation grows due to strategies used by TNCs

1) TNCs often develop and control **supply chains** that extend across **multiple countries**. They connect these countries through the **movement** of raw materials, capital, labour, part-assembled components and manufactured products.

2) TNCs are likely to focus FDI on some countries and not others, which can create **uneven** levels of globalisation. E.g. a TNC is more likely to invest in a country that offers a **potential economic advantage** (such as a cheaper labour force or a market within a tariff-free trade bloc) which would allow the TNC to maximise their profits.

3) **Glocalisation** is when a TNC **adapts** to **local** markets, for example by changing the design of a product so it complies with local **laws** or fits better with local **tastes** and **customs**. TNCs use glocalisation to make their products **more appealing** to new, local markets — e.g:
   - **Clothing** may be sized differently for people from different countries (e.g. longer or shorter leg lengths).
   - By law, **tobacco** companies have to use plain packaging in some countries.
   - **Fast food** companies like McDonald's® may change their menu to suit local tastes and religious observances.

TNCs can take advantage of economic liberalisation

offshoring
when a TNC moves branches of the company to other locations overseas

outsourcing
when a TNC moves parts of its operations to local companies overseas

- **Economic liberalisation** (having freer trade and welcoming foreign investment) has allowed many TNCs to invest in other countries by **offshoring** elements of their company, e.g. factories and distribution centres. This has helped TNCs to grow rapidly and generate large profits, since those countries can often provide cheaper labour and lower running costs. This contributes to the **international spatial division of labour** (when the labour force is split between different countries to take advantage of the cheaper wages for particular skill sets in each country).

- TNCs can also invest in other countries by **outsourcing** parts of their operations to local companies. For example, Apple employs the component manufacturer Foxconn® in Shenzhen, China, to complete the assembly of many of their products. This creates a **global production network** for many well-known products.

- TNCs may **acquire** or **merge** with local companies to be more competitive within their markets. The expansion of TNCs into new countries can also increase the **international recognition** of their brand which can increase the market for their products.

Some countries, such as China, have set up Special Economic Zones (SEZs) to encourage TNC offshoring.

Wal-Mart
- Wal-Mart is a chain of **discount department stores** that has taken advantage of economic liberalisation in other countries. It divides its labour across different countries.
- Its headquarters are still in Arkansas, USA, but most **manufacturing** is carried out where costs are lower. E.g. electronic goods are made in China and clothing is made in India.
- This **offshoring** has opened up new markets to Wal-Mart. For example, Wal-Mart and an Indian company called Bharti Enterprises are opening new retail outlets together in the style of Wal-Mart stores.
- As well as owning more than 5300 stores across the USA, it's starting to have a more **global** presence through the acquisition of other retail companies in other countries (e.g. Lider in Chile).

Warm-Up Questions

Q1 Define the term 'glocalisation'.

Q2 Explain the difference between 'offshoring' and 'outsourcing'.

Exam Question — AS and A-Level

Q1 Assess the role of TNCs in driving globalisation. [12 marks]

I'd much rather be lounging on the beach — I call it onshoring...

Make sure you learn the different ways that TNCs can affect globalisation — often they do things that increase it.

Global Winners and Losers

Both people and the physical environment can benefit from or lose out to globalisation.

The **global economy** has shifted to **Asia**

1) In different phases of history, **different regions** of the world have held **economic dominance**. E.g. for much of the nineteenth century, Europe controlled the global economy. After the Second World War, the USA rose to **greater economic power**.

2) Since the **1990s**, **Asia** has played a larger role in the global economy. The percentage of the **global GDP** now generated from that region is **increasing** and this trend looks set to continue.

3) This is called the **Global Shift** — the majority of industrial activity has shifted from one part of the world to another. This shift is seen through levels of investment — in 2020, **East Asian and Pacific countries** received 51% of all global **FDI**. Some analysts say that this represents a shift in the **global economic centre of gravity** to Asia.

4) The Global Shift has come about due to:

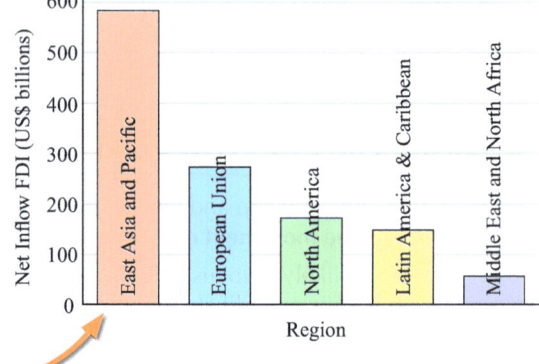

Destination of global FDI in 2020.

① **Global shift in manufacturing**
• In the 1950s, **low tech products** such as **toys and textiles** started to be manufactured in the first **Newly Industrialised Countries** (**NICs** — Taiwan, Singapore, South Korea and Hong Kong). • This happened as **TNCs** were looking to cut production costs, so they moved their operations overseas to make use of cheaper labour. • With further **investment** in the 1980s, NICs started to manufacture **consumer electronics**, e.g. computers and mobile phones.

② **Outsourcing of services**
• As more businesses began to operate globally, they required **business services** to support them. • These were needed **24 hours a day** and without large increases in **labour costs**. This means that business services, such as **IT support** or **customer services**, became more **widely distributed** around the world to meet these needs. • **Improvements** in telephone and IT **communication** allowed workers in distant countries to communicate with customers.

The **global shift** in manufacturing has created jobs and **driven economic growth** in China

- China's **large population** offered an abundant supply of **cheaper, often well-educated labour** to the global manufacturing sector. This **lowered production costs** for businesses that had previously used labour in Europe or the USA.

- Initially, **working conditions** in Chinese factories were often very poor and dangerous. Due to their regular and attractive wages, jobs in factories caused large-scale **rural-to-urban migration**. These migrants sought new jobs in **manufacturing hubs** such as Shenzhen.

- In the 1980s, Chinese goods tended to be cheap to produce and became **highly competitive** in the global market. China became known for producing **'throwaway' consumables**.

- As more **FDI** flowed into China, it was able to invest in greater **industrial infrastructure** (e.g. coal-fired power stations and railways) and higher-level **training** for workers. This made China an attractive manufacturing destination for **higher-order goods** such as **consumer electronics**. Major brands began to set up factories in China to make use of these facilities and the country's workforce (e.g. Apple).

- Increased FDI led to an improvement in working conditions, though some **labour practices** (such as asking people to work more than the legal maximum number of hours) have been criticised by human rights campaigners.

- In the late 2000s, China started to manufacture **home-grown products** to rival other large international brands, such as HUAWEI mobile phones.

- The global shift has also created a **growing middle class** in China who have become an important global **market** for the goods now produced there.

CASE STUDY

China became an attractive destination for manufacturing following the 1978 Open Door Policy. See p.91 for more on China's Open Door Policy.

An electronics factory in Shenzhen, China

The number of people living in extreme poverty in China has declined by 800 million between the early 1980s and 2022.

Global Winners and Losers

India has become a hub for **outsourcing** services

- Indian cities have attracted large companies, such as Barclays, wishing to outsource the labour needed to run their **back-office operations** or technical **call centres**.
- In 1990, a series of three **technological parks** were set up on the edge of the city of Bangalore. These hubs had **low set-up rates** as a means of encouraging technology firms to locate there.
- India has a **large, youthful population** and a rapidly growing number of **graduates** (9% of all 25-year-olds in 2011 had a degree). Many of these graduates have **qualifications in IT and engineering**, making them an especially attractive workforce to some companies.
- India has the second largest **English speaking** population in the world and these language skills enable communication with English-speaking countries.
- Call centre work can be desirable to Indians as it often earns **three times more** than the average **income**.

Poster advertising for workers for a call centre in India.

The **Global Shift** has created a number of **benefits**

Infrastructure investment

- The global shift has led to greater **investment in transport and energy infrastructure** in Asian countries.
- For example, the **capacity** of the Shenzhen port has **increased** so that it can now have more ships berthed.
- New **power stations** are also being built in China to meet the energy demands (there are approximately **1100 active coal power stations**). China is also investing in renewable energy, e.g. the **Longyangxia Dam Solar Park** is one of the largest in the world.

Education, training and health care

- Increased **national revenues** through industrial growth have allowed increased **investment in schools and health care**.
- More children are able to complete schooling, resulting in **higher literacy rates** and more people seeking and gaining **graduate level qualifications**.

When Dom finished his shift he felt like a winner.

Waged work and poverty reduction

- Asia, and particularly China, has seen a rise in the **wealth** of its citizens.
- Greater numbers of opportunities in secure and relatively well paid **waged work** have allowed people to have greater levels of **disposable income**.
- This has created a rise in the number of people in the **middle classes** and fewer people are now living in **poverty**.
- For example, the percentage of people in China living on less than US $2.15 a day dropped from 72% in 1990 to 0.7% in 2017.
- **Remittance payments** have spread this wealth beyond the immediate earner.

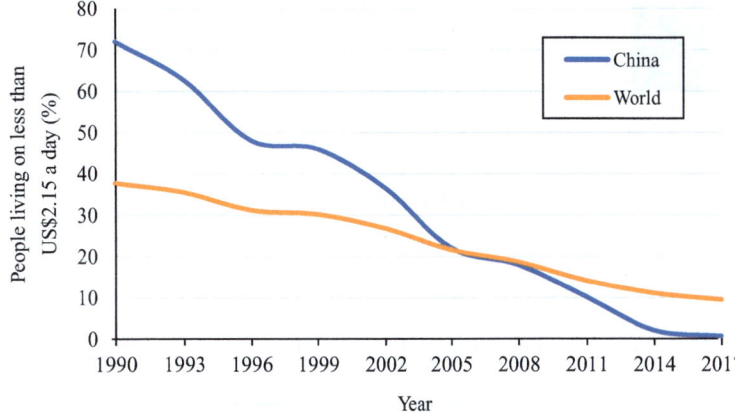

Graph comparing the percentage of people living on less than US$ 2.15 a day in China and worldwide.

Global Winners and Losers

Developing countries have experienced social and environmental problems

Land degradation
- The growing pressure on **land resources** is causing **land degradation**. The demand for industrial space has led to increased **deforestation** and a **loss of productive land**.
- Where agricultural land remains, it's often **farmed more intensively**, which can cause **soil erosion** and **desertification**.

Over-exploitation of resources
- Resources that are required by growing industries such as **minerals, energy and water** are often **over-exploited** and used at an unsustainable rate.
- E.g. China is now looking to countries in Africa for **new supplies** of key resources.

Loss of biodiversity
- With the loss of natural **habitats** comes the loss of **species** and **biodiversity**.
- **Water pollution** is also responsible for the loss of species as pollutants enter and disrupt **food chains**.

Urbanisation
- Widespread **rural-to-urban migration** has led to **rapid urbanisation**, putting pressure on urban infrastructure and resources.
- Limited housing in some cities has led to illegal **unplanned settlements** (or shanty towns) being built.

Dharavi, Mumbai

urbanisation
the growth in the proportion of people living in urban areas

Pollution
- There may be high levels of **air and water pollution**. Reduced environmental monitoring in some countries makes them attractive locations for companies who wish to focus on profits rather than **environmental accountability**.
- In some Chinese cities, such as Guangzhou, air pollution is frequently above healthy levels due to high amounts of **particulates** emitted from factories. In 2008, around 33% of the Yellow River in northern China was considered so polluted with chemicals from **industrial wastewater** that it was unfit to use as drinking water.

Developed countries face different problems

Developed countries may have seen periods of **deindustrialisation** due to the Global Shift:

deindustrialisation
the closure of manufacturing industries due to automation and increased competition from overseas

Unemployment and depopulation
There are likely to be **high levels of unemployment** as industries close. People may **move out** of the city to find work, causing **depopulation**. Residential properties may lose value as demand decreases. This may open areas to lower income groups with reduced economic potential. This can cause a **spiral of decline** (see p.122) as the skills required by new industries are higher than those held by local people, so **unemployment continues**.

Crime
Post-industrial areas are more likely to experience **crime**, increased **drug use** and **anti-social behaviour**.

Dereliction and contamination
Former industrial sites are often **left empty** as no new industries can be found to occupy them. This can lead to high numbers of **derelict buildings** and large parts of cities becoming an **eyesore**. Much of the land that remains in post-industrial areas may be **contaminated** due to poor industrial practices in the past when **regulations** were not as strict.

E.g. following the collapse of the car industry in Detroit in the 1970s, 30% of buildings in the city became vacant. In 2020, around 27% of residential buildings were still vacant.

Derelict factories in Detroit

Warm-Up Questions

Q1 Define the term 'Global Shift'.

Q2 Explain why developed countries may face social and environmental problems as a result of the Global Shift.

PRACTICE QUESTIONS

Exam Question — AS and A-Level

Q1 Assess the impacts of the Global Shift. [12 marks]

Globalisation isn't all good or all bad...

...it's like my vision without glasses on — a bit blurry. Make sure that you know the benefits and problems of globalisation inside out so that you can assess its consequences and form an opinion about them — it will pay dividends in those 12 mark questions.

Migration and Urban Growth

Cities can grow through both migration and natural increase. However, urban growth can cause issues...

Migration is changing as the world becomes more interconnected

1) **Voluntary migration** is the movement of people based on a decision made of their own free will. It's sometimes termed **economic migration** as people often voluntarily choose to move in order to improve their economic status. It's different to **forced migration** where people do not have or do not feel like they have a choice in their movement.

2) As the connections between distant countries have **increased**, the scale and pace of economic migration has **increased**. This migration further strengthens the **interdependence** between countries.

3) Globalisation means that cities have tended to develop rapidly as they are at the centre of economic activity, while rural areas remain more '**switched off**'.

Migration is closely linked to increased urbanisation

1) In 2018, 55% of people lived in urban areas and this is predicted to rise to 68% by 2050. There were 33 **megacities** globally in 2018 and this is likely to increase to 43 by 2030 as more people choose to live in cities.

2) The increasing number, size and location of megacities is closely linked to **trends in migration**:

   - There has been an increase in **rural-to-urban migration**.
   - These migrants tend to be **young** and are more likely to start a family after migrating, adding to the **natural population increase** of the city.
   - Rural-to-urban migration is more evident in **developing and emerging economies** rather than developed ones. This means that future megacity growth is likely to take place in developing economies.

megacity
a city with a population of at least 10 million people

● <u>Megacities by 2030</u>

Nine out of the ten cities predicted to become megacities by 2030 are in developing countries.

3) Rural-to-urban migration can be explained by **push factors** (reasons people want to leave an area) and **pull factors** (reasons people are attracted to a certain area):

Push Factors (from rural areas)	Pull Factors (to urban areas)
• **Lack of suitable employment opportunities** that reflect the increased skill-levels and aspirations of young people. • **Poor working conditions** and dirty, dangerous and difficult employment (the '**three Ds**'). • **The mechanisation of farming** causes unemployment as fewer people are needed to work the land. • **Land reforms** or **land grabbing** activities by agricultural TNCs can leave local people with unproductive land to farm. • **Changing climate conditions** (e.g. floods or droughts) make farming too challenging or cause **land degradation**. • **Short supply of key resources** (e.g. energy or water). • **Natural disasters** or **conflict** which destroy farmland.	• **Increased availability of employment** and **higher average wages** in urban jobs due to increased investment through FDI. • **Wider choice of schools** and more opportunities to train or study beyond the compulsory schooling age. • Larger and better equipped **healthcare facilities** that can deal more easily with specialist health issues. • The '**rural rich**' (e.g. plantation owners) may wish to invest in urban properties for their children while they complete their education.

- **IT communications** and **information sharing** means that prospective rural-to-urban migrants are more connected to and more aware of urban spaces. This can increase the occurrence of '**bright lights syndrome**' — when people perceive urban areas as being full of opportunity and offering a better lifestyle than where they currently live.

- Improvements in **transport** and **communication technology** have made rural-to-urban migration easier by removing **intervening obstacles** (barriers to migration).

- Urban and rural areas are connected in ways other than migration. **Rural peripheral areas** supply **urban core areas** with **resources**. The urban core provides the rural periphery with economic stability through **investment** and **remittance payments**.

Sitting on her chair, Flora could feel the pull of a mega-settee.

Migration and Urban Growth

Increased **urbanisation** creates **social** and **environmental** challenges

- Rapid rural-to-urban migration means that migrants with nowhere to live may build **illegal informal housing**, especially in low-income countries. These are often on **marginal land** which has been ignored by developers because it is too **difficult** or **unsafe** to build on (e.g. areas at risk of flooding).
- Cities can struggle to provide **essential public services** for large numbers of migrants. **School places** may not be available or there might not be enough **medical staff** to provide good levels of health care.
- Authorities may not be able to provide **safe drinking water**, **sanitation** and **power** to increased numbers of people.
- Without enough **formal employment opportunities**, some migrants may turn to work in the **informal economy**, which is unlikely to contribute **taxes** to the local authorities. This informal economy also has fewer healthy and safety regulations as well as lower wages.
- An increased population may mean larger volumes of **untreated sewage** and toxic waste (from factories) enters river systems. **Waste disposal** becomes less manageable and **plastic pollution** is more likely to end up in water courses.
- **Traffic congestion** and **air pollution** are likely to increase and public transport is likely to become **overcrowded**.
- With more people occupying **marginal land**, there could be an **increased flood risk** during high rainfall seasons (such as the monsoon) as there are fewer drainage areas for flood water.

CASE STUDY

- **Mumbai** (in Maharashtra state, India) has witnessed a rapid **rural-to-urban migration**. Between 1950 and 2020, the population increased by more than 560% to 20.4 million people. Migrants continue to come from all over India to seek **employment opportunities** in Mumbai. The main sources of migrants are the states of Uttar Pradesh in the north of the country, Karnataka in the south and Gujarat in the north west.
- Rural-to-urban migration in Mumbai has created one of the largest **slums** in the world. Dharavi, which occupies a 2 km² space in the heart of the city, is estimated to be home to over **1 million people**. **Informal employment**, e.g. plastic recycling, thrives in the slum. There are 15 000 **single-room workshops** which are estimated to create an annual turnover of US $1 billion.
- Conditions within the slum are very poor. **Open defecation** may occur due to limited access to sewage systems. The limited availability of **clean drinking water** means diseases such as **dysentery** are common. As the price of land in Mumbai rises, there is increased pressure on city officials to **develop** Dharavi for wealthier residents.

International migration creates **global hubs** and interdependence

1) In 2020, 281 million people lived **outside their country of birth** worldwide (3.6% of the global population) due to **international migration**. Rural-to-urban migration tends to happen at a **quicker pace** than international migration.
2) While this includes increased numbers of **refugees** and **displaced people**, there has also been an increase in people voluntarily choosing to live in other countries for economic reasons.
3) International migration has turned some cities into **global hubs**. These are cities that are **highly connected globally** and a focal point for global activities. They are often home to the headquarters of **TNCs** and have **large**, **multicultural populations**.

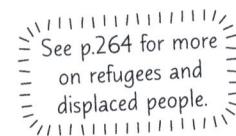 See p.264 for more on refugees and displaced people.

4) Two examples of global hubs are **London** and **Qatar**:

 **1 London**

CASE STUDY

London has experienced notable '**elite migration**'. Many of these migrants would be seen as '**global citizens**'. Elite migrants may aim to make use of the **tax breaks** and **investment opportunities** certain countries offer. The migration of 'elites' to global hubs can **spread wealth** as they are likely to employ many people in service jobs to manage their various interests and properties.

elite migration
the movement of highly skilled, influential people (usually into a city)

global citizens
people who, due to their status and skills, experience few obstacles to living and working wherever they choose

Russian oligarchs moving to central London

Wealthy Russian oligarchs have **bought properties** and **invested in businesses** in London, which benefits the UK by contributing to the economy. However, in 2022 around **100 properties** in London were owned by Russians believed to be involved in corruption. Collectively, these properties are worth more than **£1.1 billion**. Owning UK property gave Russian elites access to **UK bank accounts** and **private schools**, which are seen as highly desirable. These investments, among other factors, have led to **price inflation** and made living in some parts of London (e.g. Belgravia) **unaffordable** to most people born in the city. **Sanctions** placed on oligarchs following Russia's invasion of Ukraine in 2022 have aimed to **lessen their influence** in London.

Migration and Urban Growth

2 Qatar

CASE STUDY

Qatar has experienced '**mass low-wage economic migration**' which tends to happen mainly in **emerging global hubs**. Most of this migration takes the form of formal employment, but there is growing concern that it also encourages **illegal workers** to migrate.

mass low-wage economic migration the movement of low-paid workers to global hubs in order to fill particular employment quotas, especially in cities or countries that have low population levels and are unable to fill those positions themselves

Mass low wage economic migration of Indians to Qatar

- In Qatar, Indian migrant labour has been used to build **tourism and recreation facilities**, **infrastructure projects** and the **stadia** required to host the 2022 Football World Cup. Migrants are also employed as domestic servants.
- There are up to 700 000 Indian people working in Qatar. Workers often work for long hours in hot and difficult conditions. There's been a high **death rate** among workers building World Cup facilities. Workers often send their wages home as **remittance payments**.

International migration has implications for both host and source countries

Migration has economic, social, political and environmental **benefits** and **costs** for both host and source countries:

	BENEFITS	COSTS
HOST COUNTRY	• **Skill shortages** get filled in certain employment sectors. • Jobs that local people are **unwilling to do** can be filled. • Migrants generally spend their wages locally, adding to the **local economy**. • Migrants help to restore the balance of an **ageing** and **dependent population**. • Stronger **political and social understanding** exists between countries. • Migrants are more likely to become **entrepreneurs** and future job creators. • New cultures can change the **identity of a place** and open up **new experiences** to local people.	• **Social tensions** due to housing and/or job shortages. • **Language barriers** can prevent migrants fully integrating. Certain areas can be associated with migrant communities (**ghettoisation**) which prevents integration. • Migrants add **pressure** to the **education** and **healthcare** systems, and pressure to provide housing may lead to building on **greenbelt land**. • New cultures can change the **identity of a place** and traditional ways of life may be lost. • Public services, e.g. police and health services, may have to invest in **translation facilities**. • Greater levels of **pollution** and **waste** produced.
SOURCE COUNTRY	• **Remittance payments** can form a significant part of a country's GDP. • Authorities can spend less on **public service provision** and there is **less pressure on resources** (such as water and farmland) as population growth declines. • Migrants may return home with **new skills and qualifications**. • **Employment pressures** are eased and people can more easily find work. • Stronger **political and social understanding** exists between countries.	• People with high-level skills are more likely to migrate (known as the **'brain drain' effect**). The youngest and fittest are most likely to leave, creating a gap in the labour market for manual labourers (the **'brawn drain' effect**). • A reduced need for '**youth services**', e.g. universities, may force them to close. • Families can be split, causing **poor social stability** for children. • **Population imbalance** due to certain groups (children and the elderly) being left behind. These groups tend to have greater **healthcare** needs, adding pressure on the services. • Funding may be cut to **schools and colleges** as the birth rate declines and demand for school places goes down.

Warm-Up Questions

Q1 Explain why the global elite may be attracted to certain cities.

Q2 Describe the challenges that can come with increased rural-to-urban migration.

PRACTICE QUESTIONS

Exam Question — AS and A-Level

Q1 Assess the extent to which international migration has negative impacts on the source country. [12 marks]

I'm feeling the 'brain drain' effect from these pages...

Time to get up, stretch your legs, do 10 star jumps and boil the kettle. Then reward yourself by drinking a nice cup of tea. Ahhh.

Global Culture

As globalisation has gained strength, local and national cultures have changed and adapted to its influence.

There are **different types of culture** in a globalised world

1) A **culture** is the **shared customs of a group**. It's likely to evolve through the influence of different events and processes.

2) A culture is made up of:

- traditions
- behaviours and values
- religion and beliefs

- languages
- clothes
- music

- art and symbols
- accepted norms
- dance

- shared history
- architecture
- aspirations

3) Some people believe that globalisation causes **homogeneity** in cultures. A **homogeneous culture** is one where everyone follows the same set of cultural norms and there is a distinct lack of **cultural diversity**. **Cultural pluralism** and **hybridism** describe cultural states.

cultural pluralism
a state in which more than one culture coexists and each culture retains its defining characteristics

cultural hybridism
a state in which more than one culture has mixed together to form an entirely new culture

Different factors have caused **cultural diffusion**

1) Globalisation is thought to have created a more **westernised global culture**, based on **western norms and values**. This is because western countries can be seen to be the **drivers of globalisation** and so have spread their own version of culture more widely.

2) **Cultural diffusion** refers to the way in which western culture has spread around the world, e.g. the economies of countries being built on **capitalism** and a strong belief in the power of **democracy**. Globalisation has **accelerated the rate** of cultural diffusion. Where a person lives may no longer be the main source of their cultural identity, as different people can **adopt different cultures** at different times.

3) Cultures might **change** through:

Force — the use of **hard power** (using military might) to impose a change of culture on people.

Coercion — the use of **soft power** (using education and political persuasion) to pressure people into adopting new cultures.

Free will — people **choose** cultures that they find attractive and that suit them

4) Some of the players in cultural diffusion include:

The spread of US culture is called "Americanisation".

TNCs

- The largest **TNCs** (in terms of wealth and global spread) tend to come from the USA and Western Europe. This means many of these companies, sometimes unintentionally, spread **western values** through the distribution of their goods, services and working practices.

- Generally, products are **uniformly manufactured** regardless of the market they are targeting, though some TNCs will modify their goods to suit local tastes and needs.

This is known as 'glocalisation' — see p.93 for more.

E.g. Coca-Cola® is a global product that is uniformly manufactured but packaged to suit different markets.

© ton koene / Alamy Stock Photo

Media Corporations

- **Global media corporations** produce visual, audio and online content that is quickly and easily spread around the world. There is a wide geographical market for this content which means that it is easy to distribute cultural or political viewpoints to a large number of people.

- It is easy for an audience to believe there is a **dominant culture** if that is all they see in the media. In these cases, they change their cultural outlook to match what they experience through media. A small number of corporations, most of whom are from **developed** countries, produce the majority of mainstream media content — this could lead to the promotion of a **more homogeneous culture**.

The same media corporation produces many magazine titles.

Global Culture

Tourists

- **Tourism and hospitality** are a large part of the economy of many nations. Resorts compete to attract tourists and use culture as a selling point.
- Many tourist destinations have chosen to occupy a **cultural middle ground** — they embrace western cultural traits in order to seem familiar to **western tourists**. E.g. western food may be used in restaurants and the built environment may be designed to mimic facilities found in other destinations. Tourists themselves may also play a role in cultural diffusion as their behaviour can influence local people.

Souvenir shops may try to 'sell' local culture to tourists.

Migrants

- As **migrants** move within and between countries, they take their cultural traits with them. Many migrants will adapt to and adopt local cultures once they settle in a new location. This may be helped by marrying into new cultures or forming relationships with different people. Other migrants will live in a unique **hybrid of cultures** that embraces both their home and host cultures.
- Where the immigration of large numbers of people from one culture occurs, the host culture might alter to suit the **needs of migrants**. Migration allows people with differing viewpoints to come into contact with one another and **diverse communities** can become places of extensive **cultural influence**.

Global culture can have a negative impact

1) The movement towards a global culture based on western values represents a move towards greater levels of **consumption**.
2) This may have a negative impact on the **environment** through **unsustainable resource use** and **ecosystem degradation.**
3) A global western culture is likely to promote the continuation of global TNCs. This may increase **air and water pollution** if production methods are **poorly regulated** and it increases the need for **rapid transportation** of goods around the world.

Asia's changing diet

- Asian diets traditionally consist of **high levels of vegetables** and fish and low levels of meat. The spread of western culture (centred around more **processed** and **fast food**), along with a steady rise in the number of people in middle-and-high-income groups, has seen Asian people consume larger amounts of **red meat** and **sugar**. This has led to increased levels of **obesity**, **diabetes** and **cardiovascular diseases**. In 1990, 0.9% of adults in South East Asia were considered obese compared to 4.3% in 2015.
- **Agricultural land** in Asia is increasingly being used for **raising livestock** rather than growing crops for human consumption — this is less efficient for the amount of food produced. This change in land use has also accelerated **carbon dioxide and methane emissions** and contributed to a changing global climate.

CASE STUDY

Global culture can have a positive impact

1) Some people believe globalisation creates the wealth and the means to **understand**, **conserve** and **protect** the environment, which can lead to a **greater understanding** of **sustainable development**.
2) A global western culture also may provide opportunities for the voices of **marginalised people** to be heard. This can create a **more diverse** society and encourages marginalised people to take a prominent role within their society.
3) This has been reflected in the law. Many governments have passed acts to ensure equality is achieved — e.g. the **1995 Disability Discrimination Act** in the UK made it illegal to discriminate against anyone for their disabilities.

2016 Paralympic Games in Rio de Janeiro

- The Paralympic Games is an example of how a global culture can help bring **equality** to **marginalised people**, and particularly those from developed and emerging economies. Since the 1980s, participants and followers of parasports have increased. The 2020 Games had the highest number of parasports and athletes to date (more than 4400 participants compared to just over 3000 in 1988).
- Increased **media coverage** of the Games has led to some athletes becoming household names — this helps to draw more attention to the Games and will **inspire** more people to take part. Although some countries lag behind others in removing the **stigmas** around disability, the Paralympic Games are seen as a catalyst for change.

CASE STUDY

Athletes competing in the Paralympic Games in Rio

Global Culture

Globalisation has led to cultural erosion

1) Some cultures may be **vulnerable to cultural erosion** due to the emergence of a more westernised global culture.

2) Different countries have **different levels** of vulnerability to cultural erosion. Some sites, languages and traditions are actively protected. For example, **UNESCO World Heritage Sites** are designated sites that are placed under **international protection** due to their cultural significance.

> **cultural erosion**
> *the loss or weakening of key parts of a specific culture*

Language

- **English** is the primary business language with an estimated 1.5 billion speakers worldwide. Films, songs and published materials tend to be either written in English or translated into English.
- **Translated text and speech** can lose its nuance, so the precise meanings of ideas cannot always be conveyed to the audience — this can mean that certain aspects of a culture become lost in translation.
- There are estimated to be over **7100 different languages** in the world — 40% are thought to be under threat as so few people use them regularly.

Clothes

- Traditional clothing in some countries is being replaced with more **western styles and fashions** due to increased levels of tourism and migration.
- Denim jeans and T-shirts are universal staples in most wardrobes around the world. Women increasingly wear **trousers** in cultures that have **traditionally favoured** skirts and dresses, and it is not uncommon to see people wearing English league football shirts in countries far away from the UK.

Music

- Increased **IT communication** and **shared knowledge** has led to the spread of different music cultures.
- **Pop music** remains one of the most easily marketable genres. Its symbolism often appeals to the aspirations of young people regardless of their background.
- Television shows such as *The X Factor* and *The Voice* have been franchised to many different countries globally.
- Some people argue this has **homogenised popular music**, with winning acts appearing to have the same style of music despite coming from different countries. This suggests that traditional, **national music genres** are continuing to lose popularity.

The Voice Afrique Francophone is a French-speaking African version of The Voice.

Food

- Globalisation has created a **global marketplace** for all kinds of food. Some ingredients would normally only be available in one part of the world and would form part of **endemic** dishes (those that only originate from that place). Now those foods can be eaten by anyone worldwide.
- Equally, dishes that would traditionally be eaten by certain cultural groups are often changed to make them **more palatable** to a global audience. E.g. chicken tikka masala is thought to have been designed for British palates that found the traditional Indian chicken tikka too dry.

A traditional chicken tikka kebab

Social relations — Papua New Guinea

CASE STUDY

- In **Papua New Guinea**, there were originally thousands of cultural groups who practised individual languages and traditions. With successive waves of **colonialism** by British, German and Australian regimes, the native tribal system was broken up.
- **Missionaries** from different Christian denominations worked in the country to bring different tribes together under new religious **laws and customs**. These aimed to suppress behaviours that were seen as **animalistic** (such as cannibalism for ritualistic purposes). They also promoted living in **nuclear families** (a father, a mother and children) rather than other alternatives like sex-segregated housing.
- Today, there are thought to be only 312 tribal groups remaining as **intermarriage** and **migration** have caused some tribal extinction.

A 'sing-sing' involves the gathering of different Papuan tribes in traditional dress.

Global Culture

Global western culture has changed built and natural environments

- Traditional **architectural styles** are often lost as cities seek to appear modern to prospective industries from western economies who may wish to invest there.
- In order to attract tourists, destinations often build **facilities** to cater to western tastes and requirements.
- The **pursuit of wealth** and industrial strength is often **prioritised** ahead of the needs of the natural environment, both locally and globally.
- The **exploitation of natural resources** is often seen as a **necessary evil** in order to sustain a global western culture.

Looking up at the skyline of London's financial district

Some groups oppose the spread of globalisation

1) There is **concern** amongst some groups that globalisation and the spread of a global western culture is now an **unstoppable force**. They see the negative aspects of global culture as ever increasing **threats** to human life.

2) Others believe the **positives** of a global culture **outweigh the negatives** and so actively work to promote globalisation.

3) There are different schools of thought in the **debate**:

Hyperglobalists	Transformationalists	Global Sceptics
• Hyperglobalists see globalisation as a somewhat **inevitable and unstoppable** process. • They recognise that globalisation has **good and bad** impacts and accept that they will happen in a homogeneous global culture.	• Transformationalists see globalisation as the creator of **changing cultural ideas**. • The globalised world they want will be highly **interconnected** but with new cultural forms as they **change** and **adapt** to new world circumstances.	• Global Sceptics believe that there will be **no such thing** as one truly global culture. • They see the processes of globalisation as having an **unequal influence** — a 'core' will experience hyperglobalism while a 'periphery' of countries will continue to be marginalised.

4) **Anti-globalisation movements** have become more common as people start to feel the negative effects of globalisation.

5) Some environmental protection movements also have **anti-globalisation** and **anti-capitalist** language in their campaigns.

> E.g. **Extinction Rebellion** advocate immediate action on climate change and also support more equitable governance.

6) Other movements advocate the strengths of **nationhood** and **regionalism** over globalisation.

> E.g. in the run up to the UK leaving the European Union, the UK Independence Party (**UKIP**), a political party, campaigned for the UK to become more independent of larger systems of government.

Katie is a hyperglobalist — she likes to be around as many globes as possible

Warm-Up Questions

Q1 Define 'cultural erosion' and give some real world examples.

Q2 Explain why different groups have different attitudes to globalisation.

Exam Question — AS and A-Level

Q1 Assess the view that the spread of a western global culture has more costs than benefits. [12 marks]

Cultured, adj. — eating Yorkshire puddings when visiting Japan...

Interestingly, Yorkshire puddings are effectively made out of the same basic batter that is used to make Russian blinis, French crepes or American pancakes and waffles. I'm not entirely sure if it's relevant to globalisation, but it is intriguing. Hmmmm...

Globalisation and Development

Globalisation creates significant opportunities for wealth creation in some countries whilst highlighting a development gap.

Globalisation is closely **linked** to **development**

1) **Development** is economic advancement in a country or region that improves people's **quality of life**.

2) Development and globalisation are thought to have an **interdependent relationship**. The more globalised a society, the more likely it is to be highly developed.

3) The difference between the richest and poorest is called the **development gap**. Studying changes in the development gap helps to reveal the impact globalisation has had in different parts of the world.

4) Development can be **measured** using different forms of **indicators**:

- **Single indicators** measure just **one aspect** of development. They allow for a precise comparison of different countries in particular contexts.

- **Composite indicators** combine data from **multiple indicators** to give a single score. They give a broader picture of development in a particular place.

5) Indicators of development tend to be **quantitative** to allow for countries to get **compared** with each other and **ranked**.

Pierre has given several
indicators that he's single

There are **different types** of development

Economic development is measured using the financial and employment **statistics** of a country

- Countries that are **more economically developed** are likely to create and experience greater levels of globalisation. Developed countries are more likely to host the headquarters of **TNCs**. They are also more likely to have the capacity to invest in **IT and communication technology** that will allow their citizens to connect with people in other countries.

- It can be difficult to measure a country's level of economic development if a large percentage of its population are employed in the **informal sector**. Some of the **most frequently used indicators** of economic development are:

GDP per capita (US$) (single)	GNI per capita (US$) (single)	Economic Sector Balance (composite)
The total value of **goods and services** produced within the borders of a country in that year, divided by the number of people in the population.	The total value of goods and services produced by a country, including **overseas investments**, divided by the number of people in the population.	The ratio of population employed in **primary**, **secondary**, **tertiary** and **quaternary industries**.

Measuring **social development** tends to use a **wider range** of indicators

- It is more common to find **composite indicators** being used to measure social development because there are a **number of factors** that affect the **level of social development** of a country.

- Countries that are more socially developed may have populations who have embraced some of the benefits of globalisation. E.g. the spread of **gender equality** through social and political movements has been made more possible through the **knowledge sharing** processes of globalisation. Some **indicators** of social development are:

Human Development Index (composite)	Gender Inequality Index (composite)
Designed by the UN Development Programme, this index aims to measure overall development by combining indicators of **life expectancy**, time spent in **education**, **literacy rate** and **GDP** into a single score. By including GDP, it shows how wealth affects people's lives. It is measured on a scale from **0 to 1** (1 being the **most developed**).	This is a measure of the loss to society due to **gender inequality**. It combines **maternal mortality rate**, **adolescent birth rate**, females with **secondary education**, **female participation** in the labour force and **female representation** in parliamentary seats. It's measured on a scale from **0 to 1** (1 being the **most unequal**).

Measuring **environmental development** is difficult

- Countries tend to measure different variables depending on the type of **natural habitats** found in their country — this makes it **difficult to compare** environmental development globally.

- **Low scores** are desirable in **environmental indicators** as they show that the **negative impact** on the environment has been minimal. E.g. scores closer to 1 on the **Air Quality Index** show that there are **low levels of pollution** in the air while scores closer to 10 show that there are **high levels of pollution** in the air.

Globalisation and Development

- Some of the indicators of environmental development are shown below:

Air Quality Index (composite)	This index measures different types of **air pollution data** (e.g. particulates or nitrogen dioxide emissions) and combines them to create a single score. The index lacks the comprehensiveness of other indexes due to a lack of recordings in rural areas.

Ecological Footprints (composite)	These combine data to show how a population uses certain resources (such as **domestic water usage** and the **number of cars per household**). These can be used to calculate the number of planets it would take to sustain the world's population if everyone lived like the people of that country.

Globalisation has created **economic winners and losers**

- Globalisation has created greater levels of **inequality** within countries and comes in different forms. People may have different access to new opportunities, differing levels of income or have varying levels of **resilience**.

- There are increasing numbers of **billionaires**, many of whom control large **TNCs** that thrive due to globalisation, but there are still large numbers of people living in poverty. **Deindustrialisation** in developed economies has created large numbers of economic 'losers' and those who live in the least developed countries remain '**switched off**' to globalisation trends and its advantages.

> The **Gini Coefficient** is one way of measuring relative inequality within a country or region. This is a score given to a country or region between 0 (0% inequality) and 1 (100% inequality) based on the distribution of people with different levels of income. They can be used to produce **Lorenz Curves** (see p.301) which can be used to compare inequality in different countries or regions. Generally, Gini Coefficients have become larger over time, indicating that the amount of **inequality** within countries is generally increasing.

Globalisation has created **environmental winners and losers**

- The ability of different countries to **manage environmental issues** is largely unequal. Some issues, such as **climate change**, have received greater attention due to globalisation while others, such as **micro-habitat preservation**, have not.

- Many **rural dwellers** are seen as 'losers' in globalisation. These places are largely '**switched off**' from globalisation processes and suffer from **increased land degradation**

- Globalisation has led to some cities experiencing **rapid urbanisation**. This has led to environmental problems such as **air pollution** and the spatial growth of cities (**urban sprawl**) into natural areas.

See p.98 for the problems caused by rapid urbanisation.

Different global regions have made **different levels of progress**

- All regions of the world have experienced economic development since 1970. Growth in **Asia** and in the **emerging economies** has been very rapid compared to other regions. E.g. South Asia has seen a 549% increase in GDP between 2000 and 2021 (compared to 149% in Europe and Central Asia over the same period). This indicates that globalisation has had more of an impact in some regions than others.

- Policies in these economies have meant that many rich people have been able to get even richer very quickly. However, in some developed countries and regions, increases in wealth are felt more universally due to mechanisms that **redistribute wealth** (such as higher taxation on the richest and welfare payments to the poorest).

- This means the **development gap between countries** has got smaller, while the **development gap within countries** has got larger. E.g. in China in 2015, the top 10% of earners earned 42% of the wealth, while the bottom 50% around 14%.

- The amount of **absolute poverty** (poverty caused by a lack of **basic human needs** being met) globally has **decreased** while the amount of **relative poverty** (poverty that's below the average **standard of living**) globally has **increased**.

Warm-Up Questions

Q1 Describe how the development gap is different within and between countries.

Q2 Explain how developing countries might experience globalisation differently to developed countries.

Exam Question — AS and A-Level

Q1 Explain how indicators of development may provide different understandings of a country's level of development. [6 marks]

The Gini Coefficient — no magical lamp required...

It's the same old story — globalisation has costs _and_ benefits. Except now you also need to consider development. Lucky you.

Attitudes to Globalisation

Globalisation has created different layers of tension over socio-political and environmental issues. Different groups experience these tensions and manage them in different ways.

Globalisation has caused societies to become **culturally mixed**

1) Globalisation has seen the spread of cultures through processes such as **migration** and certain economic policies. Societies often welcome these changes as they make sizeable contributions to a **local or national economy**.

> **diaspora**
> *the large-scale spread of people away from their homeland*

2) There has been a greater recognition of **diasporas** due to globalisation. E.g. in 2021, around 31.5 million people claimed to have Irish ancestry in the USA. This contributes to a widespread celebration of Irish culture in cities like New York for festival events such as St Patrick's Day.

3) Part of diaspora is the clustering of migrant groups in certain areas, known as **ghettoisation**. E.g. Boston in Lincolnshire had one of the highest populations of residents born in the EU (around 12%) in the 2011 census.

4) Large numbers of migrants from a certain culture may start to change the **built environment** around them. E.g. there are English pubs and supermarkets in Spain to suit the needs of British expats.

Certain **policies** have **increased the speed** of cultural mixing

Open Borders
Agreements such as the European Union's Schengen Agreement gave EU workers freedom of movement. This increased the spread of different people and their cultures around Europe and a more competitive job market in many countries. This was particularly true after 2004 when the 'Ascension Eight' countries, including Poland and Hungary, joined the EU. Migrants from these Eastern European countries tended to migrate to Western European economies in search of better economic opportunities.

Deregulation
This is the removal of restrictions by the government of a country. Deregulation can affect how companies operate at home or overseas. For example, it may encourage foreign companies to invest in and set up branch offices in the less regulated country, or remove barriers that would otherwise prevent a home-grown company from employing foreign labour. This allows industry to recruit from a wider pool of talent and may increase competition for jobs. FDI is also encouraged to flow more easily into a country where there are fewer regulations in place.

Globalisation has created **social, political** and **environmental tensions**

1) Tension stems from people experiencing or perceiving **injustices**. E.g. some people would like governments to place **stricter controls** on the level and speed of international migration due to concerns about a perceived link between migration and:

- pressures on housing stock.
- a shortage of school places.
- increased competition for jobs in some sectors.
- a loss of a sense of national identity.

2) These tensions sometimes mean international migrants are subjected to **harassment** and **workplace discrimination**.

3) **Political decisions** and reactions to **environmental issues** can also create tension. E.g. some people believe there has been a lack of meaningful progress at regulating big industries' contribution to **climate change**.

4) These tensions have **strengthened the participation** of people in environmental debates and protests. An estimated 6 million people took part in the global Climate Strike in September 2019 to coincide with the **UN Climate Summit**.

The rise of nationalism in Europe

- Nationalist political parties such as the *National Rally* (formerly the *National Front*) in France and the *Freedom Party of Austria* have become more popular in Europe. They have been making gains in elections and are increasingly obtaining representation in national parliaments. These parties favour nationalism over globalism and, in some cases, focus on national independence. They reject the idea that multiculturalism can bring positive outcomes to a country, and call for a drastic reduction in immigration. These parties gain much of their support from people in lower-income groups who feel they are increasingly being oppressed and marginalised due to international migration.

- There has also been a rise in more extreme groups that support neo-fascism — an authoritarian, xenophobic movement which promotes ideas of racial supremacy. Some believe the campaign messages of nationalist and neo-fascist parties are responsible for the rise in reported racial and religious hate crime over recent years.

Marine Le Pen, leader of the National Rally

Attitudes to Globalisation

Transboundary water conflicts in southeast Asia

CASE STUDY

- With rising demands on **limited water supplies**, tension can arise between countries that share a **transboundary** water resource (one that crosses an international border). The **Mekong River** runs through six countries — China, Myanmar, Thailand, Laos, Vietnam and Cambodia, impacting over 300 million people. As of 2020, those countries (with the exception of Myanmar) have commissioned plans to build over **200 dams** on the river and its tributaries for **irrigation**, **flood control** and the **production of hydroelectricity**. Cambodia and Vietnam, which host the lower course of the river, are concerned as they're likely to receive much lower **flow rates** if the planned dams go ahead.

- A set of international guidelines known as the '**Helsinki Rules**' can help to manage these tensions. Though not enforced in law, these rules state that all countries that border a water resource should have an equal share in it and have an equal responsibility to protect it from **pollution**. All structural changes, such as dams, need to be **agreed by all countries** before they go ahead. The 1995 Mekong Agreement, based on these principles, was signed by all countries except China.

There are different ways to **control the spread** of globalisation

Some people believe that globalisation has **shifted power** from state level to a global level.
This means national states have sometimes **taken action** to control the pace and extent of globalisation in their country:

Trade protectionism

Trade protectionism is the process by which trade between countries is purposefully **limited**, sometimes through **trade sanctions** or penalties. This may be achieved by using **higher tariffs** on goods, regulations on the type and number of foreign companies able to operate in a country or through **quotas** for the maximum quantity of goods that can be traded. **Domestic goods** may be sold on the international market at more competitive prices because the state is able to **subsidise** their production.

Resource nationalism

Resource nationalism seeks to give priority access to **domestic users** of a resource. E.g. in Peru, industries have priority access to the country's copper reserves over international mineral companies. In extreme cases, this may create a **trade embargo** between countries where no trade takes place at all. These often take place on **political grounds**. E.g. in 2017, Canada placed sanctions on Venezuelan goods and services as a response to the Venezuelan government violating the human rights of protestors in the country.

Censorship

Censorship is when a government limits rights to free speech, the access of information and the spreading of ideas. Many countries limit the **freedoms of their press** in an attempt to reduce the impact of globalisation. The 2019 **Press Freedom Index** ranked the press freedom in 72 out of 180 countries 'difficult' or 'very serious'.

Censorship in China

CASE STUDY

- China was ranked 177th due to its **strict censorship** of media content. Chinese authorities have used censorship to control the population, **promote the Communist state** and **limit global western influences**. Journalists can be labelled as **dissidents** and imprisoned for promoting any anti-communist message. Most content from overseas is heavily monitored and much of the mass media produced within China is **state-owned**.

- The internet proved to be a challenge for Chinese censors. A strict 'firewall' is in place which blocks a large volume of **international web content**, though many people, especially the younger generations, have found ways to get around this. Google withdrew its services from China in 2010 after the state repeatedly blocked users from viewing content that Google felt was in the best interests of the Chinese people.

Limiting migration

Different national governments might operate a range of **migration policies**:

- Some countries (e.g. the USA) operate a **quota system** where only a certain number of entry visas are granted annually.
- Some countries have **stipulations** about who can apply for entry — language proficiency, particular skills or a guaranteed employment contract may be needed. Some countries (e.g. Australia and the UK) combine these to form a **minimum points score** a person must achieve if they are to enter the country.
- Other countries (e.g. Uruguay and Ecuador) have **almost no migration policy** and welcome migrants providing they have a small nominal value in their bank account.

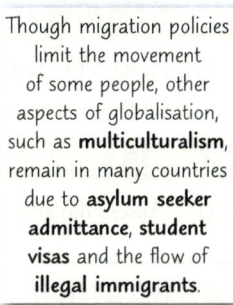

Though migration policies limit the movement of some people, other aspects of globalisation, such as **multiculturalism**, remain in many countries due to **asylum seeker admittance**, **student visas** and the flow of **illegal immigrants**.

Attitudes to Globalisation

Migration policies in Japan

1) In the past, Japan has enforced one of the strictest **immigration policies** in the world. Until 2019, Japanese companies:
   - could not employ **foreign manual labourers**
   - required very particular and **high-level qualifications** for professional labourers
   - could not be owned by non-Japanese (and one could not migrate to Japan to become an **entrepreneur**).
2) The criteria for **asylum applications** was very strict, e.g. in 2016, only 28 out of 10 901 applications were approved.
3) The Japanese attitude to migrants has not been hostile but instead places Japanese nationals first. The Japanese concept of *minzoku* (one ethnicity) remains a strong part of Japanese culture and migrants to Japan, even from a young age, may never be considered truly Japanese.
4) In 2019, the Japanese **Immigration Control and Refugee Act** began to soften some restrictions and admit foreign manual labourers for up to a maximum of five years each. Measures have been put in place to speed up asylum applications and change the approval criteria. This was largely in response to the realisation that Japan's **ageing population** was unable to fill gaps in the labour market, so **younger workers from overseas** were needed.

Some groups are working to **resist globalisation**

- Globalisation has led to the **privatisation of natural resources**, often by TNCs. Movements against globalisation are often movements against **TNCs** too as countries struggle to return the **balance of economic power** to the state. TNCs often **support globalisation** as its mechanisms allow their businesses to grow and expand into other countries.

- Some people have formed strong **pressure groups** to resist the actions of TNCs in order to protect the cultures, habitats and the natural resources found in their **ancestral lands**.

Ogoni Tribe

In 1990, the **Ogoni tribe** that occupy the oil-rich Niger Delta formed MOSOP (Movement for the Survival of the Ogoni People) to campaign for **social justice** for the people who had been **removed** from the land. They also sought to **protect the Niger Delta** from the destructive practices of well drilling.

The First Nations in Canada

1) Some groups of indigenous people wish to reject the manner in which TNCs control their ways of life and their land. At the same time, these groups may seek to take **economic advantage** of the resources found in their ancestral lands in a **sustainable** and **self-governed** way.
2) The First Nations consist of a collective of more than 630 **indigenous groups** working to self-govern and regain control of their ancestral land across Canada. They are **opposed** to the large oil, gas and timber companies looking to capitalise on the resources found there. They are concerned that:
   - the **exploitation** of resources will threaten **traditional ways of life** (e.g. the artisan salmon fishing methods of the Coast Salish people) and destroy vulnerable natural habitats — much of which are part of **national reserves**.
   - oil and gas workers may bring **new behaviour** (e.g. drinking alcohol) into **indigenous communities** that is not beneficial.

What do the First Nations do?

- **protest**, raise awareness and be a strong voice in public meetings
- set up **companies** on their land to provide **back-office** and **consultation** services for large oil, gas and timber companies in order to steer the way that the land is used and the way it is protected
- set up **schools** and **community centres** to ensure that **languages** and **traditional skills** and cultures are kept alive
- **teach tourists** about different cultures
- run **festivals** to celebrate the connection different groups have with their land

Warm-Up Questions

Q1 Explain the link between globalisation and the recent rise of nationalism in Europe.

Q2 Outline why some states may try to resist the spread of globalisation.

Exam Question — AS and A-Level

Q1 Assess the attempts that have been made to control the spread of globalisation. [12 marks]

##################################### *

This gag has been censored for expressing a criticism of A-Level Geography.

Make sure you know real world examples for these attitudes — "some people don't like globalisation" won't impress examiners.

Responses to Globalisation

A variety of different groups have reacted to the rise of globalisation by thinking more locally about consumption.

Rising concerns about **sustainability** can lead to **increased localism**

1) **Sustainability** is managing the ways we **meet the needs of today** without compromising the ability of **future generations** to meet their own needs. It is comprised of economic, environmental and social sustainability.

2) Increasing numbers of people are **worried** about the degree of sustainability that is possible through **globalisation**:

Ethical concerns

- A growing **development gap** between the world's richest consumers and its poorest producers of goods.
- Poor health is seen in communities dealing **with industrial pollution** from manufacturing businesses.
- **Exploitation of workers** in primary industries (in particular mining) as pressure to meet demand rises.
- Developing countries may sacrifice their own **food needs** in order to meet those of developed nations.

Environmental concerns

- The constant desire for newly manufactured goods creates industries that contribute to **climate change**.
- More people buy goods they don't need due to reduced production costs, increasing **wastefulness**.
- High levels of **resource exploitation** occur to create new goods, causing **habitat and biodiversity loss**.
- Increasing pressure on **water and energy supplies**.

3) The **ecological footprints** of developed countries tend to be higher than developing countries. The high demand for goods creates a need to use resources in large and unsustainable quantities. This means citizens of many developed countries are living far beyond their means.

4) There is a fear that as **levels of consumption rise**, people will become **disconnected from production processes** and lose understanding of the labour force required to produce goods and the resources that are often exploited.

5) **Localism** prioritises **local production and consumption** of goods over the **global market**. Localism also embraces **local cultures** and promotes local identities.

Ecological footprint (no. of Earths equivalent to) of the 5 wealthiest and poorest countries (by GDP per capita in 2018)

Wealthiest		Poorest	
Luxembourg	8.18	Burundi	0.53
Switzerland	2.75	Malawi	0.56
Norway	3.58	Central African Rep.	0.76
Ireland	3.32	Somalia	0.64
Singapore	3.75	Madagascar	0.60

Different groups have been involved in the **promotion of localism**

1) The demand for **greater consumer choice** means that it is normal to see **unseasonal food** in supermarkets. These products often have high food miles and generate more **carbon dioxide** emissions which contribute to **climate change** (particularly if transported by air).

2) Groups of consumers may **boycott** goods that have high-food miles, choosing instead to buy those that are **produced locally** — this is known as **local sourcing**. Supermarkets are increasingly responding to this by sourcing and marketing local food.

3) **Non-government organisations** such as WRAP (Waste and Resources Action Programme) encourage businesses to source local food as well as run campaigns (such as 'Love Food Hate Waste') to support more sustainable use of food resources.

food miles
the distance food travels on its journey from producer to consumer

local sourcing
the active efforts by communities to buy locally produced food and goods that have fewer food miles

Costs of localism

- Locally sourced goods can be **more expensive**. Local producers may not be able to utilise large **economies of scale** by having access to a large market. Some producers will create **artisan products** which increase the price too.
- Poorer nations economies may rely on exporting food. They profit from the **'throwaway culture'** of developed countries. These poorer nations may **lose income** if more people in developed countries buy locally produced goods.
- As demand for local goods increases, farmers may be encouraged to use environmentally **unsustainable farming** methods (such as **mechanisation**) so they can farm more efficiently, increase production and meet demand.
- Not all products can be bought locally.

Benefits of localism

- It allows some food producers to **diversify** into **food processing**, (e.g. producing bread from their flour) which increases their profits on certain foods.
- Local producers are more likely to use **organic production methods** which are better for soil and **ecosystem** health.
- **Profits** from selling goods stay local to the area in which they are produced, strengthening the **local economy**.
- Local food needs **less chemical processing** to remain fresh while in transit to markets.
- Producers can more easily **adapt** to changing levels of **demand** and may be able to work with greater **flexibility** in changing markets.

Responses to Globalisation

Transition towns — Totnes, Devon

The **Transition Town Movement** aims to create more sustainable settlements through strategies that put the local community, environment and economy ahead of all else. In 2006, Totnes became the world's first transition town:

- A strong drive to **reduce, reuse and recycle** waste produced in the town.
- The creation of a **local currency** — the Totnes Pound — which ensured that profits from local produce stayed local (though this ceased to be traded in 2019 as more people moved to cashless transactions).
- The promotion of **local produce** throughout supermarkets in the town and the 'Food-Link Project'.
- The creation of **community orchards and growing spaces** to encourage everyone to grow their own food.
- Regular **farmers' markets** that celebrated local specialities and showcased local producers.
- **Local transport network** improvements to encourage people to work within a short distance of where they lived.
- The creation of **industry waste networks**, where waste from one industry could be used by another.

Ethical consumption can improve people's lives and their environment

Ethical consumption is when consumers consider the impact the money they spend will have on people, animals and the environment.

- As more people become concerned about consumption, more companies promote their **social responsibility agenda**. E.g. Primark®, a company known for its cheap clothing lines, is a member of the **Ethical Trading Initiative** and requires all its suppliers and factories to meet **international standards** of hygiene and safe working conditions.
- The social responsibility of companies covers **safe working conditions** in overseas factories, **fair wages**, following **labour laws** (e.g. forbidding **child labour**) and measures to **reduce environmental degradation**. However, these regulations can be difficult to enforce as so much overseas production is done through multiple layers of **outsourced** production.
- **Fairtrade** works to reduce levels of **inequality in global trade**. Fairtrade means **fairer prices** are paid to farmer's organisations for crops, such as coffee and cocoa; and for workers, it means **better wages** and **working conditions**. The 'Fairtrade Minimum Price' acts as a **safety net** when market prices drop, and an extra sum of money called the 'Fairtrade Premium' can be invested in **environmental**, **business** or **community projects**, e.g. new schools and health clinics. High **ethical** 'Fairtrade Standards' of production need to be met to use the FAIRTRADE Mark on product packaging.

Recycling can help manage resources sustainably

Recycling can help to reduce the **consumption of resources** needed to produce new goods.

- Recycling has some **disadvantages**. It takes large amounts of **energy** to recycle products, making it expensive. Recyclables are often **shipped to countries** that can better process them, making their **carbon footprint** bigger.
- Recycling forms part of a larger **circular economy**. This means that you continue to extract value from a resource after it has had its initial use. E.g. **reusing, repairing and renting** all have a role to play in extending a resource's life.
- Engagement with recycling varies between countries. **Emerging economies** like China lead the world in **recycling infrastructure**. Of the OECD (Organisation for Economic Co-operation and Development) nations in 2020, Slovenia recycled the most waste (57%) while Costa Rica only recycled 3%.
- Some materials are **more commonly recycled** than others. There is widespread recycling of glass, paper and metals while few local authorities have the means to recycle **black plastic** that is often found in food packaging.

Local authority recycling

The **Local Agenda 21**, launched in the years following 1992, allowed UK local authorities greater control over local **sustainability** aims. The **Isle of Wight** created a far more **comprehensive recycling system**. They refurbished a large **recycling centre** where residents could drop-off items, reducing the need to travel to the mainland to dispose of waste. New **street bins** for recyclables were placed in tourist **hotspots** and **more flexible recycling collections** were introduced so the island could cope with extra waste produced in **peak tourist periods**. The island now recycles more than 55% of its waste.

Warm-Up Questions

Q1 Evaluate the costs and benefits of localism.

Q2 Explain how transition towns might create more sustainable settlements.

Exam Question — AS and A-Level

Q1 Explain how ethical consumption can have a positive impact on international workers. [6 marks]

To make this book more sustainable, all jokes will be recycled...

Phew. I don't know about you, but I wasn't sure we would ever make it to the end of globalisation. I'm off for a nap...

Places — Economic Variations

Welcome to this lovely topic on places. As well as all the theory, you need to study a local place and one other contrasting place — we've chosen to look at Liverpool and Lerwick to show you the types of things you need to know.

A **place** can be classified by its **dominant industrial sector**

1) **Industry** is grouped into **four sectors**:

> - **Primary industries create** and **extract** raw materials and goods. Farming, mining and fishing are forms of primary industry.
> - **Secondary industries manufacture** products. They take primary products and make them into new and more complex goods.
> - **Tertiary industries** are **services**. They enable people to access and use goods. Back-office (administrative) tasks and customer service jobs in retail are forms of tertiary industry.
> - **Quaternary industries research** and **develop** new products and new ways of working. Scientific research and work in hi-tech and online fields are forms of quaternary industry.

2) A place's dominant industrial sector is likely to change over time. Places almost always become more developed and move from a **pre-industrial** to **post-industrial** phase. The **Clark-Fisher Model** shows how relative levels of each sector are likely to change over time.

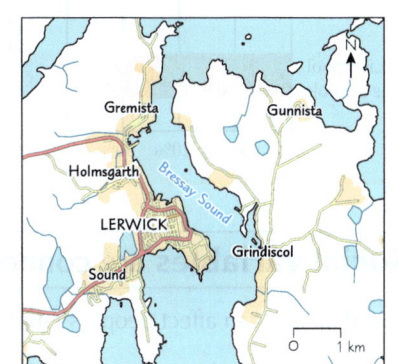

The Clark-Fisher Model

3) Certain locations have become associated with certain sectors of industry.

- **Rural locations** tend to have more **primary** industries.
- In the UK, **northern cities** like Leeds and Manchester have a history of **secondary** industries. Like many cities in other developed countries, they've since experienced **deindustrialisation** (see p.96). These areas have started to become more **industrially diverse**.
- The **tertiary** sector is the **dominant sector** in the UK. There are large numbers of tertiary sector workers in every urban settlement.
- **London** and the **south-east of England** have more recently focused on the **quaternary** sector. This has occurred largely around university towns and cities, but smaller **satellite centres** of quaternary industry (found on the outskirts of towns and cities) are starting to develop in other parts of the country, e.g. around Newcastle in the north-east of England.

4) With deindustrialisation, there's a shift away from a primary- or secondary-based economy to tertiary and quaternary industries. One aspect of this shift is the move away from the use of **outdated technology** (the '**old economy**') to the use of **cutting-edge technologies** (the '**new economy**'). The quaternary industry is now the fastest growing sector in the UK.

5) Generally, jobs in the primary and secondary sectors are **lower paid** than jobs in tertiary and quaternary sectors. Places that shift to the new economy are likely to be **more economically successful** than places that are still reliant on the old economy. Places that shift to the new economy will also become **more connected** (nationally and globally).

Liverpool and **Lerwick** are two places with **different economies**

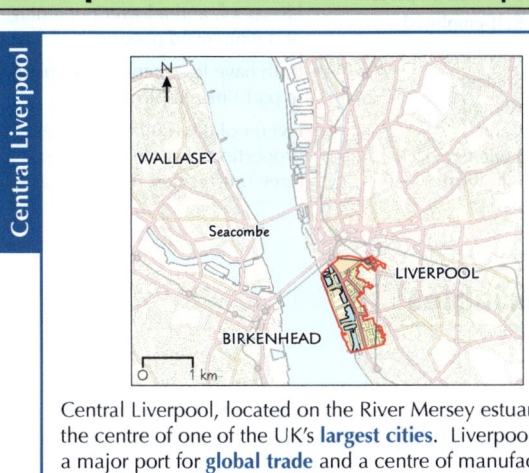

Central Liverpool

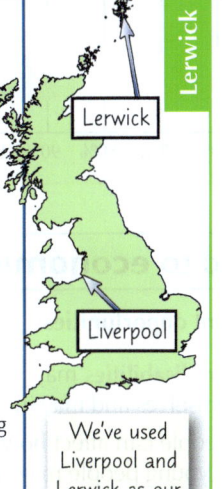

Lerwick

Central Liverpool, located on the River Mersey estuary, is the centre of one of the UK's **largest cities**. Liverpool was a major port for **global trade** and a centre of manufacturing between the 18th and mid-20th centuries. During this time, the city grew and **attracted immigrants** from around the world. The docks and factories **declined** in the 1960s, leading to **large scale deprivation**. Recently, Liverpool has attracted a lot of investment for **regeneration** and was chosen as the **European Capital of Culture** in 2008.

We've used Liverpool and Lerwick as our places of study throughout this section.

Lerwick, the **capital** of the Shetland Islands, is a small town and port located on the east of the main island. In its earlier history, the island was populated by Vikings but it's been part of Scotland since the 15th century. Its major industry has traditionally been **fishing**, but **North Sea oil** was discovered in the 1970s which, along with **increases in tourism**, has led to **economic growth**.

Places — Economic Variations

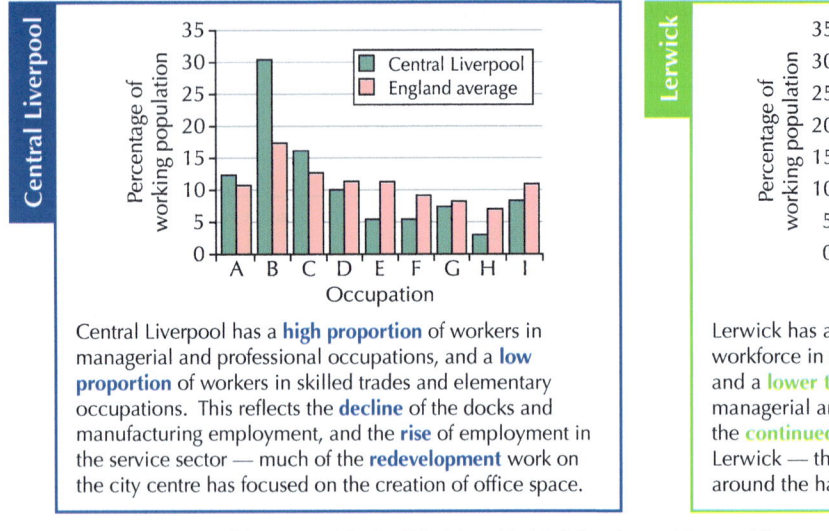

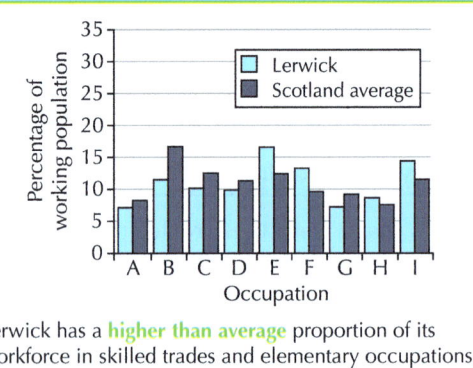

Central Liverpool has a **high proportion** of workers in managerial and professional occupations, and a **low proportion** of workers in skilled trades and elementary occupations. This reflects the **decline** of the docks and manufacturing employment, and the **rise** of employment in the service sector — much of the **redevelopment** work on the city centre has focused on the creation of office space.

Lerwick has a **higher than average** proportion of its workforce in skilled trades and elementary occupations, and a **lower than average** proportion of its workforce in managerial and professional occupations. This reflects the **continued importance** of the seafood industry to Lerwick — the catching, processing and selling of fish around the harbour **employs** many people.

Key
A Managers, Directors and Senior Officials
B Professional Occupations
C Associate Professional and Technical
D Administrative and Secretarial
E Skilled Trades
F Caring, Leisure and Other Services
G Sales and Customer Service
H Process, Plant and Machine Operatives
I Elementary Occupations (e.g. labourers)

A place can also be **classified** by the **types** of **employment** found there

The economic sector and the **nature of people's employment** are important when defining a place.

- A place can have a large number of **economically inactive** people — people who have retired, are unemployed or who have a long term sickness or disability. People may perceive places with a high proportion of economically inactive people as **less successful** than somewhere where the majority of people are in **full-time, permanent employment**. Some economically inactive people are **dependent** on **economically active people** for support.

- People who have temporary, part time or 'zero hours' employment contracts are likely to **earn less** than someone in permanent, full-time employment. They're likely to have **less financial security** and may be **more vulnerable** to becoming **economically inactive** in the future. If a place has a large number of people in this position, it is less likely to grow economically.

- **Seasonal work** is common in rural areas where there is a lot of **farming**, and in places which attract many **tourists**.

- **Self-employment** is another way gaining **financial security**. Self-employed people are more likely to succeed in places where there are **gaps in the market** for a new business, product or service. The new business, product or service offered through self-employment might be designed to meet the needs of the area and could help generate a sense of community. If successful, self-employed people may **provide employment** for others.

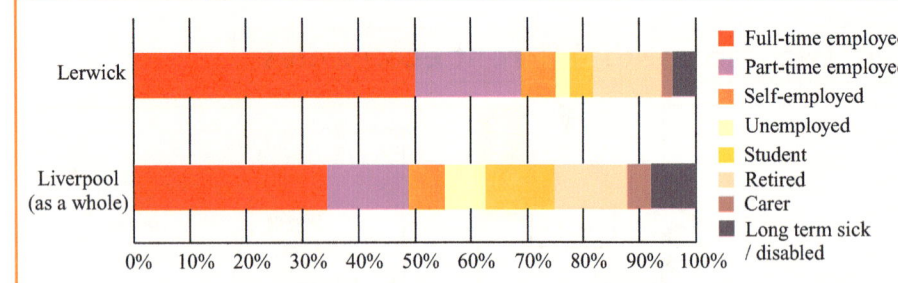

- Lerwick has a higher proportion of employed people than Liverpool, and a lower percentage of unemployed people.

- Both have less than 20% of people in part-time employment.

- Liverpool has a relatively large proportion of students — it has three universities. Lerwick has one.

Social variables are connected to economic variables

Social factors can affect people's **employment opportunities**:

1) People who have **poor health** or **certain disabilities** may have **fewer job opportunities** open to them. This may reduce their chances of earning for themselves and may increase their need for **state** and **local authority support**.

2) The type of employment available to people can affect how engaged young people are with education. If there are **few job prospects** in an area, young people may not see the point in **continuing education** beyond the compulsory years. Children whose parents work in professional or managerial jobs are more likely to achieve **higher levels of education**. They may then go on to work in similar types of jobs themselves.

Places — Economic Variations

3) A person's **level of education** affects their employment prospects. People with a **university degree** tend to earn more than someone with only **GCSEs**. Places with high numbers of graduates tend to have **higher average earnings**.

4) People who go on to further education or higher education often have greater '**personal mobility**'. This means they have **more opportunities to move** to other places in order to find better paid work.

5) There are also **location-specific factors** at play. For example, in Lerwick, although there are **relatively high numbers** of people with a degree, relatively few end up working in professional or managerial roles (categories A and B on the graph about employment on the previous page).

6) Though **apprenticeship schemes** can lead to employment, they also require students to **self-fund their living costs** while they train. This can make it more difficult for young people from **low-income families**.

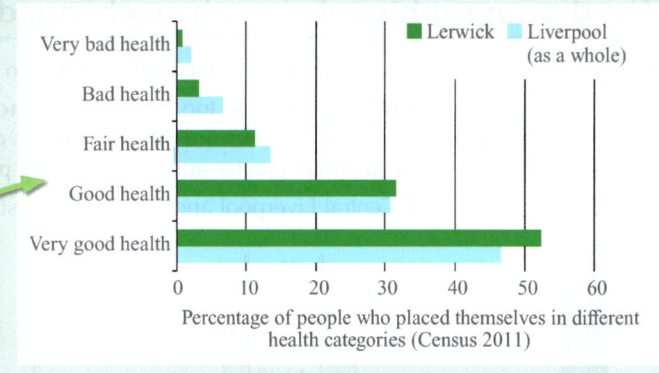

Highest level of qualification achieved (Census 2011)

- Foundation degree or higher
- A-Levels or equivalent
- GCSE (grade 4 and above)
- GCSE (grade 3 or below)
- No qualifications

A person's income can also have an impact on **social factors** such as **health** and **life expectancy**:

1) The more you **earn**, the more likely you are to live in **good health** and have a **higher life expectancy**.

2) It's also likely that people with higher earnings are able to afford **healthier foods** and undertake **preventative healthcare measures** (e.g. going for health check-ups). These options **may not be affordable** for people on lower incomes.

3) People with higher incomes may also be able to afford **higher-quality housing** (e.g. houses that are well insulated, adequately heated and aren't overcrowded) which can also improve their health.

4) However, **other factors** also affect health. For example, although Lerwick has **lower wages**, more people there say they are in good or very good health. People in urban areas such as Liverpool may be **more exposed** to air pollution from traffic, which can increase the risk of damage to the lungs and heart.

Percentage of people who placed themselves in different health categories (Census 2011)

Quality of life is connected to economic inequalities

- Quality of life measures people's levels of **health** and **living standards**, and their ability to take part in **life events**.

- Quality of life varies between different parts of the UK. Places where average earnings are **low** tend to have a low quality of life score. These places are more likely to have lower life expectancies, a higher prevalence of long-term health conditions and lower average education levels. In the **UK Prosperity Index** — an index that ranks places from 1 (best) to 389 (worst) based on a range of **quality of life indicators** — Liverpool ranks 381st while Shetland ranks 84th.

- However, a higher income doesn't always mean a higher quality of life. In many places where average earnings are high, the **cost of living** is also high, leading to lower quality of life scores. There are many places where incomes are relatively low but people's life satisfaction is reasonably high. E.g. Nigeria's GDP per capita is less than 10% of Hong Kong's, but both places report a **similar level** of life satisfaction.

Warm-Up Questions

Q1 Describe how the dominant industrial sector can vary in different places.

Q2 Give one reason why fewer educational qualifications may result in lower incomes.

Exam Question — AS and A-Level

Q1 Explain why people with low average incomes are more likely to suffer from poor health. [6 marks]

This is no plaice for a bad fish joke...

Any fin is possible, if you don't scale back your revision. Make sure you know all about how places vary economically.

Places — Changing Characteristics

Turns out it's not all about money. Geographers also like to think about other things that contribute to the characteristics of places, including their functions and who lives there. So, put your Geography hat on and get reading...

The **functions** of places have **changed over time**

1) Places have **functions** that are almost always economic in nature. The function of a place **serves the needs** of the people who live there. A function of a place could be:

- **Administrative** — providing public services, such as in schools and hospitals, through the local authority.
- **Commercial** — providing services that are profitable, such as banking.
- **Industrial** — manufacturing and distributing of goods.
- **Retail** — selling goods to the public through shops and markets.

2) One of these functions may **dominate** a place, but there's likely to be a **combination** of all of them.

3) A place's **dominant function** is also likely to change over time as its economic needs change. E.g. during the 1960s, Liverpool saw the decline of its docks and industrial spaces as this type of work moved overseas (see p.117). Over time, new **retail**, **tourist** and **commercial** offices moved into the docks, changing the function of that part of the city.

4) The increased use of **online banking** and **shopping** has meant that some places haven't been able to sustain an effective retail function and have gone through a second phase of **economic decline**.

> The COVID-19 pandemic caused a large increase in the use of online services.

The **demographics** of places have **changed over time**

- The demographic characteristics of a place are about **who** the people are and **what** they are like.

- As the function of a place changes so too might the **characteristics of its population**. New functions will attract different types of workers and different groups of people. Places that are economically developing are likely to attract greater numbers of **young people** and become **more ethnically diverse**.

- The age structure of Central Liverpool and Lerwick suggests that there are more **jobs** available in Central Liverpool.

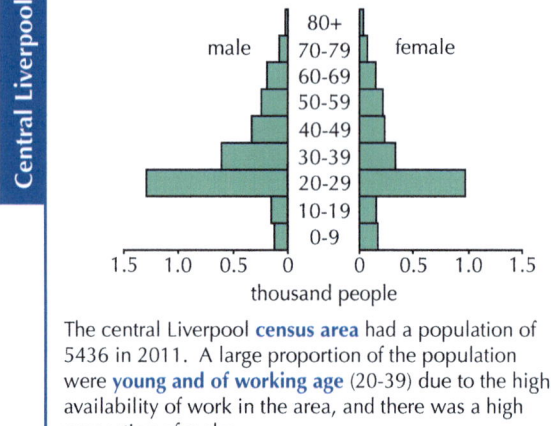

The central Liverpool **census area** had a population of 5436 in 2011. A large proportion of the population were **young and of working age** (20-39) due to the high availability of work in the area, and there was a high proportion of males.

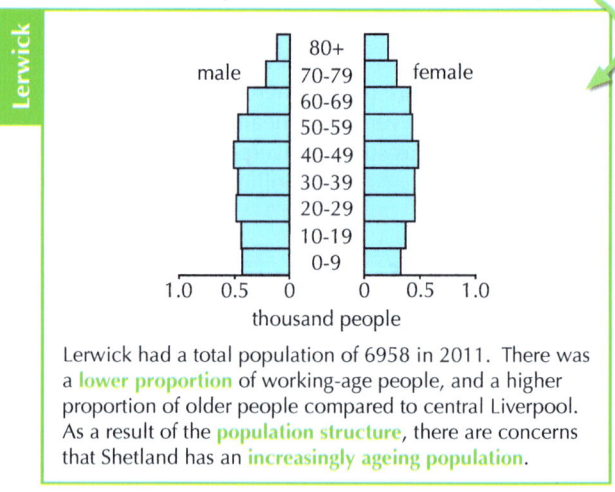

Lerwick had a total population of 6958 in 2011. There was a **lower proportion** of working-age people, and a higher proportion of older people compared to central Liverpool. As a result of the **population structure**, there are concerns that Shetland has an **increasingly ageing population**.

- The **ethnic composition** of a place is likely to change over time. In the UK, different **eras** of history have been associated with the movement of people from different **ethnic and national backgrounds** to the country. For example, after the Second World War, the UK encouraged migrants from **Commonwealth countries** to come to the UK to **fill labour shortages**. Many Caribbean migrants moved to South London, while Indian, Pakistani and Bangladeshi migrants moved into cities in the North and the Midlands, such as Bradford and Birmingham.

- Some areas have little history of **migration**. In the UK, rural and outlying areas tend to be less **ethnically diverse** than cities. E.g. Liverpool is more diverse than Lerwick.

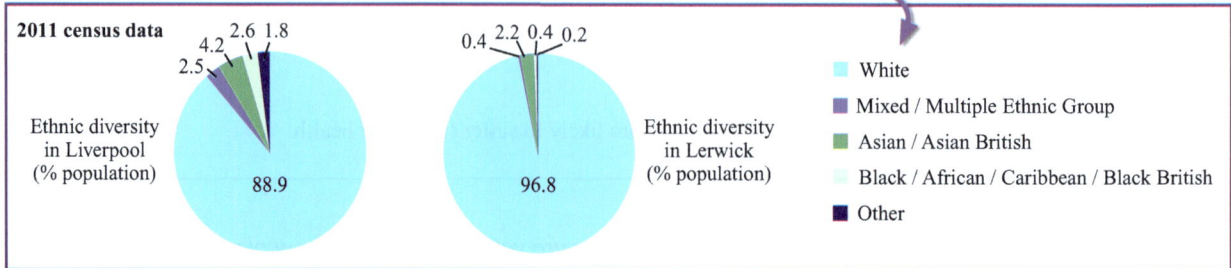

Topic 4: Option 4A — Regenerating Places

Places — Changing Characteristics

- As people of different **ages** and **socio-economic backgrounds** move into an area, the character of the area may change. In areas that are **struggling economically**, new migrants (from within the same country or overseas) may invest in local **housing** and revitalise local industry.
- Some migrants set up successful businesses that **serve their own demographic grouping**, which can transform a place and encourage further **in-migration**. This can **increase the price of housing** and raise the **cost of living** for everyone. This is one element of **gentrification**.

gentrification
when wealthy people move into a poor area, improve the housing and attract new businesses

There are many **reasons why** places **change** over time

Physical factors

- The **physical geography** of a place may change over time and some **expanding areas** might run out of space.
- The UK **coast** is in a constant state of change, with some areas **receding quickly**, such as Happisburgh in Norfolk (see p.87). Buildings and infrastructure for functions that require the coast, such as **tourism** and the **fishing industry**, may disappear as the coast recedes.
- Some parts of the UK are climatically **inhospitable**. E.g. some areas experience high levels of rainfall or very low winter temperatures. As global climates change, these may become **more attractive** and **migrants** may move to these areas for full time or seasonal employment.
- Climate change might also affect the **types of farming** that are possible in a place, and this can change the **character** of the countryside.

The Joneses weren't worried about receding coastlines. They had their defences all ready.

Historical development

Places change as they develop or decline. New **buildings** and **industries**, and the **loss** of old industries, change how people make a living. Throughout the history of a place, there are likely to be periods of **boom and bust** that reflect changes in industry.

Accessibility

- Some places may be more **accessible** than others.
- As authorities build new **transport hubs**, such as airports, people and industry can become more **connected** to places further away.
- Improved accessibility means that businesses will be able to **attract a skilled workforce** in a **competitive labour market** or **trade goods** in new ways. This may cause the place to develop economically, resulting in changes to its **function** and **demographics**.

Connectedness

- Strongly linked to accessibility is a place's **connectivity**. This might mean connectedness in a technological sense, such as the **extension of 5G mobile networks** into rural locations, or it may mean the more abstract connection places have with one another.
- For example, places may become more connected to their **cultural heritage** through the discovery of new **historical artefacts**, which in turn may increase their level of **tourism**.

Local and national planning

- How places develop or decline may be influenced by the decisions **local and national planning authorities** make.
- At a national level, **development funds** will be allocated to some locations and not to others.
- There can also be planning regulations that prevent developments from being built, such as **green belt land** (land around the edge of cities in the UK that **no development** can take place on). All local authorities have a duty to find suitable land for new housing in order to meet the **shortfall of homes** across the UK.

A map showing Green Belt land in England.

The red outline around London shows the Metropolitan Green Belt.

Topic 4: Option 4A — Regenerating Places

Places — Changing Characteristics

The **Index of Multiple Deprivation** measures change in a place

1) How a place changes can be measured through a number of criteria. Combining these criteria together can give an overall picture of whether a place is **improving or declining**.

2) The **Index of Multiple Deprivation** (IMD) measures deprivation in different places. It splits the UK into small geographical areas known as **Lower Layer Super Output Areas** (LSOAs) and gives each one a rank against 39 different **criteria**.

3) These criteria are then organised into **seven broader categories of deprivation**. Authorities in England use the following seven categories of deprivation:

> **deprivation**
> *not having access to things that people consider necessary for everyday life such as income, housing and health care*

- **Education** — the lack of attainment and skills in the local population.
- **Health** — the risk of premature death or the impairment of a good quality of life through health.
- **Crime** — the risk of a person becoming a victim of crime.
- **Employment** — the proportion of the working-age population involuntarily excluded from the labour market.
- **Income** — the proportion of the population experiencing deprivation as a result of having a low income.
- **Barriers to housing and services** — the physical and financial accessibility a person has to housing and services.
- **Living environment** — the quality of the indoor and outdoor environment for the local population.

4) The IMD compares each LSOA with others in the country through the use of **deciles**. Analysts can **rank** these scores individually or combine them together to give each area a **relative deprivation** ranking. Deciles divide the ranked areas into ten equal groups — from the most deprived (1) to the least deprived (10).

5) Greater **weightings** are given to **income** and **employment** deprivation than the other five categories (they represent 45% of the total score).

6) There are also two supplementary indices which show the ranks for deprivation amongst the **elderly population** and **children**.

Type of deprivation	Liverpool (decile rank against rest of England and Wales, 2019)	Lerwick (decile rank against rest of Scotland, 2020)
Education	1	6
Health	1	6
Crime	1	4
Employment	1	5
Income	1	6
Housing access	9	4

Liverpool having a decile rank of 1 in income deprivation means that at least 90% of places in England and Wales earn more than people in Liverpool.

There are **other ways** of **measuring change** in a place

- Changes in the relative amounts of **employment** in each of the **four sectors** (primary, secondary, tertiary and quaternary) can indicate the position of a place on the **Clark-Fisher model** (see p.111). Data on the numbers of **part-time** and **temporary employees** can also indicate wider changes in a place's **labour and employment market**.

- Increases in **house prices** may indicate changing economic fortunes for an area.

- Quantitative changes in demographic data, such as the **age** and **ethnicity structures** of a population, can also indicate the changing functions of a place. In the UK, the population of a place becoming **younger** and **more ethnically diverse** is a key indicator of increasing economic security.

- Studying **land use maps** and **aerial photography** can show how a place has changed over time. Many UK cities have former industrial sites. The length of time these stay **abandoned and derelict** can be an indicator of economic success or failure. Analysts can record what percentage of the land is used for **different purposes**, e.g. industrial or residential. How these percentages change over time is a good indicator of the **new needs** and **priorities** places have.

Warm-Up Questions

Q1 Explain how accessibility might change a place.
Q2 Describe how the Index of Multiple Deprivation works.

Exam Question — AS-Level

Q1 Explain how changing demographics might indicate a change in a place's function. [4 marks]

Changing places — when the seating plan just isn't right...

It'd be so much easier if places didn't change... but then we'd never get 5G. Hmmm... Guess you're just going to have to knuckle down and learn the ins and outs here — how and why places have changed, and how change can be measured.

Places — Influences

Now for a page about why places are the way they are. Fascinating stuff if you ask me. Oh, you didn't ask? Never mind...

Places are affected by **regional** and **national influences**...

- **Physical characteristics** — These include being located near **fresh water sources** or being in **strategic trade locations**. E.g. Liverpool developed around the River Mersey, which allowed easy trade access to Ireland and a suitable stretch of **deep water coastline** on which to build docks.

- **Transport connections** — Cities are often the **intersections** of multiple transport routes within the region. Smaller towns, e.g. Lerwick, have fewer **road connections** and might not have **rail connections**. Although ferries and aeroplanes connect Lerwick with mainland Scotland, there's a **lack** of transport infrastructure. This creates physical isolation and relative independence.

Liverpool Lime Street station has interconnecting bus routes.

- **Institutional connections** — E.g. Liverpool is situated just 35 miles from Manchester, and universities in the two cities work together as part of the **N8 Research Partnership**, which aims to create **collaborative research teams** across the region.

- **Industry** — Places can have long associations with certain industries. E.g. Lerwick's main industry is **fishing**. This has changed little since the 17th century when the town began trading fish with the Dutch.

- **National policies** — **Different migration policies** have allowed different groups of people to bring their customs and cultures to places (see p.120). E.g. in 2011, 3% of Lerwick's population consisted of migrants from Eastern European countries that joined the EU in 2004. The **Northern Powerhouse initiative** was a scheme designed to **boost economic growth** in northern cities. E.g. in Liverpool, the government is investing in a new **rail connection** called HS2. It aims to attract new business to the city and **decentralise industry from the South East**.

- **Local enterprise organisations** — E.g. the **Shetland Charitable Trust** funds local initiatives in Lerwick (such as Ability Shetland). **Local tourism boards** work to promote areas to visitors. Places are also affected by local **community groups** who target specific local needs and start small-scale developments.

...and by **international** and **global influences**

- **TNCs** — **Global brands** can dominate **retail spaces**. In some cases, **TNCs** may set up part of their business in an area, **creating jobs** and adding to the local economy. E.g. Princes Group, a TNC that produces tinned and bottled foodstuffs, have their headquarters in Liverpool.

- **Deindustrialisation** — TNCs have contributed to deindustrialisation in some areas. E.g. the decline of the Liverpool docks began after textile factories in other northern cities were bombed during World War Two — many of the **docks** had to **close** as they were no longer needed to import raw materials. Full-scale deindustrialisation in the docks occurred when textile TNCs began to move their operations to **South East Asia** during the 1960s.

- **Conflict and competition** — **Conflict and competition** may affect a place. E.g. the mackerel fishing industry in Lerwick faced **strong competition** from Iceland in 2010 — Iceland's **quota** was **increased**, allowing them to catch higher volumes of fish.

- **Tourism** — A place may try to **attract international visitors** by promoting its unique features. E.g. 310 000 people visit The Beatles Story exhibition in Liverpool every year and 60% of those come from overseas.

- **Political groupings** — As part of the EU, the UK had access to **EU Development Funds**, which were used to **stimulate economic growth**. E.g. Shetland has used more than £18 million of these funds to invest in **fish processing plants**.

- **Connections** — **transport and internet links** connect places to other countries. E.g. Liverpool John Lennon Airport connects the city to 25 countries and handled over 5 million passengers in 2019. Places that have **fast broadband networks** are better connected than those that don't. E.g. 9% of residents in Lerwick did not have access to superfast broadband in 2020.

Warm-Up Questions

Q1 Describe how national policies can influence a place.

Q2 Describe an international influence Lerwick has experienced.

PRACTICE QUESTIONS

Exam Question — A-Level

Q1 Evaluate the role that global connections have had on a local place you have studied. [20 marks]

Influences, connections — it's all about who you know...

...what you know is essential for the exam though. There's no getting out of revision. You've got to be clear on the factors that have shaped your chosen places, from the regional to the global, so go back and apply the points made here to your examples.

Places — Representations and Identity

If you thought Geography was a bit black and white, now's the time to get your coloured paints out. People's connections to places can be both real or imagined. I like to imagine a brightly coloured carousel that I can spend endless hours on...

There are lots of different ways to **represent** a place

1) A place can be represented in many different ways. How a place is represented is often dependent on what **message** the representation is trying to **convey**. When investigating places, it's important to look at a range of **different sources** to build up a complete picture of what a place is like.

2) Some forms of **representing places** are **qualitative**. This means they can't have a numerical score and the representation may be more **descriptive** or **creative**.

Maps

- Maps can be used to show any sort of data that involves a location, e.g. they can show where physical features are. They can also show quantitative demographic and economic data, e.g. different levels of income by location when used with GIS.

- Maps can also show qualitative information, such as the type of vegetation that can be found in an area — this can help to give the reader a sense of what the landscape of a place is like.

- Maps can show you reliable data, but can also be misleading. E.g. historical maps may be inaccurate — the 14th century Hereford Mappa Mundi is part map, part artwork and depicts scenes from biblical events and classical mythology.

- Many maps are drawn with a particular political purpose in mind, designed to influence the reader.

Films, photography and art

- Visual representations show what places look like and can give some sense of the character of places. However, they only represent what the artist wants to show you, and can be misleading.

- Photographs show what a place looks like in a given moment — photographs taken at different times of day can make a place look and feel different. Photographs can also be altered to make places look different to the reality.

- Films and television give a sense of place that is dependent on the nature of the story being told, e.g. a crime drama set in a city might give a different sense of place to a romantic drama set in the same city.

- Paintings or sculptures can be less reliable than films and photography at showing what a place looks like as they're an artist's interpretation. They can be more effective at conveying a sense of place and character though.

Imagery of employment in Lerwick tends to focus on the fishing industry. However, a 2015 survey of young people in Shetland found that 58% of them had an interest in working in the growing creative industries in Lerwick.

Stories, articles, music and poetry

- Written representations can be used to describe places and can give a sense of how it feels to be in that place. They usually only offer the perspective of the author though, and may not show a complete picture.

- Newspaper articles can give lots of detail about places but they may be biased, e.g. newspapers may focus on topics and ideas that are likely to sell more copies, rather than give a balanced perspective of a place.

- Stories, music and poetry can give emotional impressions of places, but only from the writer's perspective.

3) In all forms of **qualitative representation**, the **purpose** of the representation and the **agenda of the presenter** will influence the image presented. There's **subjectivity** in all forms of qualitative data.

Some representations of place are purely **quantitative**

1) A place can be judged by **number scores** and **statistics**.

Statistics

- Statistics, such as **census data**, can give lots of **quantitative information** about what places are like, e.g. population size, population structure, average income, crime figures, etc. They can be in the form of **raw data**, or visually represented through things like charts or graphs.

Places — Representations and Identity

Composite Indicators

- Composite indicators, such as indexes, **combine more than one set of data** together. This gives a broader picture of what a place is like. E.g. the **OECD Better Life Index** looks at how people rank their place using criteria such as work-life balance, civic engagement and safety. It then converts these scores into an **overall**, **weighted score**.
- The **Index of Multiple Deprivation** does something similar for small areas in England, Scotland, Northern Ireland and Wales (see p.116).

2) **Quantitative spatial data** (a range of quantitative data about a **specific place**) can be useful for creating a picture of a place. **Crime data** includes information about safety, deprivation and number of victims. **Voting and election data** provides information about public engagement, **political activism** and what **political ideals** are important in different places.

3) Statistics themselves are **objective** (based on **facts**). However, data is often used **subjectively** (based on **feelings** or **opinions**). For example, people might select data which best highlights their particular viewpoint. Statistics can sometimes **hide the real geographical picture** of a place. For example, a place might have low unemployment figures, which would be seen as a **positive characteristic**. A **closer analysis** of the data might reveal that the majority of people are in low-paid, part-time or temporary employment, which economists view less positively.

4) Statistics often don't accurately represent the **feeling** of a place and can quickly become **out of date** (see p.127).

People can be affected by continuity and change in a place

Changes (or a lack of change) in a place affect people in **different ways**.

Deindustrialisation

- The **connection** between people and local industries often remains strong even once the industry has been lost — particularly when the area has been **famous for a particular industry**, such as sugar refining in Liverpool.
- Former employees are likely to feel a sense of **nostalgia** and **romanticise** the **post-industrial landscape**.
- If there's no further industrial change and the post-industrial landscape continues, **younger people** or **new migrants** may view **abandoned buildings** as a symbol of **despair** and **fewer economic opportunities**.
- Those with **older industrial skill sets** may not welcome new forms of industry.
- However, the development of an area may **attract** migrants in search of new **employment opportunities** that involve a wider set of skills.

In-migration

As places attract people who are **younger** and more **ethnically diverse**, it can change local people's **political viewpoints**. Some people may learn more about **different cultures** and adopt more liberal viewpoints. Others may not like the change in culture of their local area and may **oppose** future in-migration.

Diversification of rural areas

- Many rural areas are moving away from **traditional farming**.
- Some farmers are diversifying into **energy production** (through solar and biofuel farming) or into **tourism** (e.g. through setting up campsites).
- This means that outsiders may perceive rural areas as **multi-functional places**, which might bring more people into rural areas to work.
- Diversification may **increase the average income** of rural residents.
- Rural economies that continue to be based around **traditional farming** sometimes **fail** to attract new workers to an area. This may lead to these farms going **out of business** and the local economy becoming **weaker**.

Development of the internet

- The **development of the internet** has changed the retail landscape of settlements.
- **Online shopping** is in direct competition with **high streets**. This has led to their demise in some smaller towns, creating **unemployment** and **reduced household incomes** there.

A boarded-up shop on a high street in the UK

Changes in the law

Changes in the law can make places **more accessible**. E.g. the Disability Discrimination Act (1995) made it unlawful for a place to deny someone who uses a wheelchair access to public premises. This changed the **streetscape** of many towns and cities. It made a big difference to how wheelchair users live and work.

Places — Representations and Identity

Local people's identity can be affected by economic and social changes

- Where multiple people share similar ideals and values, a **community** can form. These people often have **shared beliefs**, **shared ways of life** and may form a **collective identity**.

- When that **collective identity** of a community connects to the place where they live or work, changes to that place can sometimes impact how people view themselves. This can **change** the sense of belonging people have to a community.

- The amount of **migration** to a place can change depending on the **strength** of its **economy**. As a place becomes more **prosperous**, it can become home to **more diverse communities**. Migrants bring their **own cultures** and ways of life with them. These can **enhance** the local community. The **merging of cultures** can sometimes make people more **outward-looking**.

In Liverpool, Chinatown celebrates the long history of **Chinese immigration** to the city, through its architecture and restaurants.

Though Lerwick has limited ethnic diversity, it regularly celebrates its historic **Nordic connection**. The retention of Nordic street names and Viking festivals enriches the culture of the town.

- However, the addition of new cultures might lead to feelings of **cultural erosion**. If a place has a strong identity, some people feel that the influx of migrants with different identities can **weaken** this. Feelings of cultural erosion can lead to **tension** between different ethnic groups.

Riot police arrive at a fire during the Toxteth Riots.

The 1981 Toxteth riots highlighted the **economic division** existing between white Liverpudlians and **first and second generation migrants** to the city. Another factor in these riots was how black people **felt targeted** by insensitive policing (e.g. being stopped and searched), resulting in **anger** towards the police.

In 1981, riots happened in multiple cities and towns across England. As well as in Toxteth, riots took place in Brixton in London, Handsworth in Birmingham, Chapeltown in Leeds and Moss Side in Manchester.

- When forms of industry change, a **growing divide** may occur between those who are able to work and prosper in the new economy and those that start to struggle financially.

- A **rise in inequality** can affect the identity of people in a place. Those who previously felt like essential workers may now feel **marginalised**.

- Economic inequality can also lead to **power inequality** in **local decision-making**, affecting the extent to which people feel they **belong** in a place.

Jack's response to the new arrivals was to play it cool.

Warm-Up Questions

Q1 Explain how the statistics connected to a place may be used subjectively.

Q2 Describe how increased internet access has affected high streets and the people who use them.

PRACTICE QUESTIONS

Exam Question — AS and A-Level

Q1 Explain why increased migration into an area might change people's sense of identity. [6 marks]

I have an imagined identity — as a cute llama with a woolly hat...

...but enough about me. Once you start thinking about it, you'll see representations (and misrepresentations) of places everywhere and you'll start to wonder if you can ever trust those statisticians — or maybe the politicians... At least you might get to rant about it all in your exams. In a controlled and relevant way of course. So get learning this stuff.

Inequalities of Place

Unfortunately, life is not all sunshine and rainbows. However successful places are, there are always people who end up at the bottom of the heap. Then there's the spiral of decline, where the whole place goes down the drain. Cheery stuff.

Places experience **economic** and **social inequality**

1) **Inequality** is the **uneven distribution** of economic and social opportunities — e.g. income, jobs, health care and education. Some people within a place will have a **high quality of life**, but others will struggle.

2) Regardless of how successful a place is, there's often a large difference between the income of the **top and bottom earners**. However, the poorest people in one place may be better off than higher earners in less successful places. The very rich and very poor often live close to one another.

3) Inequality affects how people **view their own status** in a place. People with higher incomes tend to become **more dominant** in terms of **power** and **decision-making**. This may lead to a lack of **trust** from those who suffer most from inequality, a lack of **public engagement** and little or no **participation in local politics**.

4) Inequality can cause feelings of **despair** among young people. They may feel that however hard they work, the odds are stacked against them. This makes it more difficult for them to **break the cycle** of inequality.

5) People on **each side of inequality** may perceive a place very differently. People who have higher incomes, a higher quality of life and more opportunities may view the place as somewhere **fresh** and **exciting**. Those with a **lower quality of life** may experience the place more **negatively**, seeing it as a **trap** from which they can't escape — both economically and socially.

6) A place can be perceived as being more successful if it attracts more **migrants**.

7) When a place experiences **in-migration**, there may be **inequality** in the **lived experience** of different migrants (i.e. some migrants will be successful and some won't).

"Looks, charm, dress sense... It's not just money that's unevenly distributed," brooded Sachin.

Successful places experience **population** and **economic** growth

1) Successful places often experience a **spiral of growth**:

- The population of a place will grow as people migrate (both **internally** and **internationally**) to it, and then continue to grow through **natural increase**.

- **Cumulative causation** occurs when more people come to a successful area to provide services for those already there. This makes the area even **more attractive** to people and investors.

- As more people migrate to an area, there is a **greater demand for housing** and competition can push **house prices up**. In order to keep their workforce, businesses have to ensure the wages they pay are able to meet the **higher living costs** of the area. Everyone benefits from **higher incomes**.

- This means that successful places experience **low levels of deprivation** and **services receive consistent investment** in order to continue to meet the growing demand.

> **cumulative causation**
> *e.g. increased economic activity leading to increased prosperity and further development through a positive feedback loop*

Trying to sell personified vegetables is a niche probably best avoided.

2) The growth of **rural areas** is usually on a smaller scale — they tend to see business growth in quite **niche areas** (such as artisan foods). **Improvements in broadband coverage** mean that these businesses are able to serve local customers and also attract customers from further afield.

3) Successful places may also suffer from an **economic contraction**. A **boom** in the local economy can cause a shortage in **key workers** (i.e. people who are essential to the running of a community, like teachers and healthcare professionals). They're often **priced out** of the area, as **gentrification** (see p.115) has led to housing becoming unaffordable.

> Key workers' wages are often paid by the government. They may not rise as fast as average private sector wages in areas where there is a booming economy.

Inequalities of Place

San Francisco — a successful place

CASE STUDY

1) San Francisco is a city of over 800 000 people in California. It began as a **mining territory** and grew quickly during the **gold rush** of the mid-1800s. People **rapidly migrated** to the city and the population of roughly 1000 in 1848 had grown to around 30 000 just two years later.

2) To service the gold industry, large **banks** and **financial offices** were set up in the city. This attracted more investors and other service industries developed.

3) With the development of the Pacific Railroad into the city in 1869, **goods** started to be **imported** into and **exported** out of the city and the port area on its north-east coast developed rapidly.

4) By 1900, the population had increased to more than 340 000 people. The city also expanded further away from the coastal and bay area as more **residential** areas were added to it.

5) **Banking** and **financial service industries** continued to increase the wealth of the city throughout the twentieth century.

Population change 2010-2021	+ 1.2%
People aged 18-65	66.1%
Largest ethnic group	White (43.4%)
Average household income	US $126 187
Average house price	US $1 194 500
Living with disability	5.7%
Living in poverty	10.3%

6) More recently, the city has become highly competitive in **high-tech** industries, medical research and cyber-engineering. Some of the world's biggest tech and media companies (such as Apple) have their **headquarters** in the Santa Clara Valley, also known as '**Silicon Valley**'.

7) Today, parts of San Francisco are continuing to grow and develop. The high-income earners of Silicon Valley are living in **gentrified** areas (see p.115) such as Haight-Ashbury. Other areas which have traditionally housed lower-income earners, such as Bayview, are rising in value and becoming **unaffordable** to local workers.

Some regions experience **stagnating** and **declining economies**

1) Once a region starts to decline economically and socially, it starts to shrink. It can be difficult to stop the region from shrinking without **intervention** (e.g. formal and widespread **regeneration strategies**). The **spiral of decline model** (below) shows how an event can cause a place to experience a **negative multiplier effect**.

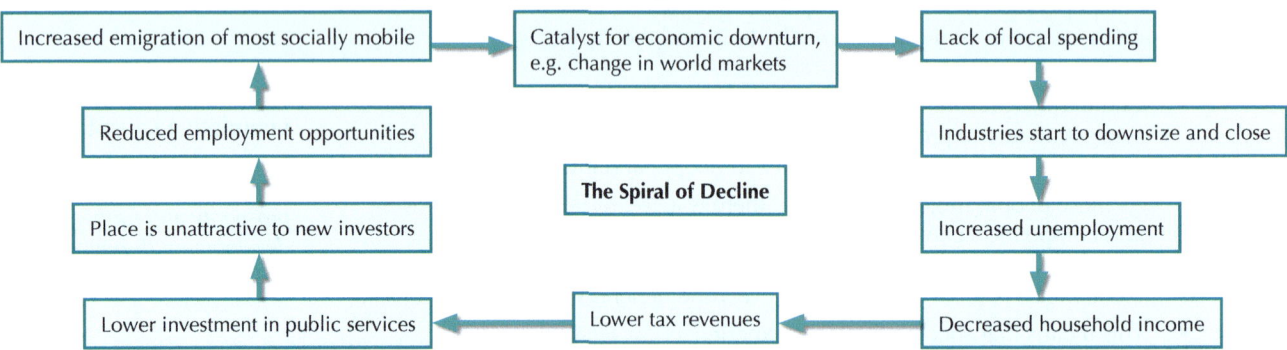

The Spiral of Decline

Increased emigration of most socially mobile → Catalyst for economic downturn, e.g. change in world markets → Lack of local spending → Industries start to downsize and close → Increased unemployment → Decreased household income → Lower tax revenues → Lower investment in public services → Place is unattractive to new investors → Reduced employment opportunities → Increased emigration of most socially mobile

2) It's common to see this kind of decline in areas experiencing **deindustrialisation**. Once businesses start to close, the resident population may start to **emigrate** in order to find work. Those who remain might not have the **right skills** to get work in other industries or may be seen as **too old** to retrain.

3) **Economic decline** can lead to **social decline**. A **lack of investment in public services** may mean that the people left behind have **poorer healthcare** and **education services**. In addition, there's likely to be an increase in **crime**, **antisocial behaviour** and **drug use**.

4) Some rural areas have also experienced declining economies. The **mechanisation of farming** has **reduced** the number of **job opportunities**, leading **young people** to **migrate out** of rural areas. As they go, they leave an **ageing population** behind. **Poor public transport links** often mean that rural areas are not well connected to local towns, reducing the chance of employment for those without a car.

5) The **decline of services** can also occur in areas where residents are from high income groups. Rail lines that pass through rural areas on their way to cities can become '**wealth corridors**' where **commuter villages** grow along the line. These become populated by **high-income groups** who want to live outside the city but still access services found in the city. These people often only live in the village and might not spend much time or money locally. This can lead to the **decline of small village services**, such as **pubs**, which may not be able to compete with the **wider choice of services** in a town or city.

Inequalities of Place

Decline in the Rust Belt

1) The **Rust Belt** is the informal name for parts of northeast USA that have witnessed **large scale decline** since the 1950s.

2) **Detroit** was the booming centre of the US car industry between 1920 and 1960. People **migrated** there from all over the country, attracted by the **high wages** and **good quality of life**.

3) **Service industries** also boomed and Detroit developed a large number of schools to accommodate the growing number of young families entering the city. Between 1900 and 1960, the population grew from 286 000 to 1.67 million.

4) **Overseas competition** from Japan and South Korea led Detroit's car manufacturers to increase mechanisation in their factories. This not only reduced the work force needed to run the factories but also cut wages to reflect the skills now required of the labour force.

5) Many white workers, who tended to be more affluent and better educated, migrated out of the city. As people left, businesses and services started to close and Detroit entered a **spiral of decline** (see previous page). This left the city with thousands of **abandoned buildings**. It's estimated that by 2010, about one quarter of Detroit's 138 square mile area was vacant.

6) The city declared itself **bankrupt** in 2013, with an estimated federal debt of over US $18 billion.

7) Since then, the city has started to **redevelop** and there are signs of a slow economic recovery. Much of this has come through private investment and tends to be based around small-scale and grassroots projects.

Data for Detroit	
Population change 2010-2020	– 10.4%
People aged 18-64	61%
Largest ethnic group	Black Af/Am (77.9%)
Average household income	US $34 762
Average house price	US $57 700
People aged under 65 living with a disability	15.7%
Living in poverty	31.8%

Poorer places are priorities for regeneration

- The role of **regeneration** is to **reverse** the spiral of decline and **spread wealth** to different levels of society.

- In most cases, there are **limited funds** available for regeneration, so authorities give **priority** to areas that have experienced **deindustrialisation** or **rural decline**.

- **Sink estates** are also targets for regeneration. These are housing estates that were originally set up by local authorities to house people with low incomes. They have become **centres of social decline**, fuelled by a **lack of local economic opportunities**.

- Other areas, such as **gated communities** and **commuter villages**, are a low priority for regeneration. Gated communities are enclosed estates in inner-city areas where walls, gates and controlled entrances separate the community from the rest of the area. People who live in gated communities tend to have **higher incomes** than people who live outside gated communities.

The entrance to a gated community

Warm-Up Questions

Q1 Explain the link between inequality and a person's perception of place.

Q2 Describe how the spiral of decline can lead to social problems in a settlement.

Exam Question — A-Level

Q1 Evaluate the extent to which the success of a place can be explained by models of economic change. [20 marks]

Spiral of decline — a poorly maintained helter skelter...

There's a glimmer of hope — places that have declined can be regenerated. Make sure you know how economic and social inequalities have changed how people see both San Francisco and the Rust Belt, and how the priorities for regeneration vary.

Experiences of Place

Turn up your emotional engagement as you read the next few pages. You've got to really live the revision experience. Warning — as your length of stay increases, your attachment may increase too. You may never be able to leave this topic.

People can gain an **attachment** to a place through their **lived experience**

- A person's **lived experience** in a place helps them to form their **unique perspective** of the place. Living in a place can give a person a better **understanding** of how it 'works' and what it needs to be **successful**.

- People often have a **strong attachment** to a place they've lived in — even after they move out of that area. They may feel a link between the place and their own **identity**.

- The attachment of people to a place can be seen in there being localised **dialects** or **colloquial terms** for people who come from certain places. For example, **Liverpudlians** are commonly known as **Scousers**. This can evoke a strong sense of **loyalty** to the place.

- This loyalty can mean that people **perceive** the place more **positively**. They may value some aspects of the place that other people might not recognise.

Joseph just couldn't understand why nobody shared his attachment to the place.

Local communities engage in a place in **different** ways

The **extent** to which people engage in a place can be measured by looking at how they choose to get **involved** in their **local communities** and how much they **participate** in local and national elections.

Elections and voting

- People can choose to take part in **elections** and referenda. Some people might even choose to become town or parish **councillors**. Analysts use **voter turnout** (the number of people who actually vote in an area) and the number of people on the **electoral register** in an area to show differing levels of **community engagement**.

> In the 2019 general election, the voter turnout was 67.3% but there were big differences in turnout **regionally**. In Liverpool, the constituency of St Helens North had a voter turnout of 62.9%, while Wirral West had 77.3%. In general, voter turnout is **lowest** in the areas of big cities where **income is low** and is highest in the **wealthier** parts. This is true in Liverpool, where St Helens North ranks 98th for deprivation in England, while Wirral West ranks 322nd.

A rank of 1 is given to the most deprived place — see p.116 for more on the IMD.

- Generally, areas that are most in need of **regeneration** are those where **community engagement** in elections and voting is **lowest**.

- There can be different levels of voter turnout when the vote decides things that affect a region **more directly**. In **local council elections**, voter turnout tends to be low — typically around 35%.

> In the 2018 local elections, Knowsley in Liverpool registered one of the lowest voter turnouts at 25.3%. For referenda that can change a local area **permanently**, the voter turnout can be much higher. In the 2014 Scottish referendum on independence, Shetland had a voter turnout of 84.4% compared to 67.7% in the 2019 general election.

The 2014 Scottish independence referendum created a significant voter turnout.

Community groups

- People can also engage with their place through **community action groups**. Though often **voluntary** and sometimes set up as **charities**, these groups of **local people** can be strong advocates for change. They often undertake development work in **small places** that might be **ignored** by local authorities.

- Community groups are often start because people are frustrated by the **growing inequality** or **social injustice** in the place where they live and/or work.

- Community groups might be a **supporting network of people** who have **similar interests** in improving places. They might also be a group of people who feel that authorities are not hearing their concerns. These groups tend to centre on **activism** and **raising awareness**. **Protesting** and **petitioning** are one way that community groups can express their **attachment** to a place.

> Voluntary Action Shetland, based in Lerwick, is an organisation that trains and supports volunteers to help with specific needs within their community (such as befriending schemes for isolated elderly residents) as well as providing administration support for local businesses.

Experiences of Place

Sports and recreation

Engagement in a place might also come through **membership of local sports clubs** and **recreation societies**, as well as choirs and music groups. These can sometimes allow people to represent their place **competitively**.

Communication

- Strong **communication within communities** can make engagement in a place easier. **Social media** can **quickly** raise awareness about issues that are affecting places, which can cause people to feel a greater sense of attachment to a place.

- On a smaller scale, community **newsletters** and **village websites** can do a lot to help people in rural communities feel a greater sense of connection to one another.

A person's **lived experience** of a place is shaped by **many factors**

There are many reasons why a person may **not** feel connected to a place or have a sense of attachment to it. If groups of people are engaging **differently** with their local area, it's an indication that they have a **different lived experience** of that place. Some people may feel that a place does not **represent** them or have a feeling of **isolation** there.

Length of residence

- Some people may have **recently moved to an area**, and there may be **language** or **cultural barriers** that make valuing every aspect of a place more challenging.

- **Transient populations** such as **students** or **second homeowners** have a different type of lived experience from someone who is there permanently and wants to spend their future there.

- **Studentification** can be frustrating for local residents as students **may not care** as much about the long term success of the area, as they won't be living there for very long.

Age

Young adults who are struggling to enter the labour market will see a place differently from **older** people who are close to **retirement**. For example, a young person may perceive a rural area as **uninteresting** and **quiet**, while an older resident may view it as **tranquil** and having plenty of beautiful places to visit.

Levels of deprivation

A feeling of **powerlessness** to change a place or to address **the level of deprivation** experienced there may create a negative lived experience.

New Students' Union, including accommodation, in central Nottingham

studentification
the process by which a place becomes heavily occupied by students

asylum
protection provided by the state to someone who has fled persecution in their home country

Ethnicity

- People from a specific **ethnic group** may feel that the place they've moved to is completely **disconnected** from their own **cultural heritage**.

- Others may have such **strong nationalistic feelings** associated with place that their **perceived experience** of it excludes those who do not share their ethnicity.

- Some communities of **international migrants**, particularly those who have sought **asylum**, can feel a strong connection to a place. Their new home may be the first place they have lived where they have **freedom** and **safety**, evoking a strong sense of **loyalty** to the place.

Differences in people's lived experiences partly explains why some people aren't interested in voting. In the 2019 general election:

- voter turnout was highest in areas that had a greater percentage of people **over the age of 65**.

- a higher proportion of **white people** voted (63%) than black and ethnic minority people (52%).

- a higher proportion of **men** voted (63%) than women (59%).

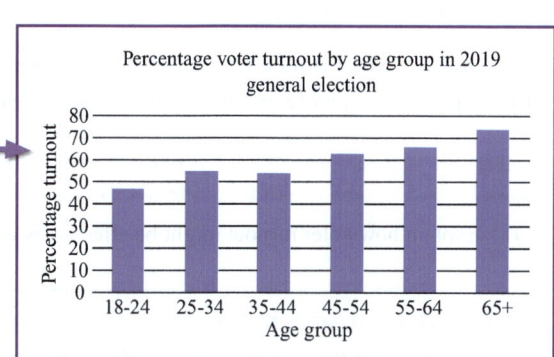

Percentage voter turnout by age group in 2019 general election

Experiences of Place

There can be **conflicting** ideas about **regeneration**

Every place will have a **mix of people** with different **lived experiences**. As they look to the **future**, these different groups may have **conflicting views** about how their place should be **regenerated**.

Age

- **Young people** may be more concerned with the future economy of the area, because it affects their **job prospects**. They may be more likely to favour regeneration that increases the number of jobs available to them.

- **Older people** who are **retired** may be less concerned with new employment opportunities. They may want to preserve the look and feel of a place where they've spent many years. For example, younger people and recent migrants may favour the **demolition** of an area, while older people may favour **renovating** existing buildings and other features of a place to **improve** them.

Older people may object to buildings being demolished as part of a regeneration project.

Inequality

- **Reducing inequality** in an area is often one of a regeneration project's aims — but the same inequality may also be an **underlying cause of disagreement** about the project.

- **Wealthy** people are often **socially well-connected** and may be able to use these connections to **influence decisions** on local regeneration projects. E.g. they may know **who they should talk to** to get their views taken into account. This may result in wealthier people's interests being **better served** by regeneration than less wealthy people's.

- Others may want to make sure that any regeneration projects don't **inadvertently** create **more inequality**, such as between different **ethnic groups** or between **able-bodied and disabled people**.

A **top-down** approach to regeneration can also lead to **conflict**

- Plans for regeneration projects are most likely to come from large **development corporations** employed by local and national authorities. The development corporations then **consult** local people, businesses, and community groups.

- Development corporations most often take a **top-down approach** to development. These agencies are unlikely to have any **lived experience** of the place they are trying to regenerate. They might try to **fit** regeneration ideas that have worked elsewhere into the place in question.

- Local people can feel that agencies **impose** ideas on them rather than having ideas come out of a **bottom-up approach**. This can lead to conflict.

- Community groups may be set up to **protest** against the planned changes.

- These reactions are **common** when there's a proposal for a new large-scale **housing development**.

Jodie was beginning to wish she hadn't taken a top-down approach...

Warm-Up Questions

Q1 Describe what is meant by a lived experience of a place.

Q2 Explain how involvement in community groups indicates engagement in local places.

Exam Question — AS and A-Level

Q1 Explain how voter turnout might be affected by a place's population structure. [6 marks]

Does your place have the wow factor — vote now...

Who'd have thought there were so many things that affect how you feel about a place. Whatever lived experience you've had where you live, these last three pages covered some important stuff, so go learn it. Then see if you can drag yourself away...

The Need for Regeneration

Here's where watching TV can be part of your revision. It's time to look at how statistics and the media present your chosen local place to see whether regeneration is required. One thing — you do have to put your geography specs on first...

Statistics can provide **evidence** of the need for regeneration...

There are various types of **statistics** that local authorities can use to assess which areas should be regenerated:

Census data

- Local authorities often assess the **need for regeneration** by using **census data** and **local area statistics**.
- This gives information about the **population structure** (age, sex and ethnicity).
- Data can also show where there are people 'missing' from the population structure and helps local authorities identify demographics that they want to **attract**.
- The population structure tells authorities about the kind of people any regeneration will affect.

Labour Force Survey

- Local area statistics might include information from the **Labour Force Survey**.
- This provides information on the average income for different jobs and the types of work in which local people are employed.
- This can **identify gaps** in the labour market. For example, a regeneration project that focuses on providing work for people in manual labour jobs might not be suitable if there is high unemployment amongst office workers.

Abdul's 'exciting role' in a regeneration project wasn't quite what he was expecting.

Index of Multiple Deprivation

- The **Index of Multiple Deprivation** (see p.116) is another way of identifying areas that might require regeneration.
- The seven criteria and **small** survey areas allow analysts to target **specific geographical areas** and their **specific needs**.

Stakeholders need to bear in mind that statistical data can quickly become **out of date**. This is particularly true of the **census** which only takes place every ten years. It's also important for stakeholders to be aware that statistics **don't** represent **how people feel** about a place.

Local councils, businesses, development companies and environmentalists are examples of stakeholders who might make use of statistical data.

...but the **media** can present a **contrasting** view

- **Television dramas**, **films** and **news broadcasting** about an area can provide viewers with a **sense of place**. This is important when considering the need for regeneration.
- Print media (both **fictional** and **non-fictional**) can give readers an idea about what a place is like. It can highlight the needs of **different stakeholders** and indicate the **relationships** that exist in an area between different groups of people.
- These portrayals are **subjective** and may not accurately represent the need for regeneration. Some writers and journalists may represent a location **favourably** to enhance a storyline. Other areas may appear **worse** than they really are to add more drama. Media evidence can be **controversial** as areas can easily be unfairly **stigmatised**.

The TV series 'Poldark' has romanticised Cornwall, where much of the series was set, creating a tourism boom.

The Need for Regeneration

Liverpool — Evaluating the need for regeneration

- Data from the 1981 census for Liverpool highlighted a number of areas for the regeneration that took place in the city over the next few decades.

  - 26% of **economically active males** were not in full-time **employment**. Those who were employed tended to work in less skilled positions, such as transporting and processing jobs.

Most common forms of employment in Liverpool (1981)		
Male	Material processing 22%	Transporting materials 13.1%
Female	Clerical 35.3%	Catering / Cleaning 21.4%

  - **Housing stock** was also in need of improvement. 5.9% of homes housed 6 or more people and 5.2% had no indoor toilet.

  - **Environmentally**, the River Mersey was one of the most polluted in the UK and in need of large scale investment. It was the dumping ground for mercury and lead from local industries as well as raw sewage from the majority of the city's residents. The river recorded high levels of heavy metals throughout the 1970s and 1980s.

- Overall, the statistical data for Liverpool seemed to suggest that the city was in need of regeneration.

- Depictions of Liverpool in the media at the time also seemed to highlight the need for regeneration. **Television programmes** such as 'The Liver Birds' and 'Bread', which depicted life in working class families, gave a good indication of what was important to local people at the time.

Lerwick — Evaluating the need for regeneration

- Some **employment data** for Lerwick suggests that the area would benefit from regeneration. However, when compared to other parts of Scotland, the need for regeneration becomes **less clear**.

  - Though employment in the town is high (less than 1% of people were long-term unemployed in 2011), the type of employment tends to be as **service providers** (who are more likely to be employed on minimum wage) and **tradespeople** (who are more likely to have unpredictable levels of work).

  - Skilled trades employ 17% of the population (compared to 13% in Scotland as a whole) and only 7% of people are employed in **managerial positions**.

  - The development of improved **transport links** to and around Shetland has caused **major environmental concerns**. An environmental quality survey in 2006 found that the proposal for a fixed link (a bridge and/or tunnel) between islands is likely to bring non-native species to ecologically sensitive areas.

See page 112 for graph showing the proportion of the population in different occupations.

- Media depictions of Shetland seemed to suggest a need for regeneration in areas relating to **community cohesion**. The television drama 'Shetland', which first aired in 2013, depicts the island as having a high murder rate, with storylines focused on the isolated nature of communities. However, data shows that only 1.3% of crimes in Shetland were violent in 2019, compared to 1.6% in the rest of Scotland.

Warm-Up Questions

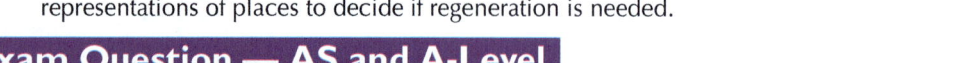

Q1 Give one reason why authorities should be cautious about using statistical data to decide on the need for regeneration.

Q2 Describe one advantage and one disadvantage of using media representations of places to decide if regeneration is needed.

Exam Question — AS and A-Level

Q1 Using a place you have studied, explain why it is difficult to assess the need for regeneration. [6 marks]

My need for regeneration is evidenced by the poor quality of this gag...

To regenerate or not to regenerate. It's not an easy question to answer. What is clearer is that you need to know there can be a range of contrasting evidence. Make sure you learn why there might be different viewpoints in your chosen local place.

Regenerating Places — Government Policies

There are a lot of technical terms along with information about governments and the world of finance on these pages — it might all seem a little bit overwhelming, but don't panic — just go slow and keep the main aims in mind.

Improving infrastructure is a form of regeneration

- Improving **transport infrastructure** systems can help continue economic growth and also act as a **catalyst for new development and regeneration**. As relatively **isolated** areas become more **accessible**, they also become more attractive to businesses, which can create jobs.

- An increase in **job opportunities** means more people might move to these areas, leading to further economic growth.

- However, infrastructure projects are **very expensive** and can take a **long time to plan** and develop successfully as they involve many **stakeholders** in multiple locations.

> To decide how regeneration should be implemented, national governments may work with charities that aim to improve the lives of local citizens.

High Speed 2 railway line (HS2)

HS2 would link London to other major cities with a faster trainline, reducing travel times. It has been claimed that this will encourage wealth to spread away from London and the South East and help to balance the economies of the North and South of England. The project would also involve the **electrification** of minor rail lines in the North of England and the **renovation** and **enlargement** of the main stations on its route.

Airport development

- Since the 1960s the capacity of UK airports has been increased to cope with rising demand for **air travel**.

- Large airports, such as Heathrow, have been **expanded**, and there has also been **investment in regional airports**. A small airport first opened in Liverpool in 1933, but became commercially viable in 1997 when a private investment company funded its expansion. Now known as Liverpool John Lennon Airport, it handles over 5 million passengers a year and flies to over 70 destinations in the UK, Europe and North Africa.

Planning policies can affect the rate and type of regeneration

Planning laws

- Both local and national governments have **regulations** in place which can affect regeneration. Some **planning laws** impact the degree of development that can take place.

- **Green Belt land** is undeveloped land that surrounds urban settlements. Green Belts prevented the **outward spread** of these settlements and have made **inner-city** areas become more **densely populated**.

- This land is usually **farmland**, but there have been instances where authorities have **reclassified** Green Belt land and have lifted **restrictions** in order to allow developments to take place.

- For example, a decision to build new homes on Green Belt land in **Maghull**, north Liverpool was unpopular among local residents. It highlighted the **conflict** between conserving land and building new homes in the UK.

House-building targets

- There are many reasons why there is a **shortage of homes** in the UK — e.g. more people living alone, people living longer, and historically high levels of **immigration** into the country mean that in many areas, and particularly in the South East of England, there is a need to build more housing.

- The **National Planning Policy Framework for England** set out the target of building 1.5 million new homes in England between 2015 and 2022.

- Between 2015 and 2019, local authorities missed the targets by around 46 000 homes a year. This has **increased the pressure on local authorities** to grant permission for housing developments in sites they might otherwise have protected. This is especially true around London, where there is most demand for housing and the least amount of land available.

Housing affordability

- Planning permission for new developments can often only go ahead if authorities set aside a proportion of the homes as 'affordable' (suitable for households with **lower than average incomes**). Sometimes there may also be requirements that developers' plans include **improvements to local infrastructure** or **investment in local facilities**. This is known as '**planning gain**'.

- In **Shetland**, the local authority was criticised when only 16 out of 62 new homes built in 2018-19 were affordable, despite the new homes being built to attract younger, working people to Shetland.

- A lack of new homes may mean that **skilled workers** are unable to move into an area. This means the economic growth and regeneration of an area may **slow down**.

Regenerating Places — Government Policies

- Planning policies can be side-stepped if authorities believe the proposed works are 'in the national interest'.

- For example, in the hope of reducing our future reliance on imported natural gas, local authorities have had certain restrictions lifted, making it easier for them to approve sites for test drilling to see if fracking is viable.

- This has caused controversy as some people believe that the environmental costs of fracking procedures outweigh the economic benefits (see p.199-200).

Izzy side-stepped fashion policies with her controversial 'frocking'.

Migration and deregulation policies also affect development

International migration policies

- International migration policies pursued by central governments can both benefit and hinder regeneration efforts.

- Increasing numbers of people migrating to an area can put pressure on housing availability. However, having more migrants can also increase the tax revenue and provide skilled labour which can kickstart economic development.

- Young migrants who start families in the UK can reduce the problems of ageing populations seen in many rural areas.

Deregulation policies

- The deregulation of the financial markets through the 1970s and 1980s led to huge changes in the financial landscape of cities. Previously only UK banks could operate in UK cities, but foreign direct investment meant areas such as London's Canary Wharf started to see the headquarters of international banks, e.g. HSBC, established there.

- For large projects, pump priming may take place. This is where national governments fund part of a scheme with the expectation that further investment will come from private sources. Many large redevelopment projects in cities have used this method to achieve their goals.

- Deregulation also brought changes to the investment landscape in other sectors. Wealthy foreign investors started to buy second homes in London and the South East, which could result in them paying less tax in their home countries. The owners often left these homes unoccupied, adding to the lack of usable housing stock available to those who needed it.

- Sometimes an increase in second homes and holiday homes can slow down economic development in an area. Though the initial purchase of the home brings funds to an area, the lack of ongoing spending means that longer term development is not always possible. People buying second homes may also price migrants and locals out of the housing market.

deregulation
the relaxation of rules and the removal of barriers

direct investment
the purchase of assets or shares which leads to direct ownership of the asset or company

indirect investment
investment which does not lead to direct ownership of the thing in question

---

CASE STUDY

Foreign investment in London real estate

- London has long been one of the most expensive cities in the world to live in. A large part is because London is a global hub that attracts businesses from all over the world.

- Between 2014 and 2016, foreign investors purchased many newly built homes in the capital.

- Overseas investors bought 13% of newly built properties in that period, 15% of which were companies situated in offshore tax havens such as the Channel Islands.

- Most of the foreign investment in London real estate used to come from wealthy Russians. Now though, most investment comes from Asia, with China, Malaysia and Singapore making up 61% of foreign buyers.

- The homes purchased are not always multi-million-pound penthouses. The City North Islington Estate had flats starting at £380 000 and foreign investors bought 78% of these.

Regenerating Places — Government Policies

Local authorities try to attract **inward investment**

1) **Local authorities** make decisions about planning and regeneration — they **choose particular areas** for redevelopment based on local needs and plan how to **fund** each scheme.

> **inward investment**
> *investment into a country from sources outside that country*

2) Almost all regeneration schemes work to attract **people** and **businesses** both from within the **UK** (domestic investors) and from **overseas**, but the balance of investors is dependent on the overall aims of each regeneration scheme.

3) Governments need to have a **comprehensive approach** to regeneration if it is to be **successful** — economic development will also require **new housing** and **infrastructure**. Local authorities may need to relax planning regulations to encourage investment in **new commercial developments** such as shopping centres.

4) One way to create a sympathetic business environment is to designate areas as 'enterprise zones' — geographical areas with **incentives** to businesses such as **tax breaks** and government support for **start-up costs**. The hope is that over time, businesses that develop in these zones become more **economically viable**, employ more people, and pay back far more into the local area than it cost to set them up there.

Science parks

CASE STUDY

- **Science parks** are clusters of offices, laboratories and design suites focused on quaternary industries. They group together **high-tech companies**, **entrepreneurs and start-ups**, who may **collaborate** and make use of expertise in **local universities**. Science Parks are often located in enterprise zones.

- One aim of regeneration is to **raise the value** of the businesses working in an area, as well as the incomes of people working there. Typically, the **quaternary sector** provides the **highest paid** employment, so regeneration schemes may look to attract these types of businesses.

- Local authorities will often manage the infrastructure and the original funding outlay of the parks before they become populated with companies and businesses in fields such as **pharmaceuticals** or **cyber-engineering**.

Liverpool School of Tropical Medicine – part of the Knowledge Quarter

- An example of a science park is the **Liverpool Science Park** in Liverpool's **Knowledge Quarter** (a group of new offices, laboratories, conference facilities and a hospital that are available to small and medium-sized enterprises as well as global companies). The Knowledge Quarter works closely with universities in Liverpool.

Local interest groups don't always agree on development and regeneration

Local authorities have to **collaborate** with a number of different groups and **stakeholders** in order to make sure the needs of a regeneration project are met and that any development doesn't have a disproportionate **negative impact** on any one group of people. There are a number of players to consider:

- **Chambers of Commerce** — These are local organisations that **represent the interests of local businesses**. They **lobby governments** to invest in infrastructure and training that will benefit the wider business community now and in the future.

- **Trade unions** — These organisations work to ensure that employers consider the **rights and needs of workers** in different decisions, particularly in relation to pay and working conditions. Trade unions offer support to workers if there are **disputes** between them and their employers. Unions can call for their members to **strike** if companies do not meet their demands.

- **Local interest groups** — These are usually groups of people who are concerned with **preserving or conserving** the **natural environment** or an **area's heritage**. E.g. 'Stop HS2' is a campaign group raising awareness and funds to fight a legal battle against the development of the High Speed 2 railway. They feel the development will do lasting environmental damage to the areas it cuts through, and that it won't be able to meet to the business needs of the northern cities.

Controversy surrounding the London 2012 Olympic Park

CASE STUDY

- The London 2012 **Olympic bid** claimed that it would transform Stratford through regeneration, but when the London Organising Committee released plans for the park, the community was not entirely happy.

- The Clays Lane Housing Cooperative was a group of **affordable homes** built in 1977. The Olympic plans required the Clays Lane homes to be acquired using a **compulsory purchase order** and for builders to **demolish** them, evicting up to 450 residents. These residents protested, but authorities moved them into homes in **poorer conditions** with £8500 in compensation. Since the Olympics, the area has been transformed through the **purchasing** and **renting** of the former **Olympic village apartments**.

Regenerating Places — Government Policies

Retail, sport tourism and rural diversification can also be used to regenerate places

Retail-led regeneration

- **Large retail centres** provide **many different jobs** to people of different backgrounds and ages.
- Retail can also **boost local spending** outside traditional shop hours with many shopping centres housing cinemas and gyms as well as restaurants. For example, **Liverpool One** was built on **redeveloped land** and the complex contains a large number of shops, bars and restaurants as well as multiple attractions.
- In smaller towns and in rural areas, smaller retail changes such as introducing **covered markets** and **pop-up shops** with reduced business rates and short term leases can provide new retailers with a foothold in the market.

Leisure and sport

- **Hosting large sporting events** can raise the profile of an area and attract **tourists** to the events themselves.
- Investing in new facilities enables events to leave a legacy, such as new buildings and transport links, for local people after the event.
- In rural locations, **holding events** can highlight the **beauty** of the area and ensure tourists return after the event has ended. E.g. the 2014 Tour de France began in Yorkshire, and people still go to cycle the route and see the bicycles that decorate many buildings.

Tourism

Making use of the **cultural heritage** of a place can boost tourism.

- The **Up Helly Aa** festival in Lerwick celebrates the Viking heritage of Shetland and attracts international visitors.
- In Liverpool, there are many tourist attractions and public art pieces celebrating **The Beatles**.

Regeneration through sport — London 2012 Olympics

The Olympic stadium, aquatic centre, and velodrome are now available for use by **both local people** and **national teams**. The Olympic Village has been converted into 2800 flats, **new office buildings**, shops, restaurants and schools. The Westfield Group bought up part of the site and created a **new shopping centre**. Transport for London (TfL) enlarged and upgraded **train services** for people who now commute into central London from the area.

Rural diversification

Diversification is when businesses start to offer **different goods or services** alongside their core business. Farms may start to grow and process **artisan foods**, or **convert their buildings** for other uses such as **galleries**, **farm shops** or **bed and breakfast accommodation**. Before 2019, diversification in rural areas in the UK was in part funded by the Common Agricultural Policy of the EU.

Rural diversification — Powys Regeneration Partnership

- Between 2014 and 2020, the One Powys Local Action Group and Powys council (in Wales) focused on multi-layered rural regeneration. This used a **combination of private and public investment**, including some from the **EU**.
- Individual projects tended to focus on key themes, such as **renewable energy developments** and **local creative industries**, that celebrated traditional Welsh crafts.
- To allow new businesses to flourish, **EU adjustment funds** were used to invest in **fast broadband connections**. This enabled people to **work from home** and helped businesses reach a wider market.

Warm-Up Questions

Q1 Describe how local governments might try to attract inward investment for regeneration.

Q2 Explain how sporting events can stimulate economic growth in both urban and rural areas.

Exam Question — AS and A-Level

Q1 Explain how planning policies can affect regeneration plans. [6 marks]

The only things I invest inwardly are cups of tea and cakes...

...it really keeps the spirits up. Make sure you've got a handle on how UK policy decisions affect regeneration, not forgetting those trusty case studies illustrating the key points. It'll really impress the examiners if you can throw in a few examples too...

Rebranding Places

Close your eyes and go to your happy place — some positive thinking is needed here. These pages are about how places try to attract people by putting up fancy new buildings and going with the 'hard sell' — "This place is seriously great!"

Public perceptions of a place can change through rebranding

- **Rebranding** a place can help to remove any **negative** image people may have of it.

- It can involve **re-imaging** — where agencies **enhance** or **change** the image of a place when **marketing** it to an audience. When re-imaging the negative aspects of the place are often **ignored**. Re-imaging is an important part of **regeneration** as it has the potential to **attract new investors** and **workers** to the place. It is often used **alongside redevelopment projects** as it's difficult to rebrand an urban space without making some changes to the **built environment**.

- Rebranding uses a variety of **media**, especially the internet, to present a new image to people. **Social media** can attract people and investors, and can be used to reach a much larger **audience**.

- Creating new **logos** or **slogans** to promote a destination is an important part of rebranding. Logos and slogans give the **impression** that investing or visiting an area makes you part of something special or exclusive. For example, Shetland previously used the slogan "Pride of Place" along with a logo that used a representation of wind, the waves and Shetland's musical heritage.

It wasn't easy talking to Herbert about rebranding his rural area...

- Rebranding can work equally well for both **rural** and **urban places**, by focussing on the particular **activities** or **features** of a place. The types of rebranding used depend on the **target-group of people** the place wishes to attract.

Deindustrialised cities can build new identities

Some **deindustrialised** cities have rebranded by **repurposing** parts of their industrial landscape.

- **Industrial heritage** — Some cities use their **industrial heritage** to their advantage. The history of a particular industry in a city, even once this industry has closed, can become the focus of a rebrand. For example, agencies **sensitively** redeveloped the former warehouses in the Albert Dock in Liverpool so that many of their original features are still present, even though their **function** is now different — there are now **restaurants** and **museums** in the old warehouses.

- **Creative arts** — Some urban areas try to move on from their industrial past. Instead, they may rebrand by capitalising on or developing a **connection to the arts**. A development agency might use a place's connection to music, theatre, paintings, photography and sculpture to highlight different aspects of a place to people and investors. This can be particularly successful if an urban place is trying to develop itself as a **creative hub**. For example, Canning Dock (next to Albert Dock in Liverpool) is almost unrecognisable from its industrial past as it now houses the Museum of Liverpool, the Open Eye Gallery and an office complex with bars and shops.

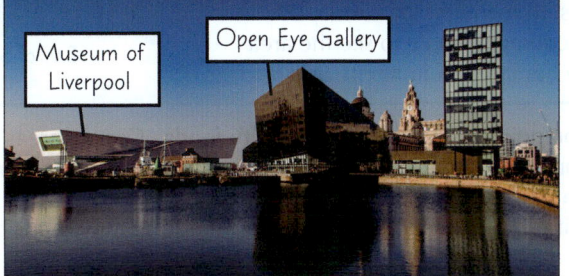

Canning Dock, Liverpool in 1910 (top left), in 2006 (top right) and in 2018 (bottom). Note the Port of Liverpool building behind the docks in all three pictures.

Rebranding Places

- Tourism — While most urban spaces aim to develop areas of **mixed use**, rebranding may focus on changing the way that a deindustrialised area is **perceived** by **tourists**. Making it more attractive to tourists can bring new forms of **investment** as well as **increased spending**.

Glasgow — 'Scotland with Style'

CASE STUDY

- Glasgow in the 1990s had an **image problem**. Though the city was starting to develop through **infrastructure projects** and more **cultural events**, many people still saw the city as suffering from the **economic difficulties** of the 1970s and 1980s. People were still living in **substandard housing** and were generally suffering from **poor health** compared to the rest of the UK.
- In 2004, Glasgow launched the 'Scotland with Style' campaign. The **slogan** aimed to promote the city as diverse, modern and forward looking. One aspect of the 'Scotland with Style' campaign was to celebrate the **city's connection** to prominent people like designer Charles Rennie Mackintosh. The city spent £1.5 million on **online** and **print advertisements** which it hoped would attract £42 million in investment. In the first two years alone Glasgow attracted £23 million in visitor revenue from the campaign.

Rural areas have different rebranding strategies

1) Some rural areas define themselves as being in a **post-production stage**. This means that the **primary sector** (e.g. farming, fishing and mining) is no longer the main source of **income** and **employment**.

2) The **isolated nature** of some rural areas can make rebranding **difficult**. Also, many rural places lack a centre or a **focus point** — e.g. they may cover large areas of farmland and be **naturally spread out**.

3) They also tend to have quite a **limited variety of attractions**. The activities that do take place in rural areas often take place **outside**, so they are **vulnerable to bad weather**. Also, some of the more popular outdoor activities **don't generate an income** for local people. For example, cycling, walking and photography on their own add little to the local economy as people may not spend money to take part.

Accessible rural areas

- Rural areas that have a greater degree of **accessibility**, such as those that are situated just outside a town or city, may find it easier to rebrand. These locations can offer greater potential for **diversification**.
- **Farm shops** and **craft centres** may have a market for their goods in nearby towns.
- Overnight visitors and people who wish to have a **quick retreat** from the city can frequently use more accessible rural areas. Farms that can diversify into **bed and breakfast** and **camping accommodation** in these areas are more likely to be successful.

A farm shop near the town of Girvan, Scotland

Remote rural areas

- Rural areas that are less accessible can focus their rebranding on their **wilderness** and the **remote nature** of the location. This can be a successful marketing tool in attracting people who wish to escape **busy urban centres**.
- For example, Shetland has successfully used imagery of the **natural environment** (often without any people or cars in photographs) to highlight the **beauty of the landscape** and encourage people to see the area as being **unspoiled by human development**.

Bay of Scousburgh, Shetland

4) Another strategy in rural rebranding is to focus on **speciality foods**. Certain areas can become **known for particular styles of food**. Some foodstuffs and beverages can have **protected status**. This means that retailers can only call certain items a **particular name** if they are made in a **specific area**. For example, lamb can only be sold as 'Shetland Lamb' if it is reared and prepared using **traditional methods** on the islands of Shetland.

Rebranding Places

Rural rebranding can attract more tourists to an area

1) The goal of rebranding is to **attract more tourists** to an area. Tourists that stay locally will spend money on **accommodation** as well as eating out in rural pubs and restaurants.

2) If done **successfully**, rural rebranding can attract people from both within the UK and internationally.

3) Rebranding can be done through:

Literary associations and film — Some rural areas focus their rebranding on the **cultural heritage** found in **literature**. They might also try to attract visitors if the **filming location** for a popular television series or film has taken place there.

Outdoor pursuits — Promoting **outdoor activities** that people can't experience in a town or city is an important part of rebranding rural areas. Rural areas on the coast might focus on activities such as **surfing** or **coasteering**. Inland outdoor pursuits might include **mountain biking**, **hiking** and **rock climbing**.

Nancy was strongly against the idea of co-steering.

Literary association — Brontë Country

- The village of Haworth in West Yorkshire lies about 8 miles West of Bradford. It is **famous** for being the place where the **Brontë sisters** wrote novels and poems during the first part of the 19th century.

- Visitors come to the village to seek inspiration from the same **landscape** that inspired the sisters — in particular the **moorland** to the west of the village. The moorland was the backdrop for Emily Brontë's *Wuthering Heights*.

- Much of the **village hasn't changed** since their time. Many of the buildings are **listed** and have a **protected status** and parts of the streets remain cobbled.

- The village has capitalised on the fame brought to it by the sisters, with multiple **tea rooms**, **bookshops** and **guesthouses** with a Brontë connection. The **Brontë museum** (in the former parsonage where they lived) is a big attraction in the village and receives around 90 000 visitors a year.

CASE STUDY

Haworth village, West Yorkshire

Warm-Up Questions

Q1 Give one difficulty with rebranding a former inner-city industrial area.

Q2 Describe the role that food might play in rebranding a rural area.

PRACTICE QUESTIONS

Exam Question — AS-Level

Q1 Assess the extent to which rebranding is only about increasing tourist numbers. [12 marks]

Scribble fests — exams rebranded...

Once you start looking for it, you'll see rebranding everywhere — logos, slogans, even fancy art museums. All these things (and more) are done to create positive perceptions of places. Don't forget, you need to know a range of rebranding strategies used in both urban and rural areas. Turns out a trip to your local farm shop for some cake could actually count as revision...

Measuring the Success of Regeneration

Right. This is the last part of this giant places topic and it's time to weigh up the pros and cons of regeneration. To make it more complicated, everyone has a different opinion and measures success in a different way. Yep, the real word is messy.

Regeneration is evaluated using **social**, **environmental** and **economic** indicators

The success of regeneration is measured using a variety of **indicators**. They're often the same as the ones that suggest a **need** for regeneration (see p.127-128).

Social indicators	**Health and life expectancy**	An improvement in **physical and mental health** can occur due to regeneration: • This may be due to the provision of **new health facilities** or the **improvement of outdoor spaces** so that people can be **more active** and exercise with greater safety. • It may be due to changes in types of **employment** — e.g. fewer people may be doing **shift work**, which can be damaging to workers' health. • **Food security** may have also improved with more people being able to access higher-quality goods.
	Demographics	• A **growth in population** suggests that people are moving into an area due to its popularity or the new economic opportunities that it offers. • A change in the **proportion** of **younger, working-age** people indicates that the area has greater economic potential or greater opportunities in **tertiary education**. • A decrease in **absolute** numbers in any age group is less positive — it may show the exclusion of some groups from the regeneration focus. It's possible that these groups have **moved out** of the area.
	Access	• Regeneration can have a positive impact on an area if people are able to access places more easily, especially if they face challenges to their **personal mobility**. • Increased **footfall** can be encouraged through improvements to **public transport** and the **upgrading of pedestrian areas**. • Increasing the amount of **wide pavements**, **pedestrianised zones**, improved **street lighting** and **cycle tracks** can make spaces more accessible and safer for people.
	Education	• Improvement in levels of **academic attainment** suggests that young people are becoming **more optimistic about their future**. • Access to higher levels of education might not have been possible before because of the need to **enter work as soon as possible** in order to support **household finances**.
	Deprivation	• Examining the **Index of Multiple Deprivation** (see p.116) scores before and after regeneration can show whether an area is generally more or less deprived than it was before — and by how much. • **Individual sector scores** (e.g. **crime**) can give more information about how a place has changed.
	Inequality	• The difference in **lived experience** between people directly affected by regeneration and unaffected people should be considered. • Regeneration can **inadvertently** create greater **inequality**, with developed areas showing huge improvements and other areas stagnating. This reveals the scope of the regeneration project — one with a very **narrow focus** is likely to allow existing levels of inequality to **continue**.

Connor's spending power was increasing by the day.

Economic indicators	**Income**	Higher levels of **personal income** are a common indicator of regeneration success. Greater incomes can generate more local **tax revenue** and will mean that local people have greater **spending power**, which will create more economic opportunities.
	Poverty	Lower poverty levels are **not usually the focus** of regeneration projects, but they are an indicator of their success. If fewer people are living on **low incomes** and more people have their **basic human needs** met, local authorities can spend more money on improving services for everyone (e.g. instead of spending the money on **support payments**).
	Employment	Increased levels of **employment** indicate that regeneration has created greater economic opportunities for people. Places with higher percentages of people in **full-time** and **permanent** employment may experience **higher levels of spending** as people are more likely to have more **disposable income**.

Measuring the Success of Regeneration

Environmental indicators	**Pollution**	Reduced levels of **air pollution** can cause people's general health to improve, and people may want to spend more time outside. Regeneration that reduces air, water and soil pollution will also improve the living conditions for plants and help **renew habitats**.
	Derelict land	An overall reduction in the amount of **derelict land** is an indication of regeneration taking place on **brownfield sites**. This improves how a place looks and may remove **contamination** from soil and ground water, where it has the potential to harm wildlife.
	Green space	An increase in the **amount or quality of green space** can be an indicator of regeneration benefitting areas of **wilderness**, **public parks** or **wetland areas**. Increased tree coverage through **planting programmes** may suggest that developers have attempted to **offset** any environmental damage done in other places.

When you're **measuring** the effects of **regeneration**, there are a few other things you need to think about:

- **Timescales** — E.g. it might take many years to see improvements in **educational attainment**, but **employment data** may change much faster.
- **Relative and absolute changes** — E.g. you wouldn't just look at the change in income in the regenerated area. You'd want to compare it to the change in average income **across the UK** over the same period of time.
- **The areas being compared** — E.g. you may want to compare how regeneration has affected the **different areas within a place**. Or you may want to compare the effects of regeneration against **a different area that has remained undeveloped**, or with other areas that have undergone **other types of regeneration**.
- **Multiple measures** — E.g. you may find regeneration has been an **economic success** for the whole region, but it may have had **negative social impacts** on people who lived directly on the redevelopment site.

Local people may have different opinions on the outcomes of regeneration

There are various factors that affect how local communities **view the outcomes** of regeneration projects:

- **Consultation** — If **consultation processes** with local communities take place **before** the confirmation of regeneration plans, local people may feel relatively **happy** with the **outcomes** of regeneration. However, when local people feel **top-down development** has been forced on them, or when **evictions** and **compulsory purchase notices** have to be issued, regeneration can feel **negative** — regardless of the eventual outcomes.
- **Integration** — If companies from **outside the local area** develop the plans, local people can be **sceptical** about the impact regeneration might have. After completion, local people may have the perception that the regenerated area is a **separate** part of the city rather than an **integrated** area.
- **Timescale** — If changes to the urban area occur **suddenly**, local people may not view regeneration positively. They may also rate the success of the regeneration on the **short-term changes**, as many will want to see an **immediate improvement** in the quality of their **day-to-day lives**.
- **Jobs** — Local people may also wish to see industrial regeneration that matches the **skill sets** of people who became unemployed from the **deindustrialisation** of urban places. Many urban redevelopments focus on **retail and leisure space**. While retail and leisure sectors might employ large numbers of people in **tertiary industries**, those who grew up working in **secondary industries** can feel **marginalised**.

Regeneration of Salford Docklands

In the 1980s, the landscape of Salford Docklands in Greater Manchester was dominated by **derelict warehouses** and **infrastructure** associated with the former industrial **port**. A series of redevelopment plans saw Salford become a large, modern residential area with improved **leisure facilities**. This included the Lowry Theatre and Gallery as well as the Imperial War Museum North. The BBC and ITV acquired **office and studio space** as part of the £550 million Media City development. The area also had a **rebrand** and was renamed Salford Quays.

Most people agree that there has been a **visual improvement** since the 1980s. However, local people question some aspects of the regeneration. **Post-modern architecture** now dominates the skyline and important buildings from Manchester's industrial heritage have been lost. Many of the new apartment buildings are **too expensive** for local people.

Some people miss the locally iconic Salford Cranes.

Two-bedroom flats that overlook the waterfront had a starting price of £400 000 in 2020. Many of the new jobs, like those in Media City, were taken by people who **relocated** to the area from elsewhere. When companies moved into the newly developed area, they often just **moved offices** rather than creating brand new jobs.

Measuring the Success of Regeneration

Rural and urban regeneration can be judged differently

- **Rural** and **urban** regeneration is often judged using very **similar criteria**. For example, both locations are likely to evaluate national and local regeneration strategies by how much **income** is generated and what **economic opportunities** it may bring to local people.

- In **rural** areas, economic growth is more likely to be measured against any **environmental costs**, such as the loss of unspoilt landscapes, noise pollution and use of natural resources.

- In **urban** areas, where population density is higher, regeneration strategies are more often judged against **social indicators** such as the well-being of residents and the **narrowing of any gaps in inequality**.

Lived experiences affect people's perception of urban regeneration

Stakeholders will have different opinions about which groups of people any regeneration is intending to **benefit**. They may also have different opinions about the **purpose** of the regeneration. Much of this is due to differences in the **lived experience** of stakeholders in a place.

Local authorities

- **Local authorities** judge the success of regeneration by using statistics such as census data, the Labour Force survey and the IMD (see p.116). They are likely to focus on **employment data** and the **long-term ability** of the area to attract further investment.

- They may seem removed from the **lived experience** of the place. However, some people working in the local authority are local residents who work for the council because they want to make a difference to the area.

National governments and authorities

- **National governments and authorities** monitor the level of **internal and international migration** to fill the job opportunities in the long term.

- They may see the regeneration as successful if industries become more **spread out** around the UK (as part of the **decentralisation** of industries from London and the South East).

- They may also see regeneration favourably if **TNCs** (see p.93) start to move into areas traditionally associated with **decline**.

- As national governments and authorities don't always have a **lived experience** of the places they regenerate, the regeneration is viewed at a **distance** and in the context of the **wider national economy**.

Local residents

- **Local residents** may feel a level of **affection** for where they live and how it **looks and feels** both before and after developments have taken place — this isn't easy to measure.

- People who have a **long-term connection** to an area (which may go back several generations) may have **different perspectives** to people who have less of a connection to the area (e.g. recently moved there).

Local residents who know an area well can feel strongly about new developments.

Local businesses

- **Local business owners** are likely to welcome regeneration because it may **boost their own businesses**. E.g. they may find it easier to **recruit skilled workers** or it may increase the number and spending power of their **customers**.

- Local businesses that have a long history in a place may feel **threatened** by any **new businesses** created. E.g. **family-run** businesses may feel **concerned** by competition from new shops that are part of a **chain**.

Local businesses that have been in a place for a long time may themselves become part of the lived experience for people living in a city.

Developers

- Other stakeholders may analyse the success purely in **monetary terms**.

- The **development corporation** in charge of the regeneration, or individual **property development companies** that have built on the site, are likely to measure success by the amount of **profit** they make and the **value of their shares**.

- Their **emotional attachment** to the site is likely to be minimal as they may have no lived experience of the area.

Measuring the Success of Regeneration

Different **factors** can affect people's views on **rural regeneration**

Large-scale developments are less common in rural areas, but regeneration and rebranding can still be **supported or opposed** within local communities. There are various issues that are contested:

Renewable energy — **Renewable energy developments** such as wind and solar farms are often the focus of **NIMBYism** ('Not In My Back Yard'). This is characterised by local people who are **in favour** of renewable energy, but don't want to have wind turbines near their house or to have solar panels covering fields near them. The same is true of **new housing developments**, which local residents often **oppose** even though they understand the need to provide more housing. Calling people NIMBYs can be an attempt by developers to **dismiss** the **concerns** of people who are against developments near where they live.

An attempt to stop a housing development in a rural area

Derelict land — **Derelict land** in rural areas can be viewed differently to derelict land in urban areas. Many local people are happy when **nature** starts to take over **former industrial sites** in rural areas (such as quarry pits and mining developments). Therefore, regeneration plans that seek to build on **rural brownfield sites** might not always get such a positive response from people in rural areas.

Access — **Access** to a redeveloped site is important regardless of location — but in rural areas, there may be **fewer road and rail links**, which can pose specific problems. For example, if a retail park in the countryside is **only accessible by car**, local people without a car won't be able to use it. Regeneration projects often place extra pressure on **road infrastructure**, which may not have been designed with higher volumes of traffic in mind — many rural roads are single lane and can easily become congested. As a result of their location, roads may also be of a **lower quality** and receive lower maintenance budgets.

Conservation — The **conservation** of natural spaces can be more contested in rural areas. There is an expectation that regeneration sites should have **minimal impact** on **local wildlife** and **ecosystems**. Local people are more likely to perceive these projects negatively if they fail to meet these standards. **Rebranding** that takes place as part of regeneration can sometimes be seen as **out-of-place** in a rural setting, especially where it aims to **commercialise** a natural, non-commercial space (such as a National Park).

Restructuring along the North Antrim coast

CASE STUDY

This section of the Northern Irish coast is most famous for the Giant's Causeway — an extended formation of basalt columns that attracted over 1 million visitors in 2018. To try and spread the **economic potential** of so many people visiting the area, and to develop industries outside of tourism and farming, there have been a number of **bottom-up initiatives** that have **restructured** the way that investments and redevelopment funds are spent. For example, the Causeway Speciality Market only allows the sale of local produce and crafts, promoting local businesses.

Between 2014 and 2020, the **Causeway Coast and Glens Borough Council** have encouraged a number of local development projects in the Moyle area of the North Antrim Coast. Under the **LEADER approach**, a **Local Action Group** (LAG) was set up to distribute **EU redevelopment funds** of £9.6 million and decide how best to regenerate the rural area in the long term. With 280 members from a **wide range of business** and **local interest** backgrounds, the LAG promoted devolution of decision-making away from **stakeholders** such as development corporations and the local authority.

They were able to provide up to £30 000 of **start-up investment** to new, small, rural, **non-agricultural businesses** as well as up to £100 000 to expand the provision of **broadband into internet 'dead zones'**.

The Causeway Speciality Market in Coleraine promotes local foods and crafts.

LEADER Approach
The LEADER approach was an EU initiative that sought to give local people the power to make decisions about the future of rural areas. They were allowed to set up partnerships between public and private sectors.

Measuring the Success of Regeneration

Lived experiences affect people's perception of rural regeneration too

Local residents

- **Local communities** and **residents** in rural areas may have **long-standing lived experiences** of a place. Multiple generations may have lived in the same village (especially where farmland is family-owned).
- This might mean that they will have seen and accepted **many changes over time** and so welcome regeneration.
- It may also mean that they have a deep **affection** for the place **as it is** — sometimes to the point of **nostalgia** about how **it used to be**. This makes them view regeneration less favourably, particularly if they feel it has not been done **sympathetically**.

Ah, the good old days...

Second homeowners and visitors

People who own a **second home** in a rural area, or are **regular visitors**, will have a different lived experience. They may only view the rural place in the **summer** or at a time when there are many other visitors like them. This means they may view the priorities of rural regeneration differently and favour projects that work best in **particular seasons**.

Local businesses

- **Local businesses** are likely to welcome a development that allows them to **expand their market**.
- In rural areas, the lived experience for business owners is often one of **collaboration**. Small businesses in rural communities tend to help each other, both **practically** (such as by sharing the costs of transporting goods out of the area) or through **connected business ventures**.
- People may perceive a new development in a rural area more **positively** if it involves local business communities. People may view it **negatively** if it creates **competition** for the businesses already there.

> *When revenue generated by one business gets spent by that business (or its employees) at another local business, that money circulates locally.*

Local authorities

- Local authority members in rural areas tend to have a lived experience of the place they work in. They're likely to think about the **long-term health of the local economy**, as well as the **traditions** and **heritage** of the place. These two things can be difficult to balance. A local authority is likely to view a regeneration project favourably if it can maintain that balance.
- Local councils in rural areas may feel more confident about **top-down developments** than those in urban areas because they have **fewer, smaller communities** to deal with.

National governments and authorities

- National government workers are more likely to have a **perceived sense of a rural place** rather than a lived experience. They may **romanticise** aspects of rural lifestyles which may not reflect the situation in villages. This means they may favour regeneration projects that reinforce stereotypes such as the '**quaintness**' of rural areas over their more **practical** needs.
- However, national governments tend to be aware of the **affection** people have for places like National Parks, and of their **responsibility to protect them** for future generations. They're likely to **think carefully** about developments in these areas so as not to upset the **public**.

Warm-Up Questions

Q1 Describe how you might measure the social success of a regeneration scheme.

Q2 Explain why local communities may view the outcomes of rural regeneration differently to developers.

Exam Question — A-Level

Q1 Evaluate the view that the success of a regeneration project should only be measured from the viewpoint of people with a lived experience of the place. [20 marks]

How does a fisherman measure his success? Net profit...

Yep, that's the best joke I have at the moment. In my defence, it's been a long topic and I've officially lost it, but there's still hope for you. Make sure you can evaluate the outcomes of regeneration from a range of viewpoints, et voilà! You're done.

Places — Changing Populations

The size and structure of the UK's population is constantly changing, just like my hair. Should I wear it up? Or down? Maybe a mohawk... Anyway, the next few pages will tell you all you need to know about the changing population of the UK.

The **population** of the UK has **grown** in an uneven way

1) The UK **population** grew from 56 million in **1981** to 67 million in **2020** — this represents a 20% **increase** in that time. Growth throughout that time period was **relatively constant**, but this wasn't always the case.

2) After the Second World War, there was a **baby boom** when soldiers returned to the UK, which caused the **population to increase**. **Smaller** secondary and tertiary baby booms followed as the first baby boomers had children and then grandchildren. However, the population increases weren't as dramatic as the first boom. This is partly because more women in the UK started to have **careers** and **delayed having children**. Recent **improvements in health care** have increased life expectancy, causing the population to increase.

Percentage population change 1981-2020	
South West	+29.1%
Yorkshire and The Humber	+12.4%
Scotland	+5.5%

3) The various **migration policies** (see p.154-155) that encouraged people to move to the UK also contributed to the population growth. Many migrants **started families** once they settled in the UK.

4) Not all regions of the UK have had the same levels of population growth. **Most growth** has been in the South and South West. There's been **some growth** in the Midlands, Wales and Northern Ireland. In the North of England and Scotland, there are areas which have had **no population growth** or even a **decline**.

London and the South-East

CASE STUDY

- There's been **rapid population growth** in this region (29.5% increase between 1981 and 2020). This is partly due to the strong growth of **knowledge, financial and service industries** which have attracted people from around the UK (and overseas) to work in them.

- The South East offers many options for **tertiary education**. In London, there are over 30 institutions of higher education that attract **young people** from all over the UK and from overseas. After their courses have finished, many graduates stay in the area to try and find **work** in London.

- These tertiary institutions also attract **high-tech businesses** wanting to capitalise on the availability of **highly-skilled labour**, which then attract even more people seeking employment opportunities.

University College London has a student population of over 40 000.

North East England

CASE STUDY

- There's been **slower population growth** in this region (1.7% increase between 1981 and 2020). This is partly due to the high levels of **deindustrialisation** that have occurred in the region and the **global shift** (see p.94) in manufacturing. Heavy industries such as shipbuilding, coal mining and steel production have **declined** and, in some cases, **closed down**. A factor in these industries closing down was the cheaper labour available overseas — this resulted in those companies shifting to countries such as China. This has caused **large-scale unemployment** in some towns and cities and there are **fewer economic opportunities** for younger people entering the workforce.

The Redcar steelworks closed in 2015.

- As a result of fewer employment opportunities, people have **migrated** out of the area to find employment. Many of these people have **moved** to the South East. Although economic, social and environmental **regeneration** has occurred in much of the North East, it's been a slow path to recovery.

Places — Changing Populations

The **rural-urban continuum** is one way of understanding places

Places can be arranged on the **rural-urban continuum** (a model showing how rural or urban places are).

This is the general trend in population change over the **last 50 years**.

This is the number of people who live in an area **per square mile** (or **kilometre**).

			Population Change	Population Density	Service Provision	Land Use
Rural	**Remote hamlets**	*Isolated communities, just a few properties*	Decline	Low	Low	Primary industries
	Commuter villages	*Made up of people who work in a nearby town or city*				Residential
	Overspill towns	*Built just outside a city to accommodate commuters*	Growth	High		
	Suburbs	*Residential area of a city, furthest from the city centre*				
	Inner city	*Residential areas closest to the city centre*	Decline			
Urban	**Central Business District (CBD)**	*The city centre*		Low	High	Commercial

This is low because there are lots of **commercial properties** in the CBD.

Places with **high service provision** have many services like hospitals, schools, public transport, retail and leisure facilities.

There are many causes of **uneven population levels**

Pollution

Rural areas can be attractive to people if they're seeking a **quieter** and **healthier** lifestyle. Rural areas have lower **noise** and **air pollution** levels. Prevailing winds blow pollution away from the south-west and towards the east. This can affect the **population density** on each side of a town, e.g. 'east ends' tend to be **more densely populated** as wealthier people, who can afford larger homes, move to the west side.

Access

Many people wish to live in **suburban** parts of urban areas as the location is **convenient** for their place of work. This is especially true when an urban area has an extensive **public transport** network. Access to places of work can be more challenging in rural areas, with **travelling** into urban areas taking longer.

Physical Barriers

Physical features, such as **mountains**, can make access in and out of an area difficult. **Flat land** is easier to build on than **mountainous land**, so settlements might **expand more rapidly** in flatter, rather than in hillier locations. Extreme rural climates may not be attractive to some people as **stormy and snowy weather** can cut access or services off.

No barrier can stop Jamal

Land Use

Rural areas are often used for **activities** that spread out over large areas of land, such as **farming** and **mining**. This means that the land use causes rural populations to be more spaced out, resulting in there being a **lower population density**. In urban areas, the land use has more of a **commercial focus**. This provides employment opportunities and can encourage more people to live there.

Planning Policy

The Town and Country Planning Act (established in 1947) allows local authorities of urban settlements to protect a ring of land around their perimeter. This undeveloped land is known as the **Green Belt**. Green Belts prevented the **outward spread** of settlements and made **inner-city** areas become more **densely populated**. In inner cities, plans for **high-rise** blocks and **brownfield developments** (land that has previously been developed) were authorised. Authorities also built new suburbs outside the Green Belt and these **commuter villages** have become heavily populated.

The five points at the top of the next page give more causes for uneven population levels.

Places — Changing Populations

1) **Suburban** areas allow relatively **easy access** to the urban centre and usually have more space and greenery. Many people with children choose to live in suburbs as there's a **wider choice of schools** and they offer a **pleasant environment** to raise a family in. Suburbs generally have train lines that run into the centre of towns and cities, allowing people to **commute** more easily.

Liverpool and Lerwick (in Shetland) will be used as examples of two contrasting places throughout this section.

2) **Fertility rates** affect the population growth of an area. **Urban areas** tend to have **higher fertility rates** and **birth rates** (the number of babies born per 1000 people in the population), which **increases** the urban population each year.

3) **Remote rural areas** tend to **have higher death** rates as they're more likely to have an **ageing population**. This means remote rural areas are also more likely to have a **declining population**. However, some rural areas like Shetland have a **lower death rate** than some urban areas because the location makes having a **healthier lifestyle** easier.

Location	Birth Rate (2016)	Death Rate (2016)
Liverpool	12.4	9.2
Shetland	13.3	10.0

4) Rural areas in the UK tend to receive migrants from within the UK (**internal migrants**) while urban areas receive more **international migrants** (from overseas). International migrants tend to move to urban places due to greater **employment opportunities** and **social clustering**. This is when people group together due to their **shared nationality, religion, culture or customs**. International migrants are more likely to find people that **share their background** if they move to highly populated urban areas.

Net number of long-term migrants (staying 12 months or longer) in 2016		
Location	International	Internal
Liverpool	+3453	+1078
Shetland	+39	-54

5) **Past migration policies** have **encouraged** international migrants to migrate to some areas and not others. For example, following the Second World War, the UK government **encouraged migration** from the then British colonies to UK cities. This was to fill **labour gaps** in specialist areas, such as the London transport network. The **expansion of the EU** in 2004 to include eastern European countries meant that many people from these countries (such as Poland) **could work in the UK**.

The **characteristics of a population** can vary spatially

1) The **population structure** (the age, sex and ethnic diversity) of different places also varies spatially.

2) Rural areas tend to attract **older people** while **younger families** usually want to live in suburban and inner city areas.

3) Although **similar numbers** of girls and boys are likely to be born in a place, some places **attract more** of one sex than the other through migration.

4) **Ethnic diversity** also **varies** between different places. Urban places are **highly diverse** with people from many different ethnic groups. These places are described as being **heterogeneous**. Rural places often have **less diversity** and the population there represents fewer ethnic groups. These places are **homogeneous**.

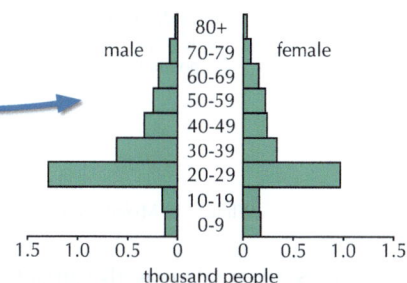

The central Liverpool **census area** had a population of 5436 in 2011. A large proportion of the population were **young and of working age** (20-39) due to the high availability of work in the area, and there was a high proportion of males.

These **spatial variations** are caused by **different factors**

1) Some areas may have higher numbers of men than women due to the **type of employment** available. In Lerwick, the sex ratio is **roughly equal** across the whole population, but there are slightly more men in the **working-age categories** (including ages 10-59) and more women in the **age 80 and over category**. This suggests that the **fishing and energy industries** of the area tend to attract more men than women. The greater number of elderly women reflects the **longer life expectancy** of women in Lerwick. In Liverpool, there are noticeably more men than women aged 20-49, suggesting that the city is **more able** to offer work opportunities to men.

2) Urban places attract **young people** because these areas have higher levels of **economic activity**. This means that it's more likely that younger people will be able to find **employment**. Compared to rural areas, urban areas have a large number of **social and entertainment spaces**, making them more attractive to younger people.

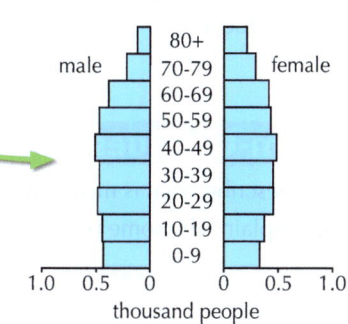

Lerwick had a total population of 6958 in 2011. There was a **lower proportion** of working-age people, and a higher proportion of older people compared to central Liverpool. As a result of the **population structure**, there are concerns that Shetland has an **increasingly ageing population**.

Places — Changing Populations

3) **Remote rural areas** have a higher proportion of older people as they're more attractive to people who wish to **retire** to quiet and beautiful surroundings. Having access to a large town or city isn't always as important for retired people, so they may choose to be **further away from economic centres**.

4) Where older people are in rural employment, it's often in jobs that might not attract as many young people, such as **farming** (the **average age** of farmers in the UK is 59). The **mechanisation** of farming has reduced the number of employment opportunities open to young people in **rural areas**, so many leave **these areas** to look for work elsewhere.

5) Migration can affect the character and structure of a population. Men tend to migrate more than women, both **internally** and **internationally**. International migration also increases the **ethnic diversity** of urban places. For example, the **increase** in international migrants from former colonies moving to the UK after the Second World War (see previous page) resulted in many areas of the UK becoming **more ethnically diverse**.

6) Cities tend to **attract international migrants** as they're better able to cater to their wider range of **cultural needs**. There are practical benefits to **social clustering** (see previous page) within cities, e.g. migrants might share a **language** or **religious practice**. These parts of the city may start to cater for these ethnic groups as migrants set up their own businesses and facilities (such as **places of worship** or markets that sell **food** from their home country).

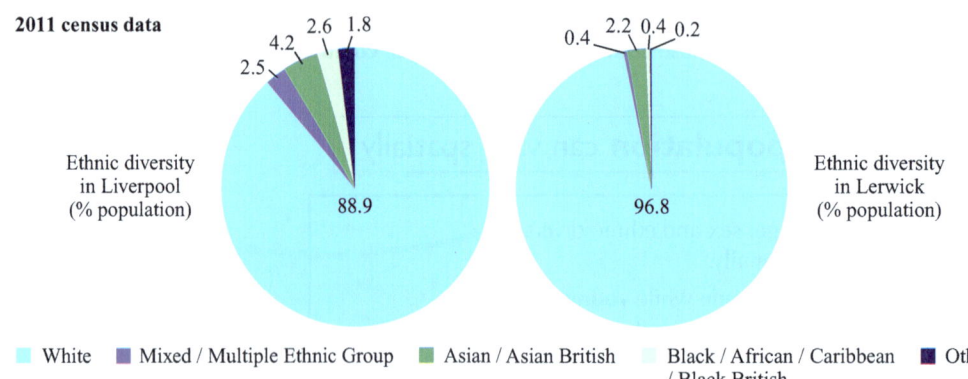

2011 census data

Ethnic diversity in Liverpool (% population)

2.5   4.2   2.6   1.8   88.9

Ethnic diversity in Lerwick (% population)

0.4   2.2   0.4   0.2   96.8

■ White   ■ Mixed / Multiple Ethnic Group   ■ Asian / Asian British   ■ Black / African / Caribbean / Black British   ■ Other

7) Some rural areas also attract international migrants. Workers from overseas often **harvest fruit and vegetables** in areas like East Anglia. The **fish processing industry** in Lerwick tends to attract Polish workers and their families.

8) The arrival of international migrants can be a catalyst for **internal migration**. Some areas of cities have experienced a wealthier population leaving an area to acquire **more expensive property** outside the city. This can leave a more **ethnically diverse population behind**, or the remaining population can become dominated by a new ethnic group.

9) In time, new ethnic groups might migrate into the area and change the ethnic mix again. Local authorities can inadvertently create **ethnically homogeneous** areas within cities by allocating **social housing** in particular areas to some socio-economic groups and not others.

Warm-Up Questions

Q1 Describe what is meant by the rural-urban continuum.

Q2 Explain why some parts of the UK have greater population densities than others.

Exam Question — AS and A-Level

Q1 Explain why the characteristics of a population may vary from place to place.   [6 marks]

I never used to like surveys but then I came to my census...

As you've discovered from these pages, population change can come about for a variety of reasons. You need to make sure that you can write about these reasons, how they impact people and places, and how they might make people feel.

Places — Influences

Places, like people, can be influenced by many things. Different forces, at a range of scales, can affect the character of a place.

Places can be shaped by **regional and national influences**...

1) Physical characteristics of the landscape can shape the character of a place. Coastal settlements will more easily form **trade connections** with other coastal settlements. For example, Liverpool's location on the River Mersey allowed easy **trade** with and **migration** from Ireland. This contributed to a history of Irish migration to Liverpool. Physically **isolated** places, such as Lerwick, can experience the opposite effect.
Fewer migrants come to Lerwick, which affects the character of the town.

2) People often have to **adapt** their lifestyles to suit the **conditions** of the place they live. This may come across in their **behaviour**, **clothing** and **food**. For example, people in Lerwick may be characterised by their strong understanding of local weather conditions. This comes from the **importance of weather** to local industries such as fishing.

3) **Transport connections** to other areas affect places. Cities tend to be extremely well-connected to other areas, as multiple transport routes (e.g. roads and railways) go through them. Smaller towns have fewer **road connections** and small towns like Lerwick may have no **rail connections**. Though ferries and aircraft connect Lerwick with Aberdeen and Kirkwall, the town remains **relatively independent** and isolated which has shaped its character.

4) Having **nationally important museums and galleries** in a place can shape its character. For example, Liverpool has the Tate Liverpool art gallery and the World Museum, which displays artefacts from around the world. These institutions add to Liverpool's character and contribute to the city being known as a **culturally rich** place.

5) Places can have long associations with certain industries. **Fishing** has been the main industry of Lerwick since the 17th century and Liverpool became an important hub for **textile** imports during the **Industrial Revolution**.

...and by **international** and **global influences**

International Migration	**International migration** can influence a place through **food**, **music**, **language** and the **religions** brought to the area. For example, The Al-Rahma mosque in Liverpool has space for up to 2500 worshippers, reflecting the religious diversity of the city. In Lerwick, the local Scots **dialect** still contains a number of **Norn words**, a language used by Vikings from Norway in around the 9th century.
International Visitors	A place may try to **attract international visitors** by promoting its unique features. E.g. 310 000 people visit The Beatles Story exhibition in Liverpool every year and 60% of those come from overseas.
Global Brands	**Global brands** can dominate **retail** spaces and, in some cases, **TNCs** may set up part of their business in an area, **creating jobs** and adding to the local economy.
Transport and Internet Links	**Transport and internet links** can connect places to other countries. E.g. Liverpool John Lennon Airport connects the city to around 25 countries and handled over 5 million people in 2019. Places that have **fast broadband networks** can more easily make connections than those that don't.
Conflict and Competition	**Conflict and competition** may affect a place. E.g. the mackerel fishing industry in Lerwick faced **strong competition** from Iceland in 2010 — Iceland's **quota** was **increased**, allowing them to catch higher volumes of fish.
Cultural and Sporting Events	Large **cultural and sporting events** can shape a place. Liverpool was designated a European Capital of Culture in 2008 which raised the **cultural profile** of the city. The city has hosted numerous international football matches at Anfield stadium and Liverpool Football Club is known around the world.

Warm-Up Questions

Q1 Describe how a place can be shaped by national influences.

Q2 Describe a global connection Lerwick has.

Exam Question — AS-Level

Q1 Assess whether places are impacted more by global or national influences. [12 marks]

Landscape — the original social influencer...

You can apply the influences on this page to any place. Have a go at using them to see what your area has been influenced by.

Places — Representations and Identity

The way that places are presented influences how people think of them. Here's a few pages to help you get to grips with all the different ways places are represented and how this impacts their perceived identity.

Places can be represented using **qualitative data**

1) Some forms of representing places are **qualitative**. This means they don't have a numerical score and the representation may be more **descriptive** or **creative**. Different forms of qualitative data can create contrasting representations of places. When investigating places, it's important to look at a variety of **different sources** to get a complete picture of what a place is like.

Maps
- Maps can be used to show any sort of **data** that involves a **location**, e.g. they can show where physical features are. They can also show **quantitative** demographic and economic data, e.g. different levels of income by location when used with **GIS**.
- Maps can also show **qualitative** information, such as the type of vegetation that can be found in an area — this can help to give the reader a **sense** of what the landscape of a place is like.
- Maps can show you **reliable** data, but can also be **misleading**. E.g. historical maps may be inaccurate — the 14th century Hereford Mappa Mundi is **part map**, **part artwork** and depicts scenes from biblical events and classical mythology.

Photography, films and art
- Visual representations show what places **look like** and can give some sense of the **character** of places. However, they only represent what the artist **wants** to show you, and can be **misleading**.
- Photographs show what a place looks like in a **given moment** — photographs taken at different times of day can make a place **look** and **feel different**. Photographs can also be **altered** to make places look different to the reality.
- Films and television give a **sense of place** that's dependent on the **nature** of the story being told, e.g. a crime drama set in a city might give a different sense of place to a romantic drama set in the same city.
- Paintings or sculptures can be **less reliable** than photography and films at showing what a place **looks like** as they're an artist's **interpretation**. They can be more effective at conveying a **sense of place** and character though.

John Constable's 'The Hay Wain' (1821) gives his impression of the Suffolk countryside.

Stories, articles, music and poetry
- Written representations can be used to **describe** places and can give a sense of how it **feels** to be in that place. They usually only offer the perspective of the **author** though, and may not show a **complete picture**.
- Newspaper articles can give lots of **detail** about places, but they may be **biased**, e.g. newspapers may focus on topics and ideas that are likely to sell more copies, rather than give a balanced perspective of a place.
- Stories, music and poetry can give **emotional impressions** of places, but only from the **writer's perspective**.

2) In all forms of **qualitative representation**, the **purpose** of the representation and the **agenda of the presenter** will influence what's presented. There's **subjectivity** in all forms of qualitative data.

Places can be represented using **quantitative data**

1) Some forms of representing places are **quantitative**. This means analysts can use **number scores** and **statistics** to judge a place.

2) **Composite indicators**, such as **indexes**, combine more than one set of data together. This gives a broader picture of what a place is like. For example, the **OECD Better Life Index** looks at what people think of where they live according to criteria such as **work-life balance**, **civic engagement** and **safety**. It then converts these scores into an overall, **weighted score**.

3) The **Index of Multiple Deprivation** does something similar for small areas in England, Scotland, Wales and Northern Ireland. Divisions (different areas within the UK) are based on population (each area has roughly 1500 people) and measured against **several indicators of deprivation**.

Type of deprivation	Liverpool (decile rank against rest of England and Wales, 2019)	Lerwick (decile rank against rest of Scotland, 2020)
Education	1	6
Health	1	6
Crime	1	4
Housing access	9	4
Income	1	6
Employment	1	5

In this data the small areas are split into deciles. Deciles divide the ranked areas into ten equal groups — from the most deprived (1) to the least deprived (10).

Places — Representations and Identity

4) Statistics, such as **census data** provide analysts with **quantitative information** about what places are like. The information might be about the population number, population structure, average incomes and household data of a place. Statistics can be in the form of **raw data** or **visually represented** using **graphs and charts**.

5) All forms of **spatial data** can be useful for building a picture of a place. For example, **crime data** includes information about safety, deprivation and number of victims in an area.

6) **Voting and election data** provides information about **public engagement**, **political activism** and what **political ideas** are important in different places.

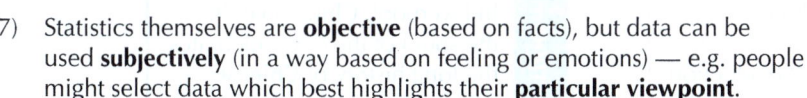

- In the 2019 general election, the voter turnout was 67.3% but there were great differences in turnout regionally. In Liverpool, the constituency of St Helens North had a voter turnout of 62.9% while Wirral West had 77.3%.

- In general, voter turnout is lowest in the areas of big cities where income is low and is highest in the wealthier parts. This is true in Liverpool, where St Helens North ranks in the 10% most deprived areas of England and Wales, while Wirral West is in the 20% least deprived.

- In local council elections, voter turnout tends to be low — typically around 35%. In the 2018 local elections, Knowsley in Liverpool registered one of the lowest voter turnouts at 25.3%. For referenda that can change a local area permanently, the voter turnout can be much higher.

- In the 2014 Scottish independence referendum, Shetland had a voter turnout of 84.4% compared to 67.7% in the 2019 general election.

The Scottish independence referendum generated a significant voter turnout and strong feelings.

7) Statistics themselves are **objective** (based on facts), but data can be used **subjectively** (in a way based on feeling or emotions) — e.g. people might select data which best highlights their **particular viewpoint**.

8) Statistics can sometimes **mask the real geographical picture** of a place. For example, a place might have **low unemployment figures**, which would be seen as a positive characteristic. However, a **closer analysis** of the data might reveal that the majority of people are in low-paid, part-time or temporary employment, which economists view less positively. Statistics also don't usually tell you anything about a **sense of place** and can quickly become **out of date**.

People can be affected by **continuity and change in a place**

1) Once rural areas have incorporated **urban features** and increased the **density of housing**, they rarely return to rurality. Urban **lifestyles** may increasingly take over village life, causing the rural character of the place to slowly become **less prominent**. This may include villages losing **traditional forms of work** such as quarrying. This changes the way people **perceive** their homes and their futures.

2) The **diversification** of farmland has created new forms of employment in rural areas. Many farms have looked to generate **renewable energy**, build **new accommodation options** for holidaymakers and **offer leisure pursuits** (such as alpaca trekking). This has created **more job opportunities** for young people, which might encourage them to stay in rural areas once they've finished their education. This means that some rural places have become **more attractive** to young people. Rural economies that continue to be based around **traditional farming** sometimes **fail** to attract new workers to an area. This may lead to these farms going **out of business** and the local economy becoming **weaker**.

3) Urban spaces that have experienced **deindustrialisation** may have **higher levels of unemployment**. People who have lost their jobs may also feel the loss of an **imagined connection** between their home and a particular industry. Former employees may feel **nostalgia** and **romanticise** the **post-industrial landscape**. If there's no further industrial change and the post-industrial landscape continues, younger people or new migrants may view the **abandoned buildings** as a symbol of a **poor economic future**.

4) The rollout of more efficient **internet services** has enabled more businesses to operate from rural areas. This has also encouraged young people to move to and stay in villages. It allows people to **work from home** and reduces the relative importance of **access** to urban spaces from rural ones.

5) **Migration policies** that have encouraged people to move to the UK from overseas have **improved urban economies**. Migrants have traditionally filled **job shortages** in urban areas. This has encouraged further numbers of migrants to seek work there. As urban centres became **more crowded**, migrants spread away from the main cities and looked for work in **more rural locations**. Wealthier people have moved out of urban areas to avoid the **increasingly dense** living conditions.

Places — Representations and Identity

(6) The growth of **commuter villages** and **suburbs** have **increased housing pressure** in these areas too. Some long-term residents of rural villages don't always welcome the arrival of new residents. This is particularly true if those new arrivals spend a lot of their **time** and **money** outside the village. Some people who move to commuter villages from cities retain their **urban-focus**, socialising in towns and sending their children to schools **outside the local rural area**.

(7) To cope with pressure on housing, UK authorities **demolished terraced housing** in many inner city areas and built **high-rise flats** in their place from the 1960s to the 1980s. This changed the **look** and **feel** of urban spaces. Many people felt they had lost a **sense of community** when this happened as people became more **isolated** from each other. By the late 1990s, new developments removed many of these tower blocks. **Urban development corporations** have now sought to create more **community-oriented** urban housing.

Local people's **identity** can be affected by **economic** and **social changes**

1) A person's **identity** is a set of **qualities** and **values** that are unique to them. A **community** can form where multiple people share similar identities. These people often have **shared beliefs**, **shared ways of life** and may form a **collective identity**. When that collective identity becomes **connected** to a place, changes to that place can affect how people view themselves. This can alter the **sense of belonging** people have to a community.

2) As the economic fortunes of a place change, so does the amount of **migration**. As an area becomes more **prosperous**, diverse communities may start to move into the area. Migrants bring their **own cultures** and ways of life with them. These can **enhance** the local community and make people **more outward-looking**.

| **Lerwick** | Though Lerwick has limited ethnic diversity, it regularly celebrates the **Nordic connection** it has developed. The retention of Nordic street names and Viking festivals **enriches the culture** of the town. | **Liverpool** | In Liverpool, Chinatown (in the south of the city) celebrates the long history of **Chinese immigration** to the city through its architecture and restaurants. |

3) However, the addition of new cultures might lead to feelings of **cultural erosion**. If a place has a strong identity, some people may feel that the influx of migrants with a different identity can **weaken** this. Feelings of cultural erosion can lead to **tension** between different ethnic groups. For example, the 1981 Toxteth riots (see p.150) highlighted the **economic division** existing between white Liverpudlians and **first and second generation migrants** to the city. Another factor in these riots was how black people **felt targeted** by the police (e.g. being stopped and searched), resulting in **anger** towards the police.

© Homer Sykes / Alamy Stock Photo

Riot police arrive at a fire during the Toxteth Riots.

4) When forms of industry change, a **growing divide** may occur between those who are able to work in the new economy and those that can't. A **rise in inequality** can affect the identity of people in a place. Those who previously felt like essential workers may now feel **marginalised**. Economic inequality can also lead to **power inequality** in **local decision-making**, affecting the extent to which people feel they **belong** in a place.

Warm-Up Questions

Q1 Explain how the statistics about a place can be used subjectively.

Q2 Describe how improved internet access has affected people who live in rural areas.

PRACTICE QUESTIONS

Exam Question — AS and A-Level

Q1 Explain why increased migration into an area might affect different people's sense of identity. [6 marks]

Exam halls — places of stress and intense concentration...

Perceptions of places aren't only restricted to urban or rural areas. People form perceptions of all sorts of different places. For example, one person's perception of school could be very different to someone else's — it could be positive or negative.

Perceptions of Urban Places

Urban places can be viewed very differently by different people. Read on to find out more...

A **lived experience** gives a certain **perception** of a place

1) A **perception** of place is how people view a place and how they **imagine** it to be. Different people will have different perceptions of a place due to their **relative experiences** of it.

2) A **media experience** of place is when someone only observes a place through media outlets. This might be films, social media, news, music or art. These people have an **outsider perspective** of a place.

3) People who live or work in an area have a **lived experience** of the place. This is the knowledge and understanding they may gain of a place from their **day-to-day experience** of it. These people have an **insider perspective** of a place.

4) Different factors can affect how someone perceives a place. A person's **gender**, **ethnicity** and **age** (or **life cycle stage** — the stage they're at in their lives) are all factors. A person's **socio-economic status** is also important. A person with good health, a high income and a high level of education will view a place differently to someone who experiences more **deprivation**.

5) People may value different aspects of a place. Some people may want a place to be highly **connected**, through **transport** and the **internet**. Other people may value the **natural environment** and place green space and clean air high on their list of priorities. Most people value **employment opportunities** and **affordable living**.

Daisy took her efforts to gain an outsider perspective very seriously.

Urban places can be seen as **dangerous**...

1) Urban places tend to have a **high population density**. This means that people frequently live among a large number of people they don't necessarily know. This can create the **impression** of a city being full of **strangers**, which some people view as **dangerous**. High levels of **poverty** in some UK cities has created **crime hotspots**.

2) Many cities face **air pollution** problems. **Traffic congestion** and **polluting industries** can release a lot of **particulate matter** into the city air. This is bad for **human health** and so some people view cities as less safe places to live.

Victorian London

- London developed rapidly during the **Industrial Revolution** at the start of the nineteenth century. Poorer people **migrated** to London from the surrounding **rural areas** to try to find work. The population of London grew from just over 1 million people in 1801 to 6.5 million in 1901.

- The **city authorities** weren't able to cope with the **pace of migration** and often refused to provide housing for workers. This meant many of the poorest workers lived in **slum terraces** close to the factories. These terraces were often **inadequate** and **unsanitary**.

Queuing for water in Bethnal Green, London (1863)

- Few residents had access to **running water** or a **sewerage system** and there was **no legislation** in place to ensure that their basic needs were met. **Extreme poverty** led some children into crimes such as **petty theft** and **pickpocketing**. By the mid-nineteenth century, Victorian London was characterised by high levels of **noise**, **pollution** and **squalor**, things the government felt the migrants had '**chosen**' for themselves when they decided to move to the city.

- The dense living conditions among the poorest Londoners created a **negative perception** of the capital. The poverty stricken areas were quickly associated with **high crime rates** and **high levels of disease**, e.g. cholera and typhoid. This made them appear **threatening** and **dangerous** to outsiders. **Wealthy residents** moved out of the **inner-city** areas to the **suburbs**. This created a clear **class divide** between the two areas.

3) Many UK cities still have **cramped** and **poor quality housing stock**. Though there have been improvements since the Victorian era, 8% of households in London are considered **overcrowded** and around 56 000 households are living in temporary accommodation that doesn't fit their needs.

...and **attractive**

1) Some people find the idea of living in the **inner city** very **attractive**. Some people value living close to their place of work. This means that they're likely to spend less time (and money) **commuting** to work. This allows them greater amounts of **leisure time** and a better **work-life balance**.

2) Inner-city living also allows people easy **access** to social and leisure activities. **Commercial** and **transport** services often work later hours (if not 24 hours a day) in urban areas. This means that people who live in the city can enjoy a more **flexible use of their free time**.

3) Although **living costs are higher**, urban areas often attract people because they usually pay **higher wages** than rural areas.

Perceptions of Urban Places

Outsider and insider perspectives of urban places can be negative

1) If negative things happen in an area of a city, that area can quickly gain a **negative reputation**. Cities, and especially the **inner-city** areas, tend to have **higher crime rates** which can lead to the city being viewed as a **dangerous place**.

2) Liverpool struggled to shift its negative reputation following the **1981 Toxteth riots**. News reports at the time depicted inner-city Liverpool as being **out of control**. The **media** in the 1980s portrayed the area inside the 'Toxteth Triangle' (a section of the city created by the intersection of three roads) as a **no-go area** for local police. Television series like the sitcom *Bread* depicted life for **working-class families** in Liverpool in the 1980s. It focused on the daily struggle many Liverpudlians had to find work and pay bills. It also featured an **overcrowded terraced house** in the area. All these factors made Liverpool appear to be an **undesirable** place to live.

3) Local people's **lived experience** can sometimes justify an outsider's perspective of an urban place. For example, data in the **Index of Multiple Deprivation** (see p.146) shows that Liverpool has **high levels of deprivation**. At least 90% of places in the UK experience less deprivation than Liverpool in six out of the seven measured categories.

4) Urban places are more likely to have **low environmental quality**. **Air pollution levels** in cities tend to be high and **noise** and **light pollution** also negatively affect people's lives. Though these are real lived experiences for people in cities, many urban dwellers get used to them.

Different groups perceive urban places differently

Groups of people urban spaces don't appeal to	Groups of people urban spaces appeal to
• People who are at different life-cycle stages have different priorities. For example, people who are retiring from paid work may wish to leave the city and move to a more rural location where houses are often larger and the environment is less polluted. • People entering retirement may not wish to live near the city as they may perceive it as too noisy and dangerous. People who have limited choices and continue to live in urban places can sometimes feel socially isolated as their peer group die or move out of the area. Elderly people who have lived their whole lives in a city can sometimes feel marginalised as the feel of the urban landscape changes around them. • Some people who belong to a particular ethnic group may feel isolated if there's no one of the same ethnicity living near them. This may be one of the reasons why UK cities experience social clustering (see p.143) in inner city areas. The proximity of particular places of worship and specific retail outlets can also cause social clustering. Racism is still prevalent in the UK and people with racist views may view urban spaces negatively because they're generally more diverse spaces. • Some international migrants may distance themselves from other migrants and integrate as quickly as possible into UK life. In these cases, they may avoid living in cities.	• Urban spaces often appeal to younger people because they tend to offer a wider range of entertainment options and job opportunities. There also tends to be a greater concentration of young people in urban spaces, which encourages other young people to move there. Easy access to public transport in urban spaces might also appeal to younger people who are less likely than older people to have their own car. • People with families are more likely to want to live in the suburbs of a city rather than the city centre. They might accept the extra length and cost of their commute because of the extra space available in the suburbs. There's generally a wider choice of schools available to families in the suburbs too, and public transport routes (such as fast train lines heading directly into the city centre) make commuting easier. • Migrants often see the inner city as having the most economic opportunities and as the cheapest place to live. This appeals to migrants who have a limited budget for setting themselves up in their new home. Cities are likely to be more ethnically diverse than rural spaces, which may appeal to international migrants.

I always thought 'suburban' areas were underneath cities and towns...

It's really important to know how perceptions of urban places change over time, so make sure you learn all this info.

Perceptions of Rural Places

The lived experience of rural places can be very different to the perceptions of outsiders.

Rural places are often seen as idyllic

1) The 'rural idyll' is the idea that rural places are attractive to a lot of people. This perception is based on the picturesque nature of rural areas. Images of green fields, hills and lakes create a stereotyped view of rural areas. They're also associated with peaceful surroundings, tranquillity and a slow and unchanging pace of life. Rural places are often associated with nostalgic and romantic views.

Some idyllic views of rural places that influence how people perceive them.

2) Idyllic rural buildings include thatched cottages, a village church and a country pub. There's a perception that quaint farm buildings are spread across the landscape and that most buildings have existed for hundreds of years. Outsiders may believe that rural areas have strong communities and that everyone in a village knows each other due to the low population densities. There's also a perception that people who live in rural areas have a worry-free lifestyle. People believe crime rates are low and that the crimes that do take place are petty and non-violent.

3) Many people see rural places as out of touch with modern living and being unconnected from the wider world.

Some rural places can be seen as undesirable by insiders

It's important to remember that these are perceptions of rural places.

1) Although the lived experience of some rural places matches the perceptions of outsiders, it can also be quite different.

2) Some rural areas can be quite isolating places. Sparse population densities mean that sometimes people might live quite far apart from each other. This means inhabitants might not have a close connection with their neighbours. The island of Shetland has a sparse population density of 16 people/km². Many of the villages outside of Lerwick are small collections of houses separated by large areas of farmland.

3) There are many challenges to living in Shetland:

- **SOCIAL ACTIVITIES** — Villages might not have a wide range of social activities or facilities for younger people. Village primary schools can be the centre of village life for children under the age of eleven. After this, there might not be things for teenagers to do in their spare time. In 2015, a survey of young people in Shetland found that young people rated gyms and leisure centres as their most available social facilities.

- **PUBLIC SERVICES** — Rural areas frequently suffer from a lack of public services. Remote villages are unlikely to have a doctor, bank or library. Rural post offices are difficult to justify keeping open as so many of the services they provide have moved online. The only banks in Shetland are in Lerwick, meaning some people have to travel long distances to access their services.

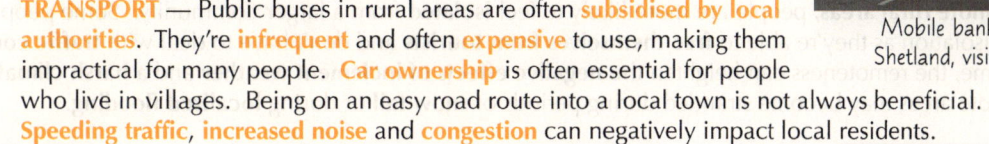

Mobile banks, like this one in Shetland, visit remote locations.

- **TRANSPORT** — Public buses in rural areas are often subsidised by local authorities. They're infrequent and often expensive to use, making them impractical for many people. Car ownership is often essential for people who live in villages. Being on an easy road route into a local town is not always beneficial. Speeding traffic, increased noise and congestion can negatively impact local residents.

- **CONNECTIVITY** — Phone and internet connections in rural areas are often not as reliable as in urban places, which can affect how easy it is to run businesses successfully. In Shetland, only 31% of young people felt that the speed of their broadband connection was good enough and just 33% rated the reliability of their mobile phone network positively.

- **BUILDINGS** — Some homes in rural areas are old and need maintenance. Older buildings have different levels of heritage protection on them, or they may exist within protected areas. The fixtures and fittings of listed buildings can be difficult to upgrade. For example, the heritage listing may prevent the upgrade of windows and doors to more efficient ones.

- **TOURISTS** — Rural areas that hold a particular interest for tourists can become inundated with visitors in summer months. This can leave local people frustrated with the levels of disruption this causes. Traffic congestion is likely to increase which might spoil the tranquillity that visitors are looking for.

Perceptions of Rural Places

Perceptions of rural areas might be **supported by media representations**

1) **Media representations** of rural areas often support the idea of **small communities** and **quaint lifestyles**. Programmes like the BBC comedy *This Country* highlight the **limited social opportunities** there can be for young people in rural areas. Television series such as *Poldark* and books like *Wuthering Heights* feature the rural landscape heavily in their narratives. Both emphasise the **remoteness** of rural places and the **beauty of wild environments**.

2) National newspapers and news channels **rarely report** events that happen in rural places. This gives the **impression** that rural areas are quiet, crime-free places to live.

3) Media representations sometimes **highlight alternative views** of rural areas. Radio soap opera *The Archers* highlights many of the problems that can come from living in rural areas. Storylines have focused on the domestic abuse that can happen on **isolated farms** and how political events, like Britain leaving the EU, can affect **employment** in rural areas.

Hardy's Wessex

- Thomas Hardy was an English novelist writing in the late nineteenth century. His novels *Tess of the d'Urbervilles* and *Far from the Madding Crowd* were set in the fictional county of Wessex. In these novels, Hardy describes what people today think of as a **traditional rural landscape**. Hills, forests and farms form the backdrop for most of the stories.

- Hardy chooses characters that appear to defy the **drive towards industrialisation** that defined the late nineteenth century. Instead, his characters have **rural occupations** and their storylines highlight the **divide between farm workers and wealthier landowners**. Hardy portrays the Wessex countryside as a place of great beauty and tradition. At the same time, the **harsh realities** of rural life play an important role in the stories.

The cottage in Dorset where Hardy wrote Far from the Madding Crowd

CASE STUDY

There are **contrasting views** of what rural places are like

1) The **type of rural place** a person lives in affects how they **perceive** rural areas in general.

Commuter villages
Commuter villages can provide people with the advantages of rural life but with the **convenience** of being close to a town or city. People with **young families** often want to have a **cleaner living environment** and more space. They find commuter villages to be a good **compromise between urban and rural living**.

Retired people
There's a perception that rural places are filled with **retired people**. These people are often attracted to rural places because of how **quiet** they are and the **slower pace of life** they can offer. The perception that rural places are mainly populated by retired people might put off **non-retirees** from living there.

Development opportunities
Developers might see rural places as just **undeveloped land**. **Housing shortages** mean that developers often target **greenfield sites** in **accessible rural** locations. Where these create **affordable homes** for local people, **young families** are likely to welcome these developments. Other people view these developments with concern as they often **remove natural spaces** and can spoil the **look and feel** of the village. The developments can also become hotspots for **second home ownership**, which makes it harder for local people to buy homes in the area. This can create tension between people who live in the village and people who visit it occasionally.

2) In the most **remote rural areas**, people are more likely to feel **isolated** from a larger community. Some people welcome this isolation as they're able to **free themselves from tourists** and don't have to deal with **traffic congestion**. At the same time, the remoteness can heighten the **negative effects** of **inclement weather** and a **harsh climate**. People in remote areas may become stranded during periods of **snow fall** or during **localised flooding**.

Warm-Up Questions

Q1 Describe a typical image of the 'rural idyll'.

Q2 Explain how media outlets might create a certain perception of rural places.

PRACTICE QUESTIONS

Exam Question — A-Level

Q1 Evaluate the extent to which a lived rural experience may be similar to the 'rural idyll'. [20 marks]

What does the Neighbourhood Watch do in rural areas? Lawn enforcement...

A good way of testing whether you know how different people perceive rural places is to draw a table comparing all the views.

Perceptions of Living Space

Like all things in life, perceptions might change over time — this page will tell you about what influences these perceptions.

Statistics can be used to evaluate a place

1) There's limited data which assesses how people **perceive an area**. Instead, local authorities use **census data** and **local area statistics**, which give information about the **population structure** of a place. The population structure gives authorities information about the **people** living in an area and how the population structure is **changing**. This helps authorities to see which groups of people tend to occupy different parts of the **rural-urban continuum** (see p.142).

2) Local area statistics might include information from the **Labour Force Survey**. This provides information on the average income for different jobs and the types of work in an area. This survey can **identify the types of lifestyle** people in rural and urban areas might enjoy and whether these lifestyles are **specific to particular areas**.

3) Statistics can go **out of date** very quickly. This means that data should be used with **caution** as outdated data can **misrepresent** the place it's about. For example, the **census** only takes place every ten years.

> Statistics also fail to represent how people feel about a place.

Media can provide contrasting representations of place

1) **Television shows**, **films**, **news broadcasting** and **social media** can affect a viewer's **perception of a place**. These types of media have the potential to reach a **large audience** and can easily **challenge** how people view somewhere. Media can give outsiders an idea about what a place is like by highlighting the **lived experiences** of real people. However, media may reinforce **stereotypes** through both non-fictional and fictional representations.

2) The way media portrays a place is **subjective** and doesn't provide the whole picture of how people perceive a place. Writers and journalists may have their own view of a place because of their own **lived experience** and may **favourably or unfavourably** present an area. Places may be made to appear worse than they really are to add more drama and **sensationalism** to a piece. This means the media can be a **controversial** way of creating a perception of a place.

Influencing the perception of Liverpool

- Statistics from the 2011 census for Liverpool highlight the difference in the **demographic makeup of households** in Liverpool's inner city and suburbs. This data **supports the perception** that the **suburbs** of Liverpool are more suited to people at the family stage of their **life cycle**.

- The media can promote a different view of Liverpool. Local people in Liverpool believed the **television programme** *Desperate Scousewives* created an unfair and **unrepresentative** image of Liverpudlians.

Household composition		Inner City	Suburbs
	One person household	53.4%	29.6%
	Parents with dependent children	14.5%	32.2%
Average age of residents		29.9 years	37.2 years

CASE STUDIES

Influencing the perception of Lerwick

- Statistics from the 2011 census don't reinforce the **perception** of **retirees** dominating rural places. People aged 65 or older occupy 17.4% of Lerwick households. This is close to the number occupied by **people with children** (19%). Though the number of retired households is high, it may not be high enough for Lerwick to be a considered a **retirement hub** (a place where retirees would outnumber young families).

- The media representation of Shetland is contradictory. Ann Cleeves is an English novelist best known for writing crime literature. Her first book set in Shetland, *Raven Black* focuses on murder and the idea that **close-knit communities** keep secrets from each other. The book depicts the Shetland **wilderness** as **haunting and suspicious**. The **lived experience of crime** in Shetland is quite different. Data from 2019 shows that 1.3% of crimes in Shetland were violent (compared to 1.6% in Scotland). Most crimes in Shetland are **motor offences**.

Warm-Up Questions

Q1 Suggest why a person should be cautious about using data to justify a perception of a place.

Q2 Explain why it can be difficult to form an accurate perception of a place through media sources.

Exam Question — A-Level

Q1 Explain why it's difficult to assess the validity of someone's perception of a place. [6 marks]

Our research shows that 100% of CGP's books are read on Earth...

While statistics are extremely useful, it's important to remember that they don't always tell you everything you need to know.

Cultural Diversity in the UK

The next three pages are here to explain how changing cultural diversity has affected the UK.

Internal migration has created diverse populations

1) Most of the northern cities in the UK had economies based on **heavy industry** and **manufacturing**. Since the closure of many of these industries, there's been a tendency for workers to **migrate** to southern cities — known as the **North-South drift**. This has created higher growth rates in London and the South-East compared to the rest of the country.

2) The movement of people through **internal migration** is often **selective**. Many migrants from the north and other areas of the UK migrate in order to find new employment opportunities. They tend to be **younger** and relatively **well educated**. Once internal migrants move into the South-East and become more established in a job, these people are more likely to start **families**. This means that the population of London and the South-East has a **lower average age** and a **higher level of educational attainment** than the North of England. This is an example of **regional disparity**.

3) The South-West of England is also a **migration hotspot**. This area generally has a **quieter lifestyle** and a **mild climate**. This attracts people who are **retired** and **increases** the average age of the population of the region.

London and the South-East

- Internal migration to London and the South-East has created **regional disparities** in the UK. In London in 2018, 39.4% of the population were under the age of 30 compared to 32.4% in Blackpool and 35.5% in Sunderland. 6.7% of working-age Londoners held no formal qualifications compared to 11.1% in Blackpool and 9.2% in Sunderland.

- The **rate of net internal migration** to London and the South-East is slowing down. This is partly due to **regeneration projects** in many of the northern cities creating improved **employment opportunities** in those areas. Another reason is the rapid increase in London's house prices. In February 2022, the average London property cost over £520 000 compared to around £150 000 in the North East of England. Young people moving to the capital are likely to face the **highest rents** in the UK. Many people move back to where they grew up when they want to **buy homes**.

- People are also leaving London and moving into the suburbs surrounding the city (in the area known as the **home counties**). This process is known as **suburbanisation**. Some of these people are retiring internal migrants while others are likely to be young families who require **more affordable** living space, but still need to be close to the city for work.

International migration has created a more diverse society

Post-colonial migration

The **British Nationality Act** (1948) gave British citizenship to people who moved to Britain from crown colonies, British overseas territories and Commonwealth countries. The majority of these migrants settled in cities, with London being the most popular choice. Some migrants took **low paid public-sector work**, such as in public transport or the National Health Service, while others were employed in **manufacturing industries** and **engineering**.

Migration from the Indian Subcontinent

Many international migrants from countries such as India, Bangladesh and Pakistan came to the UK between 1950 and the late 1970s. Many came to find work in the growing British economy and settled in cities such as Manchester, Bradford and Birmingham. Other migrants moved to areas such as Glasgow, Slough, Luton and boroughs in eastern and north east London. Many migrants aimed to move to cities with **manufacturing industries**. Some migrants **started their own businesses** once they were in the UK while others found jobs in existing businesses. Many migrants also found jobs within the **NHS**.

Migration from the West Indies

- One of the first large-scale international migrations to the UK included people coming from the **Caribbean islands**. The UK government encouraged these international migrants to come to the UK in order to **fill the labour shortages** left after the Second World War.

- The migrants are known as the '**Windrush Generation**' in reference to the name of the ship on which they sailed to the UK.

- The **vast majority** of these international migrants settled in south London boroughs such as Lambeth.

© GL Archive / Alamy Stock Photo

The Empire Windrush which brought migrants to the UK from the Caribbean

Cultural Diversity in the UK

Asylum seekers	People who are fleeing **persecution** in their own countries often apply for **asylum** in the UK. Between September 2018 and September 2019, around **19 000 people** received some form of asylum in the UK, the largest number since 2003. Most applications came from Iran and Iraq. 15% of asylum seeker resettlement in 2019 was in London and the South-East.
International migrants	More recently, international migrants have come from Eastern European countries, such as Poland. These are known as the **Accession 8 (or A8) economies** in reference to the eight countries that joined the EU from 2004 onwards. Migrants from these countries have **spread themselves more widely** across the UK in both urban and rural places.
Second generation migrants	Many international migrants had **children** once they'd settled in the UK. These children are **second-generation migrants**. These generations tend to live in the same or similar places to those of first-generation migrants. Although there's been some **internal migration** of second-generation migrants, **settlement patterns** created by different first-generation migrants largely remain in place today.

> The A8 countries are the Czech Republic, Slovakia, Hungary, Poland, Latvia, Lithuania, Slovenia and Estonia.

Some **international migrants** move to **rural areas**

1) International migration isn't something that's only occurred in **urban areas**. Migrants from the EU, particularly those from the **A8 countries**, have also settled in **rural areas**. This is often to fill particular **labour shortages**. For example, in Shetland, migrants are employed in the fishing and fish processing industries. The nature of this work can mean these job vacancies can be **difficult to fill**. The movement of international migrants to fill these positions can help rural areas to fulfil their **economic potential**.

2) In rural areas where there's an **ageing population**, international migrants can **lower the dependency ratio**. As earners within the UK, migrants contribute **taxation** and play an important role in the economy.

Eastern Europeans in North Lincolnshire

CASE STUDY

- In Lincolnshire, there's a large population of Polish nationals who migrated to work in **farming and food processing**. In towns such as Boston, Eastern Europeans represent a significant percentage of the population. More than 10% of Boston's population in 2011 was born in the **A8 countries**. Though some of the migrants have sought work in Boston itself, a large number work on farms in the **rural areas** surrounding the town.

- Few local people are interested in working on local farms as it can involve physically demanding work for **minimum wage**.

- Increased numbers of European migrants have resulted in new, **specialist shops** opening and **services** being provided for those migrants, which has added value to the local economy.

Ethnic segregation is linked to **economic and social factors**

1) Ethnic segregation is where **groups of people** from the same ethnic background live in the **same area**. This can lead to the **social clustering** (see p.143) of different nationalities and ethnic groups within a larger area.

2) There are **social and economic factors** that cause ethnic segregation:

Social Factors
- **Migrants** are more likely to feel **supported and protected** by people who share their **home language** and **cultural viewpoints**.
- Clusters of populations of a certain ethnicity may be able to gain **political power** and play a greater role in local decision making.
- Segregation may also occur due to the movement of middle-income groups out of some areas, leaving a particular ethnic group behind. This may occur due to **prejudice and racism** in the population.

Economic Factors
- Both areas of **deprivation** and **affluence** can become inhabited by different groups of people.
- **Racial discrimination** and the challenges of learning a new language can mean some international migrants are less likely to have **high levels of income**.
- This means they're more likely to occupy areas of **poor quality housing** and with **poor investment in health care** and **education**. As a result, some ethnic groups experience high levels of **deprivation**.
- Some ethnic groups might belong to a **higher socio-economic group** and be able to afford to **move to larger properties** outside the city, something that might not be possible for other ethnic groups.

3) Over time, social and economic segregation can lessen, as **second- and third-generation migrants** overcome the **initial challenges** experienced by their parents and grandparents.

Cultural Diversity in the UK

Russian oligarchs living in central London

Wealthy Russian oligarchs **bought properties** in areas of London like Belgravia. It was estimated that 7% of London's properties valued at £1 million or more were owned by Russians in 2014. Owning UK property gave Russian elites access to **UK bank accounts** and **private schools**, which were seen as highly desirable. These investments, among other factors, led to **price inflation** and made living in some parts of London **unaffordable** to people born in the city. Officials struggled to build enough affordable housing for **lower income groups** to meet demand. **Sanctions** placed on oligarchs following Russia's invasion of Ukraine in 2022 have **lessened their influence** in London.

Ethnic segregation can affect the **built environments** of **urban places**

1) Specialist buildings, such as **places of worship**, cater for the needs of different cultures in an area.

2) Ethnic segregation also affects the **retail landscape** of a place. In areas dominated by a particular ethnic group, shops are more likely to reflect the tastes and cultural needs of that group. For example, **clothing** and **food retailers** will sell goods that residents would be able to find in their home countries. International migrants are likely to have started these businesses themselves in order to serve their communities.

3) There may also be **community buildings** and **social spaces** that appeal to certain ethnic groups. People might run restaurants serving the food of their particular ethnic group.

4) This means that the built environment in some areas **reflects the ethnicity** of the people who live there. In some instances, **street signage** may be written in the **language** of the migrants who live in the area.

Southall, in the London borough of Ealing has changed as a result of the ethnic groups and communities found there. Almost all of the shops and services on *The Broadway*, the main road running through Southall, cater for the **Indian** and **Pakistani community**. There are numerous places of worship in Southall including around ten **Gurdwaras** serving the area's Sikh population (which is over 20 000). The signs at Southall station display information in both **English and Punjabi**. Desi Radio is based in Southall and broadcasts in the Punjabi language and plays **Bhangra music**.

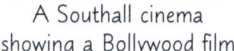

A Southall cinema showing a Bollywood film

Perceptions of culturally diverse places **change over time**

1) Many first-generation international migrants experienced **discrimination** and **prejudice** in the UK. This affected their ability to **gain employment** as well as **integrate** into the local communities.

2) Over time, international migrants have **set up their own businesses**. Many of these businesses have been successful and have **raised employment levels** in their areas. Migrants are now taking positions within **local politics**, giving them opportunities to serve everyone in their local area. As their affluence has grown, international migrants have moved to new areas, reducing the impression that certain ethnic groups are **segregated** into specific areas.

3) The position of different ethnic groups within communities has evolved as **second- and third-generation migrants** have embraced **cultural hybridisation**. This is where mixing tastes in **food**, **clothing** and **music** creates a new variant of a culture. Aspects of different cultures have **integrated** into the British way of life. For example, the **Balti curry** that's popular in the UK is a **hybrid** of Pakistani and English ingredients.

4) In many UK towns and cities, there are events that **celebrate the heritage** of international migrants. For example, the **Notting Hill Carnival** began as a way of celebrating **Caribbean culture** in London. It now involves people from many different ethnic backgrounds and focuses on the **wider diversity** of the capital.

Warm-Up Questions

Q1 Describe how internal migration has created different rates of population change across the UK.

Q2 Explain why international migrants may live in segregated communities.

Exam Question — AS-Level

Q1 Explain how cultural diversity can change the built environment of an urban place. [4 marks]

Top Tip #86 for budding geographers: Go to the Notting Hill Carnival...

The cultural diversity of a place is constantly changing. The area that you've grown up in might feel very different in 50 years. Anyway, make sure you know all about cultural diversity in the UK and how it's changed over the years.

Changing Places

Places are constantly undergoing some sort of change. These changes may bring about new challenges or opportunities.

Opportunities can come from changing land uses

1) **Investment** in urban and rural areas may come in the form of **regeneration projects** put in place by **regional and national governments**. These projects can create **new employment opportunities** for local people. They can also improve the look and feel of a space, potentially leading to further investment by **private organisations**. **TNCs** can benefit from **initial rent reductions** at a regeneration site and get **priority** over other, smaller businesses.

2) Along with regeneration projects, there may be improvements to **local infrastructure** such as **transport systems**. **New housing projects** and **public facilities** such as parks and leisure complexes may also benefit local people.

3) The creation of **new retail spaces and services**, like those created by entrepreneurial migrants to an area, can encourage **cultural hybridisation** (the exchanging and adopting of ideas and behaviours between cultures). Any form of new business in an area will also generate higher **tax revenue**.

The Gurdwara Sahib in Birmingham shares a meal with the community.

4) **New places of worship** built by migrants can benefit the wider community. For example, Sikh Gurdwara temples **provide free meals** to anyone, regardless of their religious background and can be used for things such as **food banks**.

Changing land uses also create challenges

Increasing populations	• Increasing **population** levels and increased levels of **urbanisation** mean that there's a **rising demand for land** to build on. Where authorities set aside plots of land for development, large **TNCs** have the advantage of being able to afford them over **smaller, local businesses** that may struggle. • **Local governments** may wish to create areas of **affordable housing**, which can be difficult to justify since they might make less profit than if the land was used commercially. When granting **planning permission**, local authorities may choose to develop a site **commercially** to ensure they're able to make a profit.
Regeneration	• Regeneration projects can lead to **gentrification**. This is when new developments attract **more affluent people** to work and live in an area. This increases **property prices**, which can price local people out of the area. • Many areas of regeneration include increased numbers of apartments and housing stock. The **cost** of these developments may **divide a community** between people who can afford to live in the redeveloped area and those who remain in **substandard housing**. • Regeneration can also attract people who wish to buy a **second home**. • New housing may not be to everyone's **taste** or **practical** for their needs. For example, some people consider **high-rise tower blocks** to be unattractive and they may not be suitable for elderly people or people with limited mobility.
Local people	• Local people and community groups may have a **limited say** in how new developments could meet local needs. • Authorities often approach development through **top-down methods**. This means they may **displace** local people through **compulsory purchase orders** when they build new **transport routes** and look for new parcels of land on which to expand the development.
Migrants	• **International** and **internal migrants** are increasingly moving to urban areas. This can put **pressure** on existing housing stock. • Migrants may be living in **overcrowded conditions** due to a lack of suitable housing.
Rural areas	• In rural areas there might be competition for space. Housing and industrial developments may compete with the needs of local **wildlife** and their **habitats**. • Some areas may have a **protected status**, which means that there's limited space within a rural setting for new developments to take place.

The built environment may create hostility and social exclusion

1) Changes to a **built environment** can lead to some groups feeling like they're not part of the same **community**. Some groups can feel **marginalised** by additions to the built environment which cater to the needs of a certain **group** of people and **not** to other groups in the population, e.g. a building used by a particular faith.

2) People who have resided in an area for a long time might develop **feelings of hostility** towards migrants, and migrants might begin to feel hostile in turn. As a result of this hostility, a migrant might feel **socially excluded**. This feeling can be made worse as they're in an **unfamiliar** place and may be **isolated** from their family and friends.

Changing Places

- Some parts of Glasgow represent areas of **multiple deprivation** while others have **redeveloped** and are starting to become more affluent. Much of Carntyne, to the east of the city centre, ranked in the most deprived 10% in 2012 alongside Broomielaw, to the south of the city centre. The **regeneration** of Broomielaw has seen it **economically improve** and in 2020 ranked in the least deprived 30%. Whereas Carntyne has become **more deprived**, with more areas ranked in the most deprived 5-10% in 2020.

- Some local people have **criticised** Glasgow City Council for not **prioritising** some areas outside the city centre, such as Carntyne, for investment and redevelopment. There are also fears that the development has **widened the inequality** between some groups of people rather than strengthened relationships between them. Members of some ethnic groups appear to be unfairly facing **employment discrimination**. In 2011, 32.3% of African people in Glasgow were unemployed compared to the city average of 12%. The feelings of lifelong residents and the discrimination faced by some ethnic groups suggests that some groups of people may feel **socially and economically excluded** in Glasgow.

Diverse living spaces can create tension

1) In the UK, not all people welcome **diverse communities**. Some UK citizens view the **pace** of change as concerning and as a **threat to 'British culture'**. Migrants often face **racism** and are the victims of **hate crimes**. Some people see **ethnic segregation** as a sign that migrants don't want to **integrate**. This can lead to feelings of **mistrust** between different ethnic groups.

2) The deeper levels of **deprivation** and the **lack of opportunities** experienced by some ethnic minorities can result in feelings of **anger** and **resentment** over the way they are treated. On **rare occasions** this anger has escalated into protests, where **deprived ethnic groups** have taken to the streets to **voice their frustrations** (see p.148).

Tariq's change of pace was frighteningly quick.

3) Ethnic minorities can sometimes face **deeper levels of prejudice** in rural areas. Long-term rural inhabitants might not be welcoming to people who have recently left urban areas in search of an **idyllic rural lifestyle**.

4) The movement of international migrants into rural areas can create particular **social challenges**. Once migrants settle and start to have families, there can be **increased pressure** on small rural schools. This pressure might increase if schools have limited experience of teaching children who don't use **English as their first language**.

Luton

- Luton is a town in Bedfordshire. In the **2011 census**, 24.6% of the population identified as **Muslim**.

- In 2009, a small group of **Muslim protesters** demonstrated against the **occupation of Afghanistan** by British troops. Some local people and the national media responded unfavourably to this, interpreting the action as an **anti-British protest**.

- Several people with links to Islamic extremism have come from the town, which may add to rising **Islamophobia** among residents.

- The English Defence League (commonly referred to as the EDL) is based in Luton and has been accused of stoking **tension** between the Muslim and non-Muslim community, leading to increases in the fear of crime among residents in most parts of the town.

Warm-Up Questions

Q1 Describe how people might benefit from land use changes.

Q2 Explain why regeneration projects might not be welcomed by all local people.

Exam Question — A-Level

Q1 Evaluate the view that social exclusion and ethnic tension are caused by the built environment. [20 marks]

Revision is starting to miGRATE on me...

Doing endless hours of revision in one go might seem like a good idea, but it can often work against you. Taking a break will do wonders for your concentration levels and your ability to retain info. Go outside, frolic for a while and then return to studying.

Managing Cultural Issues

There have been attempts to manage the cultural and demographic issues that may come about in rural and urban areas. There are a variety of techniques that can assess how successful these attempts have been.

Different **criteria** are used to assess **demographic issues**

1) **Employment data** can be used to measure the extent to which different demographic groups have **been successful** in integrating into urban and rural areas. For example, data can be collected to show the percentage of people that are **employed** and the percentage of those who are in **full-time** and **part-time work**.

- The percentage of people earning the minimum wage, or who rely on state benefits, can also provide an indication of economic status or inequality.

- Though measuring the average income of internal and international migrants gives an impression of how economically successful any integration has been, it doesn't give the whole picture. Some migrants become economically successful but may still think of themselves as outsiders in other ways.

- It's important to consider both absolute and relative changes in data, e.g. income and employment data. Absolute change shows the difference in raw values of an indicator between time periods (e.g. actual increase or decrease in average income). Relative change shows the same difference as a percentage change (e.g. percentage increase or decrease in average income). You can then compare absolute or relative changes between areas.

- The data in the table to the right shows that while there's little difference in the absolute change in income for Cities A and B, the relative change for City B is almost 7% higher. Looking at both absolute and relative change for different areas gives you a fuller picture of how places compare.

	2011	2021	Absolute change	Relative change
City A	£45 210	£53 800	+ £8 590	+19.0%
City B	£31 580	£39 750	+ £8 170	+25.9%

Average per capita income in two cities

2) **Demographic indicators** include different measures of the population. **Economic indicators** like employment and income can be analysed in **combination** with other indicators, such as age and ethnic diversity, to give you **more information** about a place.

- Unemployment data for Glasgow in 2011 shows that unemployment varies greatly between people who were born in the UK and those born in other regions.

- Migrants from African and Asian countries are more likely to be unemployed in Glasgow than those born in the UK or European countries. This data suggests that migrants from these countries have found it difficult to economically integrate into the city.

Region of birth	Unemployment rate
UK	11.6%
Europe	9.6%
Africa	28.5%
Asia	15.9%
The Americas and the Caribbean	9.8%
Oceania and Other	6.4%

Different criteria can be used to **assess social progress**

1) Analysts use different aspects of the **Index of Multiple Deprivation** (see p.146) to show how different groups have made **social progress** in an area. Once analysts establish the characteristics of an area (such as **age** and **ethnicity structure**), they can **compare deprivation scores** to assess the progress that different groups have made.

2) **Reducing inequality** is a form of social progress. It can be seen through a reduction in inequality between **different** groups of **people** and **different areas**, and also **within** a group of **people** that are in the same **area**.

3) For example, between 2014 and 2019, the Townhill area of Swansea became more deprived in terms of **educational achievement** (dropping 53 places in the rankings), while the nearby Brynmill area also became more educationally deprived but only dropped 15 places. Brynmill is more **ethnically diverse** than Townhill, (84.5% white population compared to 91.6% in Townhill) suggesting that being from an ethnic minority group in Swansea **may not** play a significant part in educational equality.

4) **Demographic changes**, such as improvements in **life expectancy** and changes to **birth rates** and **death rates**, are also indicators of social progress.

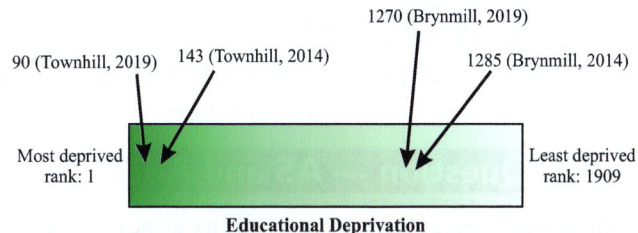

90 (Townhill, 2019)   143 (Townhill, 2014)   1270 (Brynmill, 2019)   1285 (Brynmill, 2014)

Most deprived rank: 1 — Least deprived rank: 1909

Educational Deprivation

The rankings for educational deprivation in Wales stretch from 1 to 1909. Townhill and Brynmill in Swansea occupy different parts of the ranking spectrum.

Topic 4: Option 4B — Diverse Places

Managing Cultural Issues

Different criteria can be used to **assess cultural assimilation**

1) **Cultural assimilation** is the process where one cultural group begins to take on some of the **characteristics** and **attitudes** of another cultural group. E.g. in the UK, proficiency in the English **language** can indicate a greater degree of cultural assimilation.

2) **Social integration** is another form of change. This means that minority groups feel comfortable **within** their communities. They don't need to change their identity in order to feel equal and accepted.

3) A reduction in the number of reported **racist attacks** and **incident of prejudice** towards **ethnic minority groups** and **religions** suggests that people have **assimilated** into their communities.

4) **Crime data** for Cumbria and London, areas with contrasting levels of **ethnic diversity**, shows different levels of reported racist incidents. Although there are far **fewer reported racist incidences** in Cumbria than in London, the data suggests people who are non-white in Cumbria are **more likely** to be **victims of racist incidents** than those in London.

	Percentage of population who identify as non-white (2011)	Reported racist incidents, 2010-2011	Reported racist incidents per 10 000 people who identify as non-white, 2010-2011
Cumbria	1.5%	219	288
London	40.2%	9 405	29

5) The degree to which different groups have **segregated** themselves in their place of residence is a measurable marker of **how comfortable** they feel amongst people who are not like them. A **low level** of segregation indicates that people have started to **integrate** with other groups.

6) The development of **community groups** by minorities is a strong indicator that certain groups of people **feel invested** in the place and the community where they live.

7) The degree to which different groups of people **engage in local and national politics**, such as **voting in elections**, indicates how strongly they may want to change their living environment for the better. Groups of people who display **apathy** towards decision-making may feel **socially excluded** from their community.

Stakeholders may assess the management of **cultural issues differently**

1) There are often many **stakeholders** involved in making decisions about the management of **cultural issues** in urban and rural areas. These may include

- local residents
- community groups
- local business owners
- landowners
- building contractors
- housing associations
- local, regional and national governments
- campaign groups
- political parties
- local authority bodies (such as the police)

2) Each stakeholder may view the issue and how to **best manage** it quite differently — e.g. a landowner might want **commercial** use of their land, but locals might want it to be **housing**. The **levels of interest** that stakeholders have in cultural issues can vary from strong to weak. The level of interest may depend on how **well connected** the stakeholder is to the local community.

3) Each stakeholder will also view the success of a management plan differently. Where there are different management objectives, there'll be different **criteria** to measure the **outcomes** and **success** of any actions.

4) Not all stakeholders have the **ability** to turn ideas into real actions. Some may be on the **receiving end** of the final decisions while others will only be involved at the beginning of the decision making.

Stakeholders can include campaign groups who feel strongly about social and racial equality.

Warm-Up Questions

Q1 Explain how employment data can be used to assess cultural assimilation.

Q2 Describe why the views of stakeholders on cultural issues may vary.

PRACTICE QUESTIONS

Exam Question — AS and A-Level

Q1 Explain how a local authority might try to measure the degree of cultural assimilation in a society. [6 marks]

I recently became a steak-holder at my local butchers...

Cultural issues are always complicated and there'll be loads of different viewpoints out there. Try to stay as objective as you can.

Managing Change in Urban Areas

Urban areas are changing all the time and those changes need to be managed. Have a read through these pages to find out more about how this happens. By the time you finish these pages, you could be a changed person.

Different groups tend to **view urban spaces differently**

Different generations

- Young people are more likely to view the city as a place of **economic opportunity**. For them, it's important that the city is able to provide for their **employment** needs.
- Older people may feel that the **pace of change** in cities and their rapid growth is **challenging**. The built environment may become increasingly incompatible with the physical needs of older people (e.g. due to increased pedestrianisation).

Ethnic groups

Segregation can mean that different **ethnic groups** view certain spaces of the city differently. The region someone lives in might feel **familiar** because it reflects their home culture. This means other areas of the city, which don't represent their ethnic group, may feel **unwelcoming**.

Deprived groups

People who suffer from **deprivation**, may not be able to access job opportunities within the city. Unfortunately, these groups often lack the means to be able to **migrate** to places where those opportunities exist, as city **living costs** can be high. This means that deprived groups can view urban spaces **negatively** and as areas that only cater for the needs of **wealthier** people.

Local and national **strategies** have tried to **manage change**

1) At a national level, there's been various **top-down government schemes** which have aimed to integrate different sections of society and make cities more productive. Examples include:

- The UK government channels funding towards northern cities as part of the **Northern Powerhouse initiative**. This aims to **decentralise** the UK economy away from London and to economically boost the North of England. The government is investing in **transport infrastructure**, new **business development**, **retail space** and **tourist attractions**.
- In 1999, Prime Minister Tony Blair launched the **New Deal for Communities** initiative. This aimed to create **regeneration projects** in UK cities. It focused on **residential estates** and small **retail hubs** that were experiencing multiple types of deprivation. Local people had a say in how projects were run.
- **Urban Development Corporations** have been responsible for large scale **regeneration** projects in UK cities. These have tried to improve the **visual attractiveness** of places while also **generating jobs** and **raising income levels** for local people. They've used a combination of **private investments** and funding from the UK government and the EU.

2) **Local community groups** are often at the heart of changing how people interact with each other. Groups who share common interests, faiths and similar demographics often do **outreach work**. E.g. the **Muslim Council of Britain** is an organisation that encourages mosques and Muslim groups to engage with the wider community. Some mosques have engaged in the '**Visit My Mosque**' programme, where groups hold open days, tours and workshops in mosques.

- The 2011 census recorded Slough as one of the most **ethnically diverse** places in the UK. 34.5% of its residents identified as white British and five other groups each recorded more than 5% of its population. Unlike some other **multicultural towns** in the UK, Slough is seen as a success story in terms of cultural **integration** and **assimilation**.
- Aik Saath is a **charity** in Slough that works to bring people of **different faiths** together in mutual understanding. It works closely with key stakeholders such as **schools**, **faith leaders**, **healthcare workers** and the **police**. **Workshops** and **joint community projects** (such as multi-ethnic and generational football teams and coaching sessions) aim to bring different groups of people together and create **positive actions within the town**.

Warm-Up Questions

Q1 Describe how younger and older people might view urban spaces differently.

Q2 Explain how the UK government has tried to manage change in urban spaces.

Exam Question — AS-Level

Q1 Explain how community groups can play an important role in managing change in urban spaces. [4 marks]

I wonder if I'll get funding to redevelop my local area...

I'm thinking of replacing the floors in my house with trampolines and lining my walls with sponge. Now, about that funding...

Managing Change in Rural Areas

Rural places often undergo different types of change. Here's some information on how those changes are managed.

Different groups tend to **view rural spaces differently**

1) **Rural environments** can be places of peace and quiet for people who are entering **retirement**. Having more green spaces and lower levels of pollution than cities also make rural spaces attractive to demographic groups like **young families**.

2) Some young people who are in education, or who have just left education, may see rural spaces differently. With **fewer economic opportunities** and **lower average income levels**, they may believe that rural areas aren't able to meet their needs. This means that younger people are **more likely** to **migrate** from rural areas, leaving behind an **ageing population**.

3) Migrants in rural areas might face **prejudice** or **racism** (partly due to the lack of diversity in many rural areas). Rural areas often provide migrants with **employment opportunities** (on farms for example) but migrants may not feel comfortable in the local community.

4) Rural spaces can have high levels of **deprivation**. People in lower income groups can find it hard to maintain financial security as work may be **poorly paid and seasonal**.

National and local **strategies** have tried to **manage change**

National Strategies

1) When the UK was part of the EU, it engaged in a number of **LEADER programmes**. LEADER is an EU initiative that sought to give decision-making powers about the futures of rural areas to local people. It allowed **partnerships** to be set up between **public and private sectors**, which promoted the economic development of rural areas.

2) In 2015, the UK government set up 15 new **Rural Enterprise Zones** to encourage the development of new businesses. The zones were a response to feedback from the fourteen regional **Rural and Farming Networks**. The networks brought together included **farmers unions**, **tourism executives** and **rural business owners**.

3) Increasingly, service providers such as banks and doctors are using **mobile services** to access remote areas. **National investment** in improved **broadband** has also allowed more people to work in rural areas.

Local Strategies

1) **Local authorities** can influence the future of rural areas by ensuring that grants of **planning permission** benefit particular groups. For example, land developers might only receive permission if their plans include **affordable homes** for local people and assurance that **second-home ownership** will be minimised.

2) Landowners and farmers are **diversifying** the way they use their land. For example, farmers are using their fields as campsites or are setting up farm shops and cafes. This can create **more economic opportunities** for young people in the countryside. Local authorities have encouraged employers to provide work that isn't **exclusively seasonal**.

3) Schools play a pivotal role in helping **migrant families integrate** into rural communities. Some village schools provide opportunities for children and their parents to **celebrate festivals** from their own country. Some schools provide **additional English lessons** to students and their parents to help them **integrate**.

Lake District National Park Partnership

- The Partnership brings together 25 organisations aiming to reduce the **rural economic decline** in the Lake District region. The Partnership involves public, private and voluntary organisations to improve the management of the land and seek opportunities for **sustainable development**.

- Some of the Partnership's work is on developing the **skills** of young people so they're able to stay in the Lake District and find work. The Partnership is important for the Lake District because there are so many competing land uses. The needs of **wildlife**, **local heritage** and **visitors** are all considered in the decision making.

- Specific strategies include: identifying **skills gaps** in young people, supplying **superfast broadband**, investing in **hydroelectric schemes**, encouraging farmers to raise **local breeds** of sheep and investing in **flood resilience** measures.

Warm-Up Questions

Q1 Describe how younger and older people might view rural spaces differently.

Q2 Explain how the UK government has tried to manage change in rural landscapes.

Exam Question — AS and A-Level

Q1 Explain how different types of landowner can play an important role in managing change in rural spaces. [6 marks]

The only thing I've managed to change is the flat tyre on my car...

There are many partnerships and organisations that deal with rural change. Make sure you know about the ones you have studied.

The Hydrological Cycle

The hydrological cycle is a fancy name for the water cycle, which you've studied umpteen times before...

The hydrological cycle is a **closed system**

1) Systems are normally made up of **stores**, **flows**, **inputs** and **outputs**.

   - **Stores** are where **matter** or **energy** builds up.
   - **Flows** (or transfers) are **movements** of matter or energy between stores.
   - **Inputs** are when matter or energy are **added** to the system.
   - **Outputs** are when matter or energy **leave** the system.

2) The **global hydrological cycle** is a **closed system**. This means there are **no external inputs or outputs** of **water**. Water is **continuously cycled** between different stores. It can change state, e.g. from a **liquid** to a **gas** to a **solid**.

3) The global hydrological cycle is driven by:

   - **Solar energy** — heat moves water from the surface of the Earth (as well as from oceans, lakes and rivers) to the atmosphere (evaporation).
   - **Gravitational potential energy** — the force that causes water to **flow downhill**, both on the surface and through soil, and **precipitation** to fall from the sky.

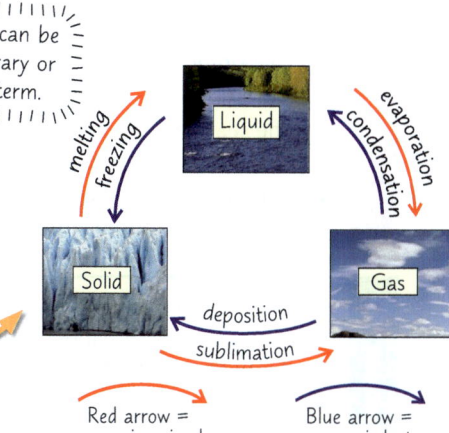

Stores can be temporary or long-term.

Red arrow = energy is gained

Blue arrow = energy is lost

There are many different **water stores**

hydrosphere includes all of the water on Earth

The hydrosphere contains 1.4 sextillion litres of water (that's 1, 4, then twenty 0s).

1) Most of this is **saline** water in the **oceans**. Less than 3% is **fresh water** (which most species, including humans, need to survive).

2) Of the Earth's **water**:

   - **96.5%** is stored in the **oceans**
   - **1.7%** is stored in the **cryosphere**
   - **1.7%** is stored as **groundwater**
   - **0.01%** is stored as **surface water**
   - **0.001%** is stored in the **atmosphere**
   - **0.0001%** is stored in the **biosphere**

3) The **amount** of water present in each store varies over a range of **scales** from **local** to **global**. The **size and importance** of each store depends on the amount of **water** flowing **between** them.

4) The ocean is an important water store. For example, around **430 000 km³** of water is **evaporated** into the **atmosphere** from the **ocean** each year. This accounts for around **86%** of **annual precipitation** that falls.

5) Over longer time scales, some stores **grow** while others **shrink**. Since the last Ice Age, rising temperatures in the **polar regions** have caused **glaciers** to melt, which is reducing the size of **cryospheric stores**. At the same time, other stores, e.g. the ocean, are receiving **meltwater** and getting larger.

The rainwater that enters rivers, lakes and groundwater is sometimes called 'blue water'.

There's a **limited amount** of water available for human use

- The **global water budget accounts for** all of the water in the whole hydrological cycle, i.e. how much water is held in different stores and flows and **how long** it stays there.

- The length of time a **water molecule** spends in a store on average is called its **residence time**.

- Water must be **physically** and **economically** accessible for humans to be able to use it. For example, **groundwater** is hard to access, so it may not be cost effective to extract it. This means only a **small amount** of water on the planet can be used by humans.

- Water with **shorter** residence times tends to contain **fewer pollutants**, making it a better source of water for human use.

- Water that's not replenished for a very long time, due to its residence time, is called **non-renewable water**. Water held in the **cryosphere** and **deep groundwater** held in fossil water (ancient, undisturbed bodies of water trapped underground) are examples of non-renewable water.

Some residence times are very short, such as just a few days in the atmosphere. Other residence times are much longer, such as thousands of years within a glacier.

The Hydrological Cycle

Fluxes between water stores **vary annually**

The Earth system is broken down into smaller parts called subsystems e.g. the atmosphere.

1) Different fluxes (flows) occur at different **spatial** and **temporal** (time) scales:

Evaporation

- Evaporation occurs when **liquid** water changes state into a **gas**, becoming **water vapour** — it normally **gains** energy from **solar radiation**. Evaporation **increases** the amount of water stored in the **atmosphere**.
- The **scale** of the evaporation flow varies by **location** and **season**. If there is a lot of **solar radiation**, a large supply of **water** and **warm, dry air**, the amount of evaporation will be **high**.
- If there **isn't much** solar radiation, little available liquid water and **cool** air that is already **nearly saturated** (unable to absorb any more water vapour), evaporation will be **low**.

Condensation

- Condensation occurs when **water vapour** changes state to become a **liquid** — it **loses** energy to the **surroundings**. It happens when air containing water vapour **cools** to its **dew point** (the temperature at which it will change from a gas to a liquid), e.g. when temperatures fall at **night** due to heat being **lost** to **space**.
- Water droplets can stay in the atmosphere or **flow** to other subsystems, e.g. when water vapour condenses, it can form **dew** on leaves and other surfaces — this **decreases** the amount of water **stored** in the atmosphere.
- The **scale** of the condensation flow depends on the **amount** of water vapour in the atmosphere and the **temperature**. For example, if there's **lots** of water vapour in the air and there's a **large or rapid drop** in temperature, **condensation** levels will be **high**.

Precipitation

- Precipitation and cloud formation are **essential** parts of the water cycle. Precipitation is the **main flow** of water from the atmosphere to the ground.
- Clouds form when **warm** air **cools down**, causing the **water vapour** in it to **condense** into **water droplets**, which gather as **clouds**. When the droplets get **big** enough, they fall as **precipitation**.
- Water droplets caused by condensation are **too small** to form clouds **on their own**. For clouds to form, there have to be **tiny particles** of other substances (e.g. dust or soot) to act as **cloud condensation nuclei**. They give water a **surface** to **condense** on. This encourages clouds to form, rather than allowing the moist air to **disperse**.
- Precipitation and cloud formation can vary **seasonally** (e.g. in the UK, there's normally more rainfall in winter than in summer) and by **location** (e.g. precipitation is generally higher in the tropics than at the poles).

Cryospheric processes

- Cryospheric processes such as **accumulation** (the build-up of snow and ice) and **ablation** (the melting of snow and ice) change the **amount** of water **stored** as ice in the **cryosphere**. The **balance** of accumulation and ablation varies with **temperature**.
- During periods of **global cold**, **inputs** into the cryosphere are **greater** than **outputs** — water is transferred to the cryosphere as snow, and less water is transferred away due to melting. During periods of **warmer global temperatures**, the scale of the cryosphere store **reduces** as losses due to melting are **larger** than the inputs of snow.
- The Earth is **emerging** from a **glacial period** that reached its maximum around **21 000 years ago**. Antarctica and Greenland, as well as numerous alpine glaciers, have extensive stores of ice on **land**. There's also a large volume of **sea ice** in the Arctic and Antarctic. However, climate change is **shrinking** these ice stores (see p.176).
- **Variations** in cryospheric processes happen over **different timescales**. As well as the changes in global temperature that occur over **thousands** of years (glacial and interglacial periods), variations can also occur over shorter timescales. For example, **annual temperature fluctuations** mean that more snow falls in the winter than in summer.

2) The **largest** and **most important flux** is the **evaporation** of **ocean water** to the **atmosphere**.

3) A far **smaller** and **less important flux** is the **evaporation** of **water** held in **surfaces** and **vegetation** to the **atmosphere**.

Warm-Up Questions

Q1 Describe how the size of different stores vary within the hydrological cycle.

Q2 Explain the role of condensation in moving water between different stores.

Exam Question — AS and A-Level

Q1 Explain why the amount of non-renewable water in the hydrological cycle is decreasing over time. [6 marks]

Sit back, relax and just go with the flow — maybe just learn this page first...

There's lots of technical words here, and it's worth learning them — examiners will love it when you use them correctly.

Drainage Basins

And now, bandage raisins. Sorry, got my letters in the wrong order there — I meant drainage basins...

Drainage basins are open, local hydrological systems

1) A river's **drainage basin** is the area **surrounding** it where the rain falling on land **flows** into that river. This area is also called the river's **catchment**.

2) The **boundary** of a drainage basin is the **watershed** — any precipitation falling **beyond** the watershed enters a **different drainage basin**.

3) As drainage basins are **open systems** with **inputs** and **outputs**, the total amount of water in the system changes over time. Water comes **into** the system as **precipitation** and **leaves** via **evaporation**, **transpiration** and **channel flow**.

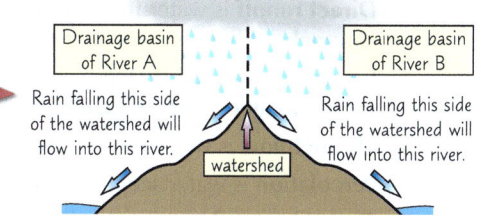

Drainage basin of River A

Drainage basin of River B

Rain falling this side of the watershed will flow into this river.

Rain falling this side of the watershed will flow into this river.

watershed

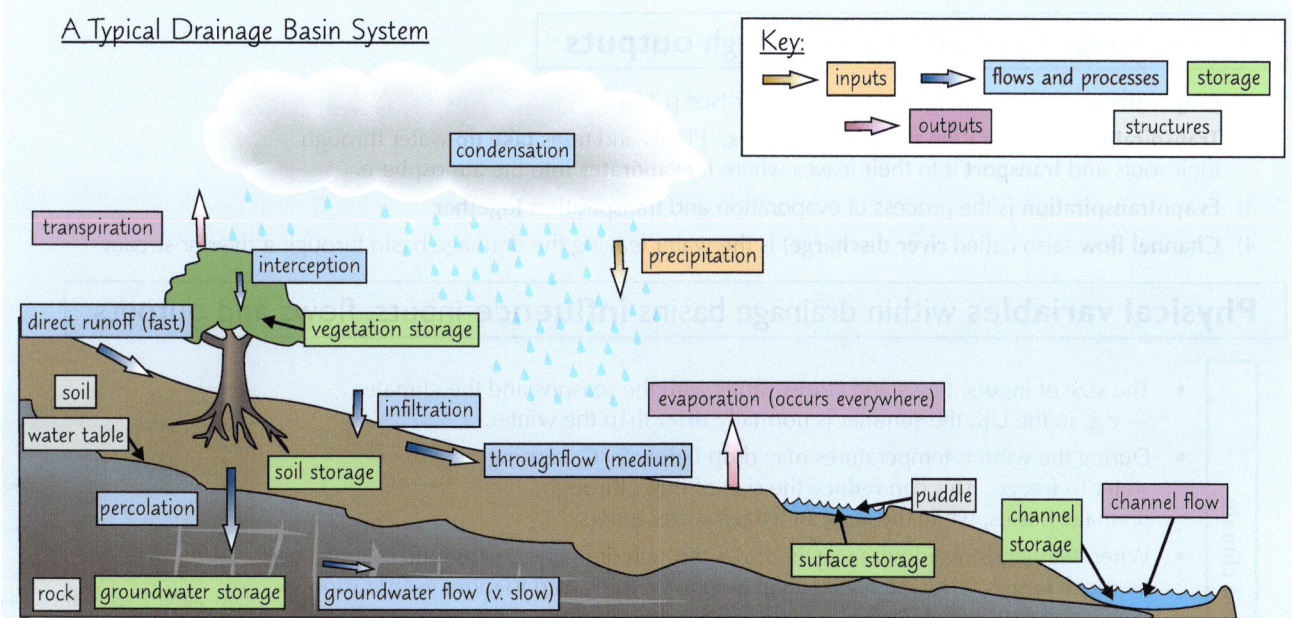

A Typical Drainage Basin System

Key:
- inputs
- flows and processes
- storage
- outputs
- structures

condensation

transpiration

interception

precipitation

direct runoff (fast)

vegetation storage

soil

infiltration

evaporation (occurs everywhere)

water table

soil storage

throughflow (medium)

puddle

channel storage

channel flow

percolation

surface storage

rock

groundwater storage

groundwater flow (v. slow)

4) The **size** and **shape** of a drainage basin affect the **processes** that take place within it.

- **Larger** drainage basins with **steep sides** have the capacity to catch precipitation and move it **quickly** into river channels. These basins are likely to have a large number of **streams**, and therefore a **high drainage density**.
- **Smaller** drainage basins, or ones with **fewer streams**, have a **low drainage density**.

Precipitation is the main input into a drainage basin

1) Precipitation includes **all** the ways moisture **comes out** of the atmosphere. It's mainly **rain**, but can also be snow, hail, dew and frost.

2) The cooling of air increases the **rate of condensation**, which causes **more precipitation**. There are **three** different types of precipitation which result from the air cooling in a particular way.

Ayo seemed surprisingly upbeat about atmospheric moisture inputs.

- **Frontal precipitation** — warm air is **less dense** than cool air. When warm air meets cool air, the warm air is forced up **above** the cool air. It **cools** down as it rises.
- **Orographic precipitation** — when **warm air** meets mountains, it's forced to **rise**, causing it to **cool**.
- **Convectional precipitation** — when the sun **heats** up the ground, moisture on the ground **evaporates** and rises up in a column of warm air. As it gets higher, it **cools** down.

3) Different **areas** experience differences in precipitation. For example:

- **Convectional precipitation** is more likely to take place in the **tropics** and produces effects like **monsoon seasons**. This type of precipitation also occurs in the **summer months** at **high latitudes** (e.g. Arctic tundra).
- **Coastal areas** often receive **more precipitation** than **inland areas** because water is evaporated from the **oceans** and falls as precipitation.

Drainage Basins

Different **flows** move **water** from **one place** in the **drainage basin** to **another**

1) **Interception** is water landing on **vegetation** or other **structures**, like buildings or concrete surfaces, before reaching the soil.

2) **Infiltration** is water **soaking** into the soil.

3) **Direct runoff** is water **flowing over** the **land**. It can flow over the **whole surface** or in **little channels**. It happens because rain is falling on the ground faster than infiltration can occur.

4) **Saturated overland flow** is water **flowing over** the **land** because the soil **no longer** has the **capacity** to allow any more **water to infiltrate**.

5) **Throughflow** is water moving slowly **downhill** through the **soil**.

6) **Percolation** is water **seeping down** through soil **into the water table**.

7) **Groundwater flow** is water slowly flowing through **permeable rock** below the **water table**.

Water **leaves** a drainage basin through **outputs**

1) **Evaporation** is water turning into water vapour (see p.164).

2) **Transpiration** is evaporation from within **leaves**. Plants and trees **take up** water through their roots and **transport** it to their leaves where it **evaporates** into the atmosphere.

3) **Evapotranspiration** is the process of evaporation and transpiration **together**.

4) **Channel flow** (also called **river discharge**) is the water leaving the drainage basin through a river or stream.

Physical variables within drainage basins **influence** inputs, flows and outputs

Seasons and climate

- The size of **inputs**, **flows** and **stores** varies with the seasons and the climate — e.g. in the UK, the summer is normally **drier** than the winter.

- During the **winter**, temperatures may drop **below 0 °C**, causing water to **freeze**. This can **reduce** the size of flows through drainage basins, while the **store** of **frozen water grows**.

- When **temperatures** increase again, flows through drainage basins (and outputs) can be **much larger** as the ice **melts**. Higher temperatures lead to **more evaporation** which can cause **convectional precipitation** (short periods of intense rainfall).

- Climate also has an impact on **vegetation** and how much **interception** and **evapotranspiration** takes place.

- **Intense storms** generate **more precipitation** and **greater peak discharges** (see p.169) than **light rain showers**.

- The **larger input** of water causes flows (e.g. **runoff**) and stores (e.g. **groundwater**) to **increase** in **size**.

- Some flows, like **infiltration**, may not be able to occur fast enough for the size of the input, increasing **direct runoff**.

Low water level in the Craig Goch reservoir during the 2018 heatwave, Powys, Wales

© Keith Morris / Alamy Stock Photo

Soil

- **Infiltration rates** are influenced by **soil type**, **soil structure** and how much **water** is **already** in the soil. For example, **larger air spaces** in sandy soils allow **more** infiltration compared to clay soils.

- **Throughflow** is faster through **openings**, such as cracks in the soil or animal burrows.

- **Soil type** can influence the type of **vegetation** that's able to grow in a location.

Vegetation

- Vegetation **intercepts** precipitation and **slows its movement** into the river channel. Interception is **highest** when there's lots of vegetation and **deciduous trees** have their **leaves**.

- The **more vegetation** there is, the **more water** is **lost** through **evapotranspiration** before it reaches the river channel, **reducing runoff** and **peak discharge**.

Geology

- Water flows **slowly** through most rocks, but rocks that are highly **permeable** (e.g. sandstone) or rocks with lots of **joints** that allow water to flow through them (e.g. limestone) can cause faster **groundwater flow**.

- **Impermeable** rocks **stop water** from flowing through them, which means there are **higher rates** of **direct runoff**.

Relief

- The **steeper** the gradient of the landscape, the **faster** processes such as **direct runoff** and **throughflow** will be.

- This means that **surface stores** have a **shorter residence time** (see p.163) in steeper areas.

- **Gradient** can also affect the amount of water **discharged** from a river to the ocean.

Drainage Basins

Human actions disrupt processes in drainage basins

Deforestation

- **Deforestation** reduces the amount of water that is **intercepted by vegetation**, increasing the amount that reaches the **surface** and increasing **direct runoff**.
- In forested areas, **dead plant material** on the forest floor helps retain the water, allowing it to **infiltrate** the soil rather than **run off**. When forest cover (and dead material) is removed, **less infiltration** can take place.
- **Soil erosion** is more likely to happen on exposed land, reducing the soil's ability to store rainwater. As there's **limited infiltration**, groundwater levels decline, and the **water table drops**.

Land use change

- Ploughing **breaks up** the surface so more water can infiltrate it, reducing the amount of runoff.
- Farming livestock, such as cattle, means that the soil is **compacted** and **trampled**, decreasing infiltration and increasing direct runoff. **Buildings** and **roads** create an **impermeable** layer over the land, preventing infiltration.
- This increases **direct runoff**, resulting in water passing through the system more rapidly and making **flooding** more likely (see p.173).
- **Urban drainage systems** can feed rivers with rainwater more quickly and increase **discharge**.

Water storage reservoirs

- Creating new **storage reservoirs** by building **dams** increases the amount of freshwater that humans can access.
- It also reduces the level of river discharge into oceans. Creating large areas of **standing water** increases the level of **evaporation** that can happen there.

Water abstraction

- More water is **abstracted** (taken from stores) to meet demand in areas of high **population density**. This reduces the amount of water in stores such as lakes, rivers, reservoirs and groundwater.
- During **dry seasons**, even more water is abstracted (especially groundwater and from reservoirs) for **consumption** and **irrigation**, depleting stores further.
- In some places, humans abstract water from **underground aquifers** (see p.172). This often happens at a **faster** rate than the water is replaced and the aquifer is recharged.

Amazonia

Amazonia covers 40% of the South American land mass.

CASE STUDY

- Amazonia is the world's largest **tropical rainforest**.
- The water cycle causes the Amazon to be very **wet**. There's a lot of **evaporation** over the Atlantic Ocean, and the **moist air** is blown **towards** the Amazon. This contributes to the Amazon's **very high rainfall**.
- **Warm temperatures** mean that the **evapotranspiration rate** is **high** in the rainforest itself, increasing the amount of **local convectional precipitation**.
- The rainforest has a **dense canopy**, meaning **interception** is high. This **reduces** the **amount** and **speed** of **water** flowing into **rivers**.
- Lots of **deforestation** takes place in Amazonia, e.g. to harvest **timber** or to use the land for **farming**. In **deforested areas**, there's no tree canopy to intercept rainfall, increasing the amount of water reaching the **ground**. As there's too much water to soak into the soil, the water moves to rivers as **saturated overland flow**, increasing the risk of **flooding**.
- Deforestation also reduces the rate of **evapotranspiration** (see p.166). This means that less water vapour reaches the atmosphere, **fewer clouds form** and **rainfall** is **reduced**. This increases the risk of **drought**.

Warm-Up Questions

PRACTICE QUESTIONS

Q1 Describe the three main types of precipitation.

Q2 Explain why a drainage basin is an open system while the global hydrological cycle is not.

Exam Question — AS and A-Level

Q1 Assess the relative importance of physical and human factors in influencing the drainage basin cycle. [12 marks]

If your basin takes an age to drain, you'll want to get a plumber in...

It might seem like a pain, but if you learn all this stuff now the rest of the section will make a lot more sense. Promise.

Water Budgets and River Discharge

If you like runoff, the water cycle and drainage basins (who doesn't?), these pages are going to blow your mind.

Water budgets indicate how much water is stored in drainage basins

soil water
water within the soil

- A water budget shows the **balance** between **inputs** and **outputs** in a hydrological system.
- When recorded over **many years**, water budgets are useful to help **farmers** plan the **frequency** and **amount** of **irrigation** needed by their crops (based on the level of **groundwater recharge** possible and the availability of **soil water**).
- The water budget can be shown using the formula:

$P = Q + E \pm S$, where P is precipitation, Q is channel discharge, E is evapotranspiration and S is the change in storage.

The general water budget in the UK shows seasonal patterns

1) In **wet seasons**, precipitation **exceeds** evapotranspiration. This creates a **water surplus**. The **ground stores fill** with water so there's **more direct runoff** and **higher discharge** — this means **river levels rise**.

2) In **drier seasons**, precipitation is **lower than** evapotranspiration. **Ground stores** are **depleted** as some water is **used** (e.g. by plants and humans) and some flows into the **river channel**, but **isn't replaced** by precipitation.

The water budget in a temperate zone

- From January to April there's a **water surplus**, with greater levels of **precipitation** than evapotranspiration. **Soil water stores** are likely to be **full** and can be used efficiently by crops.

- Between April and September, the soil water store starts to deplete. Precipitation levels are **lower** and, with increased atmospheric temperatures, there are higher levels of **evapotranspiration**. As water stores are used up, the water budget moves from a water surplus to a water deficit and farmers may need to **irrigate** their land.

- Between September and December, atmospheric temperatures start to **drop**, along with levels of **evapotranspiration**. The amount of precipitation **increases**, which leads to the **recharge** of soil water stores.

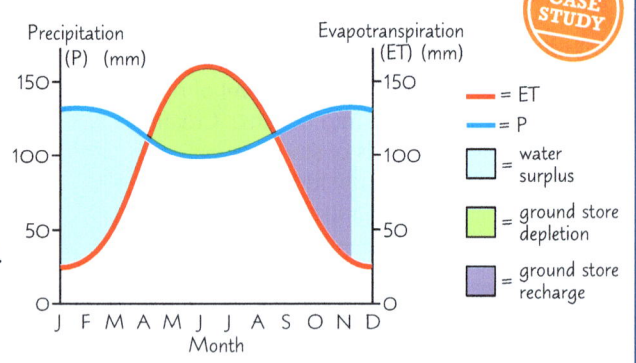

CASE STUDY

River regimes vary with physical influences

Discharge is measured in cubic metres per second (m³/s) or cumecs.

1) A **river regime** refers to the **variations** of river **discharge** over a year.

2) The **discharge** of a river is the **volume** of water that moves past a certain point in the river channel **per second**.

3) High levels of **runoff increase** the discharge of a river because **more water** enters the river, increasing its volume.

4) Changes in **seasons** and **climate** influence the amount of **precipitation** that enters a river basin, affecting river regimes:

- In **glacial** areas, an **increase** in **temperature** can create higher levels of **meltwater**, while a drop in temperature will lead to higher levels of **glacial accumulation** (see p.36).
- Higher levels of **evaporation** will happen in summer compared to winter months.

Omar didn't like to complain, but he was unhappy with the allocation of the water budget.

5) **Geology** and **soil structure** affect the amount and speed of **infiltration**:

- Geological structures that are highly **porous** (such as sand) and **permeable** (e.g. sandstone) will allow water to **percolate** (water seeping through the soil into the water table) through to the river quickly and increase the **baseflow** (see p.169).

- **Impermeable** rocks that lack joints and cracks will cause greater levels of **direct runoff** and higher levels of **saturated overland flow** into the river.

- Some soil types retain water more easily than other soil types. For example, sandy soils have **larger air spaces** between grains than clay soils. This means water will infiltrate **quickly** through sandy soils but may either pool in **groundwater storage** in clay soils or rise to the surface in puddles.

Different river basins will display different river regimes. A basin with a large river will record higher levels of discharge than basins that have a more seasonal surface flow.

Water Budgets and River Discharge

The Yukon, Amazon and Indus Rivers

CASE STUDY

- The **Yukon River** in Alaska and Canada has very **low discharge** in winter but a **rapid rise** in discharge in the early spring. This is due to rapid **snowmelt** as temperatures rise. Soils in the Yukon basin are prone to permafrost, which are impermeable all year round. This means that overland flow will happen quickly once surface snowmelt occurs.

- The **Amazon River** in South America has a more gradual rise in **discharge**. Although there's a wet and dry season, the **difference** between the two isn't as pronounced as the Yukon or the Indus. This is due to high levels of **rainfall** in most months which infiltrates the ground quickly due to the thin soils and sedimentary rocks that dominate the Amazon basin.

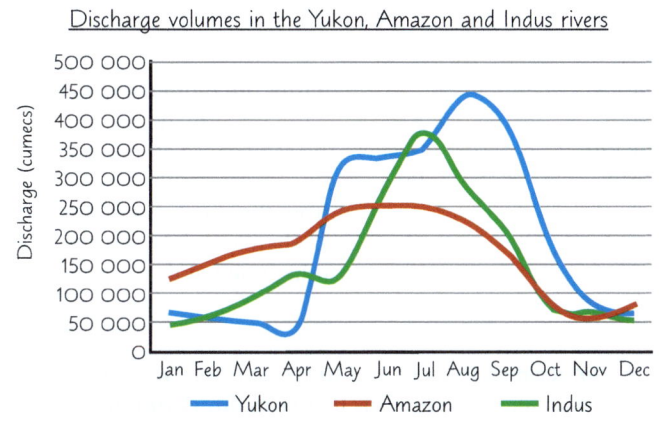

Discharge volumes in the Yukon, Amazon and Indus rivers

- The **Indus River**, which flows through South and Central Asia, is fed by **snowmelt** from the Himalayas in spring. During the monsoon months (June to September), the river level is fed by **regular** precipitation. However, the dry season from October to March means that the Indus generally flows at a **lower level** than the Yukon as it's in a water deficit for 5 to 6 months of the year.

- The Indus river basin is largely comprised of **igneous and metamorphic rocks** which are relatively hard and less permeable. Therefore, the river can **quickly swell** in volume during the monsoon season.

Storm hydrographs show river regimes

Hydrographs are graphs that show how the **volume of water** flowing at a certain point in a river **changes** over time. **Storm hydrographs** (also called flood hydrographs) show river discharge around the time of a **storm event**. They only cover a relatively **short time period** (hours or days rather than weeks or months).

The main components of a storm hydrograph are:

1. **Peak discharge** — this is the **highest** point on the graph, when the **river discharge** is at its **greatest**.

2. **Lag time** — this is the delay between **peak rainfall** and **peak discharge**. This delay happens because it takes **time** for the rainwater to **flow** into the river. A **shorter** lag time can **increase peak discharge** because more water reaches the river during a **shorter period of time**.

3. **Rising limb** — this is the part of the graph **up to** peak discharge. The river discharge **increases** as rainwater flows into the river.

4. **Falling limb** — this is the part of the graph **after** peak discharge. **Discharge** is **decreasing** because **less water** is flowing into the river. A **shallow** falling limb shows water is flowing in from **stores** long after it's **stopped raining**.

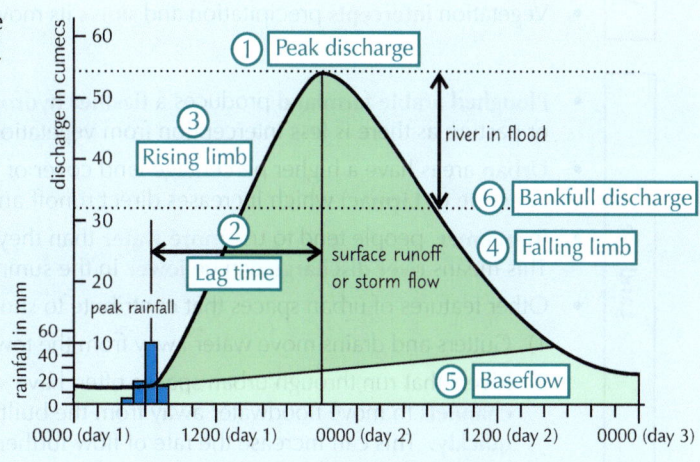

A typical storm hydrograph

5. **Baseflow** — this is groundwater flow that feeds into rivers through **river banks** and **river beds**. This is the **normal discharge flow** of the river.

If bankfull discharge is exceeded then the river will flood.

6. **Bankfull discharge** — this is the point when the **water level** reaches the **top** of the river channel without flooding.

Hydrographs can be flashy or flat

- A basin with **rapid runoff** and limited **storage** capacity gives a hydrograph with a **short lag time** and **high peak discharge**. This is called a "**flashy**" hydrograph — the graph has **steep**, roughly **symmetrical** rising and falling limbs.

- Hydrographs with limbs that rise **more steadily** are called **flat** (or delayed) hydrographs.

Water Budgets and River Discharge

The **shape** of storm hydrographs depends on **many factors**

Size of drainage basin
- **Larger** drainage basins catch **more precipitation**, so they have a **higher peak discharge** than smaller basins.
- **Smaller** basins generally have **shorter lag times** because precipitation has **less distance** to travel, so it reaches the main channel more **quickly**.

Due to his short legs, Clive preferred a smaller basin.

Drainage density
- Water enters a river **more quickly** in drainage basins with a **high drainage density**. This is because there are **more rivers** and **streams** for the water to **flow into**. This means that basins with a high drainage density are more likely to produce **flashy hydrographs**.
- Drainage basins with a **low drainage density** are more likely to produce a **flat hydrograph**.

Shape of drainage basin
Circular basins are more likely to have a **flashy hydrograph** than **long, narrow** basins. This is because all points on the **watershed** are roughly the **same distance** from the point where discharge is measured. Therefore, lots of water reaches the measuring point at the **same time**.

Relief
Water flows more quickly downhill in **steep-sided** drainage basins, shortening the **lag time**. This also means that water has less time to **infiltrate** the soil, so **runoff** is higher and a **flashy hydrograph** is produced.

Permeability of rocks and soil
- **Impermeable** rocks and soils don't store water or let water infiltrate. This increases **direct runoff**.
- **Peak discharge** also increases as more water reaches the river **faster**, making the hydrograph more **flashy**.

Vegetation
- **More vegetation** in a drainage basin means more water is lost through **evapotranspiration**. This means that less water reaches the river channel through **reduced runoff**, so there's a **lower peak discharge** and a **less** flashy hydrograph.
- Vegetation **intercepts** precipitation and **slows** its movement to the river channel.

Interception is highest when there's lots of vegetation and deciduous trees have their leaves.

This factor is a human factor. The other factors on the page are physical factors.

Land use
- **Ploughed arable farmland** produces a **flashier hydrograph** than forests or pasture as there is less interception from vegetation.
- **Urban** areas have a higher percentage land cover of **impermeable** materials (e.g. **concrete** or **tarmac**) which increases direct runoff and produces a **flashy hydrograph**.
- In **summer**, people tend to use **more water** than they do in winter. This means **river discharge** can be **lower** in the summer months.
- Other features of urban spaces that contribute to **shorter lag times** include:
  1) Gutters and drains move water away from the town and into river systems more **quickly**.
  2) Rivers that run through urban spaces often have **straightened channels** to move floodwater away from the built up area more **quickly**. This can increase the rate of flow further **downstream**.

Warm-Up Questions

Q1 Describe what is meant by river discharge.
Q2 Explain three physical factors that influence river regime.

Exam Question — A-Level

Q1 Explain why the discharge patterns for two rivers may be different for the same storm event. [8 marks]

Oops... I left the cage in the rain and now I've got a water budgie...

Make sure you learn all the bits of the storm hydrograph on the previous page. The best way is to close the book and sketch it with all the labels. Then, check back to see what you missed. In fact, that's a good way to learn the factors on this page too.

Drought

If a place has too much or too little water, it's a problem. There are lots of things that can lead to there being too little...

Droughts occur when conditions are **drier than normal**

1) Droughts can have **significant impacts** on people's **lives** and their **livelihoods**.

2) It can be **difficult to predict** when droughts will start and end. They tend to be **slow-onset hazards** — there may be **many years** of below average water levels before an impact is felt. In the **UK**, a summer **hosepipe ban** can be enforced following a year's worth of low rainfall.

3) Drought **durations** vary — the worst drought in Britain since records began lasted for **sixteen months**, whereas droughts in some African countries can last for more than a **decade**.

4) Water stores, such as lakes and rivers, become **depleted** during a drought because they're not replenished by rainfall and people keep using them. Droughts are often accompanied by **high temperatures**, which increase **evaporation**, depleting water supplies further.

5) **Farmland** is particularly vulnerable to periods of drought. If soil loses its moisture, then farmers have to **irrigate** their land. This may increase the **price** of food in the long term, or cause **food shortages**.

> **drought**
> *a shortage of water over an extended period of time*

There are **meteorological** and **hydrological** causes of droughts

- A **meteorological drought** means that there's **less precipitation** entering a local system.
- A **hydrological drought** means there's **less water** in **groundwater stores**, **streams** and **rivers**.

> A meteorological drought can cause a hydrological drought.

Short-term droughts

- Short-term droughts that affect a localised area are usually caused by **precipitation deficit** (**lower** than normal **precipitation levels**). These lower levels are usually caused by **changes** in the **frequency** of **frontal precipitation**.
- **Localised droughts** can also occur when areas of **cool descending air** create a **high-pressure** weather system (called an **anticyclone**). This can **block** the movement of warm ascending air, which would otherwise create **rain clouds**, leading to **less** rainfall and drought conditions may happen.

Longer-term changes in drought frequency

- **Climate change** causes longer-term **warming** and **cooling** of oceans, leading to **changes** in the **frequency** of droughts. This means that the recovery time between droughts may be much shorter.
- It also affects **seasonal rains** in the **tropics** — the rainy season is **shorter** and more **unpredictable**. This results in the long term depletion of groundwater and reservoir stores.

The El Niño-Southern Oscillation (ENSO) cycle can cause droughts

The El Niño-Southern Oscillation (ENSO) cycle is a **natural change** in the patterns of **ocean temperatures** in the **Pacific Ocean**.

1) Under **normal conditions**, there's an area of **low pressure** over the western Pacific and high pressure over the east. This causes the trade winds to blow from **east to west**.

2) In an **El Niño event**, pressure increases in the western Pacific and decreases in the east. This causes the trade winds to **weaken** in strength or **change direction** and blow from **west to east**.

3) The **sinking of air** in the high pressure areas can lead to drought conditions in the west (e.g. Australia) as **less precipitation** falls, increasing the risk of **wildfires**. El Niño events occur every 3 to 4 years and each event typically lasts for 9 to 12 months.

4) A **La Niña event** is when the normal conditions become more **extreme**. Trade winds increase in strength causing more **cold water** to rise in the eastern Pacific. This results in **drier conditions** in the east and wetter conditions in the west. La Niña events occur every 2 to 7 years.

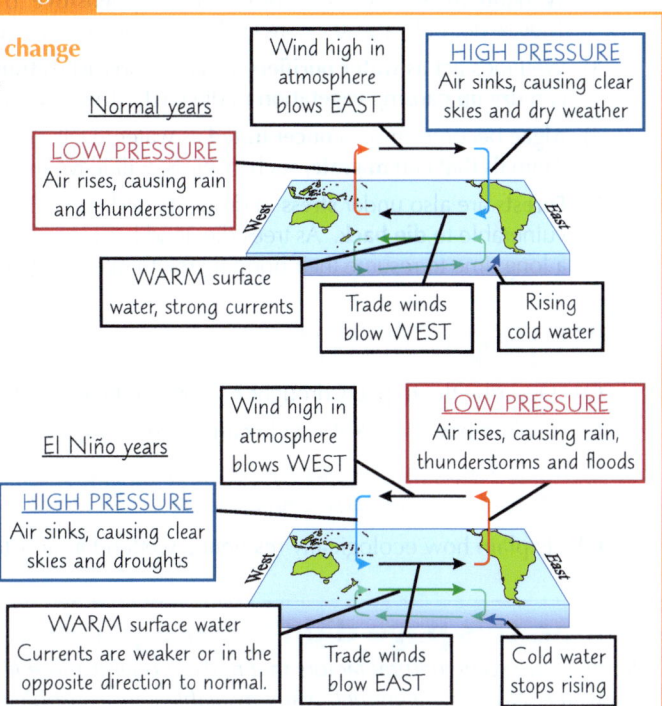

Normal years
Wind high in atmosphere blows EAST
HIGH PRESSURE Air sinks, causing clear skies and dry weather
LOW PRESSURE Air rises, causing rain and thunderstorms
WARM surface water, strong currents
Trade winds blow WEST
Rising cold water

El Niño years
Wind high in atmosphere blows WEST
LOW PRESSURE Air rises, causing rain, thunderstorms and floods
HIGH PRESSURE Air sinks, causing clear skies and droughts
WARM surface water Currents are weaker or in the opposite direction to normal.
Trade winds blow EAST
Cold water stops rising

Drought

Human activity can increase the risk of droughts

aquifer
a saturated layer of permeable rock below the water table

- Humans **can** make **low rainfall** more of an issue and **intensify** a hydrological drought.
- If the **demand** for water is **greater** than the supply by **precipitation**, surface water resources (e.g. surface lakes and reservoirs) dry up and humans often abstract water from aquifers. Abstracting water from an aquifer at a **faster rate** than the aquifer recharges is called **over-abstraction**.
- Humans also **divert** water out of **natural systems** into storage within homes and businesses. This water isn't subjected to natural processes, such as evaporation, **reducing** the supply of water to **other stores** in the hydrological cycle.

The Millennium Drought — Australia

Southeastern Australia experienced a **severe**, **long-term meteorological drought** from roughly 2001 to 2009.

1) The worst-hit area was the Murray-Darling Basin, an important **agricultural** region (it's responsible for 40% of Australia's agricultural produce).

2) Several factors contributed to the drought:
   - Australia has **naturally low rainfall** due to the **30° S high-pressure belt** passing through it, which causes **cooler air** to **sink** and **low levels of precipitation**. In many of the drought years, **El Niño** (see p.171) events caused weather fronts (that normally bring rain to southeast Australia) to move further **south**. This increased drought conditions in the region.
   - Temperatures were **higher** than normal, resulting in more **evaporation** than normal.
   - Years of **over-abstraction** for farming, industry and domestic use **depleted groundwater levels** and made the area more susceptible to **drought**.

3) **Water levels** in lakes and rivers (particularly the Murray and Darling) fell, so water supplies ran **low**. This had a large impact on **farming**:
   - **Crop yields fell** and **food prices increased**, particularly for crops that relied on **irrigation**, such as rice.
   - The number of **sheep** in Australia fell by around 7 million between 2002 to 2003, due to dehydration and starvation through lack of pasture. This caused farmers' **incomes** to fall and over 100 000 people employed in agriculture **lost their jobs**.

Drought conditions in New South Wales, Australia

The increased temperatures were possibly due to climate change.

Rice yields fell to just 2% of pre-drought levels.

Droughts affect how ecosystems function

1) The capacity for a natural area to **survive** and **recover** from a disturbance, e.g. a drought, is called its **ecological resilience**.
2) **Wetland areas** are ecosystems that are very **sensitive** to droughts and have **limited resilience**. They're not only a water store for humans but are also a **habitat** for birds, fish and other aquatic organisms.
3) Wetlands act as **water purifiers** as reeds and rushes **trap pollutants**. However, in drought conditions, wetland areas can **dry up** causing **vegetation** to die back. This also means that there are fewer **nesting spaces** for wading birds.
4) **Algae** becomes more **concentrated** as water levels drop, reducing **oxygen levels** for aquatic life. Animals that can **migrate**, such as birds, will leave the area in search of deeper water.
5) **Forests** are also under **stress** in drought conditions. Even in a mild drought, tree **growth slows** and young trees are vulnerable to **die back**. As trees die, land is exposed to **warmer temperatures** and **wind erosion**. Trees that survive can take a long time to resume their normal growth rate. **Organisms** that rely on eating leaves for water will also be under stress.

Warm-Up Questions

Q1 Describe the impacts drought had on southeastern Australia during the Millennium Drought.
Q2 Explain how the ENSO cycle can cause droughts.

Exam Question — AS and A-Level

Q1 Explain how ecologically resilient a wetland area is to a drought. [6 marks]

I hope this page didn't make you too thirsty...

So there are some natural factors that make droughts more likely, and then the dreaded humans came along with their need for water. Not having enough water available causes problems, as does having too much, as you'll see on the next page...

Flooding

If the hydrological cycle can't cope with increased levels of water in its system, there'll be a flood.

Floods occur when conditions are **wetter than normal**

1) After periods of **prolonged or heavy rainfall**, drainage basin systems may not be able to deal with the **excess water**. This means that water moves quickly as **direct runoff**.

2) During periods of heavy rainfall, river channels can reach **bankfull capacity** quickly and can **overflow** and **inundate** a flood plain. Lakes and reservoirs can also **grow** and **flood** new areas of land. This is a **natural process** and regular flooding of the flood plain can **improve** farmland as **silt deposits** in floodwater **fertilise** the land.

3) People's **health** and **livelihoods** are impacted by flooding. This is often made worse when humans **interfere** with the natural flow of the river or build on floodplains.

flood
an overflow of water onto land that is usually dry

floodplain
the low-lying land either side of a river

Missouri River flood plain

Floods are caused by different **meteorological factors**

Prolonged rainfall

- **Prolonged periods** of **heavy rainfall** can cause **slow-onset** flooding.
- This can occur when a series of **low-pressure systems** (depressions) continually move across the UK. If the time gap between different depressions is **short**, flooding is more likely to happen.

Intense storms

An **intense storm** is a **fast-onset** event. Higher levels of **rainwater** combined with possible **storm surges** from the sea can leave high levels of **standing water** on the land. This is called **flash flooding**.

Extreme monsoonal rainfall

- In some parts of the world, **prolonged rainfall** may come from **extreme monsoonal conditions**.
- A **monsoon** is a **seasonal change** in the **prevailing wind**. This change can bring **wetter** weather to some areas in **subtropical** regions. As the global climate changes, the seas in monsoon areas get warmer and often these seasonal rains get **heavier** and **last longer**.
- If the monsoon rain is particularly intense (often linked to La Niña events, see p.171), increased runoff and saturated overland flow can **inundate waterways** and cause **flooding**.

Monsoons affect South Asia, Australia and parts of Africa and Central America.

Monsoon floods in Sri Lanka

Early spring flooding from melting glaciers. Green Lake, Austria

Snowmelt

- In **upland glacial areas**, flooding can occur in the **valleys** due to the **snow** and **ice melting**. This increases **overland flow**, which can **flood upland reservoirs** and swell river channels.
- This might occur due to **abnormally high temperatures** in the spring or a **sudden change in temperature**.
- It can also come from **volcanic activity** in glacial regions, which warms the ice.
- A flood of glacial meltwater is called a **jökulhlaup** (see p.14).

Other physical factors

- **Shape** of the land — **flat** land is more **likely** to flood since there's no gradient to allow water to **flow away**. However, **steep valley sides** may also channel water into **one area** — increasing the likelihood of flooding.
- Land **elevation** — **low-lying** land next to the **coast** is more likely to flood as a result of **storm surges**.
- **Geology** of the land — if the **lithosphere** is made up of **impermeable rocks**, water will take **longer** to drain away and flooding is more likely. **Soils** of **greater depths** also have a **greater capacity** to store water so take longer to become saturated.
- **Drainage density** — the more **rivers** and **streams** in an area, the higher the likelihood of **flooding**.

Topic 5 — The Water Cycle and Water Insecurity

Flooding

Human activity can increase the risk of floods

Floodplain use

- Due to rising population levels, an increased number of floodplains are being built on. Rapid urbanisation has increased the need for land to be used for housing and industry.
- Floodplains act as temporary natural stores for water and slow the movement of water in the drainage basin. However, when floodplains are built on, an impermeable layer covers the land adjacent to a river. The concrete and tarmac stop or reduce the rate of infiltration increasing the risk of flooding.

Land use change

- Removing vegetation or deforesting an area reduces the amount of water that is intercepted by vegetation as well as exposing the soil to the greater erosive power of the water. This can move soil sediment into river systems and reduce the channel's carrying capacity, making flooding more likely.
- Construction of new buildings and roads creates an impermeable layer over the land, preventing infiltration. This means the water collects and floods, or it runs off and affects other areas.
- Changing land use for farming practices can also increase the risk of floods. For example, agricultural land is more permeable than urban spaces, but the way it's used can still increase flood risk. Areas of bare soil have reduced levels of interception. This means water moves more quickly into rivers and can allow it to build up in certain places.

Hard engineering

- Humans frequently mismanage rivers. For example, removing meanders and creating straight and unnatural river channels by hard engineering can reduce the risk of flooding in some areas but increase the risk further downstream as the rate of flow increases.
- Dredging rivers to make them deeper can cause river water to flow quickly through a built up area and prevent flooding there. However, it also increases the risk of flooding downstream beyond the dredged section.

Floods can have many socio-economic impacts

Economic activity

1) Flooding can restrict the way that businesses operate. It may be unsafe to open to the public, or they may not be accessible to customers.
2) When farmland floods, the deep saturation of soil can be harmful to crops. They can drown due to a lack of oxygen and it's more likely they'll be affected by diseases. This can lead to reduced incomes for farmers.

Road closure due to flooding

Infrastructure

1) Transport and communication networks can be affected by floodwater:
   - Trains and vehicles may not be able to move through standing floodwater.
   - Telephone lines may be damaged by floodwater.
   - Road surfaces, bridges and tunnels can be damaged by floodwater, making them unusable by traffic.
2) Fresh water and wastewater pipelines can become inundated with floodwater and cross-contamination can occur. Water supplies might need to be cut off while the system is drained.

Settlement

1) Standing floodwater can cause extensive damage to property:
   - As water soaks into the walls and floors of a building, it can leave them structurally unstable and prone to collapse.
   - Flood sediment, pollutants and mould must be removed from buildings.
   - Walls, fences and trees can get washed away.
   - Few UK homeowners living on flood plains can get insurance against flooding. This means they bear the cost of replacing and rebuilding after a flood.
2) Flooding can cause loss of life and injury. People can die through drowning and by contracting waterborne diseases, such as cholera.

Mould growing in a house after water damage caused by a flood.

Flooding

Flooding affects the **environment**

1) **Groundwater stores** near urban areas can become **contaminated** with sewage water and oil from roads.

2) **Eutrophication** (see p.182) can affect aquatic ecosystems (e.g. lakes and ponds) due to **runoff of fertilisers** from fields. Algal blooms **remove oxygen** from the water, which can cause some aquatic life and insects to die.

3) Floodwater can cause extensive **soil erosion** as water flows over the land.

4) **Soil** can become **contaminated** and **microorganisms** may die.

5) The environment can also **benefit** from flood events:

- **Lakes, wetlands** and **groundwater stores** can be **recharged** with water, making them a habitat for an increased number of plant and animal species. This enhances their **biodiversity**.

- Farmland can be **fertilised** by **silt** deposited after floodwaters have receded.

A heron visits the flooded banks of the River Thames in Oxford in 2007.

UK floods in 2007

- In the **summer of 2007**, many parts of the UK experienced extreme levels of floodwater caused by exceedingly **high levels of rainfall**. In many parts of the country, rainfall was more than **double** its normal level.

- It's believed that the heavy rainfall came from a series of unseasonably **low-pressure systems** that were caused by the **polar jet stream** being unusually further **south** than normal.

A jet stream is a rapidly moving current of air. The polar jet stream that affects the UK is a movement of warm air away from the Earth's surface in the 30° N to 60° N latitudes. The further south this jet stream is, the more warm, wet air it will contain and the heavier the rainfall will be.

CASE STUDY

Flooded land in Upton-upon-Severn

- Large amounts of flooding occurred along the **River Severn**, the **River Thames** and the **River Don**. In total, 13 people died and 48 000 homes were affected.

- The total economic cost of the floods was calculated at being around **£3.2 billion**.

- **Upton-upon-Severn** was one of the worst hit villages. Most of the village was cut off by the floodwater and there were **few flood defences** to stop this from happening. This also made it difficult for **rescue services** to access the area and support vulnerable people.

- Many **waterfront businesses** flooded including pubs and cafes that overlooked the river. In some cases, it was **months** before they were able to **reopen** for business.

Flood barriers in place in Shrewsbury

- Around **£4.5 million** has now been spent on **flood defences** (including a flood wall) to protect the village from future flooding.

- The level of **government intervention** into flood protection measures and the management of river systems was heavily criticised. However, some argue that because the floods consisted of such **unusually high levels of water** that it's debatable whether any form of flood defence could've resulted in fewer impacts.

Warm-Up Questions

Q1 Describe how floodwater can have a negative impact on an ecosystem.

Q2 Explain the role of meteorological factors in flood events.

PRACTICE QUESTIONS

Exam Question — A-Level

Q1 Evaluate the view that the damage caused by flooding is made worse by human actions. [20 marks]

Floods — good conditions for walking your goldfish...

Floods can be a lot more visually dramatic than droughts — especially fast-onset ones. Flooding is a natural event though — it's why the area around a river is called the floodplain. The problems start when humans decide to build stuff on it.

Climate Change and the Hydrological Cycle

The hydrological cycle is one of the many things that climate change wreaks havoc on. Learn about the havoc here...

Climate change will affect **precipitation** and **evaporation**

1) The hydrological cycle is a **closed system**, so climate change won't affect the **total amount** of water moving around it. However, movements **within** the system will show different **responses to climate change**.

2) Some meteorologists believe that **climate change** will affect **precipitation patterns**:

- Wet areas (such as the tropics) are expected to get **wetter**. Rapid and increased levels of **evaporation** will create more **low-pressure systems** that will create more **rainfall**. This is because a **warmer atmosphere** will have an **increased capacity** to hold water vapour and produce more **intense rain events**.

- Dry areas (such as areas on the 30° latitude) are expected to get **drier**. There are likely to be increased numbers of **high-pressure systems** operating in dry areas which will **reduce** the amount of **water vapour** in the air and the amount of rain that can fall.

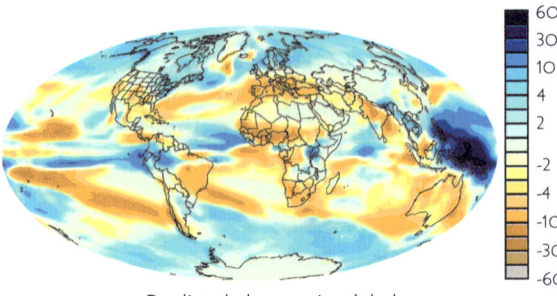

Predicted changes in global precipitation over the 21st century

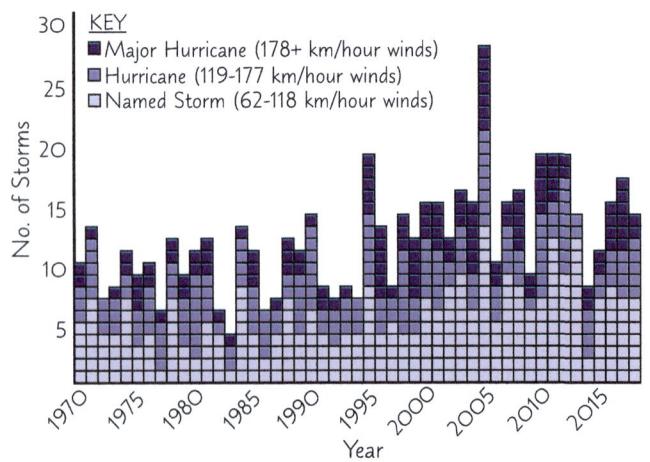

KEY
- ■ Major Hurricane (178+ km/hour winds)
- ■ Hurricane (119-177 km/hour winds)
- □ Named Storm (62-118 km/hour winds)

Number and intensity of tropical storms in the Atlantic Basin

- **Extreme weather events**, such as intense storms, floods, heatwaves and droughts, are expected to get more **frequent** and more **intense**.

- The **form** of **precipitation** is likely to change. In glacial and polar areas, precipitation is more likely to fall as **rain** rather than snow.

- Rates of **evaporation** are likely to **increase**, creating denser **cloud cover**. As air temperature gets **warmer**, there will be more **energy** in the atmosphere to create **water vapour**. There'll also be higher levels of **transpiration** from vegetation and soil.

Climate change will alter the **size of stores** within the hydrological cycle...

1) **Sea levels** are expected to **rise** as **meltwater** from glaciers is added to the store.

2) **Oceans** will also **increase** in size due to **thermal expansion**.

3) The size of **glacial ice stores** is likely to **reduce** as a warming climate creates higher levels of **ablation** (see p.164).

> Warmer seas take up a larger volume, causing sea levels to rise further (relative to the land).

4) **Snow cover** will be **shallower** and there'll be a shorter snow cover season due to **rising temperatures**.

5) **Permafrost** will also be affected. The **active layer** will **deepen** as warmer temperatures will cause more permafrost to thaw, producing more **meltwater** in the soil.

Glacial retreat shown at Chilkoot Pass, Skagway, Canada between 1906 (top) and 2014 (bottom).

6) Initially, increased levels of **meltwater** will increase the size of **water stores**. Over time, as glaciers and ice sheets **reduce in size**, the amount of meltwater feeding these stores will **decrease**.

7) Lakes, reservoirs and wetlands may see **greater fluctuations** between being at **full capacity** (after an intense rain event) and **low levels** (during heatwaves which cause high evaporation levels and times of no precipitation).

... and also change **rates of flows**

1) Rates of **direct runoff** and **streamflow** will vary according to precipitation levels.

2) **During** and **after** an intense period of rain, these flows will be **fast** and of **high volume**.

3) **Between** periods of rain, soil moisture levels may decrease as the rates slow and the ground becomes **drier**. The ground surface may also **harden**.

4) When the rain **returns**, the **hardened soil** will be **less permeable** and water will flow **over** the surface at greater **speeds**.

Climate Change and the Hydrological Cycle

Climate change may lead to **uncertainty** in the hydrological system

1) The hydrological cycle is **complex** and the **interactions** between different parts of the cycle are only **partially understood**. This means there's a **high degree of uncertainty** about many aspects of hydrology and climate change.

> Incomplete data records from different parts of the world make this uncertainty greater.

2) It's difficult to separate changes resulting from **human causes** with **natural climate change**:

> The climate **reacts** to human interference, and humans then react to a changing climate. Some human reactions aim to **restore the balance** of the climate (e.g. by reducing the dependence on fossil fuels) while others **heighten the imbalance** (e.g. by artificially irrigating barren regions).

3) The causes of climate change can come from **short-term climate events** and **long-term climate adjustments**. This makes **forecasting** exact changes difficult as the short- and long-term causes **interact** with and **influence** each other.

4) Exact changes are also **difficult to predict** because the hydrological cycle contains several **feedback systems**. Changes that might affect a water store in one direction may be **balanced** by changes in the opposite direction. This means that the hydrological cycle has a degree of **self-regulation**.

Examples of feedbacks in the hydrological cycle

Positive feedback

Temperatures rise → Evaporation increases → Amount of water vapour in atmosphere increases → Greenhouse effect increases → Temperatures rise

Negative feedback

Temperatures rise → Evaporation increases → Amount of water vapour in atmosphere increases, causing more clouds to form → Increased cloud cover reflects more of Sun's energy back into space → Temperatures fall

5) The **El Niño-Southern Oscillation (ENSO)** is a **short-term climate event** (see p.171). It's difficult to predict the exact impact it will have or when it'll happen as the intensity and speed of the onset of the ENSO changes each time. The ENSO creates **increased rainfall** and **flooding** in some areas while leaving other areas in **drought**.

6) The **warming** of the global climate is a **long-term climate event**. The impacts of this are difficult to measure and forecast as the **rate of change varies** in different locations. Knowing exactly when changes will occur is difficult because of the **lag time** between changes in global atmospheric temperature and actual changes in the hydrological cycle.

7) The ENSO is likely to be affected by climate change in the long term. El Niño and La Niña events may last **longer** and be more **intense** as the global climate continues to change.

A satellite view of a cyclone

Climate change affects the **security** of water supplies

- Rainfall patterns will become more **unpredictable** and **unreliable**, making it difficult to plan **water use** in agricultural areas. Communities that rely on **seasonal rainfall** (such as monsoons in Asia) to replenish depleted water stores may experience greater **water stress** (see p.178).

- An increase in **large tropical storm and cyclone events** may result in greater frequency of **storm surges** and inland **flooding** by seawater. Saline water **contaminates groundwater supplies** and affects water security for communities.

- Communities that live in **upland areas** and rely on **glacial meltwater** for their water supply will also be affected. There's an increased likelihood of meltwater flowing into these areas as **deluges** (a large amount of water) rather than as a constant and manageable feed. This means that some water may be **wasted** as it can't be stored. Additionally, once the glaciers have **melted** the meltwater supply will be lost — decreasing water security further.

- Increased **evaporation** from stores that supply communities can reduce their volume, **decreasing water security**.

Warm-Up Questions

Q1 Describe how global precipitation patterns may change because of climate change.

Q2 Explain why it's difficult to predict how climate change will affect the hydrological cycle.

PRACTICE QUESTIONS

Exam Question — AS and A-Level

Q1 Explain how climate change may affect the security of future water supplies. [6 marks]

So basically, it's really complicated...

All the different bits affect the other bits when they change, and then there are feedback loops too, and the ENSO which pops up now and then. Not even the best scientists or computer models have the whole thing sussed, which isn't very convenient.

Water Insecurity — Causes

Humans need freshwater. Unfortunately, there isn't always enough water in the correct places to satisfy their needs.

Water insecurity occurs when water **supply** cannot meet **demand**

Unfortunately, the global population and the sources of freshwater are **not evenly spread** throughout the world. This means that some people **receive more water** than they need while others **don't get enough**.

Water stress

- When demand for water **exceeds supply**, or when water **quality** isn't good enough to use, places experience **water stress**. This means there's **less than 1700 m³** of water available per person per year.
- Water stress is most likely to occur in areas with a **high population density** and **unreliable** or **low** water supplies.
- Places with high water availability and/or low demand have **low water stress**, such as Brazil and Russia.
- Places with low water availability and/or high demand have **high water stress**, such as Mexico and India.
- Some places have **low availability** and **low demand** which limits water stress, such as Kenya.

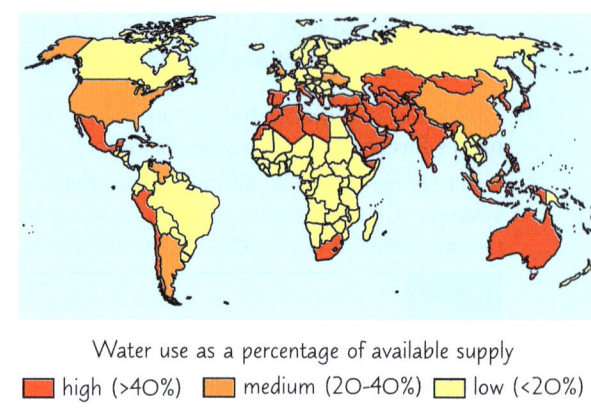

Water use as a percentage of available supply

■ high (>40%)   ■ medium (20-40%)   ■ low (<20%)

Water scarcity

- When the available water per person per year goes **below 1000 m³**, the country or region is experiencing **water scarcity**.
- **Physical water scarcity** is when there isn't enough freshwater held in rivers, streams, lakes and in groundwater to meet demand. Countries that experience physical water scarcity tend to be those that have **low rainfall levels** and **high levels of evapotranspiration**, such as Saudi Arabia.
- **Economic water scarcity** is when water held in rivers, streams, lakes and in groundwater is inaccessible to a population because there's a **lack of investment** in the technology needed to extract and transport it. E.g. developing countries in Sub-Saharan Africa.

Water insecurity is caused by **physical factors**...

1) The **climate** of a region, and particularly the amount of **rainfall** it experiences, is a key cause of water insecurity.

- Places that receive **more rainfall** will have **greater access to freshwater** and less chance of water insecurity.
- Rapid rates of **direct runoff** mean that river water quickly discharges into the sea. This means it may be **more difficult** for humans to extract the water and use it.
- If lakes and reservoirs experience high rates of **evaporation**, water doesn't reach a **sustainable level** at which humans can use it — causing water insecurity.
- Some parts of the world, such as the Sahel region of Africa, are more vulnerable to water insecurity because of the **unpredictability** of **seasonal rainfall**.
- Where stores, such as **lakes** and **reservoirs**, are experiencing lower water levels, the water they do hold is likely to be **warmer** as it takes **less energy** to heat it when it's shallower. This means that **bacteria** in the water can multiply quickly. If this water store is then used for human consumption, it may be **harmful** to human health.

2) **Rising sea levels** have increased the chances of **saltwater** encroachment on freshwater aquifers (e.g. Bangladesh). This is where saltwater is drawn into **aquifers** to replace water that humans have extracted, meaning that **wells** dug into the aquifer become contaminated with salt.

Climate variability in California

CASE STUDY

- California has experienced a greater range of precipitation levels throughout the 21st century. Californians also have a **high demand** for water (mostly to **irrigate crops**) which has led them to experience **droughts**, as supply has not been able to meet demand.
- **Climate climate** has caused an increase in atmospheric **temperatures** which has increased the rate of **evaporation** from California's main reservoirs, such as Shasta Lake in the north of the state. For example, in **2022** Shasta Lake was at **31% capacity** — its **lowest level** since records began. Of the 20 warmest years on record in California, 11 of them have been since 2000.
- Low levels of **rainfall** have also increased the risk of **wildfires**.

In 2020, the August Complex wildfire stretched for over 1 million acres.

Water Insecurity — Causes

... and **human factors** too

1) **Chemical pollution** can contaminate freshwater sources. It may come from:

   - **runoff** from agricultural land (which may contain pesticides or livestock waste).
   - **industries** where water is used as a coolant.
   - people's **homes** where wastewater is connected to river systems.

Runoff from agricultural land can result in algal blooms in water courses.

2) In countries and regions where management of wastewater has been **poorly developed**, even greater levels of pollution can occur. As water becomes **more polluted**, the supply of safe **freshwater** is reduced.

3) Groundwater supplies can also become **contaminated** with heavy metals such as **arsenic or lead**. These tend to pool there when **industrial waste** is either buried or dumped and precipitation washes through it.

4) Populations commonly abstract **too much** water from rivers, lakes and aquifers. **Over-abstraction** means that these water stores can't be replenished at a rate fast enough for water stress to be avoided. Where populations are using a water source, such as a river, people who live **downstream** may have reduced capacity to meet their water needs.

5) **Hydroelectric power plants** may hold water back in a reservoir to generate electricity. This means those downstream may face water stress.

Over-abstraction from the Aral Sea

- The Aral Sea is a **former large lake** on the border between Kazakhstan and Uzbekistan — it used to be the world's **fourth largest** inland lake. In 1960, its area was 60 000 km^2. However, between 1960 and 2010, the Aral Sea **shrunk to around 10%** of its original size.

- Mountains to the east of the Aral Sea basin feed the lake with **snowmelt** in the summer months.

- The Aral Sea has been **shrinking** since the 1960s when **irrigation canals** were built to **divert** water from the main rivers to the surrounding cotton fields. Most of the diverted water was **lost** through leaks in the canals.

- In July, temperatures around the Aral Sea frequently reach around **30 °C**, so **evaporation** means further water is lost. The diversion of water away from the Aral Sea means it isn't **recharged**.

CASE STUDY

© NASA's Earth Observatory

Satellite image of the Aral Sea in 1989 (left) and 2014 (right)

Global demand for water is rising for several reasons

Population growth
- More people means **more water** is needed for drinking, washing and preparing food.
- **Demand** for food, electricity and other goods also **increases**. Producing these uses water.

Industrialisation
- As countries develop, **energy** use increases and **manufacturing industries** grow — energy production and manufacturing use a lot of water.
- It is predicted that the demand for water by industry will **increase by around 5%** every year.

According to the UN, over 2 billion people worldwide live in countries that are classified as 'water-stressed'.

Living standards
- As people's **wealth** increases, they can afford flushing toilets and showers. Improved living standards and an increase in the number of people in middle-income groups means more homes are fitted with appliances that require more water.
- Middle-income groups often consume more **meat**, which creates an increased demand for water in **agriculture**.

Agriculture
- As countries develop economically and their demand for **food increases**, the amount of **water** needed for agriculture also increases.
- Agriculture is the largest consumer of water worldwide. Globally, **70% of fresh water** is used for agricultural **irrigation**.

There are other factors too, for example:

- New forms of **energy resource extraction**, such as **fracking**, also require large amounts of water.
- The growing world population is becoming increasingly **urbanised**. Urban spaces contain **dense living conditions** where the demand for water is centred around a **limited number of water stores**.
- The rising demand in **some locations** is causing increasing levels of **water stress** and **water scarcity**.

Water Insecurity — Causes

Water prices vary around the world

1) Water has become an **expensive commodity** in some countries and regions, despite most people viewing clean water as a **human right**. In many countries (e.g. the UK), water is a **product** like any other and is supplied by **private water companies**.

2) As with many products, the ability to **buy in bulk** can **reduce** the price for consumers. This means that **farmers** might receive their water at a **lower price** than householders.

3) Local governments may **subsidise** the cost of water use for different industries to encourage their growth.

4) The **cost of water** depends on several **different factors**:

Empty buckets waiting for fresh water from a mobile tank truck, Vietnam

- The cost of transporting the water from its source to where it's needed. This usually uses water infrastructure, such as pipelines and canals. Sometimes water is transferred long distances by road tanker or by ship.
- The price of water is market-driven in some areas. This means that as demand goes up and water becomes scarcer, the price also goes up.
- Accessing the water at its source can also be challenging and expensive. Digging wells and creating standpipes (vertical water pipes) can be costly for some communities in developing countries.
- The more polluted the water is, the more costly it is to purify it and make it suitable for use.

5) The growing costs of water means that poorer people in **developing countries** are more likely to rely on **insufficient**, **poor quality** water.

6) In addition to the cost of water itself, there's a cost for a household to **utilise it efficiently**. Families may have to invest in pipework, taps and sinks, water storage units and toilets.

7) Making water a **potentially profitable commodity** can be beneficial in meeting water deficits:

- If people must pay for their water, they're more likely to conserve it.
- Water companies may compete with each other to produce a higher quality product. These companies have an incentive to maintain water supply infrastructure to a high standard so that minimal quantities are lost through leaks and they are more likely to attract customers.

The **Water Poverty Index** is an international score of water security

The Water Poverty Index (WPI) uses **five different measures** to give an **overall score out of 100**. The five measures are:

1) **Access** — how easy it is for humans to access **safe water** for different uses.

2) **Capacity** — how effective water management measures are in making sure water continues to be **affordable**.

3) **Use** — how water is used for different purposes such as **agriculture**, **industry** and within **households**.

4) **Environment** — how sources are managed to have high levels of **environmental** and **ecological protection**.

5) **Resources** — how much surface and groundwater is **available** and what the **quality** of this water is.

> WPI scores are often shown using a radial graph, like the one on the right, which compares the WPI of two settlements

Developing countries tend to have **lower** WPIs than developed countries, which usually have established water infrastructure and management networks.

Warm-Up Questions

Q1 Describe the key differences between water stress and water scarcity.

Q2 Explain why the price of water varies throughout the world.

PRACTICE QUESTIONS

Exam Question — A-Level

Q1 Explain how human actions have contributed towards water scarcity in some areas. [8 marks]

Increase your water security — put locks on your taps...

Learn all the points on these three pages, and then you'll be well on your way to knowledge security for the exam.

Water Insecurity — Consequences

Water insecurity has a lot of knock-on effects on people's lives, and can trap the poorest in poverty.

Water supply is important for **economic development**

1) Water is **essential** to humans. People use water **domestically** for things like drinking, washing and getting rid of sewage. People also **indirectly** rely on water because of its importance in many different **industries**.

Energy Supply
- In many types of power station, water is **heated** to generate **steam**, which drives **turbines** that generate electricity. It's also used to **directly** drive turbines in **hydroelectric power plants**.
- Water is also used in **newer energy technologies**. During the **fracking process**, it's pumped into the ground to release **natural gas**.
- Water is important in the development of **biofuel crops**, as growing them requires high levels of water.

Industry
- **Factories** use water in the production of all kinds of **products**.
- Many industries use water for **cooling machinery** as moving parts can overheat easily. Water may also be used to **clean raw materials** before they enter the production process.
- The **demand** for water from industry is **growing** rapidly in **emerging economies such** as China and India. This is due to the **global shift** in the **manufacturing industry** from developed to newly emerging economies.

For example, paper factories use it to produce the wood pulp that's used to make sheets of paper.

Trade
Waterways, such as large rivers and man-made canals, enable the **transport** of goods within and between countries.

Agriculture
- Farms use water to **irrigate** crops and raise **livestock**.
- Irrigation, for crops and pastures, mainly comes from **groundwater supplies**.
- Agriculture can account for **over 25%** of GDP in some **developing** countries.

Globally, agriculture is the main user of water.

2) Some countries with rapid rates of **population growth** are in areas facing **severe water shortages**. This could hinder their **development** and have severe impacts on people's **health** and **quality of life**.

Water supply is important for **human wellbeing**

1) Health and wellbeing relies on a **safe** and **plentiful** supply of water. In 2021, nearly 800 million people worldwide still didn't have access to a clean water supply close to their home.

2) Lack of an **effective sewage system** in a community means **human waste** enters water courses and diseases such as cholera, typhoid and dysentery can be passed on to humans. This **limits** the amount of usable water. **Faecal bacteria** in drinking water can sometimes be fatal. Dirty water can also cause eye, ear and skin **infections**.

3) **Raw sewage** can also enter drinking water through **contaminated soil**, e.g. in areas without sanitation where **open defecation** occurs — threatening water security further. This can also attract **flies** that might travel from faeces to food, which will make people sick.

4) A clean water supply is needed to **prepare food** in a safe manner, e.g. to wash fruit, vegetables and kitchen utensils.

Men stand in the River Ganges washing clothes. Levels of faecal bacteria in the river are up to 12 times higher than the permissible limit.

Problems with water supply can cause **economic problems**

- Poor countries in **dry** areas can struggle to afford to **import water** or to build **heavily engineered solutions**. This means that they often can't obtain enough water to **meet demand**. This may hinder the development of certain industries or make those industries less profitable or globally competitive.

- High levels of **sanitation** means more usable water can be provided to schools, which can **improve** attendance (particularly among menstruating girls). As a result of this, young people will be **better educated** and better able to work in **skilled employment**. This will further improve the **economic development** of the country.

In 2021, around 30% of schools globally didn't have access to clean water.

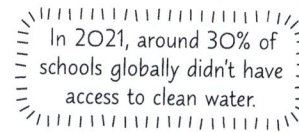

- The same is true of **water access**. In developing countries, women and girls within a family are often tasked with the **daily collection of water**. This can limit their ability to attend school on time and gain employment. Having more **piped water** into a home would allow **more** girls to focus on their education, increasing their chance of having a career as well as also improving the **economic development** of a country too.

Topic 5 — The Water Cycle and Water Insecurity

Water Insecurity — Consequences

Problems with water supply can cause **environmental problems**

- A supply of water is vital for plants and animals. Water also forms some **habitats**. If this water supply is **diminished**, this can lead to the death and even **extinction** of some **species and habitats**.
- Ponds and lakes help to **regulate local atmospheric moisture levels** and **temperatures**.
- Certain wetland features, like **reedbeds**, **purify water** as it enters the surface store. Reeds remove **toxins** from the water and also provide **nesting materials** for birds.
- **Fertilisers** that run into lakes and reservoirs cause **eutrophication** — a process where nutrients from fertilisers cause an **algal bloom** on the water's surface. This **deprives** animals and plants of **oxygen** and also of **light** at deeper depths.

> Wetland areas combine the needs of terrestrial organisms with aquatic ones. They can be breeding grounds for migratory birds as well as hold large fish stocks.

Water insecurity can cause **conflicts within a country**

1) **Conflicts** over water occur where **different groups** of people within a country share the **same supply** and either:

   - the water supply decreases, for example, if rainfall in the area has decreased
   - or, water demand increases, for example, where the population grows or industry expands.

2) There may be conflict between different stakeholders. For example, between those who are able to pay for an enhanced water supply to their homes and business and those who can't afford to install infrastructure for a domestic supply.

2) The competition for water might be between different **uses** rather than different **individuals**. Although **agriculture** commonly uses most of the available water in a local area, it may be in competition with recreation, manufacturing industries and domestic users.

3) In a drainage basin, **changes upstream** can have a major impact **downstream**, so regions upstream must **manage rivers carefully**.

> For example, over-abstraction causes lower flows downstream whereas industry can increase pollution levels. Farming can increase soil erosion and raise the concentration of sediment in rivers.

Water insecurity can cause **conflicts between countries**

- Many rivers cross **international boundaries**. This can give upstream countries **power** over downstream countries. Upstream countries may be able to **divert water** away from the main channels so that a **reduced volume** crosses the border. They may also **pollute** the water, leading to **unsafe** water entering the downstream country.
- **Lakes** can also cross boundaries, making it difficult to manage a fair distribution and use of the water.

The River Nile

CASE STUDY

- The Nile River flows through **eleven countries** in North East Africa. However, only Egypt and Sudan have full rights to the water.
- There's **conflict** between the **downstream countries**, e.g. Egypt and Sudan, that use **most** of the water, and the **upstream countries**, e.g. Uganda, Rwanda, Ethiopia and Tanzania, that **want** to use **more water**.
- A 1959 agreement gave Egypt and Sudan **priority use** of the river water and increased each country's share of water from the Nile. Egypt was also allowed to **stop developments** along the course of the Nile that threatened its water supply.
- In 2020, Ethiopia finished building the **Grand Ethiopian Renaissance Dam**. The dam will be a source of **hydroelectric power** and controls the river's flow to **prevent drought** and **flooding**. Egypt was concerned that the lake formed behind the dam would lead to the loss of billions of cubic metres of water by **evaporation**, meaning that **less water** would flow into Egypt. This led to **disputes** between the two countries, as well as Sudan, but a **partial agreement** was reached in 2015.

A charity trucks water into rural Ethiopia when local supplies run out.

Warm-Up Questions

Q1 Describe how industry uses water to create greater levels of economic development.

Q2 Explain how transboundary water stores can be a source of conflict.

> **PRACTICE QUESTIONS**

Exam Question — A-level

Q1 Explain how economic development can be influenced by water security. [8 marks]

Water pistols at the ready — it's a water fight...

Water is a valuable commodity. Everyone needs it to survive — plus it's pretty much impossible for a country to develop economically without a good supply of it. It's no wonder that it can be the source of so much conflict between countries.

Managing Water Supply

You need to know how to make sure that people have enough water. Luckily, that's what the next few pages are about.

Water supply can be managed through **hard engineering**

1) There are several things that can be done to **increase** the amount of **water available** in an area and **increase water security**. Some of these involve **hard engineering** or a **technological fix**:

Water transfer schemes
- Water transfer schemes involve **moving water** from areas of **surplus** to areas of **deficit**.
- This is normally done by pumping it through **pipes**, **channels**, **canals** and **aqueducts**.
- Water can also be moved by **redesigning the entire river course** so that it flows into the required area. There has been an increasing need to transfer water between **whole regions**, such as the **South-North Water Transfer** scheme in China (see the next page), or even between **whole countries**.

Water transfer canal, India

Mega dams
- Water can be **stored** during times of surplus to ensure there's enough water during times of **deficit**.
- Building a **dam** across a river traps water, creating a **reservoir**. The reservoir is filled during periods of **extended rainfall**, and the water is released during **drier periods**. This ensures a **consistent flow** of water in the river all year round.
- There are many large scale dams in the world, e.g. on the **Yangtze River, China** and on the **Colorado River** (see p.186).

Three Gorges Dam, China

Desalination plants
- Desalination is the **removal of salt** from seawater, so that it can be used as a water source. Seawater is heated until the water **evaporates**, leaving the salt behind. The water vapour is then **condensed** to produce freshwater. Alternatively, seawater can be passed through a series of **membranes** to remove the salt. This is called **reverse osmosis**.
- Seawater is **abundant** in many areas that lack sufficient freshwater but have a coastline, so desalination can significantly **increase water security**.
- **Saudi Arabia** has over 30 desalination plants — more than any other country.

Desalination plant

2) Choosing the right type of water management scheme is made more difficult because there are many different **stakeholders** and **players** that need to be considered.

There are **advantages** and **disadvantages** of hard engineering schemes

Water transfer schemes

Advantages	Disadvantages
• **Water security** may be increased in the **receiving area**. • The volumes of water being transferred can be controlled to suit **seasonal needs** if necessary. • Water transfer schemes can be highly effective when designed for **small-scale** use. • They can result in greater levels of **political cooperation** between regions and countries. • There's greater security against **minor imbalances** between **supply** and **demand**.	• **Water stress** may be increased in the **source area**. • The receiving area may begin to **build developments** such as golf courses, which require a lot of water. • The **infrastructure** needed is **expensive** to construct and maintain, as well as **environmentally damaging**. • **Decreased discharge** in rivers can alter freshwater **ecosystems** and increase **saltwater intrusion**. • Rivers can also become **heavy with sediment**, which can further **change the rate of flow**.

Managing Water Supply

Mega dams

Advantages	Disadvantages
• Large dams have the potential to **increase** water security for **many people**. • The construction of dams **employs** many people over many years. • The **reservoirs** created by dams can be used for **recreational activities**, like sailing. • Large dams can generate **hydroelectric power** for local communities. • Dams can **control flooding** by letting water flow at a more gradual rate.	• The reservoir created by a dam often **floods agricultural land** and causes the **displacement of settlements**. • Their construction often causes **local conflicts** and **mistrust** between displaced people and national authorities. • Dams are **extremely expensive** to build and often require loans from other countries. • **Evaporation** can take place on the large surface of the reservoir, reducing the amount of water available for human use. • Dams can hold back **silt** which is an important **fertiliser** for farmland further downstream.

Desalination plants

Advantages	Disadvantages
• This uses a **plentiful** resource. • It's suitable for **island countries**, such as the Maldives, that need high amounts of water to support tourism. • It's also suitable in areas that have **few river systems**.	• It's an extremely **expensive** process to set up. • The processes require **high amounts of energy**, most of which is likely to come from fossil fuels. This causes pollution and contributes to climate change. • As higher levels of salt re-enter the sea, the process can have a negative impact on **marine life**. • It's a **less viable option** of increasing water security in landlocked countries.

Water transfers in China

CASE STUDY

1) The **South-North Water Transfer Project** (SNWTP) is a $79 billion scheme intended to increase water security in China.

2) It involves a network of tunnels and canals that will divert about **45 billion cubic metres** of water **every year** from the Yangtze River and its tributaries in the **south** of the country to the Yellow River in the **drier north**.

3) **Two** out of **three** planned routes have been **completed** — the **Central** and **Eastern Routes**. The **Western Route** completion is planned for **2050**.

4) The completed parts of the SNWTP have had **pros and cons**:

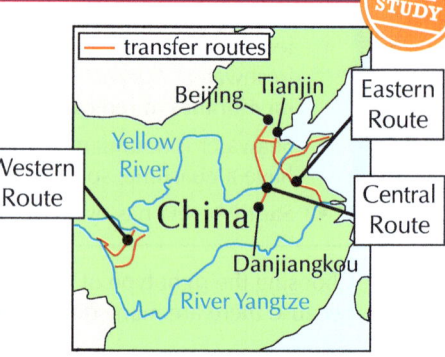

Advantages

• The project provides **clean water** to over **20 cities**, including Beijing and Tianjin. The Chinese government estimates that up to **100 million people** have benefited from this project.

• Water is important for industry, so an increased water supply means that **industrial development** can continue in the north. This can increase the country's **wealth**. The scheme provides water that can be used to **irrigate** farmland, so crops can be grown.

• It should prevent **over-abstraction** in the north, helping to stop **land subsidence**.

The start of the central South-North Water Transfer canal, China

Disadvantages

• As part of the **Central Route** construction, the **height** of the existing **Danjiangkou Dam** on the **Hanshui River** (a tributary of the Yangtze River) was increased by about 15 metres to allow **more water** to be **stored** in the reservoir behind it. This **flooded** more land, which **destroyed habitats** and **reduced biodiversity**.

• Increasing the height of the dam also **decreased the velocity** of the Hanshui River **downstream**. This contributed to an increase in **algal blooms** (thick blankets of algae on the water surface), which **prevent light** from penetrating below the surface. This causes **plants** in the river to **die**, resulting in **oxygen shortages** in the water and the **death** of organisms that **rely** on **oxygen**, such as fish and amphibians.

Managing Water Supply

Water supply can be managed through **sustainable** projects

River restoration in the Lake District, UK

- Sometimes attempts are made to **restore water** supplies. In places where **groundwater** has been **overexploited**, aquifers can be artificially recharged. For example, **injection wells** allow water to be put back into depleted **aquifers**. **Infiltration structures** can also be used. These include water-filled basins, ponds or trenches, which water **infiltrates** through to the aquifer. **Rivers** can be **restored** to their **natural course** to help improve their ability to **hold more water** and prevent flooding.

- **Habitat restoration** can also take place. The use of a lake to provide water is **paused** to allow plants and animals to return.

Water conservation makes supplies more **sustainable**

Conserving water is one way that water supplies can be made more **sustainable**:

In 2001, the UK government banned the installation of toilets that used more than six litres per flush.

Reducing domestic water consumption

- The amount of water used to flush the toilet can be reduced by adding a **displacement bag** into the cistern or installing a **modern water-efficient toilet**.
- People can take showers instead of baths. **Water-efficient shower heads** also reduce water consumption.
- **Modern appliances**, such as washing machines, are more water-efficient.
- **Water meters** can be fitted. These mean people pay for the exact amount of water they use, encouraging them to use less.
- Homeowners (and water supply companies) can fix leaks to minimise water waste.
- Homeowners are also increasingly using rainwater harvesting techniques. Rainwater is fed into large containers, such as water butts, and used to water gardens and even flush toilets.

In the UK, houses built since 1990 have been fitted with water meters.

Reducing agricultural water consumption

- **Smart irrigation** monitors **local weather** to **direct water** to exactly where it's needed.
- In the most advanced farming systems, the soil moisture around **individual plants** is monitored and when it reaches a low level, the irrigation of that plant is **automatically triggered**.
- **Stormwater** can also be collected for irrigation.
- Crops can be watered early in the **morning** to reduce water loss through **evapotranspiration**.
- **Contour ploughing** (ploughing across a slope rather than down it) can also reduce **direct runoff**.
- Genetic modification has produced **drought-resistant varieties** of plants (e.g. important food crops such as maize, millet and wheat) that would otherwise require high levels of water.

Recycling water

- Used water can be **treated** to make it safe to reuse **straight away** (rather than returning it to a river or the sea). Most recycled water is used for **irrigation**, **industry**, **power plants** and **toilet flushing**, although it can also be treated enough to make it safe to **drink**.
- **Greywater** is mostly wastewater from homes and businesses. This water is **relatively clean**, so it can be used to water gardens, flush toilets and irrigate farmland without being treated first.

All water was grey before colour film was invented.

Water management in Singapore

CASE STUDY

1) The island nation of Singapore has a **high demand for water** — they use **over 400 million gallons** of water per day.

2) Singapore has few natural sources of freshwater. Therefore, they must rely on **other means** to meet demand. Singapore has **five desalination plants** which can meet around 40% of the country's water demands.

3) The Singapore government has also led a strong campaign to encourage citizens to **use water wisely**. There's also been the introduction of a **scaled water pricing system**. This means that people who consistently use high levels of water are charged **more** than those who use less.

4) Until 2011, Singapore had a **water trade agreement** with Malaysia and parts of the agreement will exist until 2061. However, the agreement has become a source of **tension** between the two countries. Singapore has tried to become more **self-sufficient** regarding water.

5) Marina Bay is a **large freshwater reservoir** next to the coast. The **Marina Barrage** has been built to prevent saltwater entering into the bay.

6) 40% of the country's water comes from a **high-tech water recycling plant** that uses new techniques, such as UV treatment, to create what's branded 'NEWater'.

NEWater brand recycled water

Managing Water Supply

Cooperation is needed to manage water supplies

- Part of sustainable water management is to make sure that **stakeholders** and **users** are **consulted** and their **views** are **heard**. Greater levels of **cooperation** between users and stakeholders may create new ways of distributing water **fairly**.

- There are **additional players** to consider when water supplies cross borders. As well as individual users, larger **TNCs** may be consulted alongside **national governments**, **international charities** and **organisations** — each with their own aim.

- One way of achieving international cooperation to manage water supply is through **integrated drainage basin management**. This approach looks at the river basin as a **geographical unit** rather than being split by **country borders**. This means that different users, often from different countries, must **work together** to manage large river systems. Integrated drainage basin management focuses on using water **efficiently** and **fairly** while maintaining high levels of **environmental quality**.

- Integrated drainage basin management can work well within a country where there are just a **few different users**. On an international level, it can be more **challenging** and other frameworks may be necessary.

Collaborative management of the Colorado River

1) The **drainage basin** for the Colorado River extends into **seven states** in southwest USA.

2) It supplies over 40 million Americans with **drinking water** through a series of **dams**, **reservoirs** and **irrigation canals**. Problems have been caused by different states trying to use the river in **different ways**.

3) In 2007, in an update to the **Colorado River Management Plan**, the states settled on a new way of managing the water resource during a drought. Rather than divide water usage **equally** between the states, water from the river was allocated according to the size of the **water deficit** of each state. Resolving this issue involved a greater level of **collaboration** between the states.

4) Approving the plan involved trying to get several stakeholders (including National Park authorities, leisure users, farmers, water utility companies, city mayors and state governors) to **work together**.

The Hoover Dam on the Colorado River

Countries often reach agreements

- There are several international treaties, agreements and frameworks that aim to encourage **cooperation** between countries that share water sources. This is one aspect of **hydropolitics**.

- The **United Nations Economic Commission for Europe (UNECE) Water Convention** (see p.280) is a set of guidelines agreed between a group of countries that encourage **collaborative management** and **conservation** of shared water sources. They're guidelines, so aren't written into international law.

- The **Water Framework Directive and Hydropower** in Berlin in 2007 was a convention designed to promote **sustainable practices** in **hydroelectric power** generation. These included careful **management** of **water levels** in reservoirs and feeding rivers, as well as guidance on best practices in **environmental protection**.

The Helsinki Rules are based on the principle that all states containing river basins have the right to receive a fair share of water resources that cross their boundaries. They also cover environmental sustainability — they say that all involved countries should look to prevent pollution of the water resource. Importantly, every country has a duty to ensure that no users actions can harm another user.

Warm-Up Questions

Q1 Explain how integrated drainage basin management can be used to manage water supplies.

Q2 Describe what the UNECE does to help manage water supply.

Exam Question — AS and A-Level

Q1 Assess the view that water conservation is a more effective strategy to manage water supply.

[12 marks]

Management of revision — exam success now and forever...

It's a lot better for all concerned when everyone gets along (as with many things in life). It isn't always easy, but cooperation can help get everyone working towards the same aim. Well anyway, you've got through the water section. Yippee!

The Carbon Cycle

Ah, carbon — burnt toast, pencils, trees, cake... Read on to find out all the other zillions of places carbon turns up.

The carbon cycle is a **closed system**

- Carbon is a fundamental **building block** of life — all **living things** contain carbon.
- The carbon cycle is the process by which carbon is **stored** and **transferred**. It's an example of a **closed system** — there are **inputs** and **outputs** of energy, but the amount of carbon in the system **remains the same**.
- It's often called a **biogeochemical** cycle, highlighting links between the **biological**, **geological** and **chemical** processes.
- Carbon is a really important element. It's found in **organic stores** (living things) and **inorganic stores** (rocks, gases and fossil fuels). Carbon **moves between** these stores as **fluxes**:

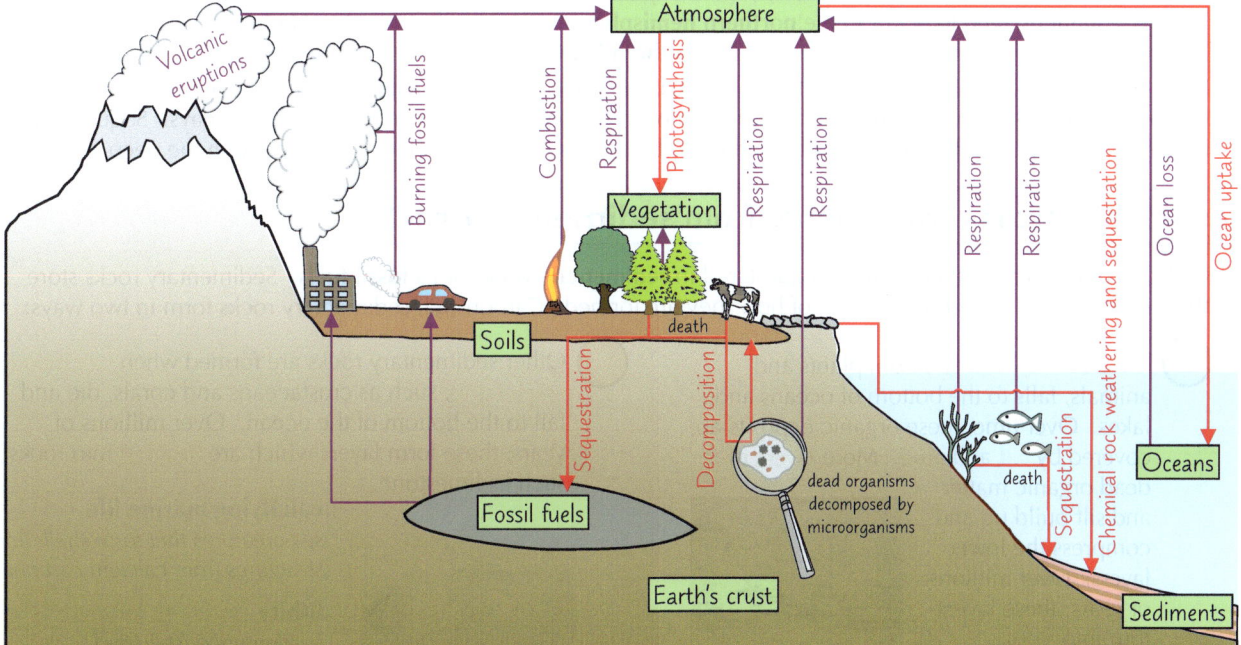

- The **carbon budget** is the difference between the inputs of carbon into a subsystem and outputs of carbon from it. For example, in the **atmosphere**, inputs of carbon come from **volcanic eruptions**, burning **fossil fuels**, **respiration** and **ocean loss**. Outputs occur through **photosynthesis**, **sequestration**, **decomposition** and **chemical weathering**.

> **sequestration**
> *the capture of carbon from the atmosphere by the ocean, vegetation or sedimentary rocks where it is stored*

- The **balance** of the inputs and outputs in a subsystem determines whether it acts as a **carbon source** or a **carbon sink**. In a **carbon source**, the carbon outputs outweigh the inputs, so it releases more carbon than it absorbs. In a **carbon sink**, the inputs of carbon outweigh the outputs, so it absorbs more carbon than it releases.
- There can be **sudden releases** of carbon (such as during volcanic eruptions or major wildfires) which can upset the balance of the carbon cycle — it can take a long time for the balance to be restored.

Most carbon is locked in **stores**

- Some carbon is locked away in long-term stores, such as in fossil fuels deep underground. Carbon can be found in each of the Earth's systems in some form or another:

Lithosphere — Over 99.9% of the carbon on Earth is stored in **sedimentary rocks** such as **limestone**. About 0.004% of the carbon on Earth is stored in **fossil fuels**, such as coal and oil, in the lithosphere.

Atmosphere — Carbon is stored as **carbon dioxide** and in smaller quantities as **methane** in the atmosphere. The atmosphere contains about 0.001% of the Earth's carbon.

Hydrosphere — Carbon dioxide is dissolved in rivers, lakes and oceans. The oceans are the second-largest carbon store on Earth, containing approximately 0.04% of the Earth's carbon. Most of the carbon here is found deep in the ocean in the form of **dissolved inorganic carbon**. A small amount is found at the ocean surface where it's exchanged with the atmosphere.

Biosphere — Carbon is stored in the **tissues** of living **organisms**. It's transferred to the soil when living organisms die and **decay**. The biosphere contains approximately 0.004% of the Earth's total carbon.

Cryosphere — This contains less than 0.01% of the Earth's carbon. Most of the carbon in the cryosphere is in the **soil** in areas of **permafrost** (permanently frozen ground) where decomposing plants and animals have frozen into the ground.

The Carbon Cycle

Carbon fluxes happen over different time and spatial scales

1) Annual carbon **fluxes** are measured in **petagrams** (Pg) or **gigatons** (Gt) of carbon per year. One of each is equivalent to 1 billion tonnes of carbon.

2) The largest fluxes are between the **oceans** and the **atmosphere** and between the **land** and the **atmosphere**. These fluxes involve both **photosynthesis** and **respiration**.

3) The carbon fluxes taking place occur at different **spatial** scales. For example:

   - At a **plant scale**, respiration and photosynthesis are the main fluxes.
   - At an **ecosystem scale**, carbon fluxes such as combustion and decomposition also occur.
   - At a **continental scale**, all carbon fluxes, including **sequestration** occur. More photosynthesis and respiration tend to occur in the **northern hemisphere** because it has a greater **landmass**. More ocean sequestration takes place in the **southern hemisphere** where there is more **sea**.

4) Fluxes occur over different time scales too. Some carbon fluxes **quickly transfer** carbon between sources. Photosynthesis, respiration, combustion and decomposition might take only minutes, hours or days. Sequestration is far slower. It takes **millions of years** for carbon to be sequestered in sedimentary rocks and in the deep ocean.

> **carbon flux**
> *the flow of carbon between the different carbon stores*

Most of the Earth's carbon comes from sedimentary rocks

- Carbon from the atmosphere can be sequestered in **sedimentary rocks** or as **fossil fuels**. Sedimentary rocks store the largest amount of carbon because of how they are formed. Carbon-rich sedimentary rocks form in two ways:

1 **Dead organic matter**, from plants and animals, falls to the bottom of oceans and lakes. Over time, these organic deposits are covered by **silt** and **mud**. More **layers** of dead organic matter and silt build up and compress the lower layers. Over millions of years, these layers turn into **shale**.

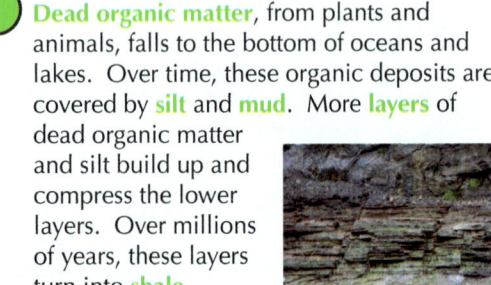

2 Other sedimentary rocks are formed when **calcifying marine life**, such as crustaceans and corals, die and fall to the bottom of the ocean. Over millions of years, these form layers which are **lithified** into rocks such as limestone.

> **calcifying marine life**
> *sea creatures that form shells and structures from calcium carbonate*
>
> **lithify**
> *harden unconsolidated material into stone*

- Fossil fuels are formed by the **deposition** of **organic material**. Without oxygen, full **decomposition** doesn't take place and the organisms decay **anaerobically**. If the organic matter **builds up faster** than it can decay, oil and natural gas (from small animals and plants) and coal (from plants) are formed.

Carbon is released into the atmosphere by geological processes

Chemical weathering

Chemical weathering transfers carbon from the atmosphere to the hydrosphere and biosphere. Atmospheric carbon reacts with **water vapour** to form a **weak carbonic acid**. On condensation, **acid rain** is formed. When this acid rain falls onto rocks, a chemical reaction occurs which **dissolves** the calcium carbonate in the rocks. This reaction produces **carbon ions** which wash into rivers and then get carried out to sea.

Outgassing

- Carbon dioxide stored within **magma** and in the Earth's **crust** is released during volcanic eruptions at plate boundaries or hot spots (see p.12). This release of trapped or stored gas in known as **outgassing**.

- Gas is also released at **ocean ridges**, **subduction zones** and through **geysers**. Rocks in these areas undergo a process called metamorphism, where sedimentary rocks change into metamorphic ones under **intense temperatures** and **pressures**. Metamorphism releases carbon dioxide.

- Recent volcanic eruptions have released much less carbon dioxide than human activities, but an exceptionally large eruption could **disrupt** the carbon cycle significantly.

Strokkur geyser, Iceland, releasing water vapour and gases (including carbon dioxide).

The combination of chemical weathering and outgassing can form a **negative feedback**.

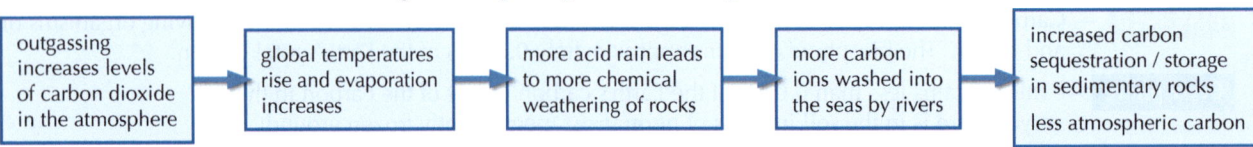

| outgassing increases levels of carbon dioxide in the atmosphere | → | global temperatures rise and evaporation increases | → | more acid rain leads to more chemical weathering of rocks | → | more carbon ions washed into the seas by rivers | → | increased carbon sequestration / storage in sedimentary rocks

less atmospheric carbon |

The Carbon Cycle

Carbon is released into the **atmosphere** by **combustion**

- **Combustion** transfers carbon stored in living, dead or decomposed **biomass** (including peaty soils) to the atmosphere by burning. Combustion releases carbon dioxide and heat which can be used to power turbines and create electricity.

- **Wildfires** rapidly transfer large quantities of carbon from biomass (or soil) to the atmosphere. The loss of vegetation then **amplifies** the increase in atmospheric carbon because wildfires cause a **decrease in photosynthesis**, so less carbon is removed from the atmosphere by plants.

Carbon can be **sequestered** into **oceans**

Carbon moves between the atmosphere, the surface water of the ocean and the deep water through **three interconnected systems** known as **pumps**.

Biological pump

- Carbon dioxide in the atmosphere can be consumed by large numbers of **phytoplankton**, single-celled microscopic organisms living on the ocean surface, through the process of **photosynthesis**.

- Phytoplankton are at the bottom of the **food chain** for marine animals, so the carbon is **transferred** up the chain as creatures feed off each other.

- **Respiration** releases some **carbon dioxide** back into the atmosphere, though most respired carbon remains in the **surface waters** and is reabsorbed by phytoplankton.

- When marine animals die they fall to the **ocean floor**, where the carbon stored in their bodies is transferred to the deep ocean.

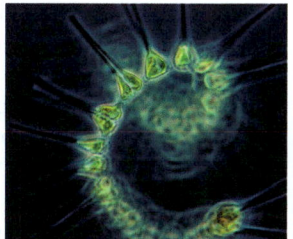

Phytoplankton are microscopic organisms that are at the bottom of the food chain for marine life.

Carbonate pump

- Chemical weathering can wash **carbon-based molecules** into the sea through river systems.

- Here, they react with carbon dioxide dissolved in the water to form **calcium carbonate**, which is used by some **marine organisms** to make shells.

- When these marine organisms die, they sink to the bottom of the ocean where they create a **sediment** rich in calcium carbonate (a process known as sedimentation).

Physical pump

- Cold water has a greater capacity to hold carbon dioxide than warm water. Cold water is denser than warm water, so it sinks and is held **under pressure** by the weight of the water above it.

- This means there is a higher concentration of carbon dioxide in the **deep water** of the ocean and at **higher latitudes**, where ocean water is cooler.

- The **thermohaline circulation** moves warmer water on the surface into cooler areas of the world. As the warm waters cools, it absorbs more carbon dioxide and sinks to the deeper waters (a process known as a **downwelling**).

- To redress the balance, cold water is forced to the surface in other areas of the ocean. Carbon dioxide is also brought to the surface where it may be released into the atmosphere or absorbed by phytoplankton. This is known as an **upwelling**.

Carbon can remain deep in the ocean for **hundreds of years** while moving around the ocean basins through the **thermohaline circulation** (the circulation of ocean currents according to their density).

- **Cooler, denser** water **sinks** to the bottom of the ocean while warmer water moves along the surface.

- Differences in **salinity** (the salt content of the water) between the **polar regions** and the **equatorial zones** (caused by freshwater melt at the poles) cause **upwellings** and **downwellings** in different parts of the ocean.

- This means the ocean acts as a giant **conveyor belt** that moves water and dissolved carbon dioxide around the Earth.

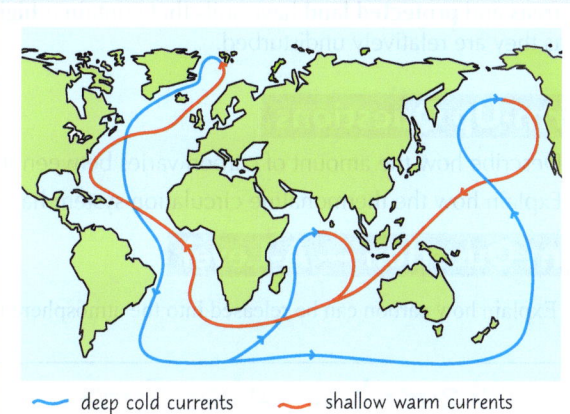

~ deep cold currents   ~ shallow warm currents

Topic 6 — The Carbon Cycle and Energy Security

The Carbon Cycle

Carbon can be **sequestered** into **plants** and **soils**

- **Photosynthesis** transfers carbon stored in the atmosphere to biomass. Plants (**terrestrial primary producers**) use energy from the Sun to change carbon dioxide and water into **glucose** and **oxygen**. This enables plants to grow as the glucose is used to build new structures.

- Carbon is passed through the **food chain** to **primary consumers** (herbivores) and **secondary consumers** (carnivores). Some carbon is converted into **proteins** and **fats** while other carbon is released through **respiration** and **decomposition**. In respiration, plants and animals break down glucose for energy, releasing carbon dioxide and water in the process.

- **Bacteria** and **fungi** (**biological decomposers**) or animals that feed on dead material (**detritivores**) breakdown animal waste and store the carbon in the soil. This is known as **organic carbon**. **Inorganic carbon** in the soil comes from the **breakdown of limestone** (formed from the shells of calcifying organisms).

- Most carbon cycling takes place in the **topsoil**, where vegetation grows and the Earth's surface undergoes the **most disturbance**. This allows the release of carbon dioxide from the soil to the atmosphere to happen more easily after decomposition. Carbon held in **deeper soil layers** tends to be more **fixed**.

- **Living organisms** are another form of carbon contained within the soil.

> Photosynthesis requires **sunlight** so only takes place during the **day**. Its rates vary with the **seasons**. When sunlight is not available a plant will turn to respiration, **changing the direction** of carbon flux.

The movement of carbon between the soil and the atmosphere.

1) Bedrock is broken up.
2) Nutrients from the rock are taken up by the roots.
3) Vegetation grows and photosynthesises.
4) Leaf litter is created.
5) Decomposition of dead plant and animal material.
6) Worms and micro-organisms break down the organic matter, returning nutrients to the soil.

Carbon can be **stored** in **dead organic matter**

- **Decomposition** transfers carbon from **dead biomass tissue** to the atmosphere and the soil. After death, bacteria and fungi break organisms down. Carbon dioxide is released to the atmosphere.

- The **rate** at which this happens depends on the **type of soil** and what is being decomposed. Decomposition in the **Arctic tundra**, where conditions are very dry and there is **little oxygen** in the soil, will take a lot longer than in a **tropical rainforest** where it's **warm** and **wet** all year round. A **large, protein-heavy mammal** will take longer to decompose than **small, soft leaves** and flowers.

- Some carbon is transferred to the soil in the form of **humus** (organic soil matter). The **lithology** of the soil determines how much carbon it can store. High levels of carbon are found in soils that are mostly **clay** and low levels are found in soils that are mostly **sand**.

- Carbon content also varies with **vegetation cover**. The regular supply of leaves and organic matter falling from **deciduous trees** will create a relatively **high carbon content** in the soil below. However, vegetation cover that is **low-lying** and **hardy** will not provide as much dead organic matter.

- Soils that are heavily worked, such as **farmland** that is regularly **ploughed** and **cultivated**, are more likely to lose carbon through **gaseous release**. **Wilderness areas** and **protected land** have soils that contain a **higher carbon content** as they are relatively undisturbed.

> Most of the carbon is stored in the topsoil layer.

Loose, partially decayed organic matter.

Topsoil with high humus (organic matter) content.

Pale layer of silt and sand, from which clay, iron and other minerals have been leached.

Zone in which minerals leached from above accumulate.

Mainly broken bedrock

Bedrock

A soil profile can show the level of organic matter and the carbon content of soil.

Warm-Up Questions

Q1 Describe how the amount of carbon varies between the different stores.

Q2 Explain how the thermohaline circulation system has an impact on carbon sequestration.

Exam Question — A-Level

Q1 Explain how carbon can be released into the atmosphere through geological processes. [6 marks]

PRACTICE QUESTIONS

The world goes 'cos carbon flows...

You've probably gathered that the carbon cycle is really important. It can be hard to get your head around the different time scales of carbon flux and the links between carbon in plants, animals and rocks, but keep going over it 'til it sticks.

The Role of the Carbon Cycle

The carbon cycle can be affected by lots of things. Some of them are natural, but surprise surprise, humans can play a big role too. There's lots to learn here, so get cracking or you'll never get to bed. Although this stuff might put you to sleep...

Atmospheric carbon affects **global climates**

The **carbon cycle** affects the amount of gases containing carbon (such as **carbon dioxide** and **methane**) in the atmosphere. These are **greenhouse gases**.

1) Some of the Sun's energy enters the Earth's atmosphere as **short-wave solar radiation** and is re-emitted as **long-wave radiation** (infrared radiation).

2) Greenhouse gas molecules **absorb** and **reflect** some of the long-wave infrared radiation, which keeps the Earth at a temperature suitable for life.

3) This is known as the **greenhouse effect** and is a natural phenomenon. Without greenhouse gases, the Earth would be in a permanent Ice Age.

4) The greater the concentration of greenhouse gases, the **less radiation** is lost to space after it has reflected off the Earth's surface.

5) Having the right balance of different gases in the atmosphere is essential for the functioning of many **natural systems**.

6) The **temperature** of the Earth's atmosphere is regulated by these gases and affects **precipitation patterns** worldwide.

7) Average temperatures differ around the world. Atmospheric temperatures determine **evaporation** rates and the amount of **moist air** available. Warm air can hold more **water vapour** than cold air and therefore has the potential for more **intense rainfall**. This is one of the reasons why **equatorial** regions are **wet** and **polar** areas are mainly **dry**.

Sunita hoped today would be a short-wave one...

The structure of greenhouse gases means that they gain energy from infrared radiation, which causes them to vibrate (non-greenhouse gases are unaffected). This traps heat energy nearer to the Earth rather than allowing it to escape into space.

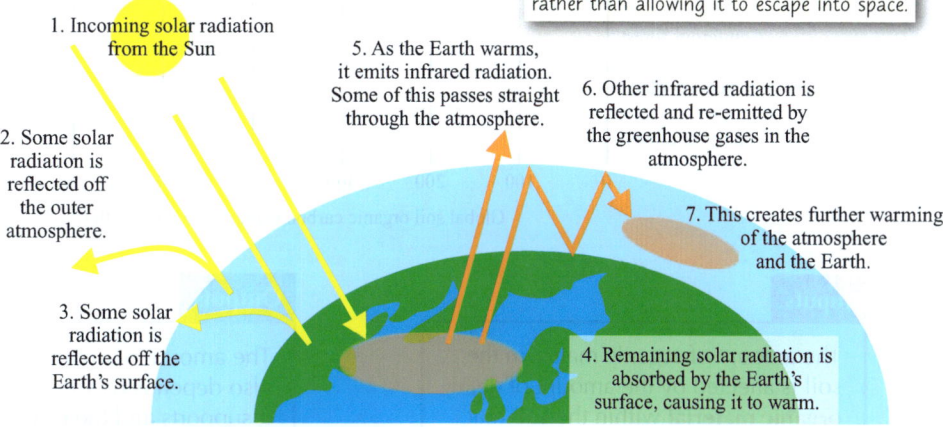

1. Incoming solar radiation from the Sun

2. Some solar radiation is reflected off the outer atmosphere.

3. Some solar radiation is reflected off the Earth's surface.

5. As the Earth warms, it emits infrared radiation. Some of this passes straight through the atmosphere.

6. Other infrared radiation is reflected and re-emitted by the greenhouse gases in the atmosphere.

7. This creates further warming of the atmosphere and the Earth.

4. Remaining solar radiation is absorbed by the Earth's surface, causing it to warm.

The enhanced greenhouse effect

- Though carbon dioxide makes up only 0.04% of gases in the atmosphere, it is the most common greenhouse gas produced by humans.
- Before the **industrial revolution**, carbon in the atmosphere was under 280 parts per million (ppm). In 2021, this figure reached around 419 ppm.
- These new, man-made atmospheric conditions have created the **enhanced greenhouse effect** — which is driving **climate change**.

Photosynthesis affects the **composition** of the **atmosphere**

- **Photosynthesis** is the process by which **terrestrial organisms** (plants) and **oceanic organisms** (phytoplankton) absorb energy from the Sun and use it with carbon dioxide and water to grow new tissue. This is known as **carbon fixation**.

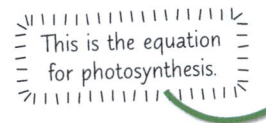

This is the equation for photosynthesis.

carbon dioxide + water $\xrightarrow[\text{chlorophyll}]{\text{sunlight}}$ glucose + oxygen

- Photosynthesis is an essential process in the **regulation** of carbon levels in the atmosphere. Sharp **rises** or **falls** in photosynthesis can have significant effects on the composition of the atmosphere.

- Different parts of the world have different amounts of **vegetation** and **phytoplankton** populations. This means that the rate of photosynthesis and the **net primary productivity** (NPP) vary globally.

net primary productivity
the amount of carbon dioxide taken in by vegetation during photosynthesis minus the amount lost through respiration

- **Biomes** that are characterised by dense vegetation, such as **tropical rainforests**, will have a higher NPP than biomes that are cold and receive little sunlight, such as **Arctic tundra**.

- In the oceans, there tend to be higher levels of photosynthesis where phytoplankton can **amass** and **bloom** — usually in **shallower** and **nutrient-rich cold water**.

The Role of the Carbon Cycle

The **health** of **soil** is influenced by **stored carbon**

1) **Healthy soil** is able to **sustain plant** and **animal life** effectively.

2) **Organic matter** held within the soil influences its **health** and its ability to provide for **organisms** and **plant life**:
   - Organic matter holds a lot of carbon — this helps increase **soil fertility**.
   - Organic matter also acts as storage for **moisture** and allows water to **infiltrate** to the roots of plants.

3) The more organic matter there is in the soil, the more growth there will be in the **biotic** (living) elements of the habitats. This increases the **productivity** and **resilience** of entire ecosystems.

The carbon content of different biomes

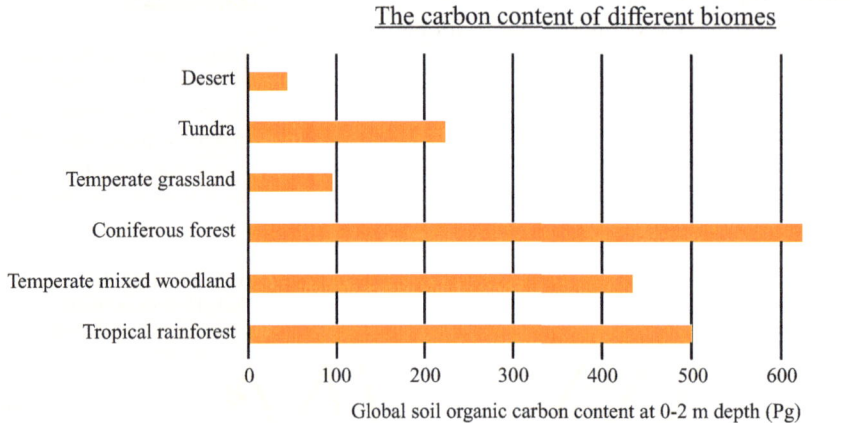

Global soil organic carbon content at 0-2 m depth (Pg)

Deserts tend to have a lower carbon content than other biomes.

Inputs

The amount of organic matter in the soil is affected by the amount of **dead organic material** within the soil that is being **decayed** by the actions of **bacteria** and **fungi**. These are referred to as **inputs** to the soil, as organic matter is being added to its structure.

Outputs

The amount of organic matter that soil stores also depends on the plant and animal life that it supports and the **rate of decomposition** that takes place within it. These are referred to as **outputs** from the soil, as organic matter is being broken down or removed from the soil through **plant growth** or **animal digestion**.

4) The health of the soil is also dependant on the organisms it can support. Soils with lots of **worms** and **microorganisms** to break up the soil and break down matter ensure that **nutrients** are available for plants.

5) A **crumbly** and **porous** soil provides the best base for plants to grow. Most organic matter is found in the **topsoil** where **leaf litter** has fallen and started to decay. This is called the **active soil carbon**.

The **balance** of **carbon** has been disrupted by **fossil fuel combustion**

- The **burning** of **fossil fuels** has increased the atmospheric concentrations of different greenhouse gases.

- Fossil fuels have been burnt in order to provide energy for both **industrial** and **domestic use**. In the UK, this began on a huge scale during the **Industrial Revolution** in the middle of the 18th century. Since then, the **UK economy** has been driven by fossil fuel use.

1 **Fossil fuels** contain high amounts of carbon that have been **stored** for millions of years.

2 The process of **combustion** breaks down the **chemical bonds** within the coal, oil or gas and releases **heat energy**. Burning fossil fuels has released a lot more carbon into the atmosphere. This has created higher concentrations of two key greenhouse gases — carbon dioxide and methane.

3 Although **carbon sinks** are **absorbing** some of this extra **carbon**, the concentrations of greenhouse gases have continued to rise **faster** than the rate of sequestration.

4 This has disrupted the balance of carbon throughout the whole carbon cycle — leading to the **enhanced greenhouse effect** (see previous page).

The Role of the Carbon Cycle

Global warming is affecting ecosystems and the hydrological cycle

Increased greenhouse gases have led to **global warming** — a rise in the average global temperature.
This rise in temperature is affecting the climate, the hydrological cycle and global ecosystems:

Climate

- It is likely the **thermohaline circulation system** (see p.189) will **alter** or **weaken** in certain areas. For example, water entering the ocean from Greenland's **melting glaciers** will **reduce** ocean **salinity** and therefore downwelling.
- This will affect the ability of **saline** and **fresh water** to mix within the ocean and, in turn, affect how air rises and falls between the **equator** and the **poles**.
- **Dry** areas are likely to get **drier**, and **wetter** areas are likely to see **increased rainfall**.
- **Rainfall patterns** have become increasingly unpredictable. Some areas are receiving more **intense bursts** of precipitation while others have experienced **unseasonal** and **extended droughts**.
- Extreme weather events such as **storms** are predicted to increase in **intensity** and **frequency**. This is due to **warmer** atmospheric conditions leading to increased **evaporation** and **humidity**.

The unpredictable rainfall had caught Caitlyn out yet again.

Hydrological system

See p.163-164 for more on the hydrological cycle.

- Higher temperatures are affecting **glacial areas**. **Ice sheets** and **glaciers** are melting at an increasing rate. This is increasing the risk of **flooding** for many glacier-fed rivers.
- This is also adding **meltwater** to the oceans and causing **sea level** to **rise**.
- Increased precipitation over time has led to a higher number of floods and, in some cases, higher rates of **evaporation** as the amount of **surface water** increases.
- This has a wide-ranging impact on the balance of the **hydrological cycle**.

Ecosystems

- Plants that are particularly sensitive to temperature have declined — and the **biodiversity** in some biomes has decreased.
- Areas of **permafrost** are no longer experiencing long seasons of **frozen ground**. This affects the ability of vegetation specifically adapted to permafrost to survive in these environments.
- Some species have had to **migrate** to other areas in order to stay within a particular **ecological niche** (the specific conditions that they need to survive).
- Species already suffering from **low population numbers** and vulnerable habitats are at **increased risk** of extinction.
- The rise in sea levels has led to **ecosystem loss** in **low-lying regions**.
- **Ocean acidification** is causing **coral bleaching** which is threatening **coral reef** ecosystems.

Warm-Up Questions

Q1 Describe the main differences between the natural and the enhanced greenhouse effect.

Q2 Explain how photosynthesis affects the composition of the atmosphere.

Exam Question — A-Level

Q1 Explain why the health of soil can vary. [6 marks]

Input: lots of nifty revision, output: increased chance of exam success...

Hooray — three more pages on the carbon cycle for you to learn. Make sure you have a crack at the practice questions to help you sequester your new-found knowledge. Then just to make sure, go over the pages again until you remember it all.

Energy Security

Energy resources are unevenly distributed — a bit like my grandad's comb-over. Time to have a look at how and why...
Before you get started, here's an advance warning about the number of giant tables coming up — I'm so sorry. Really.

Energy security depends on both physical and human factors

- Energy security means having **affordable** and **reliable access** to energy. A country is energy secure if it can meet all its energy **demands** through its own **domestic supplies**. This means energy security depends on the **supplies available**, the **size of the population** and the **amount** of energy that a **typical person** uses.

- A country that depends on **imported energy** is less energy secure because they are potentially vulnerable to changes in **geopolitics**. This can affect supply and **change energy prices**, making energy **unaffordable**.

- The world's population is growing and energy demand is increasing, so there is a need to **increase energy production** to ensure energy security.

- Producing more energy than is required by the population creates an **energy surplus**. This can then be exported to other countries. Having too little energy to meet the needs of a population is called an **energy deficit**.

- There are a number of factors that contribute to the **access** to and **consumption** of **energy resources**:

Physical availability and access	• Some countries have high levels of **domestic energy supplies**, such as sources of oil and natural gas. This means there is less investment needed to **transport** energy from producers to consumers. • **Urban** spaces tend to have an **energy infrastructure** that allows for a constant supply while **rural** areas and developing countries are more likely to have an intermittent energy supply. • The **physical terrain** of some countries makes them more suited to the production of certain types of energy. For example, countries with **large river valleys** and a plentiful water supply are more suited to **hydroelectric power** than somewhere that has a lesser **gradient**. • **Political tensions**, including **war and conflict**, can affect the trade of energy between countries. The energy source itself may also be used to **take control** of other regions and **form alliances**. • **Natural hazards** may also prevent energy supplies from reaching areas of demand.
Cost	• The limited supply of **non-renewable energy** sources globally are becoming increasingly **difficult to reach** and more **costly to extract**, e.g. deep-sea drilling for petroleum. • The cost of building new **energy infrastructure** (e.g. new nuclear power stations, wind farms, solar powered technology) can be high. If countries can't afford this investment, they may rely on imported energy.
Technology	• The **technology** needed to generate energy can be very **expensive**, and this cost is often passed down to consumers. • New forms of technology are being developed to find **long-term solutions** to energy security. Methods such as **carbon capture and storage** (see p.203) are changing the way that companies develop new forms of energy. • Investment in the technology behind **renewable energy sources** is now a priority for energy companies.
Public perceptions	• Some energy sources, such as nuclear, are perceived by the public as being **more dangerous**, **polluting** and **unsightly** than other renewable energy sources. • A person's **lifestyle** can impact how they perceive their individual **energy needs**. For example, how a person chooses to **travel** or how much energy they use in their **home**.
Economic development	• Some **developing countries** may have energy sources but cannot afford to **extract** or distribute them to homes and businesses. • People in **developed countries** often have more **disposable income** (e.g. they can afford **luxury electronic devices**), **cars** and an advanced **transport system** that requires energy, e.g. a large railway network — leading to higher energy demands.
Environmental priorities	• Some governments see **long-term environmental health** as a priority. This means they move away from more **polluting** forms of energy production and instead invest in greener energies and technologies. • Many governments have a duty to reduce their reliance on fossil fuels as part of broader **international agreements** to reduce carbon dioxide emissions, e.g. the Paris Agreement (2015).

Energy Security

The **energy mix** of a country describes how it **sources** its energy

- Many countries can't supply all of their **energy needs** from one source. Others may not want to for **energy security** reasons — e.g. countries that rely on only one energy source will be affected by **disruptions** to the supply of that source, and if the **price increases** they will have no option but to pay.

- Countries usually use a variety of energy sources instead. The **energy mix** is the composition of different **primary energy sources** that households and industries in an area get their energy from. It's usually shown as a percentage.

- Energy can be **classed** in different ways:

① **Primary or secondary** — Primary energy is released from a **direct source** as it naturally occurs, e.g. burning coal generates heat. When primary energy is converted, it becomes secondary energy, e.g. a thermal power station may use coal to generate heat, which is then used to generate **electrical energy** (a secondary energy source).

② **Renewable or non-renewable** — Non-renewable (or **finite**) energy sources will eventually run out, e.g. **fossil fuels** such as coal, oil and natural gas. Renewable energy sources (such as solar power, hydroelectric power and wind power) can be constantly **reused** and will never run out.

③ **Domestic or foreign** — Domestic energy sources are produced in the same country where the energy is consumed. Foreign energy sources are made in one country and **imported** to another country, where the energy will be consumed.

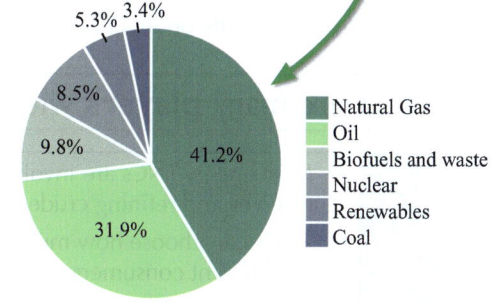

The UK energy mix (2020)

> Renewable energy can be a good option for increasing energy supply as the sources will not run out. They produce little or no waste products, and generally they require less maintenance than non-renewable power stations.

- The energy mix of different countries strongly depends on their **historical ability** to buy or source different forms of energy sources. At a global level, the energy mix is still dominated by **fossil fuels**.

Consumption of energy **varies globally**

- The ability of a country to function is based on its **energy supply** and **consumption** — transport, lighting and heating, industrial machinery and most domestic appliances rely on a source of power. The ability of a country to secure energy to satisfy these needs is strongly linked to its **level of development**.

There's a **strong relationship** between **GDP** and **energy consumption**:

1) **Wealthy countries** tend to **consume lots of energy** per person because they **can afford to**. **Most people** in these countries have **electricity** and **heating**, and use **energy-intensive devices**. E.g. **Sweden, USA**.

2) **Poorer countries consume less energy** per person as they are **less able to afford it**. **Less energy** is **available** and lifestyles are less dependent on high energy consumption. E.g. **Ghana, Mongolia**.

3) The map to the right shows the **energy consumption per person** across the world in 2014.

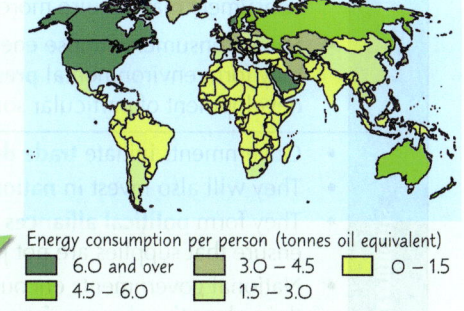

Energy consumption per person (tonnes oil equivalent)
- 6.0 and over
- 4.5 – 6.0
- 3.0 – 4.5
- 1.5 – 3.0
- 0 – 1.5

- Energy consumption can be measured by **usage per person** (per capita). This is often converted to a **single unit**, such as the equivalent tonnes of oil a year (toe/yr), gigajoules (GJ/yr) or megawatt hours (MWh/yr) per person per year. Consumption can also be measured by measuring how much of a country's GDP is spent on supplying energy. Another measure is **energy intensity**. This measures the amount of energy consumed per unit of GDP a country earns. A country with a low energy intensity is thought to be using its energy supply more efficiently.

Different countries have different levels of **energy security**

CASE STUDY

Case study continues onto the next page.

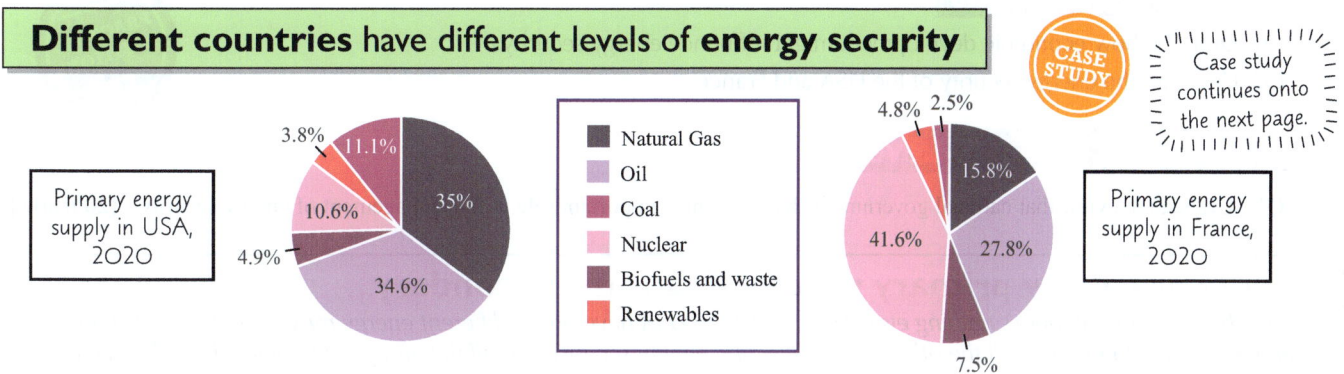

Primary energy supply in USA, 2020

| Natural Gas |
| Oil |
| Coal |
| Nuclear |
| Biofuels and waste |
| Renewables |

Primary energy supply in France, 2020

Energy Security

	USA	France
Electricity consumption per capita (2020)	**12.3 MWh**	**6.7 MWh**
Change in electricity consumption (1990 to 2020)	**+38.7%**	**+29.3%**
Annual net energy imports (2020)	**-3293 TJ**	**4188.5 TJ**

France has a relatively **diverse energy mix** and uses a lot of **nuclear power**, which is **less volatile** in terms of price than oil and gas. However, France relies on energy **imports** from other countries — including from Russia (see p.199).

The **USA** is a **net exporter** of energy and has a relatively **diverse energy mix**, which suggests high energy security. However, it still relies heavily on **fossil fuels**, which will eventually run out.

There are many **players** with **roles** in **energy security**

> **TNCs** companies that operate in two or more countries

Transnational corporations (TNCs)	• Most energy TNCs are involved in all stages of production, from **exploration** of potential reserves and **refining** crude oil to a usable product, to **marketing** it to consumers worldwide. • TNCs can choose how much to charge for energy and whether to vary their prices for different consumers. If one TNC drops its prices, a **price war** can start. • Some energy TNCs are partially or completely state-owned. For example, the Russian government controls over 50% of Gazprom and extracts, processes and sells gas. **State-owned TNCs** can exert global influence. For example, they may shut privately owned TNCs out of energy markets in some parts of the world, or they can be used to help build **political alliances** by providing low-price energy to a potential ally.
OPEC	• The Organisation of the Petroleum Exporting Countries (**OPEC**) are a group of thirteen countries that **work together** to influence the **global supply** and **price** of **oil**. Their biggest members include Saudi Arabia, Venezuela, Iran and Iraq. They controlled around 80% of the world's oil reserve in 2021 and collectively produced almost 32 million barrels of crude oil daily. • OPEC aim to **stabilise the oil market** and ensure that the income from oil is not influenced by supply and demand. They might do this by holding oil back from the market to ensure the price remains high. • More recently, OPEC have found themselves in competition with the USA and have had to reduce their prices to compete with oil that is being **fracked** (see p.197) from that country.
Consumers	• In some ways, **energy consumers** have the least influence on the energy industry. Although they influence **energy demand**, they often have little power over the prices they pay. As lifestyles change, consumers may require more or less energy to maintain their way of life. • Some consumers choose energy suppliers who use **renewable** sources. Consumers may form environmental **pressure groups** and **campaign groups** to **protest** against the development of particular sources of energy, such as wind farms or fracking sites.  An anti-fracking protestor at The Capitol, Washington DC
Governments	• Governments initiate **trade deals** with other countries to secure reliable, low-price supplies of energy. • They will also invest in **national energy infrastructure projects**, such as the building of power stations or dams. • They form **political alliances** with countries and groups, such as OPEC, to ensure that supplies are not jeopardised by **war** and **conflict**. • National governments encourage people and industry to use energy more **efficiently** through **national campaigns**. They may also **regulate** the activities of private energy companies to ensure that their citizens are set a fair price for their energy. • Governments may be under pressure to commit to **international agreements** and **protocols**, where they may have to invest in more renewable energy sources to reduce their carbon emissions.

Warm-Up Questions

Q1 Describe how economic development might influence energy security.

Q2 Compare the energy security of the USA and France.

Exam Question — A-Level

Q1 Evaluate the view that national governments have the most important role in the management of energy security. [20 marks]

Mayonnaise is my primary sauce, BBQ is my secondary...

...I only have ketchup if there's nothing else. France and the USA have pretty different energy mixes, which affects their energy security. Don't forget all the other factors affecting energy security and all the groups of people who influence it too.

Fossil Fuels

The world still relies on fossil fuels to drive economic development, enough so that people are willing to scour far and wide to find them. To be fair though, I'd go foraging for hours to ensure my biscuit supply didn't dry up...

Supplies of fossil fuels are not evenly distributed...

1) **Fossil fuels** (coal, oil and natural gas) form over millions of years from **organisms** that die and are buried. They can be burnt in power plants to generate **electricity** and oil can be **refined** to produce petrol.

2) Fossil fuel supplies are **not evenly distributed** around the world — they vary in both volume and quality. The **geology** of an area determines fossil fuel supply. For example:

   - Coal is a **sedimentary rock**. It forms when plant material undergoes specific geological processes such as **burial** and **heating**. Coal that has undergone the most change is the highest quality (**anthracite**), while coal that has been changed least is the lowest quality (**lignite**).

   - **Oil** and **natural gas** need specific geological conditions to form and be stored. When organic-rich rocks are buried, they are heated, **compressed** and begin to break down into oil and gas. Gas forms at high temperatures, often deeper underground than oil.

   - Both gas and oil travel upwards through pores in rocks until they meet a layer of **impermeable rock** and become trapped. They are found where impermeable '**cap**' rock, such as granite, overlies permeable '**reservoir**' rock such as sandstone.

   - Oil and natural gas can also form in shale. Shale is impermeable, so it is difficult to extract oil and gas from it. A process called **hydraulic fracturing** (**fracking**) can be used to extract the fuel.

Freddy had always maintained it was his impermeable cap that led to the build up of gas...

3) Some countries have **lots** of naturally occurring fossil fuels, as well as the **money** to **exploit** them:

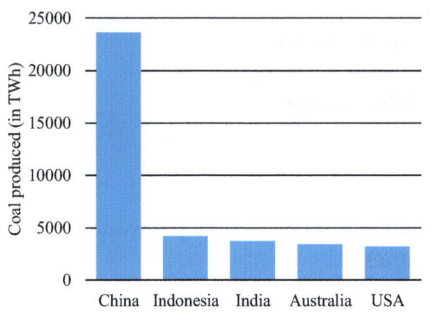
Top 5 coal producing countries (2021)

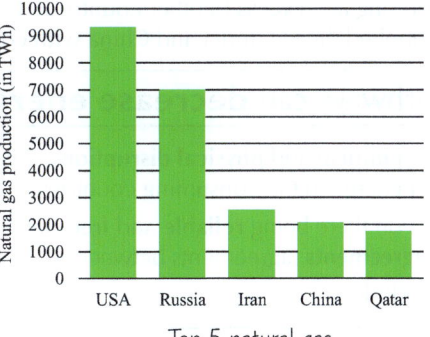
Top 5 natural gas producing countries (2021)

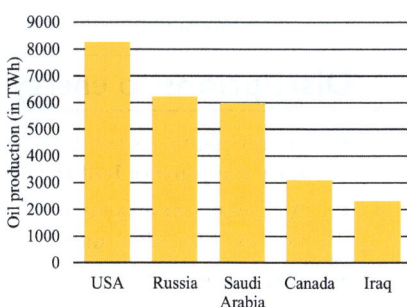
Top 5 oil producing countries (2021)

...and demand is not evenly spread

1) The countries that are producing the most fossil fuels are not always the countries that have the **highest demands** for them. This means there is a **mismatch** between supply and demand across the world.

2) The countries with the **largest populations** tend to be those that consume the most fossil fuels.

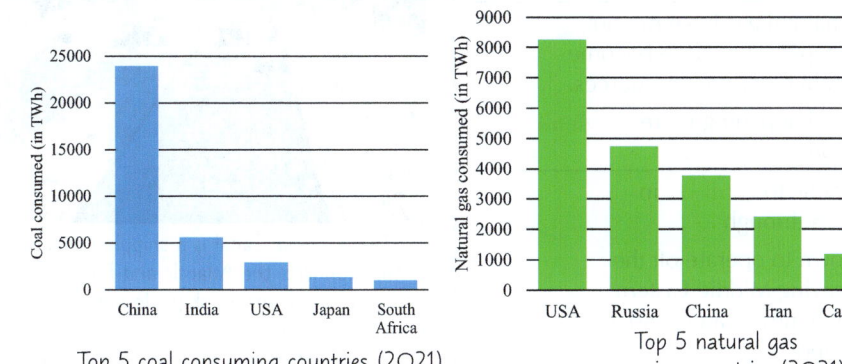
Top 5 coal consuming countries (2021)   *Top 5 natural gas consuming countries (2021)*

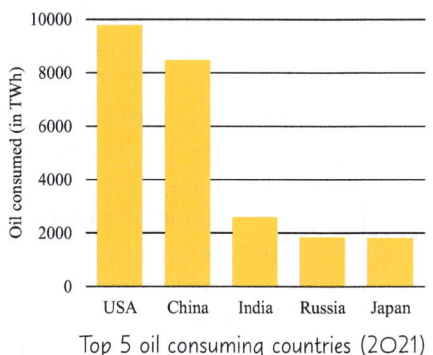
Top 5 oil consuming countries (2021)

3) Consumption of fossil fuels is now increasingly dominated by **emerging economies** (see p.215). Countries like China and India are developing **rapidly** and are now able to buy into **energy pathways** which were previously too expensive. Although more **developed countries** still account for large amounts of fossil fuel use, they are increasingly investing in **renewable energy sources** to diversify their **energy mix** (see p.195).

Fossil Fuels

Energy pathways describe ways energy is transported

1) **Energy pathways** move energy from places where it is produced to where it is consumed.

2) Energy pathways are part of a wider **energy infrastructure** that includes:

gas and oil pipelines that cross continents as well as seas.

transmission lines that transfer electricity to final users through above and underground cables.

shipping routes that transport oil, gas and coal to other countries in bulk, as well as the transportation by road and rail.

3) Energy often must travel **long distances** to reach its consuming countries. Physical energy pathways, such as pipelines and transmitters, are often **owned** by the energy company that maintains the supply.

4) Some energy pathways are extremely **well established**. For example:

- The movement of oil from the Middle East to the USA, the EU, and increasingly to South East Asia shows a long-standing **political relationship** between these regions.

- Coal is harder to move, so countries tend to use their **domestic supplies** to meet their energy needs first before exporting any **excess** to other countries. However, Australia, Indonesia and Russia have supplied the EU, India and China with coal for many years.

Disruptions to energy pathways can decrease energy security

- Energy pathways can be prone to **political** and **physical disruptions**. These can **raise the price** or **halt the flow** of energy to the consuming countries.

- A country's **energy security** relies on there being **reliable** and **free-flowing energy** through energy pathways. This often involves **multilateral agreements** (agreements between many countries).

- **Disruptions** to energy supplies include:

Hazards	• Hazards, such as earthquakes, can rupture pipelines and extreme temperatures or precipitation can damage energy pathways. • When this happens, energy resources will not reach their destination in time and the energy security of the destination country is reduced.
Time	• Wear and tear on pipelines and transmission lines can cause energy blackouts or leakage into the environment if they break.
Depletion	• As supplies of oil, coal and natural gas have run out, some well-established energy pathways have become vulnerable to competition from other energy sources. • These types of disruptions are happening more frequently, as sources of fossil fuels become depleted.
Conflict	• Energy pathways are vulnerable to conflicts in regions that the energy passes through. • For example, pirates are known to operate off the coast of Somalia. Tankers leaving Middle Eastern countries have to pass through these waters and are in danger of being hijacked and held to ransom.
Chokepoints	• Some shipping routes have chokepoints (a narrow marine passageway) such as the Panama Canal in Central America. • If these chokepoints have too much traffic, transportation can be delayed, and prices may rise for consumers and threaten energy security.

The Panama Canal is a shipping chokepoint that connects the Atlantic and Pacific Oceans. Using it means ships do not have to circumnavigate the whole of South America to get from one ocean to the other.

Topic 6 — The Carbon Cycle and Energy Security

Fossil Fuels

Russian gas to Europe

- In recent times, Russia has been a major supplier of natural gas to Europe. Producing up to 27 million TJ (tera joules) of gas each year, with around 5.5 million TJ reaching Western Europe through **several large pipelines**.

- These pipelines pass through the **borders** of several different **countries**, creating some challenging geopolitical relationships.

- To ensure **supply meets demand**, it is important that consumer countries maintain a good **geopolitical relationship** with Russia, and vice versa.

- In 2009, Russian gas company Gazprom accused Ukraine of not paying a debt for its gas supply and the company shut off gas to Ukraine. Conflict between the two countries **threatened the supply of gas** to countries in south-eastern Europe as a single supply line fed through Ukraine to those countries. These relationships became strained when Russia **annexed** the Crimean Peninsula in southern Ukraine in 2014.

- In **2022**, the conflict between Russia and Ukraine **escalated** when Russia launched a **full invasion** of Ukraine.

- The war between Russia and Ukraine has had significant implications for **geopolitical relationships**, and is likely to continue to have a **destabilising effect** on the supply of **natural gas** across Europe.

Some of the gas pipelines running from Russia to western Europe.

It's a good idea to keep up to date with current events, as geopolitical relationships are changing all the time.

The development of **unconventional fossil fuel sources** has increased...

- In 2014, 81% of energy used worldwide came from fossil fuels. The prices of fossil fuels increase as their reserves are depleted and it becomes more **economically viable** to develop **less accessible reserves**.

- Less accessible reserves are termed **unconventional fossil fuel sources**. Many energy companies are now trying to find new ways of accessing these resources. There are **four** main unconventional fossil fuel sources:

1. **Hydraulic fracturing (fracking)** is designed to extract natural gas from shale. High-pressure fluid is pumped into rock, causing it to crack and release gas which is then collected. Fracking is **controversial** because it may cause environmental issues, such as groundwater contamination. The USA has created over 1.7 million fracking wells.

2. Oil can be extracted from **tar sands**. This is **sediment** that contains bitumen (low-grade oil). It is extracted by mining the sediment and then separating the oil from it. It takes about two tonnes of tar sand to make one barrel of oil, so **mining** is **large-scale** and has major environmental impacts. Tar sand exploration is taking place in Canada.

3. Oil is also found in **oil shale**. When this rock is placed under **pressure** and **heated**, **shale oil** is released.

4. **Deep water oil** is found under the ocean floor. **Deep ocean rigs**, such as those off the coast of Brazil, allow drilling to take place.

All four sources can do lasting damage to the fragile environments in which they are found.

- As fossil fuels run out, there may be pressure to explore reserves in **remote** or **protected locations** such as Antarctica.

...with both **costs** and **benefits**

*Any further exploration for **fossil fuels** will lead to greater amounts of carbon entering the atmosphere and further **changes** to the **climate** (see p.191).*

Costs	Benefits
Tar sand and oil shale exploration takes place through **open-cast mining**. This means that large areas of ground are dug into and left open. These can leave **scars** on the landscape and can have **long-lasting effects** on plant growth.	Some economists argue that it makes financial sense for countries with an **economy** and **national infrastructure** based on **fossil fuels** to **exploit** all sources of fossil fuels while they are still available.
Deep water oil drilling increases the chance of an **oil spill** happening, which can circulate quickly through ocean currents. Oil spills can **harm marine life**, threatening the **health** of ocean **ecosystems**.	Unconventional fossil fuels can provide large-scale **employment**. Energy companies controlling the extraction may **provide funding** for other projects, such as **transport infrastructure**, **benefitting** everyone in the **local community**.
There is evidence that fracking can cause minor **earthquakes** and **land subsidence**, as well as **contamination** of **drinking water** sources.	Companies that develop unconventional fossil fuels may be in **competition** with those that develop them from conventional sources — consumers may benefit from **falling energy prices**.

Fossil Fuels

Canadian tar sands

- Canada is thought to hold the world's **largest reserves** of tar sands. The Athabasca region in Alberta has been mined for tar sands since 1967.

- It is only a viable source of energy when the price of oil is high because extracting oil from tar sands is **extremely expensive**. The process uses **large volumes of water** to help the oil separate from the crushed sands. However, this produces **contaminated wastewater** that is **toxic** to life and threatens the already **ecologically sensitive landscape**. New **vegetation growth** is almost impossible and **aquatic life** can be killed when wastewater enters **natural water courses**.

Environmental destruction in Athabasca

- Despite the exploitation of the Canadian tar sands producing **high levels of local income** (the excavation directly or indirectly **employed** over 400 000 people) the **social impacts** are felt more widely. Groups of indigenous people from Athabasca have had to move out of the area. This has caused the decline of their **cultural ways of life** (such as access to traditional hunting and fishing grounds) and removed their connection to the land.

USA fracking

- The USA has been fracking extensively for shale gas since 2005. It has been recorded in over 30 states, but mainly in Texas and Colorado.

- Fracking attempts to access **multiple**, widely dispersed, **pockets of gas** in the ground. When shale gas is detected, large areas of **vegetation** are **removed**. This can cause **habitat destruction** and the **migration of wildlife** out of an area.

- In 2014, over 3 billion gallons of **wastewater** was created from fracking sites in Colorado. This wastewater has been found to be **highly toxic** and there are widespread concerns about it entering drinking water sources.

Fracking site in Pennsylvania

- In the USA, natural gas has taken over from coal as the largest energy source, reducing the number of **coal-fired power stations**. However, fracking has caused increased **air pollution** around fracking sites as **methane** and other gases are released.

Brazilian deep water oil

- The Lapa deep water oilfield was discovered off the South Brazilian coast in the Santos Basin in 2007. Extraction of oil from the site began ten years later. It is around 270 km offshore and sits over 2 km deep. This makes the drilling of wells and oilfield platforms very **difficult** and a **more expensive process** than drilling in shallower water.

- Since then, Brazil has started to drill for oil in deeper waters. This means the oil drilling platforms are **further from the coastline** and **longer pipelines** are needed to transport the oil to shore. This has **increased the cost** of the oil for consumers. Petrobras, Brazil's state owned oil company, aims to increase production into the future.

- Brazil is looking to expand the **diversity of its energy mix** by increasing its electricity from fossil fuel sources. A large proportion of its energy is currently generated from **hydroelectric power**, but there are fears that **climate change** will bring more **droughts**, which will affect the flow of rivers and the ability of **dams** to generate power.

Warm-Up Questions

Q1 Describe the nature of the mismatch between the supply and demand for each fossil fuel.

Q2 Explain the costs and benefits of using unconventional fossil fuels.

Exam Question — A-Level

Q1 Explain why there has been a rise in the extraction of unconventional fossil fuels. [8 marks]

Oi'll refrain from making a crude joke...

Well now, it's time to dredge up your remaining energy reserves and drill these pages until they're permanently lodged in your brain. There's a few lovely case studies to help you really get a grip on the complexities of supplying energy too. Cracking.

Fossil Fuel Alternatives

I hope you've got enough energy left to consider how to cope without using dirty old fossil fuels. One day we'll have to...

There are several **alternatives** to **fossil fuels**

- As fossil fuels continue to **deplete**, there is a growing need to move towards more sustainable sources of energy and rethink how industries and domestic consumers use energy. This is known as **energy transition**. It generally involves 'decoupling' the future economy from fossil fuel use.

- One part of decoupling is a move towards **renewable** energy. Renewable energy comes from various sources:

Solar	Energy from the Sun is used to heat water and **solar cookers** or converted to electricity using **photovoltaic cells** (PV cells).
Wind	**Turbines** use the energy of the wind to generate electricity, either on land or at sea, often in large **windfarms**. There are no greenhouse gas emissions once the turbines have been built.
Hydroelectric	This uses the energy of falling water. Water is trapped by a **dam** and allowed to fall through **tunnels**, where the pressure of the falling water turns turbines to generate electricity.
Tidal	Currents or **changes in water level** caused by tides are used to turn turbines and generate electricity.
Wave	Wind blowing across water makes **waves**, which drive turbines to generate electricity.
Geothermal	Water is pumped into the ground, where heat **deep** in the Earth's **crust** turns it into steam and then drives a turbine to generate electricity. The steam may also be piped to homes for hot water and heating.

- **Nuclear power** can be used to generate a large amount of energy from a small amount of fuel. The **splitting** of **uranium atoms** releases **energy**, which is then used to generate **electricity**. Nuclear energy is a **recyclable energy source** — i.e. where the waste products can be recovered and used to make new energy.

There are **costs** and **benefits** to using **alternative energy sources**

Though renewable energy sources are **cleaner** and will **last longer** than fossil fuels, they also have some downsides:

Wind power
- Wind energy can only be generated in locations with an average wind speed greater than 5.5 m/s.
- **Very high winds** can damage wind turbines, so most stop working automatically if winds get above 25 m/s.

Nuclear power
- Nuclear power plants are very expensive to **build** and **decommission**.
- Nuclear waste is **radioactive** so must be safely stored for thousands of years and **accidents** can be **catastrophic**.

Hydroelectric power
- Hydroelectric power relies on **large flows of water** to generate electricity. In areas with **low rainfall** or **frequent droughts**, it can't produce **reliable** power. The **drainage density** (the number and size of rivers), affects the amount of energy that can be generated. The geology of an area must also be well-suited to
- To generate hydroelectric power, a dam needs to be built across a river. It works best in drainage basins with **large rivers** and **steep terrain**. However, steep terrain can also make dam construction **difficult** or **expensive**.
- Building dams can destroy local **environments** and may be too expensive for **developing** countries. **Local communities** may also be forced to **move** if their land is going to be **flooded** to create the reservoir.

Solar power
- Solar power is most effective in places with **little cloud cover** and the amount of energy produced varies depending on the time of year.
- Days are **longer** in the summer, so more energy can be generated. The need for energy in the summer is far **less**, creating a **mismatch** between supply and demand.
- In rural areas, solar farms can take up **valuable land**, though PV cells can be placed on **urban rooftops**.

> Solar energy can be produced in many countries, including in poorer ones where development of **fossil fuel based economies** have been slow.

- The use of renewable energy is **location specific** and relies on **local conditions**. This means that different countries will have different renewable energy mixes.

- The building of wind and solar farms can give **local people** concerns about the impact such developments will have on their **quality of life** or the **value** of their **property**.

> The price of fossil fuels has generally decreased as international use has declined. However, the price of fossil fuels can change due to geo-political relationships.

Topic 6 — The Carbon Cycle and Energy Security

Fossil Fuel Alternatives

The changing energy mix in the UK

- Coal used to form a **significant proportion** of the **primary energy** used in the UK (see p.195).
- Since 1990, the amount of coal used has decreased significantly and investment in **nuclear power stations** and other renewable energy sources (particularly wind power) has increased.

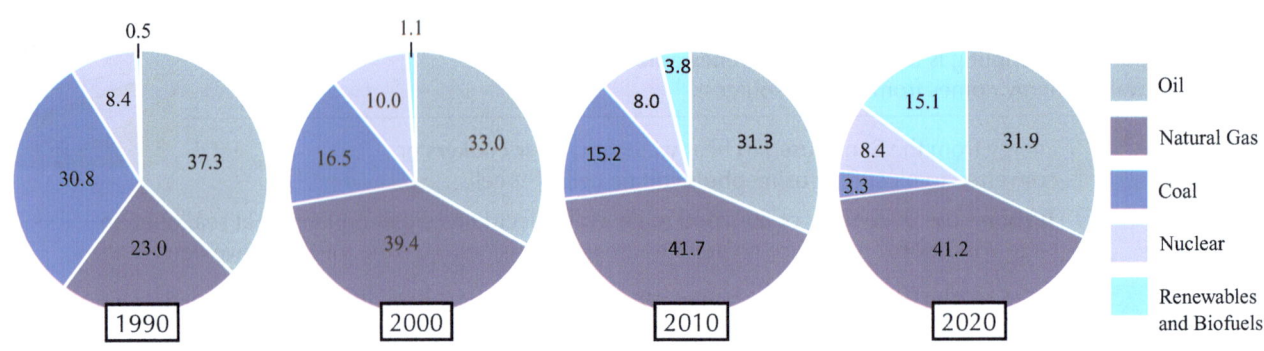

Legend: Oil, Natural Gas, Coal, Nuclear, Renewables and Biofuels

1990: 37.3, 23.0, 30.8, 8.4, 0.5
2000: 33.0, 39.4, 16.5, 10.0, 1.1
2010: 31.3, 41.7, 15.2, 8.0, 3.8
2020: 31.9, 41.2, 3.3, 8.4, 15.1

- In order to **reduce its overall energy requirements**, the UK is implementing new legislation and policies which require homes and businesses to use more **energy-efficient** appliances and machinery.
- By **2035**, the UK government is aiming to have **all** its electricity generation come from **clean energy** sources.

Biofuels are another alternative energy source

1) The combustion of **biomass** (wood, plants and animal waste) is a key energy resource in many **developing countries**. Biomass can be burnt to produce heat energy or **processed** to make **biofuels**, such as bioethanol and biodiesel.

2) These biofuels can **replace** petrol and diesel and have **lower particulate emissions**. They can be produced by growing crops such as grass, sugar cane and wheat, as well as oil based crops such as rapeseed and linseed.

3) There are **costs** and **benefits** to using biofuels:

Costs	Benefits
Growing the crops needed to make biofuels can **reduce biodiversity**. Biofuels like corn-based ethanol need a **lot of land** to grow, which means **clearing other vegetation**. This can result in the **deforestation** of **ecologically sensitive areas**, such as **rainforests**.	Biofuels rely on relatively simple processes (growing crops) which make them accessible to people in countries of all levels of economic backgrounds.
In some regions, growing crops may require the heavy use of **fertilisers** and **pesticides**, which can **pollute** water courses.	Bioethanol can be cheaper than traditional petrol if produced on a large scale.
In some countries, land is needed to grow food crops rather than fuel crops and **competition** for **land** and **water resources** means that growing biofuels reduces the ability of a country to **feed** itself.	It can bring high levels of income to rural areas from little investment.
	Biofuels do not create devastating environmental disasters if they are spilt or if there are leaks. They are biodegradable so can be cleaned up more easily and create little lasting damage.
	Although burning biofuels releases carbon dioxide into the atmosphere, the rate of release is a lot less than traditional petrol emissions.

Biofuels in Brazil

- Every year, Brazil produces around 34 billion litres of **bioethanol** from sugar cane. It provides **fuel** for **vehicles** and supplies energy to **homes**. Brazilian bioethanol isn't sold as a pure product — it is mixed with petrol to create '**flex-fuel**' which an increasing number of car engines in Brazil run on.
- Production of bioethanol has **grown rapidly** as the Brazilian economy has grown. Cleared forest land has been used for more **intense cropping patterns**, increasing the efficiency of production.
- At its peak in 2005, over half a million were **employed** by the bioethanol industry in Brazil. The Brazilian government implemented a 27% minimum of bioethanol in flex-fuel and **removed subsidies** given to petrochemical companies, making it a highly profitable industry. The 2008 economic crash reduced demand and investment in the industry but it is now steadily growing again. In 2022, the price of bioethanol cost US$ 0.84 a litre in Brazil compared to US$ 1.02 a litre for petrol.

Fossil Fuel Alternatives

Radical technologies can be used to increase energy security...

- **Radical technologies** are engineering designs and developments that either create new forms of energy or use energy in a very different way so that countries can be more **energy secure**. There are **three** main technologies you need to know about:

(1) **Carbon capture and storage** (**CCS**) — The carbon dioxide produced when fossil fuels are burnt is **captured** inside the power plant or within the factory before it enters the **atmosphere**. This gas is then **compressed** and **piped** though an **injection well** to natural underground cavities and chambers (such as aquifers).

(2) **Hydrogen fuel cells** — Hydrogen fuel cell technology combines hydrogen and oxygen to produce water and energy in the form of electricity. This means they can be used as **battery packs** for home and cars.

(3) **Electric vehicles** — Electric vehicles run off **rechargeable electric batteries** rather than conventional fuel.

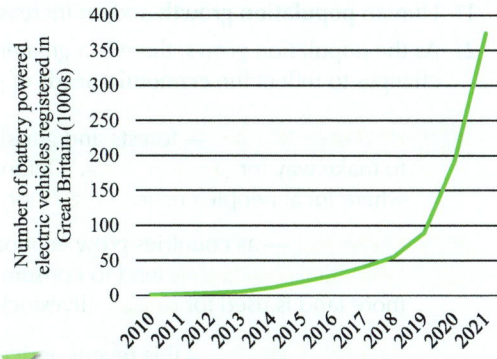

Registration of fully electric vehicles in Great Britain since 2010.

...but using them is not always feasible

Carbon capture and storage (CCS)

- Carbon capture and storage is very **expensive** and requires a high degree of mechanical engineering.
- Although CCS **removes carbon** from the atmosphere, potentially reducing the onset of a more **enhanced greenhouse effect**, results will not be felt **immediately**.
- **Leakage** from the underground stores is not always monitored.

Hydrogen fuel cells

- Hydrogen fuel cells don't produce any **dangerous waste products** (just pure water) and aren't part of the carbon cycle so can **reduce carbon emissions**. But there are still problems:

1) **Sourcing** and **storing** hydrogen for use in the cells is difficult. Although hydrogen is hugely **abundant**, it does not occur as a gas in a **natural state** from which it can be captured and used.

2) It is **highly volatile** and can cause **explosions**.

Electric vehicles

- Electric vehicles run off **battery packs** that have a **limited distance range**. Although the technology is improving rapidly, they still require a **network of charging points**. In the UK, the number of public charging points is limited and a high proportion are found in large **cities**.
- **Producing** and **disposing** of the batteries can cause **environmental problems**.
- Although owners of electric cars save money on fuel, they are **expensive** to buy **initially** and **converting** a petrol car to electric power is expensive.
- Electric vehicles are also extremely quiet. Although this creates some concerns around road safety, it does **reduce noise pollution**.

Hydrogen fuel station in Washington DC

Warm-Up Questions

Q1 Describe how the UK energy mix changed between 1990 and 2020.

Q2 Explain why alternative energy sources may not be viable for all countries.

PRACTICE QUESTIONS

Exam Question — A-Level

Q1 Explain why biofuels are not always a sustainable fuel option. [6 marks]

Where e'er you be, let the wind go free...

That's not radical like CSS or hydrogen fuel cells. Yep, electric vehicles are where the cool kids hang these days. Well, you best get learning all about renewable energy, biofuels and all those rad new alternatives — they're pretty hot topics. Sweet.

Human Impacts on the Carbon Cycle

Humans can alter the carbon cycle in a number of ways. It's complex, but read on and all should become crystal clear...

Land use is changing as demand for resources grows

1) Human **population growth** and an increase in **economic development** have changed the ways that people use land.

2) As the population grows, there is a greater **demand for resources** such as fuel and food. Land use also changes to reflect the **economic needs** of populations and their **improving living standards**.

> Industrialisation and the movement towards more urban landscapes increases the need for more living space. This is happening globally, but developing countries are rapidly advancing and their populations are growing faster than elsewhere.

1) DEFORESTATION — forests and grassland may be cleared for agriculture, logging, or to make way for developments. Deforestation might also occur in developing countries where local people create charcoal or collect wood from forests for domestic use.

2) FARMING — as countries grow economically, the type of farming required also changes. Wealthier populations tend to consume more meat than poorer ones. This means that more land is used for pasture (livestock grazing), which is less productive than crops.

3) AFFORESTATION — this reverts agricultural land back to its original vegetation cover, with varying degrees of success. For example, the same species of tree is often planted at scale. This reduces the amount of carbon stored, compared to a mixed forest where trees grow at different rates and heights.

Carbon stores are affected by changing land uses

1) The **change** of land use from **natural** to **agricultural** (or **urban**) is a major source of change to **carbon fluxes** (see p.188).

2) **Vegetation** is removed to make way for **buildings**. This can **reduce carbon storage** in the **biosphere** and reduces its ability to take in **carbon dioxide** and **remove** it from the atmosphere.

Land use changes

- Making the soil suitable for growing crops can cause moisture loss.
- Impermeable surfaces in built-up areas increase runoff and reduce evapotranspiration (see p.166).
- When forests are turned into agricultural land, the reduction in the height of vegetation cover limits the amount of shade.
- The soil is likely to become drier as rates of evaporation increase.

Deforestation

- Clearing forests, by burning, removes carbon from the biosphere and releases carbon dioxide to the atmosphere.
- Deforested land is more vulnerable to soil and wind erosion. Humus is washed away and released into the oceans, reducing the carbon content of the soil.

Farming practices

- Animals release carbon dioxide and methane when they respire and digest food.
- Ploughing can release carbon dioxide stored in soil.
- Growing rice in rice paddies releases methane.

Carbon stores may be affected by climate change

Humans have **influenced** the carbon cycle for centuries, particularly by extracting and **burning fossil fuels**. There's now over **40% more carbon dioxide** in the atmosphere than there was in **1750**. As the concentrations of **greenhouse gases** in the atmosphere continue to **increase**, **temperatures** are expected to **rise**. This is known as the **enhanced greenhouse effect** (see p.191).

Land
- Changes in the carbon cycle can reduce the amount of carbon stored in the land.
- For example, warmer temperatures, caused by global warming, are causing Arctic and alpine permafrost to melt. This releases carbon from this store into the atmosphere.

Atmosphere and Climate
- Extreme weather events (such as storms, floods and droughts) are expected to increase in frequency. These can destroy carbon stores such as forests.
- The poles are likely to decrease in size and continental arid zones are likely to increase, which will affect the size of carbon stores in these areas.

Oceans
- Organisms that are sensitive to temperature, e.g. phytoplankton, may not be able to survive at higher temperatures, so their numbers decrease. This means that less carbon dioxide is used by them for photosynthesis and less carbon is removed from the atmosphere.
- Warmer water is less able to absorb carbon dioxide. As global temperatures rise, the amount of carbon dioxide that could be dissolved in the sea decreases. This leads to increased levels of carbon dioxide in the atmosphere.

Human Impacts on the Carbon Cycle

Amazon drought events

- The Amazon rainforest plays an important role in **regulating** the local, regional and global **climate systems**. It **releases moisture** into the atmosphere, through **evapotranspiration**, affecting humidity and rainfall patterns.
- **Climate change** can alter evapotranspiration and rainfall patterns within tropical rainforests. For example, the Amazon experienced **severe droughts** in **2005** and **2010**.
- Many **plant and animal species**, which have **adapted** to live in **moist** conditions, **died** in the **dry weather**.
- During droughts, **photosynthesis slows** down and **less carbon is removed** from the atmosphere.
- **Forest wildfires** became more common, **releasing** more **carbon dioxide** into the atmosphere and **increasing** local atmospheric **temperatures** further.
- Frequent droughts could lead to the **extinction** of some **species** and the overall health of the forest is threatened as **carbon and water cycling** within the ecosystem is affected.

Standing dead trees from the drought in 2010

The rate of **ocean acidification** is **increasing**

1) The oceans **absorb carbon dioxide** from the atmosphere, making them an important **carbon sink**. When carbon dioxide is dissolved in water, it reacts to form a weak **carbonic acid**. Over time, the **acidity increases** in the ocean which **reduces the pH**.

2) **Increased** levels of **carbon dioxide** in the atmosphere, due to **fossil fuel combustion**, can **increase the acidity** of the oceans.

3) Increased acidity can have a negative impact on **marine life**. Carbonic acid **dissolves calcium carbonate**, which some **sea creatures** need to **make their shells**. Acidity also affects the ability of **marine plants** to take up **minerals**, which means that the population numbers of **primary consumers** may **decline**.

An example of bleached coral

4) Even a **small increase** in **acidity** makes it **harder for corals** to produce their **calcium carbonate skeletons**. Acidification **removes carbonate ions** from the sea, making **thinner corals** that are more likely to break up. High levels of acidity can **dissolve coral reefs**. The ocean can become so acidic that the reef is no longer able to recover, causing **permanent damage**. The point at which this occurs is known as the '**critical threshold**'.

5) **Marine algae** live in and feed on the coral, which gives the coral its **bright colours**. **Warmer ocean temperatures** cause coral to remove the algae and turn white in response. This is called **coral bleaching**. If ocean temperatures remain **too high**, the coral won't let the algae (its primary **food source**) back in, which can cause the coral to eventually **die**.

Coral reefs are estimated to benefit around 1 billion people globally, either directly or indirectly.

6) Corals provide **ecosystem services**. Ecosystem services are ways in which the natural world **benefits human populations**. Coral reefs can generate an **income** for **local people** in the **tourist industry**, who operate dive schools and boat trips. Coral reefs also provide **rich fishing grounds** for fisherman, who take their catch to local **markets**. These markets sell produce to tourists and local people.

Warm-Up Questions

Q1 Describe the impacts that increases in ocean acidification have on coral reefs.
Q2 Explain the link between population growth and deforestation.

PRACTICE QUESTIONS

Exam Question — A-Level

Q1 Explain how ocean acidification impacts ecosystem services. [6 marks]

How to influence the carbon cycle — everyone hold your breath...

There's a sure-fire way to influence your marks for questions on interactions between the carbon cycle, water cycle and climate, and it starts with you learning everything on these two pages, then having a crack at the practice questions...

Changes to the Carbon Cycle — Impacts

Ready to store some more information? Make sure you learn the facts to avoid going around in a vicious cycle...

Human wellbeing is reliant on the carbon and water cycles

1) Human wellbeing refers to the **social and economic quality** of life and **standard of living** that humans experience. Humans have a good wellbeing if they are **safe**, have a **healthy lifestyle** and have the capacity to **better themselves economically**.

2) The water and carbon cycles affect the ability to **provide food**, **form settlements** and **develop economically**. This means the **health** of the water and carbon cycles is strongly linked to the **wellbeing** of human **populations**.

3) Some people view these natural cycles as **resources** to be exploited for **human gain and wellbeing**. Others want to have a more **custodial role** over the water and carbon cycles and feel strongly about **conserving** them.

Managing forests is important for human wellbeing

1) Forests provide several **ecosystem services** to humans. Forest products such as timber, paper and foodstuffs provide **economic benefits**. Globally, it is estimated that more than **1.6 billion people** are **dependent** on forests for their **livelihoods**.

2) Pharmaceutical companies carry out **research** in tropical rainforest areas in search of new compounds to be used in **medicines**.

3) Forests regulate **local micro-climates**, particularly in relation to **temperature**. Large areas of forest also act as a buffer to **noise and air pollution**.

Deforestation, for timber, in the Amazon rainforest

An **environmental Kuznets Curve** shows the relationship between **economic development** and **environmental degradation**.

1) Countries often use **more resources** as they **develop**, which causes an **increase** in environmental **degradation**, e.g. through deforestation.

2) The country will then reach a **peak of development** and industrialisation. This is known as the '**turning point**'.

3) After this, the **demand** for some **resources** remains **high** but the country will try to act more **sustainably** as people become more aware of the role the environment plays in human wellbeing.

4) Countries will usually **increase** their **conservation measures**, or begin **importing resources** from other countries, **reducing** the amount of **environmental degradation**.

5) The **height and width** of the Kuznets Curve will **vary** from country to country. Some countries will take a **longer time** to reach industrialisation and then remain in that phase for **longer**.

This is called the 'turning point'.

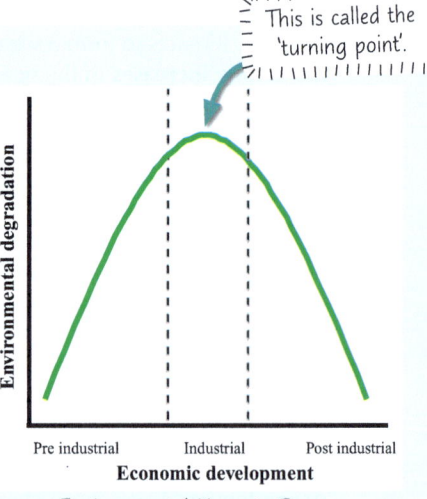

Environmental Kuznets Curve

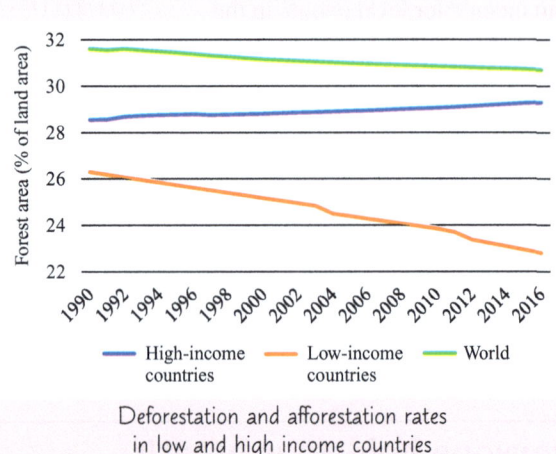

Deforestation and afforestation rates in low and high income countries

- In many **developed countries**, where the impact of deforestation is being felt, programmes of **conservation** and **restoration** are common.

- **Charities** are working towards large-scale **tree planting** schemes, as well as **raising awareness** about the value of forests.

- In **low-income** countries, where the immediate **economic needs** of a population are often **prioritised**, the **exploitation of forest** environments is **more likely**.

- More people are recognising the **emotional and spiritual value** that forests have. Walking through a woodland can **reduce stress** and improve overall **human wellbeing**. Some people strongly feel that areas such as **forests** should be **conserved** simply because they are **beautiful places**.

Changes to the Carbon Cycle — Impacts

Increased temperatures affect water stores

The increased risk of **flooding** and **droughts** will cause **social** and **economic** disruption, negatively affecting **human wellbeing**. Growing **crops** will become more **difficult** and **expensive**, as it will require an increase in **human input**. Water stores are likely to change in three main ways:

1) **Precipitation patterns** — Rising temperatures increase **evaporation**, increasing the amount of **water vapour** in the atmosphere. There are likely to be more **intense** periods of **rainfall** (leading to flash flooding) and **longer periods** of **no rainfall** (leading to increased drought risk).

2) **River regimes** — In **glacial** areas, extended periods of **snow melt** will increase volumes of **meltwater** feeding into **rivers**, increasing the likelihood of **flooding**.

3) **Freshwater stores** — Glacial snow melt will **decrease** the amount of **water** stored in the **cryosphere**. **Drier** regions will see a **reduction** in **lake** size and also **ground water storage**.

Rising temperatures in the Arctic

- **Rising temperatures** will **reduce snowfall** in the Arctic, cause longer periods of **melting** and **increase** the **global sea-level**, which will threaten people's livelihoods as coastal areas flood.

- **Ice reflects** more **solar radiation** than darker surfaces (such as water). This creates a **positive feedback**, called the **albedo effect**, where more solar radiation is absorbed by meltwater, which will further **increase melting**.

- **Migratory patterns** of some animals, such as **caribou**, are changing. They're able to **graze** in more **northern latitudes** for longer, **extending** the **hunting season** for the indigenous populations that hunt them.

- **Warmer temperatures** in the Arctic have **lengthened** the **growing season** for crops and **increased yields** as a result.

- As more **sea surface** becomes **ice free** and exposed to the Sun, the **habitat** range of **phytoplankton increases**. This **increases** the amount of **carbon sequestration** that takes place.

- The **melting of sea ice** has increased the number of **shipping vessels** using the northern sea routes, in particular the North-West Passage. This may increase the risk of **oil spills and pollution**, which can harm **marine ecosystems**.

- The Arctic is a **net carbon sink**. However, when **permafrost thaws**, the **carbon** stored in the layers can be released as **carbon dioxide** and **methane**. In northern Canada, the melting of permafrost has caused **infrastructure** to **subside** as the **ground** beneath it becomes **unstable**.

CASE STUDY

September 14, 1984   September 13, 2012

Extent of Arctic sea ice in 1984 (left) and 2012 (right)

Managing oceans is important for human wellbeing

1) As **ocean temperatures** and **acidity levels** continue to **rise**, their ability to **support marine life** will become increasingly **threatened**. To escape warm waters, some tropical **marine species** are **migrating** towards the poles, which is having a wider impact on **marine food webs**. **Competition** between species is **increasing**, causing some fish to migrate even further from the **populations** which **depend on them**.

2) Many **developing countries** rely on the **fishing industry** as a source of **food** and **income**. These communities may not have **expensive** fishing trawlers that can travel the **greater distances** needed to find fish.

3) **Coastal areas** that rely on **tourism** may also see **economic declines**. For example, dive schools and boat trips to coral reefs will be harder to operate as **corals disintegrate** and **bleach**.

4) **Rising sea levels** might threaten **low-lying** areas in both developed and developing nations. Countries must weigh up whether local **coastal defences** are a good investment, given the **global** nature of the problem.

Warm-Up Questions

Q1 Describe the different ways in which human wellbeing is connected to natural systems.

Q2 Explain the effects that rising temperatures are having in the Arctic.

PRACTICE QUESTIONS

Exam Question — A-Level

Q1 Assess the extent to which a standard environmental Kuznets curve holds true for different regions or countries. [12 marks]

With all this cycling, carbon could win gold at the Olympics...

There's plenty to learn here. At least it's all exciting stuff. Oh no, wait... Anyway, you've got to know it all, so you'd best crack on with safely storing it in your noggin. There are some technical terms in here too, which are well worth learning.

Future Changes to the Carbon Cycle

Grab your crystal ball, we're about to have a glimpse into the future and see what's in store for the carbon cycle...

Natural factors make future climate warming difficult to predict...

There are several **natural factors** which make **predictions** about future climate change very **difficult**:

Natural Factors

1) Whether there is a **maximum value** of **carbon dioxide** emissions at which **temperature increases** will **stop**.

2) The capacity of **carbon sinks** to **sequester carbon dioxide** from the **atmosphere**.

3) The amount of **carbon** stored in more **inaccessible places**, such as **permafrost**.

4) The **lag time** there may be between carbon being released in emissions and changes being felt in **global temperatures**.

5) How other **large-scale systems** such as the nitrogen cycle, Milankovitch cycles and El Niño may **impact carbon and temperature levels**.

> Oceans, in particular, are thought to have a slow response to changes in atmospheric greenhouse gas concentrations.

...and so do human factors

There are several **human factors** which make **predictions** about future climate change very **difficult**:

Human Factors

1) **POPULATION** — The amount of **carbon dioxide** released by the **largest emitters** could **decrease** if their **populations** start to **decline**. As **urban spaces expand** and **populations increase**, the amount of **carbon** emitted in some countries could **increase**.

2) **ECONOMIC GROWTH** — As countries develop **economically**, they usually use **more energy** which **increases** the amount of **greenhouse gases** emitted. But it's hard to predict the **rate** of different countries' **economic development** and whether these countries will be able to **access green technologies**.

3) **ENERGY** — The future usage of **renewable energy** sources is difficult to predict. **New sources** of energy, or new ways of using existing resources, could affect greenhouse emissions. Industry could become **more efficient** in the way it uses energy and **reduce its greenhouse gas emissions**. Industries in some areas are likely to **decline** while new industries will **grow and develop** in other areas.

Feedback mechanisms could affect future changes in the carbon cycle

1) The effects of a change to a system may be **amplified** by a **positive feedback** or **dampened** by a **negative feedback**.

2) It's **unclear** exactly how physical **feedback mechanisms** will work in the **future** and which processes may **dominate** these **mechanisms**.

3) **Peatlands** and **permafrost** could **store** more **carbon** than previously thought, creating a **negative feedback**. They may also **release more carbon** as they thaw, creating a **positive feedback** and releasing more **carbon** into the **atmosphere**.

4) **Natural processes** may reach a '**tipping point**' where they are **unable to store** more carbon and the impact of humans on the climate becomes **irreversible** and **permanent**. For example, **widespread deforestation** could reduce the moisture levels in soil, making it impossible for forests to reseed, causing **forest dieback**. An increase in **freshwater** (from glacial meltwater entering the oceans) could decrease the salinity of the oceans and cause **permanent changes** to the **thermohaline circulation** and physical ocean pump.

5) There's also a **tipping point** for many **species**. If **habitat destruction** causes numbers to decline and a reduction in breeding pairs, the **extinction** of the species may become inevitable.

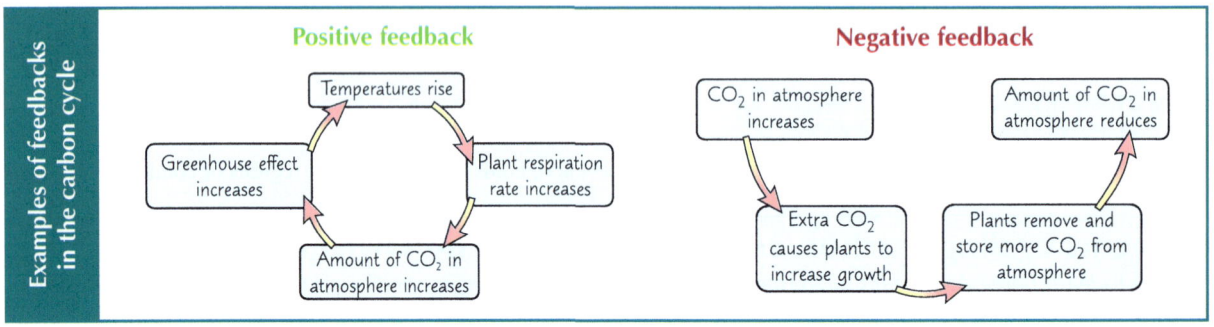

Future Changes to the Carbon Cycle

There are ways of adapting to climate change

- Adaptation means **changing the way we live** to best cope with a **new situation**.
- In terms of **climate change**, this means that humans will need to **adapt their lives** to allow them to cope with **raised temperatures**, rising **sea levels** and unpredictable **weather** patterns.
- Below are some ways that people can **adapt to climate change**, along with some **criticisms** of these adaptations:

Water Conservation

- Changing **behaviour** in **industry** and in **homes** to reduce **water consumption**.
- Using more **grey water** and **adapting home design** to collect and store **grey water** more easily.
- **Charging people more** for water to encourage them to think more carefully about their **water use**.

Criticisms

- **Campaigns** to encourage **water conservation** may be needed to ensure **maximum uptake**.
- The **demand** for water is currently **rising faster** than the extent of **conservation awareness**.
- Populations need to realise the **value of water** as a **precious commodity** as well as a **human right**.

Land use planning

- **Restricting** the amount of **built development** that can occur in **vulnerable areas** such as **floodplains**, low-lying **coastlines** and land next to **ecologically sensitive** areas.
- **Designing towns** and street layouts so that **minimal disruption** occurs.
- **Relocating** critical **buildings** (such as hospitals) away from the **riskiest areas**.

Criticisms

- High **land values**, particularly in **urban areas**, means it's difficult to abandon land to the possibility of **flooding**.
- Abandoning land is very difficult to **enforce**, given that **flooding** may happen very **infrequently**, so the **risk** isn't perceived as being that high.

Bruno has tried his best to adapt to his environment since moving to Lapland.

Resilient agricultural systems

- Using 'smart irrigation' that tailors water supply to **meet the needs of individual plants** (see p.185).
- Planting crops that have been **genetically selected** to be more **drought resistant**.
- **Rotating crop types** and using 'no till' techniques (such as not ploughing) to **replenish carbon** and **water** stores in the soil.

Criticisms

- New forms of **farming**, such as **vertical** or **indoor farming**, have a **higher energy demand** through the need for artificial lighting and temperature regulation.
- New forms of agriculture might only be **available** to those who can **afford it**, rather than those who **need it most**.
- **Genetic selection** and **modification** is not always **legally** or **morally** accepted.

Flood risk management

- Using **hard and soft engineering techniques** (e.g. flood relief channels, dredging rivers or raising embankments) to manage **floods** and prevent them from damaging **homes and businesses**.
- Stopping **deforestation** within a **drainage basin** so that trees and **vegetation** are better able to take in **excess water**.

Criticisms

- Some forms of **hard engineering** can be very **expensive** and do not work with all types of rivers or drainage basins.
- **Flood defences** can **detract** from the natural beauty of a river or coast.

The Thames Barrier flood defence, London

Solar radiation management

- **Designing** and **engineering** ways of reflecting **solar radiation** back into space, such as the use of satellites to act as giant reflectors. More broadly this is known as **geo-engineering**.

Criticisms

- There's **limited evidence** that these types of proposals will be successful and plans are in the **early stages of development**.
- It doesn't address **all of the impacts** of global warming, e.g. **ocean acidification**.
- These ideas **aren't popular** with many people as they're seen as trying to **intervene** with **natural processes**.
- It's **unclear** which countries would **pay** for these schemes. Some schemes could be used **unequally** and **benefit** some **countries** more than others.

Future Changes to the Carbon Cycle

Mitigation strategies can rebalance the carbon cycle

Mitigation strategies attempt to directly **minimise** the **impacts** of **climate change** by reducing the amount of **carbon dioxide** in the atmosphere.

Renewable switching

- Increasing the **availability** and decreasing the **cost** of **renewable energy** sources can reduce reliance on **fossil fuels**.
- To do this, national **governments** need to make **investments** in **renewable energy sources** and make them available on their **national grid**.

Energy efficiency

- People can **reduce** their **car** use or buy more **fuel-efficient** cars.
- **Homes** can be made more **energy-efficient** by using double glazing, insulation and more efficient appliances. In some countries, **government regulations** require new homes to be built with a minimum **energy** and **heat efficiency value**.
- Governments sometimes offer **grants** and **loans** to people installing **energy-efficient features** in their homes.

Carbon taxation

- Governments can **charge businesses** for their **carbon emissions** or for using **building materials** that have to be transported over **longer distances**.
- In some countries, **older cars** are **taxed more** than newer models because they have higher emissions.

Carbon capture and storage

- **Governments** can invest in **carbon capture and storage** (CCS).
- **Carbon dioxide** emitted from burning **fossil fuels** is captured and stored **underground**, such as in depleted oil and gas reservoirs (see p.203).

Afforestation

- Afforestation and **restoring** degraded **forests** can increase **carbon** uptake by the **biosphere**.
- **Grants** might be given to **landowners** to **reforest their land**.

Some mitigation strategies require international cooperation

1) The Intergovernmental Panel on Climate Change (IPCC), an **international organisation** of the **UN**, states that countries need to **reduce** their **greenhouse gas emissions**.

2) The IPCC report published in 2021 highlighted **five possible future** scenarios. These are called **Shared Socioeconomic Pathways** (SSP) and they take into account different levels of **climate change mitigation**.

3) Countries can work **together** to **reduce emissions**. For example, the **Kyoto Protocol** (1997) and the **Paris Agreement** (2015) are **international** treaties where participating countries agree to keep their **emissions** within **set limits** (see p.222).

4) There are also international **carbon trading schemes** which put a **limit** on the amount of **emissions** countries and businesses can emit. If they emit **less** than the limit, they can **sell the extra carbon credits**. If they emit **more** carbon emissions, they have to **buy more credits** (see p.224).

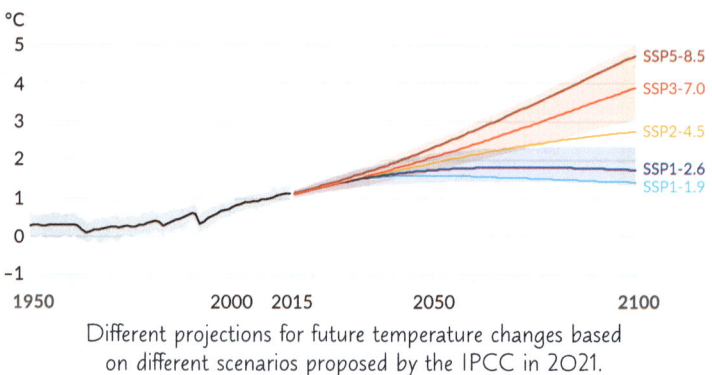

Different projections for future temperature changes based on different scenarios proposed by the IPCC in 2021.

Warm-Up Questions

Q1 Describe three different methods of adaptation to climate change.

Q2 Explain why it is difficult to predict future global temperature changes.

PRACTICE QUESTIONS

Exam Question — A-Level

Q1 Evaluate the view that mitigation is a better response to climate change than adaptation. [20 marks]

Mirror mirror on the wall, who is the greenest of them all...

Hopefully your head isn't spinning after reading that information about the carbon cycle. It's really important stuff, so make sure you've got all the info stored in your brain, it'll help to mitigate any negative effects in the exam...

Introduction to Superpowers

Welcome to 'Superpowers 101'. Today we will be learning how to fly. Well, no, not those kind of superpowers. A superpower is determined by the human and physical characteristics that can make them have more or less geopolitical power.

There are three ways of **classifying powerful countries**

Superpowers	**Emerging powers**	**Regional powers**
Superpowers (like the USA) can influence the rest of the world. They are a dominant global force and hold strength through their **economy**, **military** and **political influence**.	Emerging powers (like China) are the **potential superpowers** of the future. Their influence is relatively large compared to other countries and is growing rapidly. They hold significant **economic** power.	Regional powers (like Mexico) can influence other countries **within the continent** they are in.

- There are **six characteristics** that a country needs to have before being called a superpower:

  1) **Economic** — A country's economy is seen as the foundation of all its power. It is difficult for a country to be strong in other areas without a **strong and stable** economy to support it. A superpower will have a **high GDP** and be able to make money around the world through the **ownership** of **TNCs** (transnational corporations). Superpowers are likely to have **stable currencies**.

  2) **Political** — A superpower holds significant influence in **intergovernmental organisations**, such as the UN, and can influence how other countries **behave** and what **policies** they follow.

  3) **Military** — A superpower can use its military to threaten, invade, defend and aid other countries. Military power is an important way of **expanding influence** over other countries. Superpowers spend a significant percentage of their **GDP** on their armed forces and are likely to **export** military technology to their allies.

Annual Military Spending (US$ billion) (2020)	
USA	778
China	252
India	72
Russia	62
Saudi Arabia	57

  4) **Cultural** — Superpowers often have a specific **ideology** (a dominant set of beliefs or behaviours) that other countries may wish to follow. They may also have a strong influence on the **arts**, **food** and **fashion**.

  5) **Demographic** — Having a **larger population** that is educated and healthy can lead to a greater amount of power. For example, a larger population can create a **larger military** and a **bigger labour** force. A superpower is also likely to have more **migrants** entering the country than leaving it each year.

  6) **Access to natural resources** — A country with a **large supply** of natural resources is unlikely to be reliant on other countries for their supply. A country can also influence other countries through the trade of those natural resources for profit. Having an abundant **source of energy** (such as coal or oil) or **key metal reserves** (such as iron) is especially powerful.

- A **true superpower** dominates in **all** of the above characteristics. Currently, the **USA** is the only true superpower. A **hyperpower** also dominates the above characteristics but is **completely unchallenged** by other powers. Britain was a hyperpower from around 1850-1910 and the USA was a hyperpower from around 1990-2010.

- A country that has a large surface area of land is more likely to become a superpower — an exception to this is Great Britain before the creation of the British Empire. A country's **geographical position** and **topographical features** (such as mountains and coastlines) impact the ability of a country to **influence other countries** around them. For example, a large coastline that is easy to navigate will help a country establish itself as an **international trader**.

Countries can maintain power through **hard and soft power**

1) **Hard power** is the use of power through **military** or **economic force**. The different elements of hard power are outlined below.

Hard Power

- **Military force** involves the threat of action and the invasion of another country. Countries most effective at using this form of hard power have a **well-equipped military**, **military bases in foreign countries** and **nuclear weapons**. For example, the USA and its allies used hard power during the Iraq War (2003-2011).

- **Economic sanctions** can take the form of **trade restrictions**, **freezing of assets**, **arms embargoes** and **travel bans**. They aim to **limit opportunities** for a country to act in an undesirable way. For example, Russia's invasion of Ukraine in 2022 resulted in sanctions being applied to Russia.

- Hard power can work but it is expensive, risky and can create greater **long-term political instability**. Although hard power is still used today, its **importance has decreased** over time.

G7 leaders at the 2019 summit in France

© World Politics Archive (WPA) / Alamy Stock Photo

Introduction to Superpowers

2) **Soft power** is the use of power through attractive policies or ideologies. The term 'soft power' was first used in 1990 by Professor Joseph Nye. Countries can be influenced **voluntarily** or by **negotiations** and persuasion.

> **Soft Power**
> - It's seen through the **culture** of one country spreading around the world in the form of **films**, **music**, **television**, **social media** and **recognisable brands**. These cultural exchanges help countries to spread their **global influence**.
> - The spread and use of a **common language** — the English language has become the world's most used tool of communication, which helps **spread ideas and beliefs**.
> - **International relations** is a key element of soft power. Countries that get on well with others are able to **influence** countries with **appealing policies** and **cultural values**.
> - **Globalisation** (see p.88-89) has contributed to the spread and **growing importance** of soft power in today's world. Soft power is more likely to succeed when countries already share a **degree of cultural similarity** with one another. It may have less of an impact on countries that are more **culturally different**.

3) Some strategies **combine** elements of hard and soft power and occupy the **centre ground** of the spectrum. Some economic policies (such as **trade agreements**) are examples of soft power since they involve no direct threat. However, the nature of the trade agreement may control how the country wishes to sell its goods. They may be forced to sell at a **lower price** than they would like, which is why there are **elements of hard power** in the agreement.

Mackinder had a specific **geostrategic theory** about superpowers

- Halford **Mackinder** was a British geographer who created a theory about **territory** and **geopolitical power** at the start of the 20th century.

> **geostrategic theory**
> *the idea that controlling certain areas of land can make a country strategically stronger and have a greater influence over other nations*

- Mackinder identified the centre of Eurasia as a '**heartland**' or a 'pivot area'. The geographical position of this region means that it is centred on a **large land mass** and has an **inland position** that gives it protection from attacks via the sea. It also contains a large percentage of the world's **natural resources**.

- Mackinder suggested that whoever had control of the heartland would become a superpower. Mackinder realised that this meant **Russia** should be the world's only superpower. However, he also recognised that some of the physical geography surrounding Russia was a **disadvantage**. For example, a large part of Russia's coastline is within the **Arctic Circle** and unusable as a **trade route** for most of the year.

Mackinder argued that the heartland can only be strategically useful if the land surrounding it is made up of just a few countries. Russia, with its **multiple borders**, would have to defend **territorial attacks** from multiple directions.

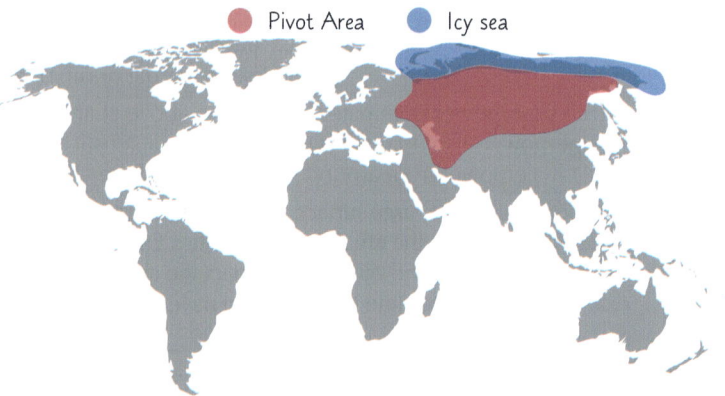

- Mackinder's theory influenced **geostrategic** thinking beyond the Second World War. Superpowers, such as the USA, tried to **restrict countries** that looked to dominate the heartland (a policy known as **containment**). For example, during the Cold War between the USA and the USSR (1945-1991), the USA was **influenced** by Mackinder's theories and wanted to **restrict** the USSR's control of the heartland.

Mackinder's Heartland Theory (1904)

Warm-Up Questions

Q1 Outline the characteristics of a superpower.

Q2 Describe the Mackinder heartland theory.

Exam Question — A-Level

Q1 Explain the idea of a spectrum of hard and soft power. [4 marks]

My superhero name would be 'Aluminium Man' — I love foiling crime...

There are so many different types of 'power' that it can be quite confusing to keep track of all of them. However, you'll be fine so long as you know the differences between 'hard power' and 'soft power' and how they are used on the global stage.

Patterns of Power

The last time I learned about patterns of power, I was tracing strange lines in an old spell book I used to own. The patterns of power found on these pages are a bit less magical than my spells, but they are more useful when preparing for an exam.

There are different levels of **geopolitical stability and risk**

There are **three** main **patterns of power** that come about through changes in geopolitics:

1. A **unipolar world** (British Empire) — This is where **one superpower dominates** the world. There are **different views** on how **stable** a unipolar world is — some scholars argue that a unipolar world is unstable because other powers are likely to **challenge** the dominant power, while others believe it can be stable if the dominant power has **sufficient hard and soft power**.

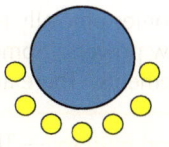

Elena finally found some stability

2. A **bipolar world** (Cold War era) — A world where **two superpowers** that have very different **ideologies** contest each other's influence and power. If both powers are content to **share control**, this can be a relatively stable system. If the two powers do not trust each other, **diplomacy may breakdown** and peaceful stability may be lost.

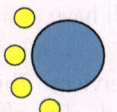
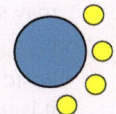

3. A **multipolar world** — This is where **multiple superpowers** and **emerging powers** are influencing other countries, often within the geographical **region** they are in. Although this might appear to be the most stable scenario as there is a greater division of global power, this stability may only occur if all powers within the system are the **same size**. The stability of a multipolar world can quickly change to an **unstable** unipolar power system if two or more powers decide to form an alliance.

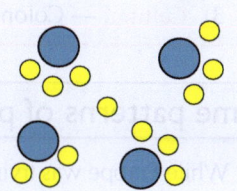

Power can be maintained by **direct colonial control**

- Colonisation is when one country takes control of another country or territory, often by force. This control is then **reinforced** by the colonising country **settling the colony** with its own people and **cultural ideals**.

- The **British Empire** created a **unipolar geopolitical system** through **direct colonial control**. At its largest in 1922 A.D., the British Empire covered about 25% of the world and contained around 25% of the world's population.

- Britain's rise to dominance was aided by its strong **navy** and established **trade routes** with its colonies.

- The British maintained their territories through **military force** and **forming alliances** with local rulers. They also installed British seats of power and **governance** in the country to govern the people according to British values. **Acculturation** (where a new, dominant culture **overtakes** an indigenous one) took place over time with the **spread** of the English language and Christianity.

- **Wealth** mostly benefited the colonisers and local people who worked within the colonial system. **Enslaved people** and **goods** from colonies (such as tea, sugar and tobacco) were traded through large British companies (such as the East India Company), whose activities were defended by the British military.

The extent of the British Empire (in maroon) in 1921

- The colonies weren't only a source of additional wealth for the British — they were also a **source of softer power**. For example, the British set up more permanent homes in the colonies such as India, Australia and Kenya. With these more permanent residences came **public facilities** (such as schools) and the extension of communication and transport lines. In the late 19th century, the British began to sell **manufactured goods** back to the colonies, with second and third generation settlers often growing up to know little of Britain at all.

- After 1919 A.D., the world became more **multipolar**. The First World War had exposed weaknesses in Britain's power, and strengths in the emerging powers of Russia and the USA. Japan also started to become a **regional power** in Asia.

- The **multipolar** nature of the world at this time may have contributed to the massive **scale** of the Second World War and the number of countries it involved.

Patterns of Power

Neocolonialism was a form of **indirect control**

- The **post-colonial era** started **after** the Second World War as there were new ways of creating and maintaining power.

- This post-colonial era began after most colonies became **independent**. For many former European colonies, independence was gained in the 1960s and 1970s. Independence partly came about because, after the Second World War, European powers **lacked the finances** needed to maintain power in their colonies while **rebuilding** their own countries.

- After many colonies officially gained **independence**, they formed new relationships with the former colonial powers and new powers. Some of these relationships contain neocolonial elements. There are several forms of **neocolonialism**:

> **neocolonialism**
> *the use of economic, political and cultural power to gain and maintain indirect control over developing powers*

1) **Political and military** — There may be military **alliances** between countries where a former colony relies on weapons and equipment from another country. Countries may still have **political parties** that represent the people and ideologies of the old colonial order.

2) **Economic** — A former colony may be in **debt** to another country or rely on **aid** to provide its citizens with basic services. This aid may be **tied aid** (aid that is donated but there are **conditions attached** to how it is used). This effectively gives control of a country's finances to another country. Development may stall without **trade deals** between countries that disproportionately benefit the former colonial powers.

3) **Cultural** — Colonial place and road names, religions and languages are often used long after independence.

Some patterns of power have become **more important over time**

- While Europe was trying to rebuild itself after the Second World War, the USA experienced rapid urban growth and industrialisation and led the post-war economic expansion. A **bipolar power era** began with the capitalist USA and the communist USSR fighting for **hegemony** (the dominance of a single superpower). This period was known as the **Cold War** (1947-1991).

> It was called a 'Cold War' because no direct conflict took place between the USA and USSR. The constant threat of nuclear weapons being used kept both powers from starting a 'hot' war.

- The USA strengthened its global influence during the Cold War era through **indirect economic control**. The **Marshall Plan** (1948-1951) was a programme of US funding that extended into western European countries. It provided financial assistance to countries trying to **rebuild infrastructure** after the Second World War.

© GRANGER - Historical Picture Archive / Alamy Stock Photo

- Some countries aligned themselves politically to the USA (like the UK) and others to the USSR (like East Germany). This made the Cold War period intense and uneasy. **NATO** (the North Atlantic Treaty Organisation) was formed by the allies of the USA in 1949, while the **Warsaw Pact** (1955-1991) was an alliance between the USSR and the socialist countries of Eastern Europe.

A political cartoon depicting the constant threat of nuclear weapons being used.

- These alliances were involved in other conflicts (known as **proxy wars**) — e.g. the Korean War (1950-1953) divided Korea between the USSR-aligned North and the USA-aligned South. In the Vietnam War (1955-1975), the USA fought directly against the USSR-backed communist government of North Vietnam.

- The collapse of the USSR in 1991 allowed the USA to become the dominant superpower. However, there is debate about whether the current era is more of a **multipolar** power system or a **unipolar** power system. **Emerging economies** like China and India are becoming more powerful, while established economies are starting to **stagnate**.

Warm-Up Questions

Q1 Describe the three main patterns of power.

Q2 Explain how the power of the British Empire was maintained.

PRACTICE QUESTIONS

Exam Question — A-Level

Q1 Assess the extent to which direct and indirect control contribute to economic power. [12 marks]

I once wrote a song about a tortilla — well, I guess it was more of a wrap...

Sorry about that one. The names of the different eras have useful clues to remind you of what was going on at the time. For example, you know there were two powers (the USA and USSR) during the bipolar era as the prefix 'bi' means 'two'.

Emerging Powers

Emerging powers are not a new group of young, up-and-coming heroes, but they might become the next superpowers.

Some emerging economies have a **growing role** in global systems

1) In the future, it is likely that the USA, as the current superpower, will have less influence. Countries like the USA are predicted to see the **growth of their economy slow down** as **finite resources** (such as fossil fuels) on which they rely start to run out and other countries start becoming more competitive.

2) Countries in **BRICS** (Brazil, Russia, India, China and South Africa) and **MINT** (Mexico, Indonesia, Nigeria and Turkey) have growing numbers of **middle-income** citizens with larger disposable incomes. The disposable incomes make these nations important **marketplaces** for **manufactured goods**.

3) The strong secondary sectors of the BRICS countries may start to shift towards more **service-based economies** in the future (something that is already seen in the growth of back-office support companies in India). The MINT economies are seeing rapid increases in population which also gives them greater potential to become emerging powers.

4) In 2014, the BRICS economies created the **New Development Bank** — an institution that provides financial **loans for infrastructure projects** within developing countries. This gave the BRICS countries more economic influence around the world as the New Development Bank now competes with the **World Bank** for bids.

5) Membership of the **G20** (a forum of twenty countries aiming to create global financial stability) is a good indicator that a country has **political value** and a **sizable economy** (the wealth of the G20 countries is 85% of global GDP).

6) The BRICS and MINT economies are also starting to have a greater say in **global organisations**. Since 2017, there has been growing support for India to become a **permanent member** of the UN Security Council. Both India and Brazil have said that they are willing to opt out of having the **veto power** that the other permanent members have.

Percentage increase GDP 2000 to 2020	
USA	104%
EU	111%
Brazil	120%
Russia	471%
India	468%
China	1115%
South Africa	121%
Mexico	52%
Indonesia	541%
Nigeria	522%
Turkey	162%

The only bank I'll ever need is right next to my bed.

Emerging economies also have a role in **global environmental governance**

1) The drive for economic growth in BRICS nations has often meant that concerns for the environment have been **less of a priority**. In the early 2000s, Beijing's **air quality** failed to meet the **World Health Organisation's** minimum standards, especially with regards to PM2.5 concentrations (airborne particulates). This was caused by the reliance that China had on energy produced by **coal-fired power stations**.

2) Although China is becoming a leader in **renewable energy sources** (mainly wind and solar power), **water pollution** of the Yangtze, Xi and Huangpu Rivers remains a problem due to the rapid development of pollution generating industries.

3) While economically emerging powers are growing in strength, they are not always able to do so in an **environmentally sustainable** way. Countries like Brazil and Indonesia have a history of **deforesting** large areas of land in order to farm **cash crops**, **build infrastructure** and **mine for resources** (such as precious metals). **Drilling for oil** in Nigeria has previously caused **oil spills** and **soil contamination** over a large area of the now **infertile** Niger delta.

4) **Changes** are happening at a local level within emerging powers and these countries have an increasingly important role in the **global governance** of the environment. Under the **Kyoto Protocol** (see p.222) developing countries (including China, India and Brazil) **did not have to commit** to reducing their emissions. However, these countries still **actively participated** in summits and conferences aiming to tackle climate change and other environmental issues.

Beijing's Forbidden City covered in smog

Emerging Powers

Emerging powers have evolved different **strengths and weaknesses**

Emerging powers, such as the BRIC countries, have the **potential to develop** into superpowers and **challenge** the status of the USA. Below are some examples of the **strengths** and **weaknesses** that will impact the potential superpower status of the BRIC countries in the future.

	Strengths	Weaknesses
Brazil	• Has large supplies of oil and biofuels • Has a growing middle-class consumer market • Agricultural and mineral heavyweight (e.g. soya, iron ore , tin and phosphates)	• Small military only capable of being involved in regional affairs • High levels of domestic inequality (e.g. income, health care and education) • Forests threatened by illegal deforestation
Russia	• Strong military power and has nuclear weapons • Has significant oil and gas supplies • Growing economic and political links to Asia (particularly with China)	• Has an ageing and declining population • Difficult geopolitical relationship with the USA and Europe and is no longer a member of the G7 • Limited amounts of cultural exports
India	• Has a young population with huge economic potential • Has a growing middle-class • Possesses nuclear weapons and sophisticated space and missile technology	• High levels of rural and urban poverty • Difficult political relationships with neighbouring countries (e.g. with Pakistan over the Kashmir region, see p.229) • Poor transport infrastructure
China	• A large and increasingly educated workforce • A growing military that possesses nuclear weapons • Investing in infrastructure projects to increase trade between China and the rest of the world	• Ageing population due to the legacy of its One Child Policy • Major air and water pollution as a result of rapid and unregulated industrialisation • Heavily reliant on importing raw materials (e.g. oil, iron ore and soya)

Development theories help to explain patterns of power

There are three main development theories to consider.

Sofia was just finishing her 'pattern of power'.

Modernisation Theory

- This theory was developed by American economist Walt Whitman Rostow in 1960. It was created to explain the **dominance** of the British Empire and the USA. Rostow identified **five stages** that he believed all countries would go through as they developed.

5) **High mass consumption** — Increasingly urban society with lots of disposable income and consumption of goods.

4) **Drive to maturity** — This is characterised by increasing levels of disposable income and a strong domestic market for goods. There are high levels of urbanisation.

3) **Take-off** — This is shown by the development of manufacturing industries.

2) **Preconditions for take-off** — Identified by the growth of export-based trade. This trade is supported by advancements in transport and communication networks, as well as central governance and financial systems.

1) **Traditional society** — Characterised by very slow development in localised ways.

- The Modernisation Theory is useful when describing the **economic development** of superpowers, but it does not consider the development of the **political and cultural characteristics** needed to become a true superpower.

Emerging Powers

Dependency Theory

This theory was developed by a number of theorists, including German sociologist Andre Gunder Frank, in 1966. This theory recognises that development is **not a simple linear progression** (as believed by Rostow). Instead, development is a **set of relationships** between developed countries (the '**core**' economies) and developing countries (the '**periphery**' economies).

1) The **core** provides the **periphery** with manufactured goods, aid and FDI, dominant cultures, political ideals (which are usually seen as desirable) as well as pollution and waste materials (seen as undesirable).

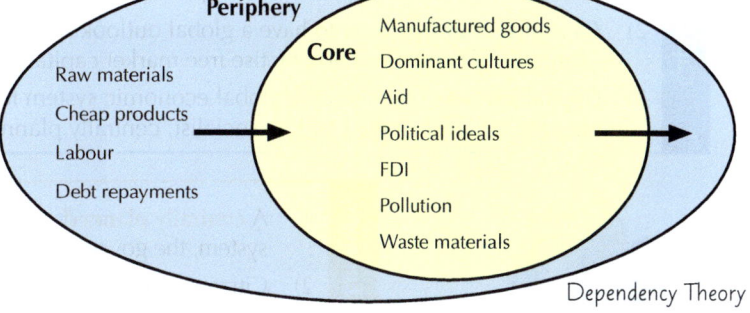
Dependency Theory

2) The **periphery** provides the **core** with raw materials, cheap products (such as sugar, cocoa and coffee), cheap labour and debt repayments. The periphery can become **exploited** by the core, **slowing its own development** and furthering the core's development. People that **migrate** from the periphery to the core tend to have a higher level of education than the workers that are left behind. This means the remaining workers may not be able to contribute as much to the economy. This is called the '**brain drain**'.

3) Superpowers occupying the core are able to control the **value and volume of trade** between themselves and periphery countries. This allows superpowers to **protect their status** as they control how much economic power developing nations gain, preventing them from becoming an emerging power.

- One **criticism** of the Dependency Theory is that not all countries belong to a periphery or a core. **Newly industrialised countries** (NICs), such as Singapore and South Korea, have been able to **develop independently** by **specialising** in particular economic fields (banking in Singapore and consumer electronics in South Korea).

- Another criticism is that the theory suggests countries are unable to move from being a **peripheral nation** to a **core nation**, and that the world can only be **split** between these two poles.

World Systems Theory

- The **World Systems Theory** attempts to model the complexity of the relationships that exist between countries. It was developed by American sociologist Immanuel Wallerstein in 1974.

  1) The World Systems Theory **expanded** the Dependency Theory to include the **semi-periphery**, which is made up of emerging economies and NICs. The semi-periphery is able to **dominate** the periphery but not the core.

  2) While the characteristics of the core, semi-periphery and periphery stay the same, the countries can **move between the regions** over time as they move through different periods of development and decline.

- A **criticism** of the World Systems Theory is that it describes which countries have **power** and **wealth** but it does not explain **why** they do.

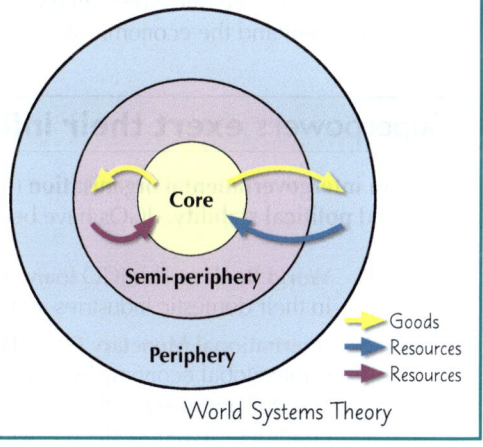
World Systems Theory

Warm-Up Questions

Q1 Describe the factors that could threaten a superpower.

Q2 Explain the barriers that could stop an emerging power from becoming a superpower.

Exam Question — A-Level

Q1 Assess the extent to which development theories explain patterns of world power. [12 marks]

What do you call a yeti with a strong core? An abdominal snowman...

It's important to know about the three development theories and all the similarities and differences between them. Having a strong understanding of those theories will help you to form your own opinions about each theory's strengths and weaknesses.

Influence of Superpowers

Superpowers have a pretty big say in what goes on in the world. Here are some pages about how they influence the world.

Superpowers influence the **global economic system**

Capitalism

1) **Free market capitalism** is one type of economic system. It encourages the **private ownership** of industries and their associated wealth. Goods and services are competitively traded on a **free market** where their value is based on **demand**, the **quality** of the good or service and the ability of the industry to **supply** these products.

2) **Capitalist countries** tend to have a global outlook and form trading connections with many other countries. Examples of countries who practise free market capitalism are the USA, Japan, members of the EU and the UK.

3) Capitalism has **dominated** the global economic system for more than a century and became **stronger** after the collapse of the USSR (a socialist, centrally planned economy) in the early 1990s.

The headquarters of Sinopec Group, the second largest TNC in the world and a state-owned Chinese company.

Centrally Planned Economy

1) A **centrally planned economy** is one alternative to free market capitalism. In this system, the government owns all industry and profits, as well as land and property.

2) Citizens work for the state rather than for privately owned industries. There is no competition between industries, so prices for goods and services remain **stable** and the state **controls the supply** of products. Examples of countries that have centrally planned economies are Cuba and Vietnam, though both have had reforms that have opened up their economies to private ownership.

3) Extreme centrally planned economies, such as North Korea, aim to be entirely **self-sufficient** and have almost no connection to other countries.

- During the **Cold War era**, the capitalist USA and the centrally planned USSR worked alongside each other in the global economy even though the two powers were ideologically against each other.

- More recently, China has moved from a fully centrally planned economy to what may be called '**state capitalism**' or a '**socialist market economy**'. This means that most industry is still owned by the state, but some private investment and business ownership is permitted. However, the state retains control over how these private businesses operate, usually by owning the **largest share** of the business. **Foreign investment** is also permitted in China.

- Superpowers influence the global economic system because the dominant superpower tends to promote one type of **economic ideology**. Other countries are influenced by this system and the economic strengths of the dominant superpower grow.

Superpowers **exert their influence** through the work of IGOs

- An **intergovernmental organisation** (IGO) is a committee of countries that work together to gain global **financial** and **political stability**. IGOs have become platforms that can help to promote international cooperation.

1) **World Bank** — This IGO **loans money** to developing countries so that they are able to invest in their domestic industries and eventually export more of their goods to the global market.

2) **International Monetary Fund** (IMF) — This organisation aims to create a more stable global economy by providing financial assistance to developing nations. The IMF are responsible for **distributing loans** from developed countries to developing countries. In return, the receiving country must agree to trade more freely.

3) **World Trade Organisation** (WTO) — This organisation **regulates** world trade and encourages countries to **remove or reduce tariffs** (import and export taxes) on trade.

4) **World Economic Forum** — This group of **business** and **political leaders** try to influence global agendas, play a role in international-decision making and promote capitalism.

Turn to p.90 for more information on the World Bank, the IMF and the WTO.

- Many IGOs promote **free market capitalism** and **free trade**. Free trade is the international trading of goods and services without countries placing **taxes**, **tariffs or quotas** on the goods and services being traded. Free trade is designed to increase the volume of world trade. This has helped countries around the world **grow their economies** while also ensuring the USA remains a superpower. A criticism of IGOs is that they tend to **support capitalist ideologies** over other economic systems since their members are mainly the capitalist nations themselves.

- Some countries prefer economic policies that promote **protectionism** (see p.90). This means that taxes are placed on imported goods so that they automatically become more expensive than goods produced **domestically**. This '**protects**' local economies as it creates a **stronger market** for local goods and services.

Influence of Superpowers

TNCs influence the global economic system

- The world's 10 largest TNCs are all either private or state-owned companies.

  1) **Public TNCs** — Privately owned companies with numerous shareholders. They are often well-known, recognisable **global brands** — e.g. Apple.

  2) **State-owned TNCs** — Companies that are owned by the governments of countries. They might be well-recognised **within their country of origin** but not outside it.

See p.93 for more on TNCs.

- TNCs can influence the global economic system in a number of ways:

  1) Their **size** means they can dominate the market of some goods and services. Some countries have little choice but to trade with them as **alternatives may not be available** (e.g. drugs made by pharmaceutical companies).

  2) The expansion of free market capitalism has opened **new markets** to TNCs. This has allowed them to grow their profits and expand on a **greater scale**.

  3) TNCs often grow by **merging** with other companies or **acquiring** smaller ones — allowing them to **reduce competition** for raw materials and labour and to pick the price they pay.

  4) TNCs can **invest** in **technology** (to develop new products), **business infrastructure** (to increase production or efficiency) and **operational infrastructure** outside the company location (such as new transport and communication systems) to increase the speed of products going to market.

- The growth in the power and influence of TNCs has mainly come from the growth of free market capitalism. They have **benefited** from globalisation while also being the reason for its **continued growth**.

- This power can **influence government policies** because developing countries often need their investment in order to develop themselves.

- TNCs dominate research and development activities, which lead to the invention of new products and technology. These developments are protected by **intellectual property law**, such as **patents**, **trademarks** and **copyright**.

- The **royalties** get paid to TNCs, most of which are based in a superpower like the USA. Payments often continue to be **paid long after** the inventions were designed and can create **high levels of income** for TNCs.

'Westernisation' is one form of cultural influence

Westernisation is often unseen and quite subtle.

- TNCs often create global brands that become known worldwide. Historically, most TNCs have come from the USA, so companies like **NIKE, McDonald's®** and **Apple** have helped to promote western culture around the world. The growing influence of **western culture** is known as '**westernisation**'.

- Westernisation is also reinforced through the **arts** (music, film, art, literature and fashion), **food** (ingredients, cooking styles and presentation) and **media** (news outlets, television and radio programming, Internet search engines and social media).

- However, the **movement of culture** means western countries can also be influenced by the cultures of other countries.

  1) **Glocalisation** occurs (see p.93) when western TNCs **adapt** their products to suit **local tastes and needs**.

  2) Some **cultural elements from emerging economies** have influenced western superpowers (e.g. the popularity of Indian food across the world and the increasing number of students learning Mandarin).

  3) Some aspects of western cultures have had a **limited impact** on the rest of the world. For example, sports like American football and cricket have not travelled far beyond North America and the former colonies of the British Empire.

Warm-Up Questions

Q1 Explain how IGOs might support developed countries in protecting their superpower status.

Q2 Describe what westernisation is and how it is reinforced.

PRACTICE QUESTIONS

Exam Question — A-Level

Q1 Explain the difference between free market capitalism and centrally planned economies. [4 marks]

What superpower will eventually replace the USA? USB...

A good way of assessing a county's cultural impact is to try and see what you know about their languages, food, fashion, literature, sport and general culture (without researching). The more you know about them, the more influential the culture.

Global Governance

Global governance is less to do with how you manage your snowglobe connection and more to do with how countries form alliances, join councils and make decisions that impact the entire world. Collecting snowglobes seems much simpler.

Certain countries play a key role in international decision-making

- Today, **international decision-making** is mostly controlled by the world's most powerful countries. This comes from the **Bretton Woods** system that was created towards the end of the Second World War.

- The **Bretton Woods Agreement** was signed in 1944, resulting in the creation of the **International Monetary Fund** (IMF) and the **World Bank**. In 1945, the United Nations Conference on International Organisation established the **United Nations**.

- Decision-making at the IMF reflects the positions of different countries in a **hierarchy of development**.

- The **formation of the United Nations** (UN) in 1945 was led by countries that were **allies** during the Second World War — the USSR, the USA and the UK. Although **193 countries** are now members of the **UN Charter**, international decision-making **still favours** the most powerful countries.

- The **UN** is linked to a large number of bodies and organisations that operate all over the world. Examples include the **World Health Organisation** (WHO), the **Food and Agriculture Organisation** (FAO) and the **International Criminal Court** (ICC).

	UN Security Council Permanent	NATO Member	G7 Member	G20 Member	WTO Member
USA	✓	✓	✓	✓	✓
UK	✓	✓	✓	✓	✓
China	✓			✓	✓
Russia	✓			✓	✓
India				✓	✓
Brazil				✓	✓

Countries are part of different international organisations

- Powerful countries often lead the global response to **international crises**. For example, the USA, UK and France sent **financial aid**, **medical personnel** and **supplies** to Sierra Leone, Liberia and Guinea during the Ebola epidemic between 2014-2016.

- Countries hit by national crises (such as **natural hazard events**) frequently receive emergency relief packages from other countries. The areas affected by the 2019-2020 Australian **bushfires** received **international aid** from 16 countries in the form of firefighting personnel, equipment and financial aid to support the rehabilitation of people and wildlife.

- The involvement of the USA in conflicts around the world has led to them being nicknamed '**the world's police**'. One aspect of some USA interventions has been to **protect their interests overseas** (such as protecting oil stocks) while also **strengthening their ties** with other countries. When a country acts alone in this way it is known as **unilateral intervention**.

The UN aims to provide geopolitical stability

- The **United Nations** was created in an attempt to prevent conflict on the scale of the Second World War happening again. Today, the UN recognises that the reasons for conflict can include issues like the provision of food, safe water supply and the protection of human rights.

- One branch of the UN is the **UN Security Council**. This is a group of **15 countries** that aims to maintain **international peace and security** by reacting to outbreaks of conflict. Member countries vote on **deploying UN peacekeeping forces** or applying **sanctions** to a country that has started a conflict.

1) The **UN peacekeepers** are a **neutral army** made up of troops from all Security Council members. They aim to **maintain order** in areas of conflict but are not allowed to use an armed response to a situation except to defend themselves from attack.

2) A **sanction** is an action taken by a country or a group of countries against another country by **withdrawing from a mutually beneficial agreement**. For example, a country may be **removed from trade deals** or **banned from competing** in international sports for having a poor record in protecting the human rights of its citizens.

UN Peacekeepers in Paris in 2008

© CHARLES PLATIAU / REUTERS / Alamy Stock Photo

- There are **five permanent members** on the UN Security Council (the USA, UK, Russia, France and China) and 10 members from other countries that serve on the council for **two years** at a time. A **criticism** of the Security Council is that any permanent member can **veto** (reject) **a decision** that the other 14 members are in favour of. This maintains the political power held by the world's strongest powers.

- Being a member of the Security Council does not prevent members from taking **military action** against each other. This has the potential to **derail** the work of the council in trying to maintain peace internationally.

Global Governance

Some countries have formed **military alliances**

Countries often form alliances in order to combine their **military power**. This can **increase the global influence** of countries and allows them to act **collectively** and **interdependently** when executing international strategies.

- **NATO** has 30 member states, including the USA and most of Europe, and is the world's largest alliance in terms of military power. Three of the five permanent members of the UN Security Council (USA, UK and France) are in NATO.
- The alliance guarantees that no member of NATO will **start a conflict** with another member. It also means that if one member is attacked from outside the group, other NATO members will take action to defend it.

- The **ANZUS Treaty** is a military alliance between Australia, New Zealand and the USA.
- The aim of the treaty is to keep the Pacific region safe and that they will act as a **collected defence** if any of the three countries is attacked.

- Though China is not part of any military alliances, it does cooperate with a number of countries (such as Russia, India, Pakistan and Tajikistan) in matters of international security as part of the **Shanghai Cooperation Organisation** (SCO).
- Similarly, Russia has military cooperation with five former members of the USSR, such as Belarus and Armenia, through the **Collective Security Treaty Organisation**.

The 2020 meeting of the Shanghai Cooperation Organisation

Alliances are also important to the **global economy**

- Countries that form military alliances often form **economic alliances** (such as **free trade agreements**) as well. Countries that are members of multiple military and economic alliances have a **strong geostrategic position** (see p.221) and tend to be superpowers or emerging powers.
- The EU has 27 member states and one of the **largest trade blocs** in the world. **Free trade** exists between its members. Free trade also exists between the EU and the **European Free Trade Association** (Iceland, Liechtenstein, Norway and Switzerland).
- The **Transatlantic Trade and Investment Partnership** (TTIP) is a proposed set of trade agreements between the USA and the EU. The negotiations for TTIP ended in 2016 as the USA and the EU have not been able to reach an agreement yet. If the TTIP goes ahead, it would create a huge free-trade zone and potentially strengthen the USA and the EU.
- Other economic alliances such as the **North American Free Trade Agreement** (NAFTA) (see p.278) and the **Association of Southeast Asian Nations** (ASEAN) have created greater economic stability and provided **greater geopolitical security** in member countries. Greater economic **interdependence** between countries strengthens the political alignment between them — countries that rely on each other for trade are **less likely** to start an armed conflict or political disagreement with each other.

Members of the EU			Members of NAFTA	Members of ASEAN
Austria	France	Malta	Canada	Brunei
Belgium	Germany	Netherlands	Mexico	Cambodia
Bulgaria	Greece	Poland	USA	Indonesia
Croatia	Hungary	Portugal		Laos
Cyprus	Ireland	Romania		Malaysia
Czech Republic	Italy	Slovakia		Myanmar
Denmark	Latvia	Slovenia		Singapore
Estonia	Lithuania	Spain		Thailand
Finland	Luxembourg	Sweden		The Philippines
				Vietnam

Other economic alliances include the Eurasian Economic Union (EEU), the Gulf Cooperation Council (GCC) and the Central American Market.

Global Governance

Alliances can strengthen approaches to tackling **global environmental issues**

- **International cooperation** is essential when tackling global environmental issues. Countries can work **interdependently** to try to strategically manage global problems. The **Intergovernmental Panel on Climate Change** (IPCC) is an organisation responsible for producing regular **scientific assessments** on climate change, its impacts and ways to mitigate these impacts.
- The IPCC creates **reports** that inform the **United Nations Framework Convention on Climate Change** (UNFCCC) and proposes recommendations that form the basis of climate change agreements between countries.
- There have been **four** notable conferences of the UNFCCC (known as **COPs** — Conferences of the Parties).

1 COP3 — Kyoto 1997

This conference established the **Kyoto Protocol**, an agreement to cut the **greenhouse gas emissions** produced by the members by a combined average of 5% of 1990 levels by 2012. It expected countries to find their own ways to reduce emissions. Developing countries, like China, India and Brazil, were not expected to sign the agreement. In response, the USA refused to sign too as they felt that these countries would **capitalise on the lack of restrictions** and **compete economically** with the USA. Despite the USA not joining the agreement, the agreed target by 2012 was exceeded with levels being **12.5%** lower than 1990 levels.

2 COP18 — Doha 2012

This conference saw members agree to extend the Kyoto Protocol until 2020. With **China**, **India** and **Brazil** still not required to ratify the agreement, some nations (such as the USA, Russia and Canada) also refused to participate. This meant that member countries only accounted for 15% of global carbon dioxide emissions at the time.

3 COP21 — Paris 2015

COP21 resulted in countries moving away from **climate change mitigation** and placed a greater emphasis on **adapting** industries and giving **financial support** to countries most affected by climate change. The **Paris Agreement** was established and countries agreed to **continue reducing** greenhouse gas emissions and to keep any rise in global temperatures below a **2 °C increase**. All countries (including developing nations) were expected to plan and report on how they would play their part in this. At the time of COP21, 196 parties (195 countries and the EU) signed the agreement.

Protestors marching through Glasgow during COP26

© Jonathan Porter / Alamy Stock Photo

4 COP26 — Glasgow 2021

This conference sought to establish a stronger set of emissions targets from all countries in order to **limit global warming to 1.5 °C**. 153 countries established new emissions targets (called the **Glasgow Climate Pact**) and it is believed that this is **unlikely to be enough** to achieve that goal. Steps were agreed to phase out the use of **coal power** and provide stronger financial support for **greener energy development**. There was a focus on the wider economic systems required to meet the needs of all countries with **changing climates**. Part of this was to financially support nations who have suffered '**loss and damage**' as a result of climate change through new funding solutions.

Warm-Up Questions

Q1 Explain why a country might try to form a military alliance.

Q2 Describe the main differences between COP3, COP18, COP21 and COP26 UNFCCC conferences.

PRACTICE QUESTIONS

Exam Question — A-Level

Q1 Assess the extent to which the decisions made by intergovernmental organisations may strengthen the influence of superpowers.

[12 marks]

Why are horses such good companions? They live in a stable environment...

The first UNFCCC took place in Berlin in 1995. The conference's approach to tackling climate change has evolved since then — it's a good idea to do some research to see how their approach has changed over the years.

Superpowers — Environmental Impacts

The next couple of pages will tell you how the rapid industrialisation of the world and the formation of modern superpowers have had a large and damaging impact on the environment. Grab a hot drink and then let's crack these pages.

Environmental degradation is disproportionately caused by some countries

1) Superpowers usually have a **high demand** for natural resources.

- **Water** is needed for human consumption, industry and food production.
- **Energy** is used to power homes and businesses (in the form of electricity) and to power transport (in the form of oil-based fuels).
- **Minerals** provide industries like manufacturing and construction with raw materials.
- **Land** is used to build settlements and grow food.

Rosina wasn't sure how the plant would give her energy.

2) Superpowers have a **high demand** for natural resources because they:

- often have **large populations**.
- often have large numbers of **high-income residents** with high-level resource needs.
- have economies that are often based on **industries** that consume a lot of raw materials and energy.
- are likely to be **hubs of trade and transportation** that have a high energy demand.
- have a plentiful **domestic supply of resources** (which encourages unrestricted use).

3) There are widespread concerns that this resource use is **unsustainable**. There are also concerns that the high levels of consumption are causing local and global **environmental issues**.

- Natural areas of land (such as forests and wetlands) are **degraded** through practices like **open-cast mining** and **deforestation**.
- **Ecologically sensitive** areas of land are exploited for **unconventional sources of energy** (see p.199).
- Water sources get **polluted** through **inefficient farming practices** (such as chemical fertilisers running off the land and into water courses) and the **dumping** of waste and industrial effluent. There has also been an increase in **oil spills** from tankers moving fuel around the world.
- **Poor air quality** due to **older energy technologies** (such as old vehicles and coal-fired power stations).

Some countries also have disproportionately **large carbon emissions**

- Historically, superpowers and emerging powers have relied on energy produced by burning **fossil fuels**. Superpowers and emerging powers are still the **largest producers** of carbon emissions despite their attempts at making their economies **less reliant** on fossil fuels.

- As societies with high levels of **consumerism**, superpowers are **large markets** for goods produced in other countries. The transportation of these goods by **container ships** and **aeroplanes** is very demanding on fuels that produce high levels of carbon dioxide.

- Countries with the largest **populations** tend to produce the greatest volume of carbon dioxide emissions.

- The graph on the right shows that the most developed superpower (the USA) was the largest producer of carbon dioxide until 2000. China was going through a period of **rapid industrialisation**, so saw a sharp rise in its carbon emissions. Other **BRICS** nations did not follow this pattern. India has seen an increase in its carbon emissions but on a far smaller scale then China. Until 1991, Russia was producing a large amount of carbon emissions. Russia reduced its level of industrialisation immediately after the end of the Cold War, but has had a **slow** rise in carbon emissions since 1994. Brazil, South Africa and the **MINT** nations all produce far fewer carbon emissions than other emerging powers, but their emissions are **still rising**.

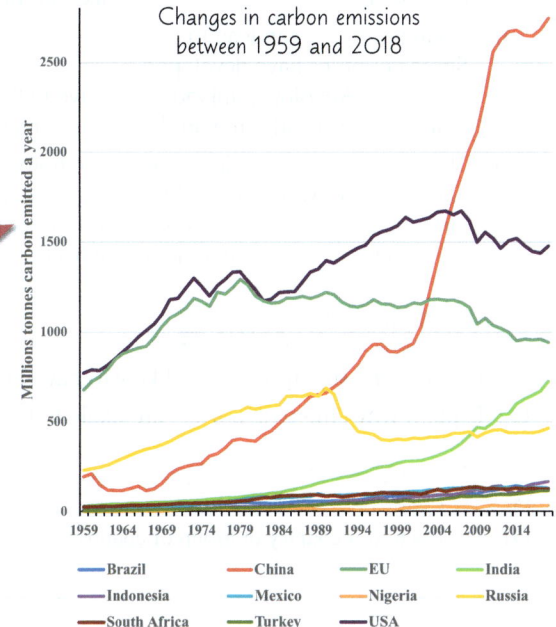

Changes in carbon emissions between 1959 and 2018

Millions tonnes carbon emitted a year

Brazil — China — EU — India — Indonesia — Mexico — Nigeria — Russia — South Africa — Turkey — USA

Superpowers — Environmental Impacts

The EU is an example of a power tackling carbon emissions

Kyoto Protocol

The **Kyoto Protocol** (see p.222) aimed to **reduce emissions** from industrialised nations and showed how countries were trying to address rising carbon emissions. EU countries, as signatories of the agreement, reduced their carbon dioxide emissions by around 32% between 1990-2020. The USA, who was not a signatory, reduced its carbon dioxide emissions by around 7% between 1990-2020. The EU **developed a target**, known as **20-20-20**, which aimed to reach a 20% increase in **energy efficiency**, a 20% reduction of **carbon emissions** and 20% of European energy coming from **renewable sources** by 2020. The target was based on **1990 levels** and the EU claimed in 2021 that the targets had been met.

European Union Emissions Trading Scheme (EUETS)

In 2003, the EU created the **European Union Emissions Trading Scheme** which set a **cap** on how much carbon an '**installation**' (like an industry or a power station) was allowed to produce annually. The size of the cap gets **reduced** each year, meaning that the emissions produced by the EU lowers every year (if the targets are met). EUETS allowed these installations to **trade** any carbon they produced above its capped level with another installation that produced less carbon than its capped level. The EUETS is the world's largest emissions trading system and has been used as a blueprint in other countries.

- A country's willingness to reduce carbon emissions is often determined by how **industrialised** it is and the **energy production infrastructure** it has in place. Some countries have industrial practices and infrastructure that can be **more easily modified** than others. For example, the European success in **emission reduction** was partly due to a more easily modified industry and identifying '**easy gains**'. These countries invested in projects that they believed would quickly and easily reduce carbon dioxide emissions, leaving the harder-to-manage sources of emissions **unchecked**. This means that in order to continue the same rate of emission reduction, significantly greater levels of willingness and investment will be needed.

- There is also evidence that some countries are more willing to address environmental issues than others. For example, there are many **laws** within the EU about acceptable **levels of pollution** and how **intensely land can be used** to grow food. These minimum standards protecting habitats, certain species and water sources are **strictly enforced**, and those who manage land can be **fined** or **prosecuted** for being negligent.

The future growth of emerging powers creates environmental concerns

1) Emerging powers are undergoing **rapid urbanisation and industrialisation** as their economies grow. These countries are also experiencing **rapid environmental degradation** and are competing more for natural resources.

2) In growing economies, the demand for **higher wages** increases as workers become more **highly skilled**. Employers are able to pay higher wages and the general wealth of the labour force increases. This creates a growing middle-class that has more **disposable income**.

3) People in **middle-income groups** consume more **manufactured goods**, such as electronics and cars, than those in **lower-income groups**. Making and transporting these goods increases the demand for **fossil fuels**, such as **oil**, and the risk of **environmental disasters**, like oil spills, occurring. It also places a greater strain on **rare earth elements**.

4) Changes in **diet** can lead to a **greater demand** on staple grains (such as rice and wheat) and **increasing prices**. Some countries have developed **intensive farming practices** that are successful but rely on **pesticides** and **fertilisers**. Keeping farmland productive in this way can cause greater levels of soil degradation, which means that more farmland is required. **Wetlands** may be drained and **ecologically sensitive** areas might become farms.

5) The **demand for water** increases as the human populations grow. There are concerns over the **quantity** and **quality** of water available. In **rural areas**, the greater demand on the land to produce food means that farmers turn to new sources of water in order to meet the demand. Previously untouched **groundwater stores** (like **aquifers**) are under threat from **salinisation**, **under-replenishment** and **complete draining**.

Warm-Up Questions

Q1 Explain why superpowers are likely to have high levels of environmental degradation.

Q2 Explain why emerging powers are likely to have high levels of carbon dioxide emissions.

Exam Question — A-Level

Q1 Assess how a country or group of countries has attempted to address global environmental issues. [12 marks]

Don't you hate it when someone answers their own questions? I do...

Powerful countries impact the environment and cause different types of environmental degradation in various ways. Make sure you know how superpowers and emerging superpowers impact the environment and what they can do to slow the damage.

Conflict Over Resources

The sharing of natural resources is as difficult for countries as it was for me to share my toys as a child.
However, the conflict over resources is much more complicated and dangerous than my inability to share Mr. Ted.

The ownership and use of some **physical resources are disputed**

- As some **natural resources** start to run out, the claims over **who owns them** becomes more important. Having **control** over these resources is crucial if a country wants to gain (or maintain) **superpower status**.

- **Disputes** over who has the **right to use** certain resources can come about in a number of ways:

  1) The reserves of resources may not sit completely within one country's **borders**. Resources such as **groundwater aquifers**, **lakes**, **coal seams** and **oil reserves** often cross international boundaries. These are known as **transboundary resources**. It is difficult to divide the resource between two countries (see p.107).

  2) The land or sea that resources are found in may be under **territorial dispute**. This is where two or more countries claim **ownership** over an area. Equally, a country may not **politically recognise** an international border and continue to operate within a territory.

  3) Some resources are seen as **basic human rights** (such as water). Countries are more likely to share these resources than those that are **commercially profitable** (such as oil). However, the difference between these two types of resources is becoming increasingly blurred as **plentiful and safe drinking water** becomes more essential to **economic development** as well as **human health and wellbeing**.

Disputes over resources have caused tension

1) Disputes over resources do not always come from two countries **wishing to use the resources** themselves.

2) A dispute may arise between one country who wishes to use a resource and another who wishes to **conserve** it.

3) An **Economic Exclusive Zone** (EEZ) is the marine area **200 nautical miles** off the coastline of any country and is included in the territory of that country. When the coastlines of two or more countries are located close to each other and have the **same EEZ** (and are unable to reach an agreement on how to share the marine area), the **UN** can be asked to broker a decision between the countries.

4) Disputes can also occur when multiple countries show an interest in **unclaimed territory**.

Arctic oil and gas

- There are **reserves of oil** and **natural gas** in the Arctic Circle (with some estimates placing the volume of the oil at 90 million barrels). Unlike Antarctica, the area around the North Pole is not that well **protected** from **resource exploitation**. This means that many countries are making **territorial claims** on the Arctic Ocean, the seabed beneath it and the **Lomonosov Ridge**.

- Canada, the USA, Russia, Norway and Denmark (through its connection to Greenland) have made territorial claims in the area as they all have land territory that borders the ocean. Most of the countries' territorial claims on the Arctic are for the areas **closest** to their country.

- However, the position of these countries in relation to the Arctic Circle means there are **overlapping claims** based on each country's **EEZ**.

- In the Arctic, there are disputes between the **USA and Canada** as well as between **Russia and Norway**. Russia planted a Russian flag on the seabed at the North Pole in 2007, which **intensified** the disputes in the region.

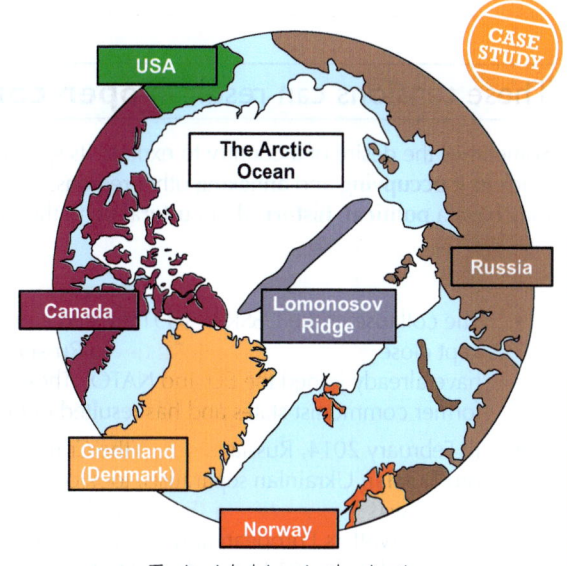

Territorial claims in the Arctic

- Both the USA and Russia are powers with a **significant military presence** in the region. As the warming Arctic loses ice (as a result of climate change) and the region becomes **easier to navigate**, naval ships can move their position closer to the disputed area. Greater accessibility also means that **test drilling** for oil and gas can take place in disputed areas.

Conflict Over Resources

Overlapping **spheres of influence** can cause disputes

1) All countries have a political **sphere of influence**. A sphere of influence includes the country itself and **other countries** that it may influence politically, economically and culturally. A superpower will have a **larger sphere of influence** than a country on the **periphery** (see p.217).

> **sphere of influence**
> *the territorial area that a country has influence over*

2) **Disputes** are likely to occur when the spheres of influence of two significant powers **overlap**. This means that the geographical area where the overlap occurs may be pulled back and forth between two **ideologically opposing influences**.

3) A sphere of influence can be created through the **claiming of territory** as well as the level of **cultural** and **political sway** they hold over other countries. This might occur due to:

- a country wishing to **expand its territory** to gain access to **new reserves of valuable resources** or **new trade routes**.

- the **disputed breakup** of a territory following the end of an **armed conflict**. Land that was formerly part of one country might be acquired by another.

- a country wishing to defend the **territorial and cultural rights** of a particular group of people (such as an ethnic group that regards a particular area as their **ethnic homeland**).

South and East China Seas

- China has looked to expand its **marine territory** into the **South China Sea** in order to make use of the **fishing grounds and the oil and gas reserves** in the area. It has done this by **claiming** a number of small and, until recently, largely **uninhabited islands** (like the Spratly Islands and Scarborough Shoal). This has **expanded the EEZ** of China, which now has a much larger marine territory.

- Other countries in the South China Sea (the Philippines, Malaysia, Vietnam, Indonesia, Brunei and Taiwan) have disputed China's claims on the islands. China has reacted by posting a **large naval presence** throughout the region. It has also started to **construct artificial islands** and expand others. These islands have supported the construction of **ports**, **military bases** and **bases for oil exploration**. The creation of new islands has caused **disruption** to **fish breeding cycles** and dredging has **destroyed coral reefs** in the area.

- Claiming and creating new islands is a **direct breach** of the **United Nations Convention on the Law of the Sea** (UNCLOS). Tensions have increased through China saying that **foreign militaries** have no right to be in the area.

- Japan and China are involved in a dispute in the **East China Sea** (centred around the Senkaku islands). This dispute is over the oil and gas resources found in the area as well as the fishing grounds.

The disputed area in the South China Sea

These tensions can result in **open conflict**

Sometimes the desire of a country to expand its sphere of influence leads to it **occupying** and **annexing** other regions. The annexed region may have a **political**, **historical** or **cultural** tie to the invading country.

> **annexation**
> *the act of one country seizing part of an established territory without permission*

The annexation of Crimea

- The collapse of the USSR in 1991 created a number of **new independent states** in Eastern Europe, which mostly kept close **political and cultural ties** to Russia. Some of these countries, such as Estonia, Latvia and Lithuania, have already joined the EU and NATO. These new **western ideologies** have created tension between Russia and its former communist states and has resulted in Russia trying to **maintain a presence** in areas through **military force**.

- In February 2014, Russia **annexed** the Crimean Peninsula in southern Ukraine. Russian forces (along with pro-Russian Ukrainian separatists) took control of the military bases and the Crimean parliament. In response, Russia was **removed from the G8** and **sanctioned**. These sanctions were applied in 2014 and included **travel bans** as well as European and American industries **stopping trade** with Russian companies. The sanctions that were applied in 2014 are believed to have **cost Russia** over US $400 billion in revenue from lost oil and gas sales.

- The **armed conflict** began in April 2014 in the Donbass region of eastern Ukraine. The conflict has killed over 3 000 civilians and more than 1.4 million Ukrainians have been **internally displaced**.

- In 2022, the conflict between Russia and Ukraine **escalated** when Russia launched a full invasion of Ukraine.

Conflict Over Resources

Counterfeiting undermines intellectual property rights

- The UN operates the **World Intellectual Property Organisation** (WIPO). This global organisation **regulates** **intellectual property** and enables creators to **protect** their work and the inner workings of their companies. This protection is known as **intellectual property rights**.

> **intellectual property**
> *the design of a product, a piece of art, a musical score or a piece of literature. It might also be a brand associated with a set of products*

- Intellectual property rights **prevent** an idea or design being used by another person or company in order to **profit** from it. **Theft** of intellectual property is **illegal** and companies can lose revenue from it.
- If a company wishes to use the **intellectual property** of another, they are required to pay **royalties**. Royalties generate **high levels of income** if the intellectual property is in **common use** and needed in the marketplace (like the way a particular drug is made). However, this can lead to theft of the intellectual property if a company can't afford to pay for it.
- **Breaches** of intellectual property rights have grown rapidly with the **expansion** of the internet.
- Intellectual property theft can result in **two outcomes**:

1) Companies that steal intellectual property may produce **counterfeit goods** that **match the real branded products** so closely that it is difficult to tell them apart. These instances represent very clear breaches of intellectual property law.

2) Companies may also steal intellectual property and produce their **own version** of a product that uses some features of the original but is a **completely different product**. These instances often go to **court** where the legal system determines whether the new product is **different enough** to be classed as a **new invention**.

Counterfeiting also **creates tension** in trade relationships

1) **Counterfeiting** has a big economic impact. Imports of counterfeit goods into the EU in 2019 amounted to US$134 billion worth of products. The global trade in counterfeit goods accounted for 2.5% of world trade in 2019.

2) Counterfeiting can also **strain economic and political relationships** between countries. A **government** may be viewed as **politically weak** if they do little to stop the production of counterfeit goods in their country. This means the country may be seen as weak in other geopolitical areas, such as in **international agreements** or in other areas of **international law**.

George has finally found a way to get out of P.E.

3) Economically, a country may not wish to create **trade agreements** with a country that turns a blind eye to counterfeiting. A country that does not stop counterfeiting might be seen as producing counterfeit goods to try to **undermine the global trade system** by producing goods that are very **competitive** in some markets.

4) Companies that produce counterfeit goods are known to operate in countries such as China and Turkey. This has caused **disputes** between the countries that own the intellectual property and those that are stealing it. It can affect the way that **TNCs** may operate in other countries:

- TNCs may reduce their **investments in research and development** facilities in a country where counterfeiting happens.
- TNCs may **withdraw wider operations** from a country where counterfeiting takes place if they feel the local government does not do enough to prevent counterfeiting from happening.

Warm-Up Questions

Q1 Describe the disputes that are happening in the Arctic.

Q2 Explain why counterfeiting can affect the operations of a TNC.

PRACTICE QUESTIONS

Exam Question — A-Level

Q1 Explain why marine territorial claims are often disputed. [4 marks]

I've always thought that a utility belt is a waist of resources...

The situations unpacked on these pages are complicated and constantly changing. Keeping up with the news (and getting your news from multiple sources) is a great way to stay updated on these tensions and conflicts.

Superpowers — International Relations

Relationships can be tricky at the best of times and international relations are certainly no different. There's a whole lot of 'give and take' and questions over who is benefitting the most. Loyalty is important too...

Emerging powers are **creating new relationships** with developing countries

- Developing countries that are former colonies often remain closely tied to the former colonial powers in a number of ways (e.g. through **trade** and **alliances**).

See p.214 for more on neocolonialism.

- The relationship between developing countries and superpowers continues to be one where the superpower often **benefits more** from the trade or transaction than the developing country. **Raw materials** and **basic crops** (like sugar and tobacco) are bought cheaply by superpowers while more complex, **manufactured goods** are sold back to developing countries at a higher price.

- This trade relationship can be **unfair** as the developing countries have **no choice** but to accept the **terms of trade** set out by the superpower. The superpower is also unlikely to agree to changes in the trade relationship that may **weaken their superpower status**.

- The situation for **emerging powers** is different. They also rely on low prices for raw materials but can usually provide **cheaper manufactured goods** than their superpower competitors. Emerging powers have **less to lose** than superpowers when forging strong relationships with developing countries.

© Dirk Renckhoff/ imageBROKER / Alamy Stock Photo

Sugar cane from Cape Verde gets exported around the world.

These relationships create **new opportunities and challenges**

The **relationship** between China and various African countries is a fairly new one in terms of **geopolitics**. This relationship has a number of **different features**:

1) China is increasing the **volume of trade** it has with African countries. Most of this trade is in raw materials (such as **coltan** from the **Democratic Republic of Congo**, which is used in mobile phones) and energy sources (such as **oil** from **Nigeria**).

2) China has become one of the **main sources** of FDI in African countries. In 2020, China invested US $4.23 billion in the African continent with five countries receiving over 50% of the FDI flow.

3) China provides **foreign aid** in the form of grants and low-interest loans to countries it has important **trade relationships** with, such as Angola.

Top five African countries receiving Chinese FDI (2020)	
Kenya	US $630 million
D.R. Congo	US $612 million
South Africa	US $400 million
Ethiopia	US $311 million
Nigeria	US $309 million

4) Chinese investment (both foreign aid and FDI) is often used for large scale **infrastructure projects** such as new roads, rail lines and improvements to ports and communication networks. It's also been used to help fund energy projects such as the Djibloho Dam in Equatorial Guinea. China has also invested significantly in **water treatment plants** across the continent.

This developing relationship has increased the **interdependence** between some African countries and China. However, increased interdependence does not mean that both African countries and China benefit all the time. The relationship creates numerous **opportunities** and **challenges** that have economic and environmental impacts on the countries involved.

Opportunities

- Chinese funds have paid for projects that will **encourage further investment** from foreign states and private companies. Countries that wish to establish new trade links with African countries will also benefit from Chinese investment in transport infrastructure.

- **Infrastructure investments** also benefit home-grown companies and encourage trade within the African continent.

- Many African countries have benefited from investment in **hydroelectric power plants**, encouraging the use of **green energy sources** on the continent. Between 2000 and 2013, China partly or wholly funded 17 hydroelectric power plants in 14 different African nations.

- Many investment projects create **employment opportunities** for local people, which can have a long term-effect on local economies.

- Investment deals are often accompanied by **aid projects** that are designed to **benefit local communities**.

- More people can access goods to consume because African countries are a **growing marketplace** for Chinese goods. This means more people on lower incomes can access them because they are **cheaper**.

Superpowers — International Relations

- China's relationship with many African countries is based on the trade of resources that will eventually **run out**. It's possible that, as these resources run out, China may **withdraw its investment** as it no longer benefits them.

- The relationship between China and some African countries has been compared to the **neocolonial relationship** between many European and African countries, which can **exploit politically weak** and **vulnerable** countries.

- The race to supply China with materials has caused **environmental degradation** in many African countries. For example, **open-cast mining** for copper has destroyed parts of Zambia. The Nchanga mine in Zambia has **denuded** a huge area of **woodland savanna** and the mine is so deep that it is unlikely the landscape will ever recover. Other countries that supply China with oil (such as Angola and Chad) have had **oil spills** that have affected **water sources** and **contaminated farmland**. **Toxic waste materials** from new industries can more easily find their way into water sources when countries are not able to control and police such activity.

- The growing marketplace of Chinese goods means that there is more competition against **African manufacturing**, which may **harm** local economies.

- Jobs generated from Chinese investment tend to be **low-paid** and **short-term**.

- The impact on the **health and well-being** of African workers in the **metal extraction industries** is also a concern in some countries.

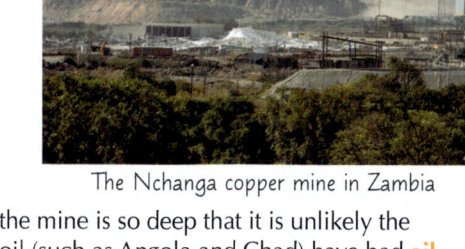
The Nchanga copper mine in Zambia

Some Asian countries are experiencing **rising economic importance**

CASE STUDY

1) The **emerging economies**, such as China and India, are developing strong trade links with the rest of the world. This means that Asia's **geopolitical influence** is increasing.

2) The economic centre of the world is shifting away from Europe and **closer to Asia**. This is sometimes called the **economic centre of gravity** and it represents the idea that a far larger proportion of the global GDP is generated in Asia.

3) Asia also represents the global centre of population growth. It is predicted that by 2030, 27 of the world's 33 **megacities** will be in Asia. This means there is an increasingly **large market for goods** in the region, mostly being consumed by **middle-income groups**.

4) As well as China and India, there are a number of **other rapidly growing economies** in Asia. For example, Indonesia and Malaysia saw **steady increases** in their GDP between 2000 and 2019, compared to Europe which has experienced a decline in their GDP.

5) In 2020, 15 Asian-Pacific countries (ASEAN members, China, Japan, South Korea, Australia and New Zealand) agreed to form the **Regional Comprehensive Economic Partnership** (RCEP). This trade bloc is the largest in history and includes almost a third of all **global economic activity**.

6) This rise in importance has created both economic and political stress, especially between countries who have past histories of tension and conflicts.

- China has strained political relationships with other countries over its **leadership of Hong Kong**, its **conflict with Taiwan** and its **control over Tibet**.

- China's attempts to control the **South China Sea** has created regional and international tensions (see p.226).

- Tensions between India and Pakistan over the future of the **disputed region of Kashmir** has the potential to have a negative impact on the political and economic prospects of each country.

- The establishment of the **China-Pakistan Trade Corridor** has increased tensions between India and China and Pakistan.

A standoff between armed guard and protesters in Hong Kong in 2019

Superpowers — International Relations

Countries in the Middle East are **allied with different superpowers**

The **political alignment** of various Middle Eastern countries is **very complex**. This complexity dates back to agreements reached after both the First and Second World Wars, when the region was **divided into new states**. Many of these states aligned themselves politically to different superpowers. For example:

- **Israel** was established in 1948. The **UK** and the **USA** supported this, and Israel remains **allied** to the UK and USA today.

- **Iraq** previously had links to the **Soviet Union** but has generally been allied to the **USA and its allies**. However, in the Gulf War (1990-1991) and the Iraq War (2003-2011), the USA and its allies aimed to end the **Iraqi occupation** of Kuwait, **overthrow** the Iraqi leader Saddam Hussein and to try to bring **stability** to the region.

- The **Syrian** government has been allied with **Russia** since 1980 and is also allies with **China and Iran**.

- Countries like the USA, Japan, the Republic of Korea and Switzerland have placed **sanctions** on **Iran** for its continued development of **nuclear weapons**.

- **Saudi Arabia**, **Qatar** and the **UAE** are allies with the USA and the UK, largely as a result of **economic ties** through trading oil.

There are **complex reasons** for the tension in the Middle East

The Arab Spring

- **Young people** looking for change have led numerous **uprisings** in the Middle East and North Africa. The uprisings have been about **economic decline**, **corruption** within governments and a desire for stronger **democratic rights**.

- Between 2011 and 2012, these uprisings were collectively known as the **Arab Spring**. During this time period, leaders struggled to create order and appease the protestors. This sometimes resulted in **armed responses** to protests and, in many cases, a return to **autocratic leadership** (being ruled under a single leader). This has increased the number of **dissatisfied people** in the region.

Protestors in Yemen during the Arab Spring in 2011

Systems of Governance

- In some cases, **newly established** systems of governance are not yet providing **stable forms of democracy**.

- The **constitution for modern Iraq** was established in 2005 following the first free elections in the country's history. However, **conflict continued** as separatist groups tried to highlight weaknesses in the Iraqi government's early strategies and **undermine their new authority**.

Division of Territory

- The division of territory between **religious and ethnic groups** can result in **tension**, particularly where one or more of the groups concerned consider an area to be their **ancestral homeland**.

- For example, since Israel was founded in 1948, there have been **ongoing disputes** between **Israelis** and **Palestinians** in the area over who has the rights to live there.

Cultural Ideologies

caliphate
a religious state under Islamic rule

- Conflicting **cultural ideologies** have created tension within the Middle East. This tension can turn into conflict when people feel their **heritage or belief system** is under threat from others. In extreme cases, people may take up arms to defend it.

- The **Islamic State** (IS) terrorist group are an extremist group whose actions are rooted in their **extreme interpretation** of Sunni Islam. They have occupied territory and committed terrorist attacks in an attempt to set up a **caliphate**. It has been argued that IS's growth is partly due to the **imbalance of power** created after the Iraq War, after which the majority **Sunni** government was replaced by one predominantly made up of another Muslim group, the **Shia**.

Superpowers — International Relations

The **legacy of alliances** can present challenges to superpowers

- The **UK**, **USA**, some **European countries** and **Russia** have a history of **selling arms** to the **Middle East** region. A combination of weapons being **traded internally** within the Middle East and **changing alliances** means that foreign allied forces sometimes find themselves fighting against weaponry that their home countries supplied.

- Armed conflict has created **refugees** across the region (more than 6.6 million Syrian refugees since 2011). Large numbers of refugees can put a **strain** on neighbouring countries, causing tension. This tension can be heightened by the refugees **seeking asylum** in a country that **may not support** their culture or religion.

The Kawergosk Refugee Camp in Iraq

- Some countries are caught between **conflicting alliances**. For example, Turkey, as a member of **NATO**, supports the fighting in Syria to remove Islamic State forces from the country. The **Kurds** (an ethnic group that lacks a recognised territory in the region) also joined the fight against the Islamic State but are in conflict with Turkey over the rights to have an **established Kurdish state** of their own.

- Some countries, while not actively supporting **terrorist groups**, do not actively work against them and allow them to grow in size. **Funding** for these groups may come from cells within different countries in the Middle East even if the group does not hold territory there.

The **abundance of resources** in the Middle East impacts international relations

- Much of the **involvement** of Europe and the USA in the Middle East is connected to the **oil reserves** found there.

- It is estimated that the Middle East holds around 48% of the world's oil, so **peace and stability** in the region have a direct impact on the **efficiency** and **stability** of the oil trade. In the **early years** of oil exploration in the region, **investment** by the UK, and later the USA, helped the region to **develop its own technology** and **expand the industry**.

- The price of oil **rises significantly** during conflicts in countries that have reserves. If a country faces **trade sanctions** (as the USA applied to Iran in 1979 and again in 1995), **oil prices may drop** as other countries compete to supply the country that applied the sanction.

- The **hot and dry climate**, and the limited capacity to grow **cash crops**, means the Middle East heavily relies on the oil trade. This reliance means that political stability across the region is beneficial to many countries as it allows them to continue trading. One of the causes of Iraq's invasion of Kuwait in 1990 was the accusation that Kuwait was '**slant drilling**' oil from Iraq's reserves by laying pipelines under the border of the two countries.

- Other resources that are **not in abundance** in the region (such as water) can also create tension. For example, the water flowing through the Euphrates and Tigris rivers became a **valuable and strategic resource** during the Turkish, Iraqi and Syrian conflicts. For example, withdrawing from agreements that allow one country to supply another with water or **damming transboundary water sources** so they favour supply to one country can also heighten tension in the area.

Warm-Up Questions

Q1 Outline the ways in which China has built relationships with different African countries.

Q2 Explain how the abundance of oil reserves in the Middle East has caused tension there.

Exam Question — A-Level

Q1 Assess the relative importance of economic, political and cultural reasons for tensions in the Middle East. [12 marks]

International relations play a big part in the status of a country...

The conflicts and tensions in the Middle East are extremely complex and emotional for many people. While it's important to know how the conflicts started and continue to develop, it's also important to be sensitive when talking about it.

Challenges for Superpowers

All superpowers want to maintain their global status for as long as possible. These pages look at the challenges they face.

Some superpowers are experiencing economic and social problems

1) The **USA**, as the **dominant superpower**, and the **EU** have maintained their **power statuses** for a considerable period of time. **Emerging powers**, such as China and India, are becoming more influential and are **challenging the dominance** of the USA and the EU. There are also a number of **challenges** faced by the USA and the EU:

- The EU was formed in 1993 and originally consisted of a small group of European countries with some of the highest GDPs. Since 1993, countries with lower GDPs have joined the EU (e.g. Poland and Hungary joined in 2004 along with six other countries), and economic analysts have suggested that this is one of the reasons for the per capita decline in the EU's GDP.

- Some cities in the USA and the EU have undergone a shift from manufacturing industries to the service sector — a process called economic restructuring. This resulted in significant job losses in these industries, causing economic decline in cities such as Detroit in the USA (see p.123) and in Turin, Italy.

- The EU and the USA are reliant on migrant labour in some sectors of work (especially those which are traditionally low paid, such as care workers and those working in agriculture). As emerging economies grow, they may become more attractive destinations for workers looking to migrate. Existing superpowers may face labour shortages in some high-skill sectors (for example, health and social care in Germany).

- Many countries in the EU are experiencing ageing populations. In 2021, people aged 65 and above accounted for 21% of the total EU population. An ageing population can increase the pressure on healthcare services and result in higher pensionable ages.

- The USA and EU may struggle to attract investment from industries looking to expand their operations. Higher back-office costs (see p.111) or higher wages mean that large TNCs are increasingly expanding their businesses in other regions of the world (like Asia and Africa), where they offshore and outsource production.

- The global financial crisis of 2008 was largely sparked by the collapse of the housing market in the USA. It prompted a global recession — between 2008-2017 global public debt doubled.

2) **Emerging powers** tend to have more **youthful populations** than the USA and the EU. This means countries like Indonesia and India have a **lower dependency ratio**. As a result, the central governments of these states have a lower social welfare bill (through payments such as **pensions** and **disability allowances**) as well as an **expanding** economically active population.

dependency ratio
the proportion of the population that has to be supported by the working population (aged 15-64)

3) The USA and some parts of the EU are becoming increasingly **energy insecure** (see p.195-196). These powers are becoming **reliant** on other countries to supply them with **energy**, particularly fossil fuels, which **weakens their power status**. This could result in them becoming less able to **dictate** the trade price of such resources.

Maintaining military power is a significant economic cost

1) In 2018, the USA and the EU spent 3.3% and 1.4% of their annual GDP respectively on their **military**. This represents a combined total of over roughly US $900 billion that many people believe should be spent on **foreign aid**, **poverty prevention measures** and **education**.

2) Established powers **maintain their power** by regularly updating their military equipment. Powers might find the increasing amount of funds spent on developing and maintaining their **military power** to be **unsustainable**.

3) Established powers have the greatest capacity to build **nuclear weapons**. Trident (the UK's submarine nuclear missile programme) has cost the UK government approximately £20 billion and there are **ongoing debates** about whether the scheme should be extended to include further nuclear submarines at even greater costs.

4) The **RAF**, the UK's Royal Air Force, plans to introduce a **new aircraft** called the Tempest into service by **2035**. The government expects each new aircraft to cost around **£2 billion**.

5) The threat of terrorist attacks (e.g. the 11th September attacks on the World Trade Centre in New York, 2001) is a key reason for countries to spend money on maintaining their military powers. In response, militaries now spend significant amounts of money on **developing and maintaining intelligence gathering** and **international operations** to **prevent** further attacks.

Challenges for Superpowers

Superpowers are becoming **less dominant** in space exploration

1) The '**space race**' of the 1950s and 1960s was an example of two powers (the USA and the USSR) investing large amounts of money into being able to claim a number of 'firsts' in space. Although the USSR succeeded in sending the **first man into space** (Yuri Gagarin in 1961), the USA placed greater importance on having the **first man to walk on the Moon** (Neil Armstrong and Buzz Aldrin in 1969).

2) The **annual budget** (2019-2020) for the National Aeronautics and Space Administration (**NASA**) was US $22.6 billion. However, some believe this money should be spent on **domestic and foreign issues facing the USA on Earth**.

3) The USA has **led the way** in terms of space exploration for decades, but there are other countries that are starting to **contest the USA's dominance** in space.

4) **India** launched its first space programme in 1962 and since then has **launched hundreds of satellites** into space, including its **first Mars orbit** in 2014. There is also competition from **private enterprises** for control of leisure-based space travel. **VIRGIN GALACTIC** and **SpaceX** have invested large budgets into space science to enable **space tourism** to be a more realistic prospect for future generations.

Atiyyah was convinced she'd seen the finish line.

Emerging powers are beginning to **challenge** superpowers

1) As countries continue to develop and decline, there is likely to be frequent **changes** in the **balance of power** that different countries and regions share. The pace of change within the process of **globalisation** (see p.88-89) suggests that changes in the ways that countries influence each other will also happen quickly.

2) **Global population growth** is starting to **slow down**, with global fertility rate standing at 2.4 in 2020, compared to 5 in 1960. This may reduce the importance of having a **large population** in order to become a powerful nation. **Other factors**, such as a nation's position in international decision-making or its ability to tackle global issues such as **climate change**, may become more important.

3) The future may hold unforeseen **global problems** (such as new wars or further pandemics similar to COVID-19) which may **disproportionately affect** the most advanced economies. Emerging powers may gain more influence and create a change in the global **power balance**.

4) The future superpowers may become the countries that are able to maintain the highest degree of **self-sufficiency** when both **renewable resources** (such as a water supply) and **non-renewable resources** (such as metal minerals and fossil fuels) are in short supply. These countries will be able to provide for themselves domestically and may also **control the trade prices** of these resources to existing powers.

Renewable energy sources (wind and solar power)

5) There are a **range** of **possible outcomes** in **global power balance**, these include:
- **Unipolar world system** — continued **dominance** of the **USA**, where no other power is able to rival that of the USA.
- **Bipolar world system** — in the **short term** (until 2050), studies are suggesting that the gradual **division of power** between the USA and China will continue until a bipolar world system is in place.
- **Multipolar world system** — the ongoing social, political and economic **challenges** faced by the USA and the EU along with the growth in emerging economies could mean **power is shared** between more countries.

Warm-Up Questions

Q1 Outline some of the economic problems experienced by current superpowers.
Q2 Describe how the balance of global power might shift in the future.

PRACTICE QUESTIONS

Exam Question — A-Level

Q1 Assess the extent to which emerging powers are challenging the dominance of superpowers. [12 marks]

Why did the astronaut go on a vacation? She needed some space...
Global affairs often move at a rapid rate, so countries can gain or lose influence in a short space of time. Staying up-to-date with global affairs will help you to write about the challenges superpowers are facing and how they are coping with them.

Introduction to Development

Most people agree that human development is a good thing — but what this actually means is a whole lot trickier...

Human development means that things are improving

1) The term **human development** is used to **describe** a country or society's **improvement** over **time** as well as increasing people's **opportunities** e.g. employment and education. Most understandings of human development are based on the **idea** that the **more developed** a country is, the **better life** is for the **people** who live there.

2) Levels of development **vary** greatly between countries:

Aisha tried her best to stop others reaching their goals.

- A **developing** country has a **low** level of human development.
- An **emerging** country has a **medium** or **high** level of human development.
- A **developed** country has a very **high** level of human development.

3) **Governments** and **organisations** make **policies**, carry out **programmes** and invest lots of **money** to try to **promote** human development. This can take place **within** a country or can be carried out by a government or organisation from **one** country (often a **developed** country) in **another** country (often a **developing** country).

4) However, human development is a **contested** idea — people disagree on what it **means** for a country or society to improve. Depending on people's **political** and **cultural** viewpoints, they can have fairly different ideas on what the **main goal** of human development should be (e.g. **wealth**, **health** or **happiness**) and **how to** reach that goal.

GDP has been the traditional way of measuring human development

- **Gross Domestic Product** (GDP) is the **total value** of **goods** and **services** a country produces in a **year**. It tells you how **wealthy** a country is.

- GDP has **historically** been the **main way** that governments and development organisations have **measured** human development. The view was that increasing a country's income would lead to **all citizens** increasing their **wealth** and therefore their **human development**.

- They sometimes use **variations** of GDP too, e.g. **GDP per capita** (a country's GDP divided by the country's **population**) and **GDP adjusted for PPP** (purchasing power parity — how much you can **buy** for that amount of money in that country, **compared** to **other countries**).

- This has meant that people have understood an **increase** in a country's GDP to be the **main goal** of human development. As a result, many development policies have tried to promote **economic growth** — e.g. by encouraging **investment** in a country's **industries**.

	GDP (US$ billion, 2021)
USA	22 996
UK	3187
China	17 734
Nigeria	441
Brazil	1609

Some people have challenged the importance of GDP to human development

1) Many people have **criticised** this economy-focused approach to development because **economic factors** don't provide information about **quality of life**. This has led people to create **alternative** measures that include a **broader understanding** of what human development means.

2) The **Happy Planet Index** (HPI) doesn't consider a country's **wealth**. Instead, it looks at 'sustainable wellbeing'. It's calculated using **three different measures** of human development:

- Wellbeing — the level of happiness people feel about their life
- Life expectancy — the average age that a person can expect to live to
- Ecological footprint — the average impact that a person has on the environment

3) The **relationship** between wealth and other measures of development is **complex**. Wealthy countries tend to have **higher** life expectancies — but they **don't** always perform **well** in terms of **wellbeing** or **ecological footprint**.

4) For example, **Vanuatu** has a relatively **small GDP** but was ranked 2nd out of 140 countries in 2021 in the Happy Planet Index, partly due to its **high score** for **wellbeing**.

5) By contrast, **Luxembourg** is ranked 143rd on the HPI. Despite its **high scores** for **life expectancy** and **wellbeing**, it performs **poorly** on the HPI because of its **large ecological footprint**.

Introduction to Development

Bolivia's socialist government proposed a different model of development

Despite **criticisms** of how development is measured in **Western** countries, the **dominant** model for development is still based around the idea of **increasing** a country's **wealth**. However, this dominance has been **challenged** by **ideas** and **models** of **development** from other **cultures** and **political systems**.

Bolivia and 'good living'

1) Between the late 1970s and early 2000s, **Bolivia's economy** fluctuated between **growth** and **financial crisis**. The country had severe economic **problems** including high levels of **unemployment**. Responding to **pressure** from the **IMF** (see p.240) and other international organisations, in the early 2000s the Bolivian government **sold off** some of its **industries** and **resources** (this is called **privatisation**). This was meant to **promote economic growth**, but it was mostly **unsuccessful**.

2) **Evo Morales** was elected as **president** in 2005 and he promoted an approach to the country's problems that **challenged** the western **model of development** used by previous governments. His party **criticised** capitalist policies that promoted economic growth and **profit**, arguing that this model of development enables only a **few people** to **live** well.

3) Morales's model of development is based on **socialist** ideas (e.g. that resources should be **publicly** rather than **privately** owned and wealth should be spread equally among the population) as well as the **philosophy** of Bolivia's **indigenous peoples**.

A map showing the location of Bolivia, South America.

A Bolivian woman selling souvenirs on the Isla del Sol, Bolivia.

4) A key example of indigenous philosophy is '**Sumak kawsay**', an idea that means '**good living**'. According to this view, the **goal** of human development should be for **everyone** to live in **harmony** with **nature** as a **community**. This involves making sure that **natural resources** such as forests and water are being used without damaging them, rather than being **privately owned** and used to make a **profit**.

5) Morales's policies that **nationalised** Bolivia's **oil** and **gas** resources, and **redistributed** the ownership of **land**, demonstrated some of these **core ideas**.

6) This development model had some **big successes** — the **percentage** of Bolivians living **below** the international poverty line **decreased** from 15.6% in 2005 to 1.9% in 2019. Although GDP growth wasn't the main aim of Morales's government, GDP **increased** by **257%** during his presidency.

7) However, there were **problems** too — Bolivia still had high levels of **inequality**, and Morales's government was **criticised** for not doing enough to promote **gender equality** or protect **human rights**, e.g. thousands of people have been held in overcrowded prisons because of delays to trials.

Warm-Up Questions

Q1 Explain how the Happy Planet Index is used to assess human development.

Q2 Explain how some Western models of development are challenged by different models.

Exam Question — AS and A-Level

Q1 Study the table below. Suggest two reasons why a country can rank highly in one measure of development but poorly in another. [3 marks]

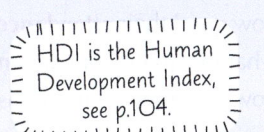

HDI is the Human Development Index, see p.104.

	GDP (US$, 2021)	HDI (2021)	HPI (2019)
Costa Rica	64.282 billion	0.809	62.1
USA	22.996 trillion	0.921	37.4
Nepal	36.289 billion	0.602	53.1

Develop your mental resources — learn these pages...

That might seem like more information than your poor, overloaded brain cells can handle. But don't panic — from here on, it all gets a bit more real and a bit easier to get your head around (well hopefully). Go on, stop loitering and be off with you.

Measures of Development

Development can be measured in a number of ways — education is a really important one, but it isn't valued everywhere...

Some people see **GDP growth** as a tool for **increasing quality of life**

1) Despite its **criticisms**, GDP has remained a **popular** way to measure development because many people believe that **increasing** a country's **wealth** will eventually **lead** to **improvements** in **other aspects** of human development.

2) **Hans Rosling** was a doctor and statistician from Sweden. He co-founded an organisation called **Gapminder**, which produces **animated graphs** of various measures of human development. The graphs show the **relationship** between a country's **income** and other **development indicators**, and how this **changes** over **time**.

3) According to Rosling, GDP growth **shouldn't** be the **goal** of development — instead, it's a **means** to other, more **important** improvements, such as **environmental quality**, **health**, **life expectancy** and **human rights**. However, Rosling argued that **economic growth** is the **best means** of **achieving** these goals.

Gapminder

- These Gapminder graphs show **infant mortality** (deaths of children before their first birthday, per 1000 births) against **GDP per capita** (adjusted for purchasing power parity), **by world region**, in 1975 and in 2015.

- The **size** of the **dot** reflects the **size** of the **country's population**, and each **continent** is grouped by **colour**.

- There's a **strong negative correlation** between **income** and **infant mortality** — the wealthier a country becomes, the fewer babies die — but there are also plenty of **exceptions** to this trend.

The graphs in this box are just examples of the kind of data available on Gapminder. These interactive graphs are best used on the website, so don't spend time trying to interpret the graphs here.

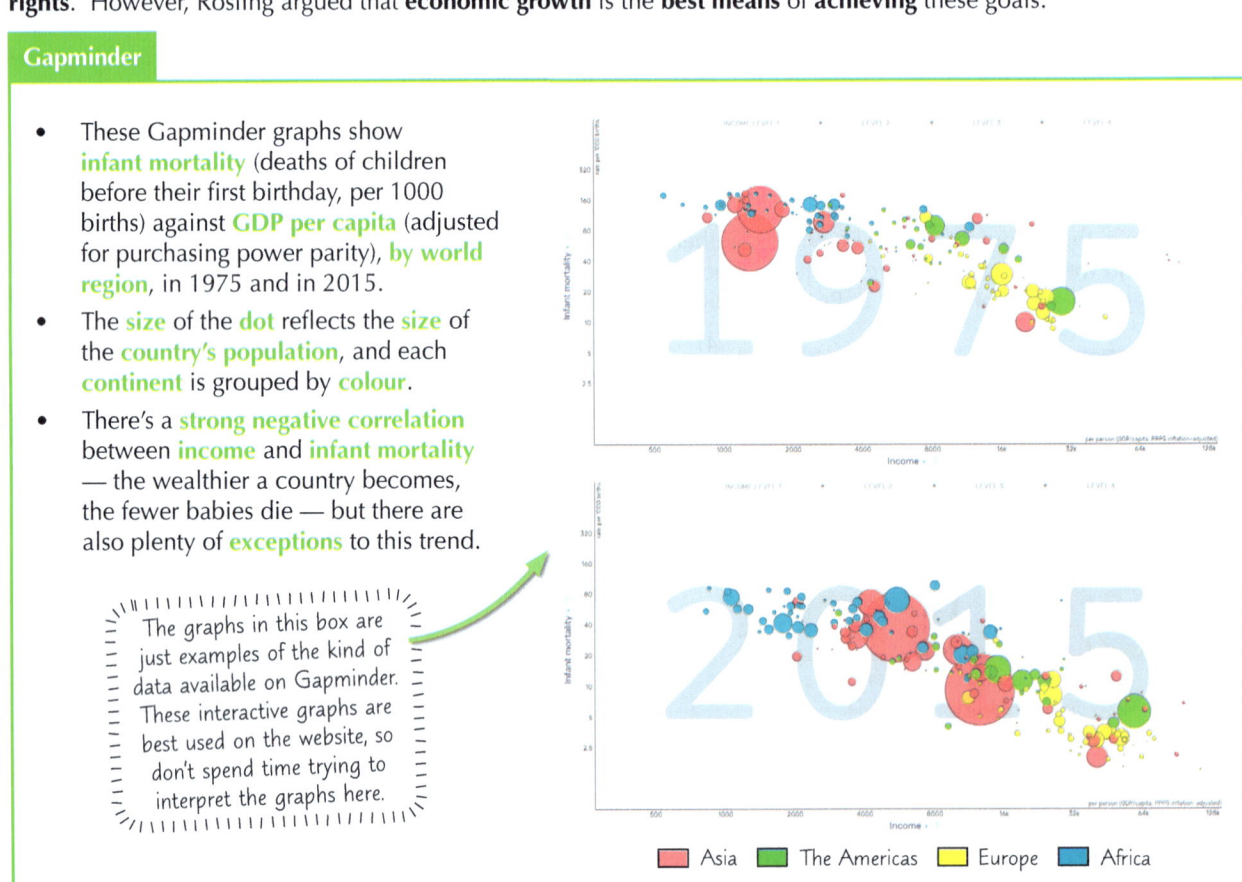

Asia   The Americas   Europe   Africa

Education is a key theme in **human development**

1) Like economic growth, **education** is seen as both a **means** to increasing human development, and a **goal** of development.

2) **Improvements** in education can be **measured** in lots of different ways, including:

- whether children have **access** to school
- how **good** their **attendance** is
- what the **quality** of **teaching** is like
- how **well** they do in **exams**
- how many **years** they **stay in school** for
- whether they go on to **further education**
- whether boys and girls have **equal access** to education
- the **quality** of the **facilities** available to pupils and teachers
- the **range of subjects** being taught

A girl attending school in Cambodia.

Topic 8: Option 8A — Health, Human Rights and Intervention

Measures of Development

Education helps develop a country's human capital...

1) **Human capital** is the **total amount** of **skills** and **knowledge** that a country's **population** has, and the **contribution** these make to the **human development** of the country.

2) Education **provides** a country with human capital by **giving** people the **knowledge** and **abilities** they need to be **economically productive** — to get **better paid**, more **highly skilled** jobs that help the economy **grow**.

3) This means that for a country to **achieve** economic growth, it also has to **invest** in **education** to make sure that in the **future**, the **workforce** will be able to **contribute** to the country's **economy**.

Human capital also includes good health.

...and allows people to **understand** and **assert** their **human rights**

1) Access to a **good education** can also help improve people's **access** to their **human rights** (see p.244). **Learning** about **rights**, **responsibilities** and the **law** teaches people about what their rights **are**, what this **means** and to **respect** other people's rights.

2) Education also gives people the **ability** to **find out more** about their rights (through being able to **read** and carry out **research**). In **situations** where people's rights are being **limited**, this combination of **awareness** and **skills** means that people are more likely to **challenge** the situation and get access to **support**, such as help from **human rights NGOs** (non-governmental organisations) and the **legal system**.

Mary was surprised to see her investment in education had led to visible signs of growth.

Not everyone **values education**, especially for **girls**

1) Children's **access** to education and the **quality** of education **varies** between and within countries. In many countries, **girls** have much **poorer** access to good quality education compared to **boys**. For example, around **one third** of primary-school-aged **girls** in **Pakistan** don't attend school, compared to **21%** of **boys**. Of the primary-school-aged girls attending school, **only 13%** of these are **still enrolled** at school by **9th grade** (children aged 14-15).

2) One cause of this is **attitudes** to **gender equality**. Many **cultures** have different **expectations** about the **roles** that men and women should have in **society**. In some cultures, it's seen as **less important** for girls to have a **formal education** than boys. This is because men are often expected to **work** and earn an **income**, whereas women are expected to provide **food** and look after the **home**. This is more **pronounced** in countries that don't provide everyone with **free** education as **poorer families** may have to make **difficult choices** about sending their children to school.

3) Organisations such as **UNESCO** (the United Nations Educational, Scientific and Cultural Organisation) are working to **promote gender equality** in access to education. UNESCO works with **governments**, encouraging them to pass **legislation** protecting girls' **right to education**. It also supports **programmes** that **challenge stereotypes** about gender, and aim to make schools **safer** and more **welcoming** for girls. UNESCO points out that gender equality in education is important for **human development** — for example, it can increase **economic growth** by adding to **human capital**.

Warm-Up Questions

Q1 Describe what is meant by human capital.

Q2 Explain how gender inequality can affect access to education.

PRACTICE QUESTIONS

Exam Question — A-Level

Q1 Evaluate the view that education is the most important factor in increasing human development. [20 marks]

All this talk of education got you exhilarated? I thought so...

Education is a very, very important thing. It can make a big difference to an individual's life and isn't to be sniffed at. I know school can seem a chore at times, but trust me, it'll be worth it in the long run. Right, pep talk over, now get learning...

Development and Health

Health is a key part of human development, but the health of populations, and the causes of poor health, are very uneven.

Health and life expectancy vary between and within countries

1) There are many things you can look at, for example how **common** certain **diseases** are. To understand how health **fits** into **human development**, you can also measure **trends** in the **population** that indicate how **healthy** people are, as well as whether they live in a **clean**, **safe** environment or have **access** to effective **health care**.

2) International organisations like the **UN** look at **life expectancy**, **infant mortality** and **maternal mortality** to measure how health **varies** between **countries** and between different **groups** of people **within** a country.

Measure	How it's calculated	UK	Zambia
Life expectancy at birth (2020)	The average age that a person can expect to live to.	81 years	64 years
Infant mortality rate (2020)	The number of babies who die under 1 year old, for every 1000 live births.	4	42
Maternal mortality ratio (2017)	The number of mothers who die from problems related to pregnancy or childbirth (up to 42 days after birth), for every 100 000 babies born.	7	213

Health is affected by medical care and people's lifestyles in developed countries

1) In **developed** countries, most people have **access to basic needs** (water, sanitation and food) and there are **lower rates** of **most infectious diseases** than in developing countries. However, there are generally **higher** rates of **non-communicable diseases** (illnesses that **can't be passed** from one person to another), e.g. heart disease and cancer.

2) There are **three main factors** that affect health patterns in **developed** countries:

Lifestyles
- **Risk** of **some diseases** (e.g. some **cancers** and **heart disease**) **increases** if you're **overweight**, eat **unhealthy food** or don't do enough **exercise**.
- These factors are **more common** in developed countries.

Deprivation
- **Wealthier** people can afford **larger** homes, more **nutritious** food, more **active** lifestyles and better **medical care**.
- This means they're likely to be **healthier** and have a **longer life expectancy** than a poorer person living in the same country.

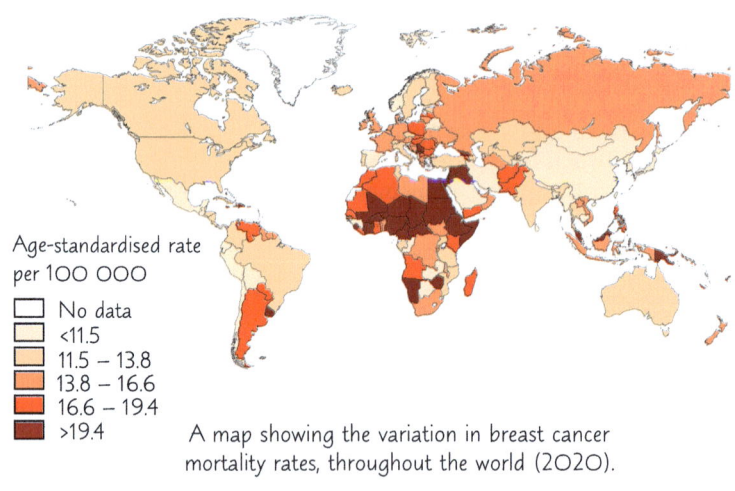

Age-standardised rate per 100 000
- [ ] No data
- <11.5
- 11.5 – 13.8
- 13.8 – 16.6
- 16.6 – 19.4
- >19.4

A map showing the variation in breast cancer mortality rates, throughout the world (2020).

Medical care
- In some countries, **free medical care** (paid for through general taxation) is available to **everyone**, whereas in others, people have to **pay** for **medical treatment** (which may be partly or fully covered by health insurance). For example, in the **USA**, receiving treatment for a **broken leg** could cost a patient without medical insurance **$7500**.
- This means that the **quality** of medical care that people receive depends on how much **money** they have. **Poorer** people might have to make **choices** about whether to get treatment at all.
- Even in countries with **free universal health care**, like the UK, the **quality** of medical care can **vary** between **local areas**. **Wealthier** people can **choose** to pay for **private** medical care, which can be **quicker** and have **better quality** facilities.

In countries without free health care, many people have health insurance that is often paid for by their employer.

Development and Health

In **developing** countries, **health** is mostly affected by **access to basic needs**

1) **Developing** countries tend to have **higher** levels of illness and death than **developed** countries. This is due to higher rates of **infectious diseases**, e.g. cholera, tuberculosis and HIV/AIDS.

2) The **patterns** of where these diseases occur can often be **explained** by **uneven access** to **basic needs**. For example, diseases like **cholera** spread **quickly** in places where people don't have access to **sanitation** (toilets and washing facilities) or a **clean water supply**. This contributes to **infant mortality** — it's estimated that **one** in every **five deaths** of **newborn babies** could be **prevented** if they were washed in **clean water** and looked after by someone with **clean hands**.

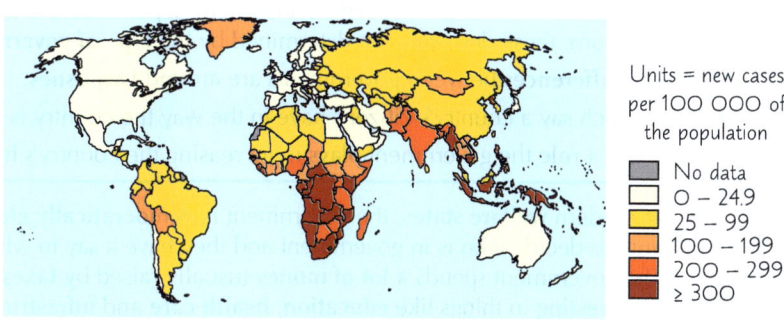

Units = new cases per 100 000 of the population

▨	No data
▨	0 – 24.9
▨	25 – 99
▨	100 – 199
▨	200 – 299
▨	≥ 300

Disease rate of tuberculosis (TB) in the population (2017)

3) People who don't have **access** to enough **nutritious food** are also more **vulnerable** to disease — they're **more likely** to **get** an illness and more likely to **die** when they become ill. Similarly, **malnutrition** can increase **maternal mortality** — **pregnant women** who can't get the **nutrients** they need in their diet are at a **greater risk** of **dying** in childbirth.

Inequality drives **variations** in **health** within countries

The UK

1) There are big differences in the **health** and **life expectancy** of people in **different regions** of the **UK** and people with different **income levels**.

2) People living in **wealthier** areas tend to **live longer** than those in more **deprived** areas. Boys born between 2012 and 2014 in the wealthy London area of **Kensington and Chelsea** have a life expectancy of **83** years, whereas boys born in the same period in **Blackpool** are only estimated to live until **75**.

3) This reflects England's **North-South divide** — the major **inequalities** in **wealth** and **investment** in infrastructure and services between the generally more affluent **south** and the more deprived **north** of the country.

4) Health and life expectancy also vary between people from different **ethnic groups**. For example, people from South Asia or of South Asian descent are **more likely** to die from **coronary heart disease** than White British people.

5) People from most ethnic minorities are **more likely** to live in **deprived** areas than **White British** people, meaning that they may be less likely to afford **healthy lifestyles**.

Australia

1) **Australia** was **colonised** by European **settlers** from 1788. The settlers took **land** inhabited by **Aboriginal peoples**, many of whom were forced to live on reserves. They were **marginalised** and faced **discrimination**, e.g. they were not allowed to vote until 1962.

2) From the 1890s to the 1970s, Aboriginal children were taken from their families and raised by **white parents** or in **children's homes**. Many were **harshly treated** and some suffered **abuse**.

3) The **impacts** of these policies are still **felt today**, e.g. **income** is far **lower** amongst Aboriginal than non-Aboriginal peoples. Tobacco and alcohol use are **more common** amongst Aboriginal peoples.

4) These factors impact health, e.g. from 2015-2017, the life expectancy of Aboriginal males was **8.6 years less** than non-Aboriginal males.

5) Many Aboriginal people have suffered mental health problems. In 2020, **52%** of Aboriginal people suffered from **depression** compared to **33%** of non-Aboriginal Australians.

CASE STUDY

Warm-Up Questions

Q1 Name three basic needs that affect variations in health in developing countries.

Q2 How does deprivation affect life expectancy in developed countries?

Exam Question — A-Level

Q1 Explain why there are variations in health and life expectancy in the developed world. [8 marks]

I hope UN-joyed revising development and health...

...what's that? You didn't? Well you've got to learn it anyway, I'm afraid. It may not seem the cheeriest of topics, but it's on the syllabus so you don't really have a choice... Make sure you've got those case studies pinned down too while you're at it.

Increasing Development — Role of Government

When it comes to increasing a country's development, the big players are its government and international organisations.

Different types of governments have different approaches to development

1) The **way** a country is **run** has a huge **impact** on how it **develops**. Whether or not a country's **economic development** contributes to its **social development** depends on the **decisions** that its **government** makes. In turn, these decisions are determined by the **type of government** the country has.

2) The main **differences** between governments are around **two issues**:
   - how much **say** a country's citizens have in the way the country is run.
   - how big a **role** the government **plays** in increasing the country's **human development**.

> ⎧ In a lot of totalitarian
> states, a big proportion
> of government spending
> goes towards building
> military strength. ⎭

   - In modern **welfare states**, the government is **democratically elected** — the citizens **vote** to decide **who** is in government and they have a **say** in what the government **does**. The government spends a **lot of money** (usually raised by **taxes**) on **social development** by **investing** in things like **education**, **health care** and **infrastructure**.

   - In a **totalitarian state** or **dictatorship**, the government is run by elites and has **complete control** over the **population** — **ordinary people** have **no say** in how the country is run. If the government chooses **not** to spend money on **social development** (**education**, **health care** and **infrastructure**) and instead use the country's wealth in the **personal interests** of the country's **leaders**, there's **not much** that people can do about it.

3) Most governments are somewhere **between** these two **extremes**, but still **vary** in terms of the **total amount** of money that a government **spends** and the **proportion** of government spending that goes towards **social development**.
   - For example, **Sweden** and the **UK** are both democracies, but in 2019 Sweden spent **7.6%** of its GDP on **education**, compared to **5.2%** of GDP in the UK. **Turkmenistan**, which is a **dictatorship**, spent **3%** of its GDP on education in 2019.

Intergovernmental organisations play a big role in development

1) Besides governments, the other **big players** in development are **intergovernmental organisations** (IGOs) that work **with** and **between** national governments. They help to create **rules** and **policies** that affect countries' **economic development**. They also provide **funding** and **advice** that **shape** governments' **decisions** about **economic** and **social development**.

2) The **three big IGOs** in development are:

1 The International Monetary Fund (IMF)
   - The IMF aims to encourage **global financial stability**. It does this by **monitoring** the global economy and **advises** governments on how they could improve their economic situation through economic policy.
   - It also gives **loans** to countries with **economic problems**.

> ⎧ The World Bank also
> offers advice and expertise. ⎭

2 The World Bank
   - The World Bank provides **loans** and **grants** to developing countries to invest in areas like **health**, **education** and **infrastructure**.
   - The money for the loans comes partly from **subscriptions** from its member countries — **all** members pay in, but **only** those who **need** it can receive a loan.
   - This means funds are **redistributed** from developed countries to less developed countries. However, countries are expected to eventually **pay back** loans.

Penny was regretting saying yes to standing in for the world bank.

3 The World Trade Organisation (WTO)
   - The WTO was set up to open up trade, help **negotiate trade agreements** and **settle trade disputes** between member countries.
   - It sets **rules** about how countries should trade with each other. For example, countries should promote **free trade**, e.g. by **removing** as many **barriers to trade** as possible, such as stopping countries imposing **unfair tariffs** on each other's companies.

Increasing Development — Role of Government

In the past, **IGOs** were keen on **economic** development and **neoliberal** policies...

1) Historically, **IGOs** have focused on **economic development**. They've promoted **neoliberal ideas** about how economic development can be **achieved**.

2) **Neoliberalism** argues that economic development happens when the **economy** is allowed to operate with **minimal interference** from government. This should create **ideal conditions** for companies to compete and create **profits**. In theory, this wealth will 'trickle down' through society leading to **development**. This reduces the reliance on government spending for social development.

3) Neoliberalism has a few **central ideas**, including:

Free Trade	• Some countries **limit trade** using **tariffs** and **other barriers** to shield their **industries** from **foreign competition**. • **Free trade** is the **policy** of **removing** these barriers.

Deregulation of financial markets	• **Deregulation** of financial markets means **removing laws** and other **controls** that **restrict** what people can do with their **money**. • For example, governments removing **limits** on how risky an investment a bank can make, to try to make the **economy** more **efficient**.

Privatisation	• Privatisation is about moving the **control** of **services** and **assets** (e.g. property or resources) from the **state** to **private companies**. • For example, the **selling** of state-owned companies or giving a private company a **contract** to run a hospital.

Barry was disappointed when he realised there was no loan on its way for all the tree sap he'd spent weeks collecting...

4) Beginning in the **1980s**, the IMF and the World Bank put **conditions**, e.g. free trade and reducing government spending, on their **loans** that meant that the country **receiving** the loan had to follow **neoliberal policies**. These conditional loans were known as **Structural Adjustment Programmes** (SAPs).

5) Many **developing** and **emerging** countries, including **Bolivia** in the early 2000s (p.235), made significant changes to their economies due to SAPs. These often involve **selling off** state-owned **resources** and **industries** as well as **removing regulations** and **restrictions** around trade and finance.

> Austerity policies are implemented by a government in order to reduce the country's debt. Examples of austerity policies include cutting public sector pay and reducing healthcare spending.

6) These policies have been **controversial** — critics argue that they've **increased poverty** and **inequality** in the countries receiving the loans (as a result of austerity policies), and have promoted the **interests** of **TNCs** rather than increasing **human development**.

...but recent IGO programmes look at **environmental** and **social development** too

1) In recent years, the IMF has **questioned** its **neoliberal approach** to development.

2) Increasingly, the IGOs' development programmes and loan conditions promote **environmental sustainability** and **social development** rather than just **economic development**.

3) They include a growing focus on **environmental quality**, **health**, **education** and **human rights**.

The World Bank

• For example, the **World Bank** now has a **Global Platform for Education Finance**, a **Human Capital Project** and a **Global Program on Sustainability**.

• The Global Platform for Education Finance aims to help make sure that countries (especially low-income countries) are receiving **sufficient funding** for education and that **resources** are being **managed efficiently**.

• The Human Capital Project aims to invest in the **productivity of people** by focusing on improving their **knowledge**, **skills** and **health**. The hope is this will also help to **end extreme poverty** as well as increasing **economic growth**.

The World Bank, Washington DC

• The Global Program on Sustainability encourages countries and private companies to create **policies** and make **investments** that promote **sustainability**. This is done by offering them **advice** and **financial incentives**.

Increasing Development — Role of Government

The **Millennium Development Goals** aimed to **reduce poverty**

- The **Millennium Development Goals** (MDGs) were created by the **United Nations** in 2000.
- They aimed to increase **equality** and **sustainability**, eradicate **global poverty** and close the **development gap** between the **richest** and the **poorest countries** in the **new millennium**.
- All UN member states **agreed** to try to achieve the **eight goals** by **2015** and **progress** towards these would be **monitored** each year:

189 countries signed the declaration in 2000.

The Millennium Development Goals

Goal 1 — Eradicate extreme poverty and hunger
Target: halve the number of people living in extreme poverty or suffering from hunger.

Goal 2 — Achieve universal primary education
Target: make sure that all children had a primary education.

Goal 3 — Promote gender equality and empower women
Target: increase the number of girls and women in education and in paid employment.

Goal 4 — Reduce child mortality
Target: reduce death rates in children under five years old by two-thirds.

Goal 5 — Improve maternal health
Target: reduce death rates amongst women caused by pregnancy or childbirth by three-quarters.

Goal 6 — Combat HIV/AIDs, malaria and other diseases
Target: stop the spread of major diseases, including HIV/AIDs and malaria.

Goal 7 — Ensure environmental sustainability
Target: protect the environment and make sure development is sustainable, while improving quality of life.

Goal 8 — Develop a global partnership for development
Target: make sure that countries around the world worked together to help developing countries.

- Each goal included several additional **targets**. For example, **MDG 7** (ensure environmental sustainability) included targets to **halve** the number of people without **clean drinking water** and **basic sanitation**.

The level of **success** of the **MDGs** varied between countries

1) **Progress** towards the MDGs was **mixed**. At a **global** scale, several of the targets were **achieved**. For example, the number of people living in **extreme poverty** fell by more than **half** between 2000 and 2015.

2) However, at a **regional** and **national** scale, progress has been **uneven**. The fall in global numbers of people in poverty was partly due to **improving conditions** in some parts of the world, especially **China** and **South Asia**, whereas the number of poor people in Sub-Saharan Africa actually **increased** (although between 1996 and 2005 the number of people in poverty in Sub-Saharan Africa **decreased** by 8%).

3) Some **countries** and **regions** performed well on **some** targets and goals but **poorly** with **others**. For example, India was close to succeeding in **halving** the proportion of its population living in **extreme poverty**, but **struggled** to reduce **hunger** — the proportion of malnourished children declined slowly, from 53.5% to 40% between 1990 and 2015.

Since 2022, extreme poverty has been defined as living on less than US $2.15 per day.

4) Below are examples from **four** different regions of the world on the **progress** they made towards meeting the MDGs:

	MDG	Southern Asia	Latin America and the Caribbean	Oceania	Sub-Saharan Africa
1	Poverty and hunger	Good	Good	Mixed	Some
2	Primary education	Good	Some	Good	Some
3	Gender equality	Good	Good	Mixed	Good
4	Child mortality	Good	Good	Some	Good
5	Maternal health	Good	Some	Some	Some
6	Disease	Good	Good	Mixed	Good
7	Environmental sustainability	Good	Good	Limited	Mixed
8	Development	Some	Good	Good	Good

Increasing Development — Role of Government

The MDGs have been replaced by the Sustainable Development Goals

1) The **Sustainable Development Goals** (SDGs) were adopted by the UN's member states in 2015 to **replace** the MDGs.

2) They're **similar** to the MDGs, but they have a bigger focus on **environmental sustainability**. This means they try to **ensure** that **human development** and **economic growth** are achieved in ways that don't cause **environmental damage**. This is to ensure goals are **met** without **compromising** the human development of **future generations**.

3) The SDGs are more **ambitious** than the MDGs. For example, MDG 1 aimed to **halve poverty**, whereas SDG 1 aims to 'end poverty in all its forms everywhere' by 2030. There's also **more of them** — there are 17 SDGs.

4) **Seven** of the goals carry on the **main aims** of the MDGs, although some of them also include sustainability:

- End **poverty**.
- End **hunger**, increase **food security** and make **agriculture** more sustainable.
- Enable people to live **healthy lives** and promote **wellbeing** for everyone.
- Provide everyone with a good quality **education**, including opportunities for **adults**.
- Reach global **gender equality**, **empowering** women and girls everywhere.
- Provide access to **clean water** and **sanitation** for everyone.
- Work together in a **global partnership** for sustainable development.

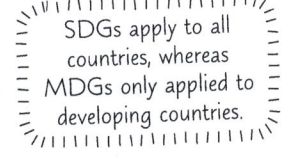
SDGs apply to all countries, whereas MDGs only applied to developing countries.

5) Unlike the MDGs, some of the SDGs talk explicitly about **economic development** and its **relationship** with other aspects of **human development**:

- Achieve sustainable **economic growth** that provides **decent jobs** for everyone.
- Decrease **inequality** between and within countries.

6) **Some of the goals** look at how **human activities** can be made **fairer** and **more sustainable**:

- Provide access to **affordable**, **clean energy** for everyone.
- Make **industries** and **infrastructure** more **sustainable**.
- Make **cities** more sustainable, inclusive, resilient and safer.
- Make global patterns of **consumption** and **production** more sustainable.
- Promote **peace**, **justice** and **strong institutions**.

I'm a big fan of renewable energy.

7) Other SDGs focus much more on the **environment**:

- Tackle **climate change** and manage its impacts.
- Protect the **oceans** and use **marine resources** sustainably.
- **Protect** and **restore ecosystems** on land, stop **biodiversity loss** and reduce **deforestation** and **degradation**.

8) Like the MDGs, **progress** on the SDGs has been **mixed**. At the start of 2020, the world was moving **too slowly** towards many of the goals to meet them by 2030. There are also concerns that the widespread impacts of the **COVID-19 pandemic** and **climate change** are **slowing the progress** towards achieving the SDGs even further.

Warm-Up Questions

Q1 Explain what a totalitarian state is.

Q2 Describe three main ideas of neoliberalism.

PRACTICE QUESTIONS

Exam Question — A-Level

Q1 Evaluate the view that IGOs have been effective in deciding the goals of development. [20 marks]

So the IMF and the WTO are both IGOs — got it...

Yep, all those acronyms can be a tad confusing. Here's an idea — write down a list of acronyms from this section, close the book and see if you can remember what each of them stands for. You'll get to know your MDGs from your SDGs in no time.

Human Rights

Human rights — they're the basic principles and values that belong to every human being in the world.

Everyone is entitled to human rights

1) **Human rights** are important for **guaranteeing** the **freedoms** that **all human** beings are **entitled** to.

   - The concept of human rights isn't a new idea. Throughout history, some societies and groups of people have fought to have their rights recognised by those in power, e.g. women's right to vote or workers' rights to fair pay and hours.
   - However, it was only in the 20th century that the idea of a universal set of rights (rights that apply to everyone, everywhere) became widespread. Human rights were agreed on through international agreements and in some cases protected by law.

2) Despite these agreements, the human rights of **many people** around the world **aren't** being **met**.

The Universal Declaration of Human Rights was the first global agreement

1) After the **Second World War**, many countries wanted to **make sure** that the **terrible abuses** that happened during the conflict never happened again. The **United Nations** (UN) was set up in **1945** to establish a **peaceful** and **fair** world.

2) The UN established a **Commission on Human Rights** to agree a set of **universal human rights** between the UN's **member countries**. The idea was that **setting out** these rights, and **protecting** them, would promote **peace** and **fairness**. The commission drew up a **list** of rights, which were **adopted** by the UN in **1948** as the **Universal Declaration of Human Rights** (UDHR).

> ### The UDHR
>
> 1) The UDHR states that **all humans** have **equal rights** that are **inalienable** (they can't be taken away from you) and **should be protected by law**. The **30 articles** that make up the UDHR include the right to **life**, the right to freedom from **slavery** or **torture** and the right to **education**.
>
> 2) The UDHR isn't **legally binding**, so it can't be used to **enforce** people's rights. However, some countries have **included** parts of the UDHR in their **national constitutions** and many people have used the Declaration to **condemn** human rights abuses.
>
> 3) However, **not everyone** views the UDHR as **universal**:
>   - Although the Declaration is described as 'Universal', not **everyone agrees** with it. When it was adopted by the UN, some member countries, e.g. the **Soviet Union** and **Saudi Arabia**, **abstained** from voting on it.
>   - **Critics** of the UDHR argue that it's **biased** towards **Western culture** and **values**. By ignoring the fact that people in **different societies** have different **beliefs** and **moral values**, the UDHR can be seen as a way of **imposing** Western ideas on the rest of the world. Saudi Arabia chose to abstain from the Declaration because it felt that some of the articles (e.g. the right to **freedom of religion**) went against **Islamic teachings**.
>   - People have also **challenged** the way that the UDHR has been used as a **political tool**. Since the 1990s, some countries have used the UDHR to **justify** intervening in the **affairs** of countries that are **infringing** on their citizens' **human rights**. These interventions have usually involved taking **military action** or applying **economic sanctions** on trade, aid and investment in the country.

The European Convention on Human Rights lists human rights in Europe

1) Like the UDHR, the **European Convention on Human Rights** (ECHR) was created in the **aftermath** of the **Second World War** to make sure that people's **rights** would be **protected** in the **future**. It was signed by members of the **Council of Europe**, which had 12 members at the time, in **1950** and came into force in **1953**.

2) The ECHR has **18 articles**, some of which were taken from the **UDHR**. They include the right to a **fair trial** and the right to not be **discriminated against** because of characteristics like **sex**, **race** or **religious beliefs**. Some articles are **specific** to **Europe**, e.g. the abolition of the death penalty.

3) Unlike the UDHR, the ECHR **is legally binding**. The Council of Europe set up the **European Court of Human Rights** in 1959. This meant that **European citizens** who felt their human rights were being **infringed** could bring a **legal challenge** to the **court** and the ECHR could be **enforced**.

4) In 2022, Russia withdrew from the ECHR. Some people in Russia felt that the ECHR **undermined** the **Russian constitution** and its **national sovereignty**.

Human Rights

The ECHR was brought into UK law via the Human Rights Act

1) The UK incorporated **most of the rights** in the ECHR into its own **national law** in 1998 when it passed the **Human Rights Act**. This made it **easier** for UK citizens to **defend** their human rights, as they could bring a legal challenge to a **court** in the **UK** rather than needing to go to the **European Court of Human Rights**.

2) Under the Human Rights Act, **every law** that's passed in the UK has to **respect** human rights. **Public authorities** like **councils** and **schools** also have to take human rights into **account**.

3) Some people are **critical** of the Human Rights Act because they feel it gives the European Court of Human Rights too much **sway** over the UK's **laws**. They believe this **undermines** the UK's **sovereignty** — its power to make its own **laws** and **decisions** as an **independent country**. This view became more **popular** during debates around **Brexit** and the relationship between the UK and Europe.

European Court of Human Rights, Strasbourg, France

The Geneva Convention protects people's human rights during conflict

1) The ideas behind the **Geneva Conventions** date back to **1864** when the founder of the **Red Cross** set up a **treaty** to protect the rights of **wounded soldiers** and **civilians** who were caring for them. The Geneva Convention was set up by the **International Red Cross** in **1949** (in the aftermath of the Second World War) and has now been **signed** by **196 countries**.

2) The Geneva Convention protects people's rights during **war** and **conflict**. It tries to **regulate** how war is carried out, to **minimise** the harm done to **civilians** and **prisoners of war**. It bans actions such as **terrorism**, **torture**, taking **hostages** and **sexual assault**.

The ICC building in The Hague, the Netherlands

3) The Geneva Convention provided the **basis** to set up the **International Criminal Court** (ICC), at The Hague in the Netherlands. The ICC holds **trials** for people accused of crimes that concern the whole international community, e.g. **war crimes** and **genocide**. For example, the former **Liberian president** Charles Taylor was tried by the ICC for **war crimes** committed in Sierra Leone's civil war, such as recruiting **child soldiers**.

4) However, **many cases** of war crimes don't come to trial at the ICC as it can be difficult to get enough **evidence** to **prove** whether the crimes were committed. **Not all** of the countries that have signed the Geneva Convention are members of the **ICC** — there are currently **123 member states**.

5) Although **most** of the world's countries have **signed** the Geneva Convention, this doesn't mean that they all **abide** by it. For example, **torture** is still reportedly used in around **140 countries**.

Warm-Up Questions

Q1 Outline two similarities and two differences between the UDHR and the ECHR.

Q2 Explain why the ECHR is controversial in the UK.

PRACTICE QUESTIONS

Exam Question — A-Level

Q1 Explain how the Geneva Convention is used to protect human rights. [6 marks]

Human rights — it's heavy stuff...

Don't get your letters in a twist. It's easy to get confused by all the different agreements. But if you remember the UDHR was the first global agreement on human rights and the ECHR is specific to Europe, you'll be part of the way there at least...

Human Rights — International Variability

People's ability to have their human rights fulfilled isn't the same everywhere, it varies a lot between countries.

Some countries are big advocates of human rights

Some countries are **advocates** of human rights — they **speak up** about human rights in **international forums** like the **G7**.

The UK — CASE STUDY

1) The UK **positions** itself as a **world leader** in **defending** human rights, especially **democracy** and **freedom of speech**. The UK government has **criticised** actions by other countries that have **compromised** human rights, such as moves by the **Chinese government** to reduce **freedom of speech** by introducing **censorship laws** in **Hong Kong**.

2) In 2020, the UK passed a **law** to allow it to impose **sanctions** on countries that are **abusing** the **human rights** of their **citizens**, called the **Global Human Rights Sanctions Regulations**.

3) At a G7 meeting in 2022, the UK and other G7 members **condemned** Russia's **invasion** of Ukraine. The G7 believes the invasion is **illegal** and **unjustifiable**.

> The G7 is a group of seven countries (that together account for over 50% of global net wealth) whose aim is to solve global problems.

Other countries favour economic development over human rights

1) For other countries, **economic development** is a **bigger priority** than protecting human rights.

2) Some countries **defend** this position in **international forums**, arguing that they need to meet their citizens' **basic needs** before worrying about meeting all of their **human rights**. They point out that some human rights (e.g. the right to **life** and the right to **education**) rely on **human development**, which is often **dependent** on economic development.

India — CASE STUDY

1) For example, in **India**, the exploitation of **natural resources** has **threatened** human rights. The development of **coal mines** in **Chhattisgarh** has involved **evicting indigenous communities**.

2) Some members of these communities were **imprisoned** for holding **protests** about the evictions and have accused the police of **mistreatment**. However, **coal** is central to India's **economic development** — it's the world's second biggest **consumer** of coal.

3) Government ministers argue that coal mining will bring **wealth** and **development** to indigenous communities, as well as to India as a whole.

Superpowers and emerging powers all have different levels of democracy

1) In some **superpowers** and **emerging powers** (see p.211), economic changes have been accompanied by a **transition** from **totalitarian regimes** and **dictatorships** (see p.240) to more **democratic** governments. Emerging powers like **Brazil**, **Mexico** and **South Africa** have become **democracies** within the last 50 years.

2) However, how **easy** it is for people to **exercise** their **rights** varies, even between democracies.

3) For example, **freedom of speech** means that people can **share** and **exchange** ideas and information **freely**, without **interference** from the government.

4) In less democratic countries, the government **censors** some information — for example, **limiting** people's ability to **criticise** the government on **social media** or in **newspapers**.

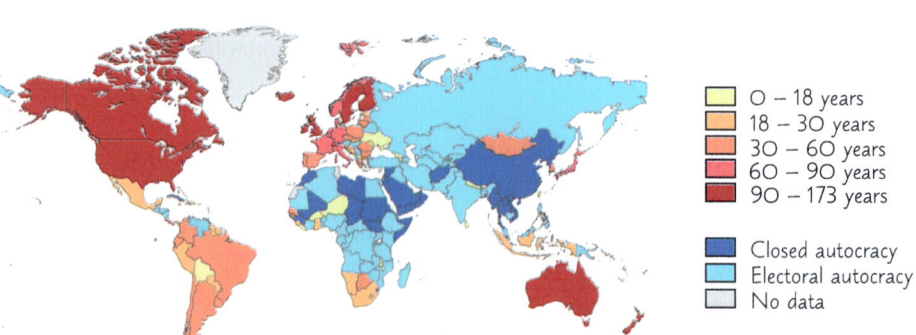

Age of democracies throughout the world, 2021

- ☐ 0 – 18 years
- ☐ 18 – 30 years
- ☐ 30 – 60 years
- ☐ 60 – 90 years
- ☐ 90 – 173 years

- ☐ Closed autocracy
- ☐ Electoral autocracy
- ☐ No data

> In closed autocracies, people don't have the right to elect the head of state. In electoral autocracies, people do have the right to elect the head of state but they still have limited freedoms.

Human Rights — International Variability

China is an example of an authoritarian system of government

1) In an **authoritarian** system, power is centralised and the **government** has complete (or nearly complete) control over people's **lives** and **behaviour**.

2) China is an **authoritarian** state. It's been ruled by the **Chinese Communist Party** since 1949. Until 1976, it was led by Chairman Mao, who used his **absolute power** to carry out huge **social** and **economic changes**. Since Mao's death, China has **opened** its **trade** to the rest of the world, but continues to be run as an **authoritarian** state. It's quite **different** from a **democratic** country, e.g. **Germany**.

	China	Germany
System	Authoritarian	Democratic
Governing party (in 2022)	Chinese Communist Party	Social Democratic Party (in a coalition with other parties)
Are elections held?	Yes — but only high-ranking officials can vote in the presidential election	Yes — everyone over 18 can vote in elections for the German parliament
Freedom of speech	Limited — the government uses its 'great firewall' to restrict access to information on the internet. Laws about criticising the government mean that many journalists and reporters censor themselves to avoid getting into trouble.	Free — freedom of speech and freedom of the press are protected by Germany's constitution.
Other human rights	International bodies have criticised China's record on human rights, e.g. the Chinese government has been criticised for its treatment of ethnic and religious minorities, such as Tibetans and Ugyur Muslims.	Human rights are strongly protected, e.g. the German government accepted large numbers of refugees in 2015-17, although there was also a rise in the number of attacks on refugees in this period.

Corruption can threaten human rights

1) **Corruption** is when people in authority **misuse** their power to serve their own **personal interests**, e.g. **politicians** giving **jobs** to their friends or family, or **public servants** taking **bribes**.

2) The **Global Corruption Index** measures the risk of corruption occurring in a given country. It's calculated by looking at several factors, including people's reports of experiencing corruption. The risk of corruption is often **lower** in more **democratic** countries.

3) Corruption can allow people to carry out **illegal activities** without facing

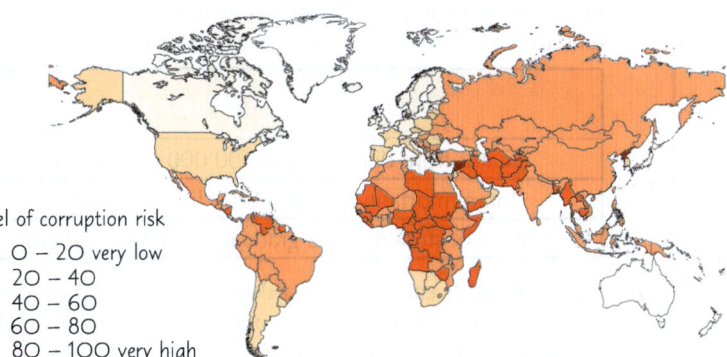

Level of corruption risk in countries throughout the world, 2022

Level of corruption risk
- 0 – 20 very low
- 20 – 40
- 40 – 60
- 60 – 80
- 80 – 100 very high

the **consequences**, e.g. by paying the police a bribe to look the other way. This can **threaten human rights** as people can infringe on others' rights without worrying about being arrested.

Warm-Up Questions

Q1 Explain what is meant by freedom of speech.
Q2 What is the Global Corruption Index?

Exam Question — A-Level

Q1 Evaluate whether governments are justified in pursuing economic development rather than human rights.

[20 marks]

Learn this lot and you're done*...
*with the international variability of human rights, that is.

There are quite a few case studies on these pages, but it's really important you learn them all. It'll give you the best chance of doing well in the exam. I know it's tempting to skip over them, but you'll be glad to have them all stored in your noggin later.

Topic 8: Option 8A — Health, Human Rights and Intervention

Human Rights — National Variability

Human rights are meant to be universal, but access to them often varies within countries. Let's take a closer look...

People's **access** to their **human rights** varies **within** countries

1) Within a country, not everyone has the **same rights**, or the same **ability** to **access** their rights. Members of certain social groups, e.g. **ethnic** or **religious minorities**, often have access to **fewer rights** than the social group that's in **power**. **Systemic discrimination** is when more **powerful** people **ignore** or **restrict** the rights of certain groups because of their characteristics, e.g. **gender** or **race**. This can be **intentional or unintentional**.

2) These **differences** in rights can be especially pronounced in **post-colonial countries**, where there's often a history of **one** social group **dominating** others. In some countries, this has led to **violence**:

   - For example, Rwanda has three main ethnic groups — Hutu, Tutsi and Twa. Rwanda was originally colonised by Germany, before being transferred to Belgium after the First World War. The Tutsi were viewed by the Europeans as superior and therefore were given more power. Under Belgian rule, special rights (e.g. education) were given to Tutsi people but not to Hutus. This led to growing resentment among some Hutu people which resulted in a rebellion that forced numerous Tutsis to flee the country.

   - In 1994, an extremist Hutu group started a genocide against Tutsis and moderate Hutu people. Twa people weren't eligible for the survivor benefits provided by the government (only Tutsi people were). Twa people are also unable to claim particular rights, resources and representation based on being an ethnic minority.

A **lack of rights** often means **lower human development**

Social groups with **fewer rights** often have **poorer** levels of **health** and **education** than social groups with more rights. The systemic discrimination that reduces their **rights** can also **limit** people's access to **services**. For example, the **government** may choose to **invest less** money in **infrastructure** and **services** that are used by these groups.

Indigenous peoples in North America

CASE STUDY

1) Indigenous peoples in North America have experienced **prejudice** and **discrimination** since European colonisation. Around **1 million** Indigenous peoples now live on **reservations** with some rights to **self-governance**. However, land in some reservations is **poor quality** or contaminated, with **few resources**.

2) **Schools** and **healthcare facilities** in these areas are **under-funded**, and the remoteness makes it harder to provide and access the facilities people have a right to. Indigenous peoples tend to have **worse** health and lower educational achievement than the **national average**:

Human Development Indicator	Indigenous American /Alaska Native	USA, all ethnic groups
Deaths from diabetes, per 100 000 people, 2019	41.3	26.7
Infant mortality rate per 1000 live births, 2019	7.9	5.6
Percentage of students who graduate from high school, 2019	74	86

3) A **lack of rights** can make it **harder** for these groups to **protest** about the lack of support for human development in their communities. It may also be more difficult for them to use their **freedom of speech** and rights to **assemble** and **protest**. For example, in the USA, Indigenous populations are **more likely** to be **arrested** or **killed** by the **police** than **white** people.

People have **campaigned** for **equal human rights**

1) In some countries, **oppressed groups** have **campaigned** for their human rights to be **recognised** and **realised** by the state. For example, the **civil rights movement** in the **USA** campaigned for **African American** people to have **equal rights** to white Americans.

2) These campaigns have been more **successful** in some countries, and for some groups, than others. In some cases, **campaigns** led to a **change in the law** — for example, in the UK, the **2010 Equality Act** made it illegal for men and women to be paid different amounts for doing the same work. However, many argue that more still needs to be done, given that there was a **gender pay gap** of **over 8%** in 2022. In other cases, campaigns have become an **important part** of the country's **history**, e.g. campaigners from the **women's suffrage movement** in the UK are now **celebrated** as national **heroes** e.g. Emmeline Pankhurst.

A civil rights march in Washington D.C, USA (1963)

© U.S. National Archives and Records Administration

Human Rights — National Variability

Women are campaigning for gender equality in Afghanistan

Gender equality — Afghanistan

CASE STUDY

1) **Afghanistan** has been involved in **conflict** for around 40 years (including civil wars and the US-led intervention in 2001). Between 1996 and 2001, a large proportion of the country was controlled by the **Taliban**.

2) The Taliban is an organisation that applies a very **conservative** understanding of Sharia Law (the law of Islam). This includes enforcing very strict **rules** about what women are allowed to do. For example, many women don't have access to an **education** and are less likely to be given senior positions in **government**. Women also experience high levels of **violence**.

3) In 2001, the Taliban regime collapsed following a US-led military intervention. In the years that followed the **removal** of the **Taliban** from **power**, some **progress** was made in improving women's rights in Afghanistan. For example, in 2001, very few girls were attending school, whereas by 2012 almost 3.2 million girls were in school. However, progress was still **slow** as it took time to remove the laws put in place by the Taliban, and many girls still **dropped out** of school **before** completing their **education**.

4) In August 2021, the Taliban **seized control** of several cities in the country, including the capital Kabul, and the Afghanistan **government collapsed**. This has led to an **increase** in the **violence** and **discrimination** experienced by women in Afghanistan, **undoing** the **progress** made towards **gender equality** in the country.

5) The Taliban imposed a **ban** on girls **attending secondary school** throughout the majority of the country. In March 2022, the Taliban announced it would **extend** this ban further, leaving the **future** of girls' education in the country **uncertain**. Human rights organisations are concerned that the continuing **political instability** in Afghanistan could **prevent** lasting progress from being made.

6) Women (and men who support them) have **campaigned** for **change**, e.g. by protesting against bans on women's education. NGOs and IGOs like the UN (see p.220) strongly support women's rights in Afghanistan.

In Australia, Aboriginal peoples have campaigned for equal rights

Ethnic rights — Australia

CASE STUDY

1) Australia's **Aboriginal peoples** have experienced **discrimination** for decades (see p.239). Health and education indicators like **life expectancy** and **literacy rates** are lower for Aboriginal peoples and Torres Strait islander people than for other Australians. Aboriginal women and Torres Strait Islander women are especially **vulnerable** to domestic violence.

2) In 2008, health authorities and NGOs launched a campaign called '**Closing the Gap**' to improve the health of Aboriginal peoples and Torres Strait Islander people and achieve **health equality** by 2030. Another campaign, '**Women's Voices**', aimed to promote Aboriginal women's rights by giving them opportunities to talk about their needs and concerns.

3) In response to the "Closing the Gap" campaign, the **Australian government** has taken **action** to improve things, with some success. For example, 20% more Aboriginal children and Torres Strait Islander children finished high school in 2020 than in 2008. However, research has shown that **racism** towards Aboriginal peoples and Torres Strait Islander people is still **widespread** in Australia. For example, in 2021 around 40% of Aboriginal peoples and Torres Strait Islander people faced racial prejudice, in contrast to only 20% of remaining Australians.

Warm-Up Questions

PRACTICE QUESTIONS

Q1 Why do post-colonial states often have variations in rights between different social groups?

Q2 Give an example of a campaign for human rights and explain the progress it made.

Exam Question — A-Level

Q1 Explain the relationship between human rights and human development indicators. [8 marks]

You're right to care about human rights...

It's very heavy stuff, I know... but human rights are fundamental for ensuring equality and it's on the syllabus, so you need to make sure you learn it. Don't forget to brush up on your case studies in this section too, they're vital for getting top marks.

Defending Human Rights

Sometimes countries and organisations get involved in the affairs of other countries to try to prevent human rights violations.

There are several **different** ways to **intervene** in **human rights**

When countries or organisations believe that human rights are being **abused** in a particular country or region, they might choose to **intervene**. A **geopolitical intervention**, in defence of human rights, could take several forms — it might provide an **incentive** for the country to protect human rights, **punish** a country for human rights abuses or **support** movements for better rights.

Geopolitical Intervention	What it is	Example
Development aid	• Countries or organisations provide **money** or other kinds of **assistance** to increase **human development** in the country. • Development can help the country to meet its citizens' human rights. • Donors might also make the aid **conditional** on the country making improvements to human rights.	• In 2010, **Malawi** prosecuted two men for trying to get married. • In response, some donors threatened to **cut-off aid** to countries, like Malawi, not recognising gay rights. • In the following years, the Malawian government **overturned the ban** on homosexuality.
Trade embargo	• Countries **restrict** or **refuse** to **trade** with the country until it **stops** human rights abuses.	• In response to worries over potential **human rights violations** and the possible development of **nuclear weapons**, **EU** member countries restricted their trade with **Iran** in 2011. • They have an embargo on the trade of **military equipment** that might be used to develop nuclear weapons and restrict the human rights of Iranian citizens.
Military aid	• Countries provide **weapons** or **military training** to help stop human rights abuses. • For example, to support an oppressed group in **overthrowing** a **dictatorship** or to fight **terrorism**. • Like development aid, military aid can also be **withheld** as a way of **protecting** human rights.	• The **USA** provides **training** and **funding** to the **Philippines' security forces**. • In 2020, the Philippines brought in laws that repress **freedom of speech**, leading to the USA threatening to withdraw its military aid.
Indirect military action	• Countries provide **financial support**, **military equipment** and **advisors** to one side in a conflict.	• The **USA** supported **Ethiopia's** invasion of **Somalia** in 2006, with the aim of bringing **regional conflict** to an end.
Direct military action	• Countries or organisations (e.g. NATO or the UN) sometimes use their **military power** to intervene in another country. For example, carrying out **airstrikes** or sending in **troops** to try to stop human rights abuses.	• In 2011, **NATO** intervened in the **civil war** in **Libya**. They carried out **bombing** and **missile strikes** in support of **rebel groups** trying to overthrow the oppressive **government**.

Western governments use human rights as a **negotiating** tool

1) Some Western governments act as **champions** for human rights by **condemning** other countries' human rights violations. They often use human rights as **conditions** for providing **aid** to other countries and use them in **negotiating trade deals**, e.g. allowing the investigation of suspected labour rights violations under the US-Mexico-Canada Agreement. Some of these countries have also used human rights as a **reason** for launching a **military intervention** in another country (see p.255), e.g. the intervention in Libya, 2011.

2) Some people have **criticised** the way that Western countries use the idea of human rights to **justify** their actions. For example, they believe it is **hypocritical** when some countries call out other countries' human rights records without addressing their own failings with human rights. For example, Australia condemned Russia's invasion of Ukraine however, it has itself been criticised for the inhumane treatment of asylum seekers. It can also be **selfishly motivated** — countries use human rights as the reason to intervene in another country, but they might have an **ulterior motive**, such as gaining **access** to that country's **resources**, e.g. the invasion of Iraq (see next page).

3) Human rights interventions also challenge **national sovereignty** — the idea that a country is **independent** and has the power to **make decisions** about what happens there, without **interference** from outside. By **overriding** other nations' sovereignty, some claim that Western countries are seen as continuing their **colonial** attitudes towards other countries.

Defending Human Rights

Different groups promote different kinds of interventions

1) Most interventions have **pros** and **cons** which can **limit** or **protect** human rights. For example, **cutting off** development aid can **limit** a country's ability to provide for its citizens' **basic needs**, while **military action** can lead to the **injury or death** of civilians.

2) Some **countries** and **organisations** favour certain kinds of interventions over others. Even in situations where it's clear that serious human rights violations are happening, these different groups don't usually **agree** about the **best way** to intervene.

3) The kinds of interventions used by **national governments** depends on their **power** and **resources**. Superpowers are more likely than less powerful countries to carry out **military interventions** in other countries.

Charlie's parents decided it was too late to intervene.

4) **IGOs** like the **UN** and the **EU** use political and economic interventions. These interventions include publicly **condemning** human rights abuses, putting **conditions** on aid or withholding **aid** and imposing **trade embargoes**. The UN has security forces that it uses to help prevent or reduce conflict — but these are usually meant to be used for **peacekeeping** missions rather than military action.

5) Human rights **NGOs** tend to promote **peaceful** interventions. **Amnesty International** and **Human Rights Watch** monitor the **status** of human rights around the world and **campaign** for the protection of those rights. They use **petitions** and **lobbying** to put pressure on oppressive governments (e.g. to free people who have been imprisoned unfairly). They also **encourage** IGOs and governments to use their **political influence** to support human rights in other countries.

6) Some NGOs argue that military intervention should **only** be used as a last resort, when there's **no other way** of preventing the most serious human rights violations, like **genocide**.

The USA used human rights to justify the invasion of Iraq

Invasion of Iraq

CASE STUDY

1) The USA launched a military intervention in Iraq in 2003 as part of the 'war on terror' (see p.256). It argued that military intervention was justified by the need to remove Saddam Hussein's oppressive regime, which had killed hundreds of thousands of Iraqis. The USA also said it believed that Iraq was creating weapons of mass destruction, which posed a global threat.

2) The USA was supported by some other governments, such as the UK, but the invasion wasn't approved by the UN. There were mass protests against the intervention in the USA, UK and elsewhere. People accused the USA of launching the intervention to help secure its access to Middle Eastern oil reserves.

3) The intervention succeeded in overthrowing Saddam Hussein, but led to a long and complex war. The war also had negative impacts on human rights and development — the conflict displaced around 1 in 25 people from their homes in Iraq and severely damaged Iraq's health care and education systems.

An anti-war protest outside Westminster, London

Warm-Up Questions

PRACTICE QUESTIONS

Q1 Name two human rights NGOs and briefly outline what they do.

Q2 How can military interventions challenge national sovereignty?

Exam Question — A-Level

Q1 Explain why some forms of intervention may be favoured more than others. [8 marks]

Read these pages to defend yourself from embarrassment...

Trying to mix up your revision routine? I thought so. Instead of reading the pages over and over and over, why not try covering up each of the different types of intervention in the table and seeing what you can remember for each one.

Increasing Development — Aid

There are several ways of increasing development, and we're starting with a classic — development aid.

Aid is given to a country to help it develop

Development aid is a form of **assistance**, usually **money** or **resources**, that's given to a country by an **organisation** or another **country** to help increase its **human development**. There are several types of development aid, depending on **who** is giving it and **what it's for**:

1 National governments

1) This is usually transferred **directly** from the **donor** government to the government of the **recipient** country (bilateral aid).

2) It's often spent on large, long-term and national-scale projects, such as promoting **economic growth** or building **infrastructure**.

3) A country's **geopolitical interests** can influence which countries it chooses to give aid to.

2 Intergovernmental Organisations

1) IGOs, like the IMF and the World Bank, provide **financial assistance** as a **loan**, to be paid back to the IGO.

2) IGOs, like the WTO, might also **attach conditions** to the loan, such as adopting specific economic policies (see p.241).

3 Non-governmental Organisations

1) NGOs raise money through **charitable donations**, e.g. from businesses and members of the **public**. They may use this money to provide **humanitarian assistance**, such as food and emergency supplies, to countries experiencing a **crisis**.

2) NGOs also work with **communities** in developing countries to increase their human development in the longer term.

- **Oxfam** is a major confederation of 21 NGOs from around the world. It began in the UK but now works in over 60 countries.

- Oxfam raises money through **funding** from national governments and IGOs, holding sponsored events, receiving donations and selling second-hand items in its charity shops.

- In 2021, Oxfam raised £105 million from **donations** and **legacies** (money left in people's wills) and received £27.9 million from the UK government.

- Oxfam's aid work includes providing people with a clean and safe **water supply** (including during **emergencies** like wars and natural disasters, e.g. the Haiti earthquake in 2010).

CASE STUDY

Development aid has had successes...

Development aid has contributed to some **major improvements** in human development, such as reducing deaths from **life-threatening conditions** and promoting **human rights**.

Tackling Malaria

1) **Malaria** is a life-threatening disease that's transmitted by **mosquitos**. It's **preventable**, e.g. by using mosquito nets and antimalarial drugs (these are the most common strategy against malaria), but still kills **hundreds of thousands** of people every year. **Women** and **children** in **developing countries** are the worst affected.

2) IGOs, NGOs (such as Nets For Life) and other donors, e.g. The Global Fund, have worked **together** to tackle malaria — it was one of the UN's **Millennium Development Goals** (see p.242). They funded **awareness campaigns** as well as the **distribution** of mosquito nets, diagnosis kits and medicine to treat malaria. For example, the UN provided 30 million mosquito nets to Nigeria. This had a **big impact** — global death rates from malaria **fell by 25%** between 2000 and 2020.

Achieving gender equality

1) Around the world, **women** generally have **fewer rights** than **men** and this inequality is often much greater in some countries (e.g. those with more conservative values) and social groups than others.

2) Development organisations and donors have run programmes at national and local scales to **empower** women around the world. For example, providing women with small **loans** to start their own **businesses** and training **community groups** to recognise and prevent **violence** against women and girls. UN Women also launched the global programme "Safe Cities and Safe Public Spaces for Women and Girls" to try and **reduce violence** against **women**.

3) In many countries, gender equality has **improved** in recent decades — **more girls** are attending and finishing school, **maternal mortality** rates have **fallen** and more women have been **elected to parliament** in many countries. The **Gender Inequality Index** has also fallen in most countries since 1997.

Increasing Development — Aid

Trends in gender equality

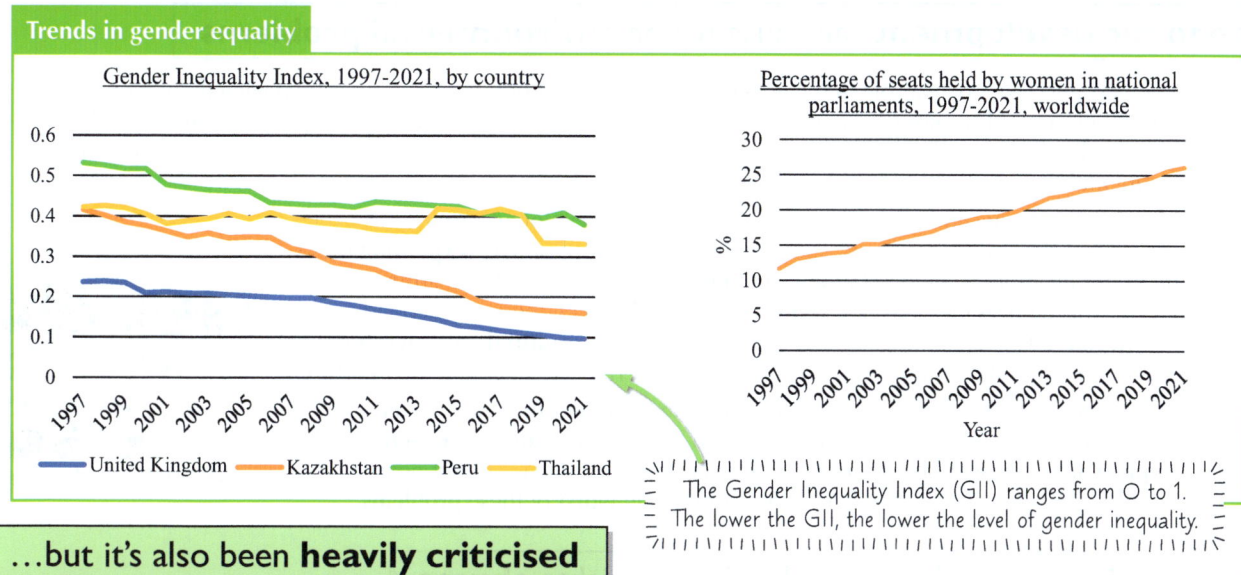

Gender Inequality Index, 1997-2021, by country

Percentage of seats held by women in national parliaments, 1997-2021, worldwide

United Kingdom — Kazakhstan — Peru — Thailand

The Gender Inequality Index (GII) ranges from 0 to 1.
The lower the GII, the lower the level of gender inequality.

...but it's also been **heavily criticised**

There are **debates** over how much aid actually contributes to human development.
Some people argue that aid can **create more problems** than it **solves**:

1) **Dependency** — Poorer countries that receive aid for a **long time** can end up **relying** on that aid to support their **basic services**, rather than taking **action** to promote development. Countries that receive **food aid**, for example, lack an **incentive** to produce their own food. Recipient countries can also get used to **responding** to donors' demands, rather than **setting their own agenda** for development. If the aid is a loan, recipient countries may become **dependent** and can end up in **debt**.

2) **Corruption** — Aid money and resources can be **stolen** by **corrupt** politicians and public servants, or spent on development projects that serve the interests of a small, powerful elite. Although some donor countries place **conditions** on the aid they give to try to **prevent** corruption, research has shown that giving aid to countries with high levels of corruption can actually **increase** their corruption. As a result, aid money **doesn't** always **reach** the **people** who need it most.

Humanitarian aid in **Haiti** had **mixed** results

Humanitarian aid in Haiti

CASE STUDY

1) **Haiti** is a **developing** country in the Caribbean — **30%** of its population were living in **extreme poverty** in 2021. The country is also exposed to a wide range of **natural disasters**. In 2010, it was hit by a severe **earthquake**. An estimated **220 000 - 300 000** people were **killed**. Hundreds of thousands of people were **displaced** and had to live in makeshift 'tent cities', which were then hit by an outbreak of **cholera** that killed over 9000 people by 2016.

2) **Governments**, **development organisations** and members of the **public** around the world **donated money** to help Haiti cope with the aftermath, raising around **US$ 10 billion**. Thousands of NGOs started work on the relief effort, e.g. the UN, the Red Cross and Oxfam. The money raised was used in a number of ways, e.g. to provide **emergency shelters**, **repair homes** and restore access to **clean water**.

Devastation from the 2010
Earthquake in Port-au-Prince, Haiti

Imposing a time limit is one way of trying to prevent corruption, making sure that money is spent on disaster relief. However, this can backfire, encouraging organisations to spend the money, even if it's not going towards development.

3) However, there were **concerns** about where all the **money** was going. Two years after the earthquake, around **500 000 people** were still living in **temporary shelters**. Foreign governments **pledged** US$ 5.6 billion, with the **condition** that it would be spent within the first 18 months after the earthquake. However, 18 months later, only **40%** of this had been spent. Haiti's **government** and the **NGOs** didn't have the **capacity** to use the money **effectively**. Instead, due to a lack of involvement from locals, it's thought a lot of donated money was spent on the **salaries** and **accommodation** of NGO workers, many of whom were from **outside** Haiti.

Increasing Development — Aid

Economic development can cause major environmental problems

1) Superpowers and TNCs play a role in the **economic development** of poorer countries by paying them money to **exploit** their **natural resources** (e.g. minerals or timber). The money received by poorer countries can help them to **develop** their own **industries** and to exploit **their own resources**. Exploiting these resources can cause **environmental damage** that has **knock-on** effects for **human rights**.

2) In many countries, **unexploited resources** are found in **less developed**, rural areas. These may be areas where minority groups, such as **indigenous** peoples, live. **Exploiting** these resources, for example by **deforestation** or **mining**, often means that these groups are **displaced**. This **ignores** these groups' **rights** to their **land**, which they may depend on for their **livelihoods**, e.g. farming. Where people have a deep **cultural** and **ancestral** connection to their land, displacement also ignores their rights to maintain their **culture**.

Deforestation in the Amazon

3) Economic development can also create problems like the **pollution** of **air**, **water** and **soil**. Minority groups and poorer people often have less **freedom of speech** to **complain** about these problems, or insist that a factory or mine isn't built near their homes. This means that they tend to experience the **worst** of these **problems**.

Oil exploitation in Nigeria led to human rights abuses

Oil exploitation in Nigeria

1) The **Niger delta** on the coast of **Nigeria** is home to large **oil reserves**. These reserves have been exploited by TNCs since the 1950s. Sales of oil have contributed to Nigeria's economic growth — it's one of the wealthiest countries in Sub-Saharan Africa.

2) The Niger delta is inhabited by several **minority** ethnic groups, including the **Ogoni** people. In Ogoni culture, the **land** and **rivers** are considered **sacred**. They are also vital to people's **livelihoods** of **fishing** and **farming**.

A crude oil refinery in Ogoni, Southern Nigeria

NIGERIA

▪Abuja

Lagos

■ Niger Delta States ● Oil fields

Map of Nigeria, showing the location of the Niger delta and the major oil fields.

3) Ogoni communities have **lost land** as a result of oil drilling. **Oil spills** have led to the **pollution** of local **wells** and rivers, threatening people's food and water supplies. Despite the growth of Nigeria's **GDP**, levels of **human development** in the Niger delta are still **lower** than the rest of Nigeria.

4) **Conflict** over the oil reserves has also led to **human rights abuses**. In 1993, local communities **protested** against the laying of a new oil **pipeline**. **Military police** and **security guards attacked** the protesters. Around 1000 people were **killed** and villages were destroyed, making 30 000 people **homeless**.

Warm-Up Questions

Q1 Outline three different sources of development aid.

Q2 Explain how development aid can lead to dependency.

Exam Question — A-Level

Q1 Evaluate the view that development aid harms more people than it helps. [20 marks]

Learn these pages to aid you in the exam...

If the last three pages have taught you anything, it's that aid can be given in a number of ways and to varying degrees of success. One thing that's almost guaranteed to give you success is if you swot up on the information from the last three pages.

Increasing Development — Military Intervention

Some countries have used human rights as a reason for carrying out military interventions in other countries, but they don't always have the best track record on human rights themselves. Read on to find out more...

Countries use human rights to **justify military interventions**

A military intervention is when a country uses its military resources to get involved in the affairs of another country. Take a look at p.251 for more on military interventions.

1) The type of **military intervention** used by a country can **vary** — it could provide **military aid** (see p.250), offer **support** to another country's military intervention or intervene **directly**, e.g. by **invading** or **declaring war** on another country.

2) Military interventions are often **controversial** and so countries need to be **strategic** about how they explain their reasons for the intervention. Some countries have **justified** their military interventions by saying that they're intervening to **defend human rights**.

3) It's generally not **acceptable** for countries to just **invade** each other — doing this leads to **international condemnation** and more military interventions from other countries. Many people believe in the idea of a '**just war**' — that military interventions should **only** happen for a **good reason**, e.g. to **prevent genocide**.

4) Countries also **use** military interventions to advance their own **strategic**, **political** or **economic interests**, e.g. to get **access** to another country's **resources**, to **support** their **allies** or **remove** a **threat**. Military interventions can be a way for a country to demonstrate its **military power** and to increase its **influence** over other countries (see p.260).

CASE STUDY

For example, **Syria's civil war**:

- Since 2011, there's been conflict in **Syria** between the authoritarian **government**, pro-democracy **rebels** and an **extreme Islamist group**, ISIL. **The Kurds** (although not actively involved in the civil war) are also fighting against ISIL. There are **global strategic interests** in the **outcome** of the war as Syria's in an **important** location in the Middle East. The **government** has been supported by emerging powers like **Russia** and **Iran**, while the **USA** and their **allies** like **Saudi Arabia** have supported the **rebels** against ISIL as well as the government.

- In 2013, evidence suggested that the Syrian government had used **chemical weapons** to attack civilians. This goes against the **Geneva Convention** (see p.245). The following year, the **USA** and its allies began carrying out airstrikes in Syria. In **2018**, after **more evidence** of chemical attacks, countries including the **UK** and **France** launched further airstrikes. They used the chemical attacks to **justify** their **involvement**, arguing that military intervention was needed to **prevent** further **human rights violations**.

Military aid can lead to the **abuse** of **human rights**

1) Military aid is a form of **development aid** that one country **gives** to another country to help it develop its **military strength**. Types of military aid include providing **training** for military **personnel** and supplying **weapons**. For example, by giving the country **credits** to buy weapons or **selling** them weapons at **reduced prices**.

2) Giving military aid is a way of building **new alliances**. It also makes sure that allies have the **capacity** to carry out **military interventions** in the **strategic interest** of the donor country.

3) Some of the biggest suppliers of **military aid** are countries that **stand up** for **human rights** on the international stage, e.g. the **USA** and **France**. However, many of the major **recipients** of military aid are countries with a **questionable record** on human rights, e.g. **Saudi Arabia** and **India**.

Largest arms exporters and importers

USA: 36%　　France: 7.9%　　Russia: 21%

India: 9.2%

Egypt: 5.8%

Saudi Arabia: 12%

● Exporter
● Importer

Top three arms exporters and importers by percentage of world trade, 2015-2019

Turn over to the next page to learn about military aid and human rights in Yemen.

4) This can **undermine** the donor countries' **position** on human rights. On the one hand, donor countries may **criticise** some countries for human rights **violations**. On the other hand, they may **provide** other countries with the **resources** to **abuse** human rights. This suggests that their support for human rights can sometimes be influenced by their **strategic interests**, rather than **good intentions**.

Increasing Development — Military Intervention

UK military aid to Saudi Arabia

- Military aid, in the form of **weapons** and **assistance** in **training** military forces, has been sent to **Saudi Arabia** from the UK. From 2015-2020, the weapons maker contracted by the UK government sold **£15 billion** worth of arms and services to Saudi Arabia's military.

- The **Saudi government** is carrying out a military intervention in **Yemen**. This intervention has been accused of committing numerous **human rights violations**, e.g. bombing attacks on **civilians**. At least **8000 civilians** are estimated to have been killed since 2015. Human rights NGOs such as Amnesty International have claimed that the military aid sent to Saudi Arabia by the UK has played a role in the human rights abuses that have taken place in **Yemen**.

- Some people believe that the UK government hasn't stopped its military aid to Saudi Arabia because it would threaten the important **economic ties** and **damage** the **trading relationship** between the two countries.

The 'war on terror' used human rights as a reason for **direct military intervention**

1) On **11th September 2001**, a **terrorist** organisation called **al-Qaeda** carried out attacks on the **USA**, killing around 3000 people. In response, the US President George W. Bush declared a global 'war on terror'. This involved launching **direct military interventions** in **Afghanistan**, where al-Qaeda was **based**, and **Iraq**. The USA and its allies also supported **indirect** military interventions and provided **military aid** to several other countries, including **Somalia** and **Yemen**.

2) The **invasions** of **Afghanistan** and **Iraq** weren't **approved** by the UN Security Council, but the USA argued that they were **justified** — partly by the need to **fight terrorism** and **defend** the **USA** and partly to **protect human rights**.

- Afghanistan was run by the very conservative **Taliban**, which had **severe restrictions** on **women's rights** (see p.249) and used brutal **punishments** to **enforce** these restrictions. The USA and its allies argued that by **removing** the **Taliban**, they would be helping to **liberate women** and **promote** their **human rights**.

> In 2021, following the removal of foreign troops, the Taliban regained power of Afghanistan (see p.249).

- Iraq was a **dictatorship** and its leader, **Saddam Hussein**, and government were responsible for many **human rights abuses**, especially towards people from the **Kurdish ethnic group**. Again, the USA argued that **overthrowing** Hussein's regime would bring an **end** to these abuses and stop the threat of Iraq having weapons of mass destruction.

However, **some** of the **countries** involved used **torture**

1) Although the USA and its allies **used** human rights to **justify** their military interventions in the 'war on terror', some of the countries involved ended up **committing** human rights **abuses** as part of the war.

2) People have **criticised** the USA for using human rights as a **reason** for military intervention, and then actually committing human rights **abuses**.

3) For example, US soldiers **tortured** and **humiliated** prisoners at **Abu Ghraib** prison in Iraq. People **accused** of being terrorists have also been **tortured** and **held without trial** for years at the USA's prison at **Guantánamo Bay**.

4) The **Universal Declaration of Human Rights** (see p.244) states that **no one** should be **tortured**. It also says that no one should be **treated** or **punished** in an **inhumane** or **degrading** way. The USA and its allies have **signed** the Universal Declaration of Human Rights and several other international agreements **banning** torture. By using torture, these countries have **undermined** their **justification** for the invasions.

Warm-Up Questions

Q1 Give an example of a conflict where countries used human rights to justify military intervention.

Q2 What was the 'war on terror'?

Exam Question — A-Level

Q1 Explain the extent to which a government's stance on human rights can be undermined by their actions.

[8 marks]

Hopefully you've developed your understanding of military intervention...

Military interventions aren't always justified and have led to severe human rights abuses. Sometimes they can be selfishly driven, but that isn't always the case. It's a complex issue, but hopefully these pages have helped to simplify it a tad.

Increasing Development — Measuring Success

Just as people disagree about how best to intervene in a country, there are also debates about how to measure how successful an intervention is. Success can be measured in lots of different ways, here are some of them...

Success can be measured using different variables

1) One way people might measure success is by thinking about whether the **main objective** of the intervention was **achieved** (e.g. to overthrow an oppressive regime). Alternatively, people might consider what the **broader impacts** of the intervention were (e.g. impacts on human rights in the country). They could also compare the **short-term** and **long-term** consequences of the intervention.

Syrian refugee camp in Atmeh, Syria

2) **Variables** that are used to measure success include:

- Indicators of **human development** — e.g. **improved health**, **higher life expectancy** and **increased access to education** can suggest that standard of living has increased.
- Indicators of how well **human rights** are upheld — e.g. **gender equality** and **freedom of speech**
- How **well** the **negative consequences** of the intervention are **managed**. For example, military interventions often **displace** large numbers of people, who become **refugees** — e.g. **6.8 million people** have left **Syria** since the start of its civil war in 2011. In some cases, governments or NGOs set up **refugee camps** with **facilities** to make sure that refugees are **safe** and **healthy**. However, many refugee camps have **poor sanitation** and **food supplies**. Many refugees travel by **dangerous routes** (e.g. in small boats) to get to safer places.

For some countries, a stronger democracy is a sign of success

1) Some countries and IGOs (e.g. the IMF) promote a **global Western culture** of **democracy** and **capitalism** (see p.101). They believe that **societies** work best when people have the **freedom** to make their own **political** and **economic choices**, e.g. choosing **political leaders** or deciding how to **spend money**.

2) For these countries and IGOs, taking steps to make a country more **democratic** is a sign of a **successful intervention**. They also believe that making a country more **democratic** leads to stronger **economic growth**.

3) **Steps** towards a **stronger democracy** usually include:

These are the ideals of a democracy. In reality, these freedoms aren't always experienced and censorship can still exist.

1 Increasing freedom of expression

- In a democracy, people are free to **express** their **beliefs** without being **censored** by the government (though there are usually exceptions for very **offensive** beliefs). However, one group of people with certain beliefs are **prevented** from **dominating** or **repressing** those of others.
- People are able to make more informed decisions as they are allowed to openly discuss a range of ideas.
- People in **power** are also meant to be **accountable** to the people who **elected** them. This means that anyone can **criticise** the government — as an **individual** or through the **media**.
- Increasing freedom of expression can involve **removing censorship laws** and supporting **independent media organisations**, e.g. **newspapers** or **radio stations**.

2 Creating, or strengthening, democratic institutions

- Democratic institutions are the **organisations** that are **elected** by the citizens of a country or region to run the government of the area. E.g. the USA's democratic institutions are the **House of Representatives** and the **Senate**.
- **Setting up** new democratic institutions can involve writing a **new constitution**, or making big changes to existing **laws**.
- Other institutions and organisations can **strengthen** democratic institutions and **protect** freedom of expression, e.g. a **judiciary** (a country's system of law courts and judges) that's **independent** from the government.

Bella was grateful that democracy allowed her the right to refuse her greens.

4) **Iraq** (see p.251) is an example of a country moving towards a stronger democracy after an intervention. During the intervention, the USA and its allies worked to set up **democratic institutions** in Iraq. In 2005, **two years** after their intervention, **elections** were held for a new **National Assembly** and a **new president** was elected. The National Assembly wrote a new constitution and Iraq held elections **again** in **2010**, **2014**, **2018** and **2021**. This was seen as a success by those who measure success through democracy. However, **further steps** towards **strengthening democracy** in Iraq were **held back** by continuing **corruption** and **threats to security** in the country.

Topic 8: Option 8A — Health, Human Rights and Intervention

Increasing Development — Measuring Success

For other countries, **economic growth** is more **important**

1) Not all countries follow the **Western model** of democratic and capitalist development. Some countries see economic growth as the most important measure of success. This can mean that they **focus** on strategies for **economic development**, rather than following a more **holistic** model of development that aims to increase **human wellbeing**.

> Holistic development looks at the whole picture of economic, social and environmental aspects rather than focusing on just one.

2) They might also **prioritise** economic growth over protecting **human rights** or developing **democratic institutions** — especially where **democratic processes** (e.g. **debating** and **voting** on a new economic policy) might **slow down** economic progress.

Vietnam

CASE STUDY

For example, the government of **Vietnam** has followed policies that have **limited human rights** and the development of **democratic institutions** but promoted **fast economic growth**.

- Vietnam is governed by the **Vietnamese Communist Party** (VCP). Citizens can **vote** in local and national elections, but they can only choose between **representatives**, not between **political parties** as the VCP is Vietnam's **only** political party. This means that people have **limited choice** in how the country is **governed**.

- In 1986, Vietnam's government brought in **reforms** called the '**Doi Moi**'. These moved away from Vietnam's centrally-planned, **socialist** approach to development towards a more **capitalist** system. Vietnam reduced **restrictions** and **tariffs** on **international trade** and eventually joined the **WTO**. This attracted **foreign investment** and led to big **increases** in **GDP**, e.g. in 1985 GDP was **14 billion** and by 2021 GDP had increased to over **360 billion**.

A map showing the location of Vietnam

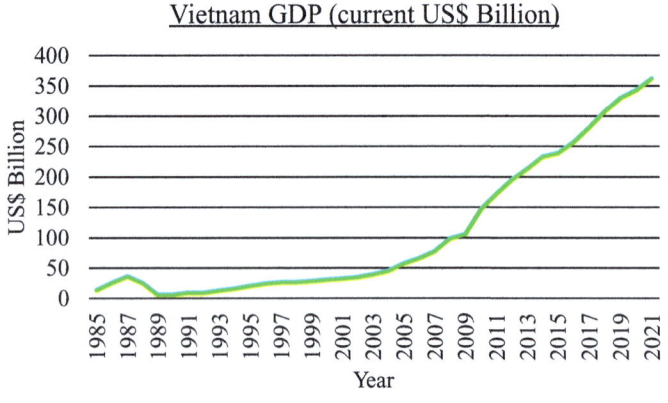

Graph showing Vietnam's GDP in US$ Billion, 1985-2021

- This economic growth has helped to **reduce poverty** in Vietnam. For example, 45% of the population were below the international poverty line in 1992 compared to 1.2% in 2018. Vietnam has also seen **improvements** in **human development**, e.g. **life expectancy** has increased by **6 years** since 1986.

- However, NGOs have **condemned** Vietnam's record on **human rights**. For example, **freedom of expression** is very **limited** — the government **controls** the **media** and people who **criticise** the government are often **intimidated**, **attacked** and even **imprisoned**.

- For example, although land is **owned collectively** by farmers in Vietnam, it is **controlled** by the **government**. This means the government has a lot of **power** over land prices and land use. If the government chooses to prioritise development, **farmers** living on the land have **little power** to change the decision. As a result of this, some **farmers** have been **forced** to give up their land for the development of an Ecopark 13 km from Vietnam's capital city Hanoi.

Warm-Up Questions

Q1 Name three variables that are used to measure the success of an intervention.

Q2 What are democratic institutions?

PRACTICE QUESTIONS

Exam Question — A-Level

Q1 Explain the relationship between freedom of expression and democracy. [4 marks]

Flex those brain cells to help you on the road to success...

As I'm sure you've gathered from the last two pages, deciding how to measure the success of different countries is a complex topic. Every country's different — there's no one glove fits all rule, so keep this in mind when you reread this section.

Success of Aid and Military Intervention

Both development aid and military interventions have had a mixed record, with failures as well as success.

Aid has led to successes **and** failures for **human rights** and **development**

It's not always **clear** whether aid **actually does** promote **development** or **improvements** in **health** and **human rights**. In some cases, aid has been **relatively successful** in achieving these aims, whereas in **others** it has been a relative **failure**. In **many** countries, the **outcomes** of development aid interventions are a **mixture** of **successes** and **failures**.

A relative success — tackling Ebola in West Africa

1) From **2014** to **2016**, several countries in **West Africa** were **badly affected** by a **virus** called **Ebola**. Ebola is very **contagious**, so it **spreads quickly** through a population, especially in places with **poor sanitation**. It's also very **deadly** — around **half** the people who get Ebola **die**. **Liberia**, **Guinea** and **Sierra Leone** had about **28 000** cases of Ebola and around **11 000** deaths.

2) The **World Health Organisation** coordinated the **international response** to Ebola. **Governments** and **NGOs** such as Doctors Without Borders sent **medical staff** and **equipment** to help the affected countries **cope** with the virus. For example, the **UK government** pledged a support package worth **£427 million**. This included sending **troops** to build **treatment centres** and **funding** the development of a **vaccine** to prevent Ebola.

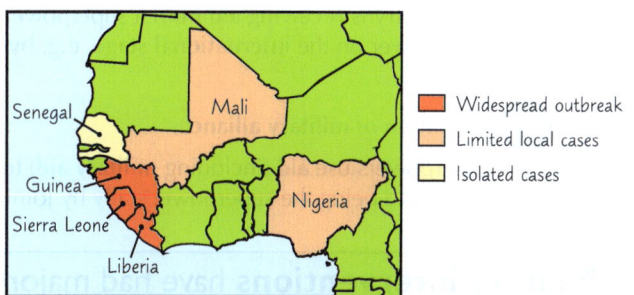

- Widespread outbreak
- Limited local cases
- Isolated cases

3) The intervention managed to **slow** and eventually **stop** the **spread** of Ebola. Scientists developed a **vaccine**, which helped to **slow** the spread and impact of Ebola in the **Democratic Republic of the Congo** in **2020**.

Map of West Africa, showing the main countries affected by the 2014-16 Ebola outbreak

4) However, some people have **criticised** the response to Ebola. **Doctors Without Borders** has argued that the **international community** was too **slow** to respond. However, there is some disagreement about whether or not the response was too slow. Doctors Without Borders also say that the response was **held up** by **local governments** refusing to **cooperate**, e.g. by **sharing** their **data** on the spread of the virus.

A relative failure — development interventions in Haiti

1) **Haiti** is a **developing country** in the **Caribbean**. It has **received** a lot of development **aid**, especially after **disasters** like an **earthquake** in 2010 (see p.253) and a **hurricane** in 2016.

2) One of the **biggest** aid projects in recent years was the development of the **Caracol Industrial Park**, in northern Haiti. This was **funded** by the **USA**, with the aim of **attracting investment** in Haiti's industry. It cost **US$ 300 million**.

3) Despite all the aid and interventions, Haiti remains **poor** and **vulnerable** to disasters. Caracol Industrial Park didn't **live up** to the donors' **promises** that it would **create** lots of jobs, and **plans** to build a **port** nearby were **abandoned**. As of 2022, farmers who had to give up their land for the construction of Caracol Industrial Park reportedly still haven't received their **compensation**.

4) Some people have argued that **aid** has left Haiti **worse off** by making it more **dependent** on other countries. They point out that donors tend to fund projects that **look good** (e.g. Caracol) rather than tackling more **difficult issues**, e.g. reducing **corruption**.

Former US President Clinton standing for photos with workers at Caracol Industrial Park, Haiti

Development aid can lead to **higher** levels of **inequality**

1) The **relationship** between **aid** and a country's levels of **inequality** is **complex**. For some countries that have received **large amounts** of development aid, levels of **economic inequality** have increased.

2) For example, **Zambia** receives over **US$ 1 billion** every year in aid from other governments, but inequality between people's incomes has **increased** and Zambia is now one of the **least equal countries** in the world. Economic inequality often leads to inequalities in people's access to **health care** and differences in **life expectancy** between the **richest** and **poorest** people in a country.

3) However, development aid has helped other countries to become **more equal**. E.g. **Mali** also receives more than US $1 billion and has **decreased** the inequality in the country. According to the **Gini coefficient** (see p.105), where a **higher** score means **more** inequality, Mali's income inequality **dropped** from 50 in 1990 to 36 in 2018.

Topic 8: Option 8A — Health, Human Rights and Intervention

Success of Aid and Military Intervention

Superpowers often use development aid as part of their approach to foreign policy

1) Development aid often forms part of the **foreign policy** of superpowers. This means that they may **measure** the success or failure of aid in terms of how much it **increases** their alliances with other countries.

2) **Three ways** that superpowers may judge the success of their aid are:

1 **Access to resources**

Superpowers provide developing countries with **infrastructure** to help them **exploit natural resources** that superpowers **need**. Superpowers can also attach **conditions** to aid to make sure that it has **access** to these resources at **low prices**.

2 **Political support**

If a country is **receiving aid** from a superpower, it has a big **incentive** to **support** the superpower on the **international stage**, e.g. by **voting** in favour of its **ideas** at **UN** meetings.

3 **Formation of military alliances**

Superpowers use **aid** (including **military aid**) to **encourage** other countries to **commit** to being the superpower's **ally** by joining a **military alliance** (e.g. NATO).

Military interventions have had major costs for human rights

1) Military interventions used to be carried out mainly for **geopolitical** reasons, but in the last few decades, there's been a **significant increase** in military interventions for **humanitarian** reasons.

2) However, **evidence** has shown that both **direct** and **indirect** interventions can have **serious costs** for the **countries** and **communities** that they're meant to **help**. When other countries **intervene** in its **affairs**, a country loses its **sovereignty** — its **control** over what **happens** there. Interventions can also **reduce** or even **abuse** people's human rights.

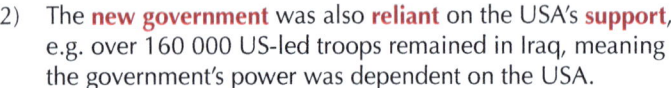

Loss of sovereignty and human rights — Iraq

CASE STUDY

1) Some argue that the US-led military intervention in **Iraq** (see p.251) undermined Iraq's **sovereignty**. For example, the USA handed over control to the Iraqi Interim Government in 2004, but this government was created by the USA and its allies. This meant the government had **limited power** and couldn't make long-term policy decisions.

2) The **new government** was also **reliant** on the USA's **support**, e.g. over 160 000 US-led troops remained in Iraq, meaning the government's power was dependent on the USA.

3) A **full-term government** was formed in Iraq in 2006, but they still had **limited power**, e.g. security was still threatened and more US troops were stationed around Baghdad. The USA **withdrew** their **remaining troops** by the end of 2011, leaving Iraq **vulnerable** to **attack**. In 2014, an extremist group called **Islamic State** (ISIL) started a violent **rebellion** and **took over** some of Iraq's major cities.

The US army in Mosul, Northern Iraq

Military interventions can make short-term gains but cause long-term problems

The USA's military intervention in Iraq

1) In the **short term**, the USA's military intervention in Iraq helped to **improve** human rights. The intervention achieved its goal of **removing** Saddam Hussein's **oppressive regime**. This improved the human rights of minority groups in Iraq, such as **Kurds**. It also gave Iraq the opportunity to start moving towards a **democracy**.

CASE STUDY

2) However, in the **longer term**, military intervention had **negative** impacts. Saddam Hussein's removal caused violent conflict between various groups aiming to fill the power vacuum created. Conflict lasted for years, killing over **200 000 civilians** and displacing more than **9 million people**.

3) The war also caused **long-term problems** for **health** and **education**. Schools, hospitals and **health centres** were **destroyed** in the conflict and around **half** of Iraq's **medical doctors left** the country.

Success of Aid and Military Intervention

Non-military interventions can be better at promoting human rights

Non-military interventions, e.g. **peace-keeping missions**, have often been **more successful** than military interventions at protecting **human rights** and promoting **development**. Sometimes a **combination** of the two are used:

CASE STUDY

A combination of military and non-military intervention helped prevent civil war in Côte d'Ivoire

1) After years of rising tension within **Côte d'Ivoire**, a civil war broke out between 2002 and 2007. A series of **attacks** were made against the **central government** by troops from the north of the country. In 2007, a **peace agreement** was signed — this marked the end of the civil war.

2) In 2010, the **president** of **Côte d'Ivoire**, Laurent Gbagbo, refused to **step down** after a contested **election**. **Violence** erupted between the **government** and **rebel groups** — this made people **worried** that war was about to **break out** again.

3) **IGOs** like the **African Union** put **pressure** on Gbagbo to **end** the violence and the **World Bank** stopped **funding** to Côte d'Ivoire, but the fighting **continued**.

4) **France** and the **UN** provided **military support** for the **rebel groups** by **bombing** Gbagbo's **weapons supplies** and **military bases**. Gbagbo was **arrested** in 2011 and **tried** by the **International Criminal Court** (see p.245). This military intervention by France and the UN **reduced** the level of **damage** but failed to bring about peace.

5) UN troops **stayed** in Côte d'Ivoire until **2017**, carrying out **peace building** activities including **teaching soldiers** about **human rights** and how to **prevent** human rights **abuses**.

Doing nothing can have negative impacts too

1) Military interventions can have **negative** impacts, but **not intervening at all** can also have **global consequences**.

2) Staying **out** of other countries' affairs and **allowing** governments to pursue their **own agendas** can have severe impacts. It can affect the **environment**, **politics** and **social development** in those countries and around the world. Ultimately this can **compromise** human rights and human wellbeing.

The international community failed to prevent genocide in Rwanda

CASE STUDY

1) Countries **chose not** to intervene in the **genocide** that happened in **Rwanda** in **1994** (see p.248). Although **UN troops** were present in Rwanda, most decided to **leave** after the **murder** of ten soldiers. Nearly **800 000 people** were **murdered** in the genocide.

2) There were **longer-term impacts** on human rights and wellbeing in Rwanda and other countries. For example:

- Hutu and Tutsi refugees **fled** to the **Democratic Republic of the Congo** and formed **armed groups**. **Conflict** between these armed groups affected **civilians**, with thousands killed, and contributed to **political instability** across the region.

- The **president** of Rwanda, Paul Kagame, **restricted** human rights and **freedom of expression** (e.g. imprisoning his critics), arguing that this was **necessary** to prevent another **genocide**.

- After the genocide, there was a lot of pressure on **land** as so many people had been **displaced**. This led to **deforestation** and the **degradation** of **ecosystems**, e.g. wetlands, to clear land for **housing** and **farming**.

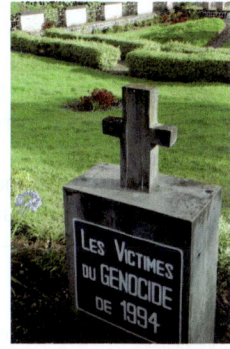
The Genocide Memorial in Kigali, Rwanda

Warm-Up Questions

Q1 Give three ways that superpowers might measure the success of development aid.

Q2 Name one short-term gain and one long-term cost of a military intervention you've studied.

PRACTICE QUESTIONS

Exam Question — A-Level

Q1 Evaluate the view that aid brings more failure than success in terms of promoting human rights and development.

[20 marks]

Revision's all about long-term gains...

Why are there SO many case studies? A fab question — I was thinking the same. Well they're great at providing real-life examples of the content you need to learn, so make sure you learn them. The short-term pain will pay off — I promise.

Globalisation and Migration

Populations don't just change because of changes to birth and death rates — migration can also have a major effect...

Globalisation has led to more migration

Winnie was only interested in the pull factor.

1) **Migration** is when **people move** from one place to another. The **reasons** for migration can be divided into **push** and **pull** factors:

- **Push factors** — these are things that **make** people want to **move out** of the place they're in. They're **negative factors** about the place they're **leaving**, e.g. **war**, **famine** or a lack of **jobs**.
- **Pull factors** — these **attract** people to a **new place**. They're **positive factors** about the place they're **moving to**, e.g. **jobs** or a better **quality of life**.

2) **Globalisation** (see p.88) has **increased** the amount of migration between countries. As it's become easier to travel and communicate, countries have become **more connected** to each other.

3) Globalisation has also caused a **major shift** in the global **economy**. For example, the different functions of **companies** have become more **concentrated** in particular areas, e.g. manufacturing has largely moved away from **developed countries** (such as North America and Western Europe) and towards **emerging and developing countries**, especially in Asia (this is called the '**global shift**').

4) Meanwhile, functions like **research and development** and **consultancy** have become focused in **developed countries**. This has created different types of jobs in different areas, affecting the pattern of **international migration**. Jobs have also become more concentrated in **cities** over time, leading to **rural-urban migration**. Wages tend to be **higher** in more **developed countries** and in **urban areas**, which **increases migration** to these areas.

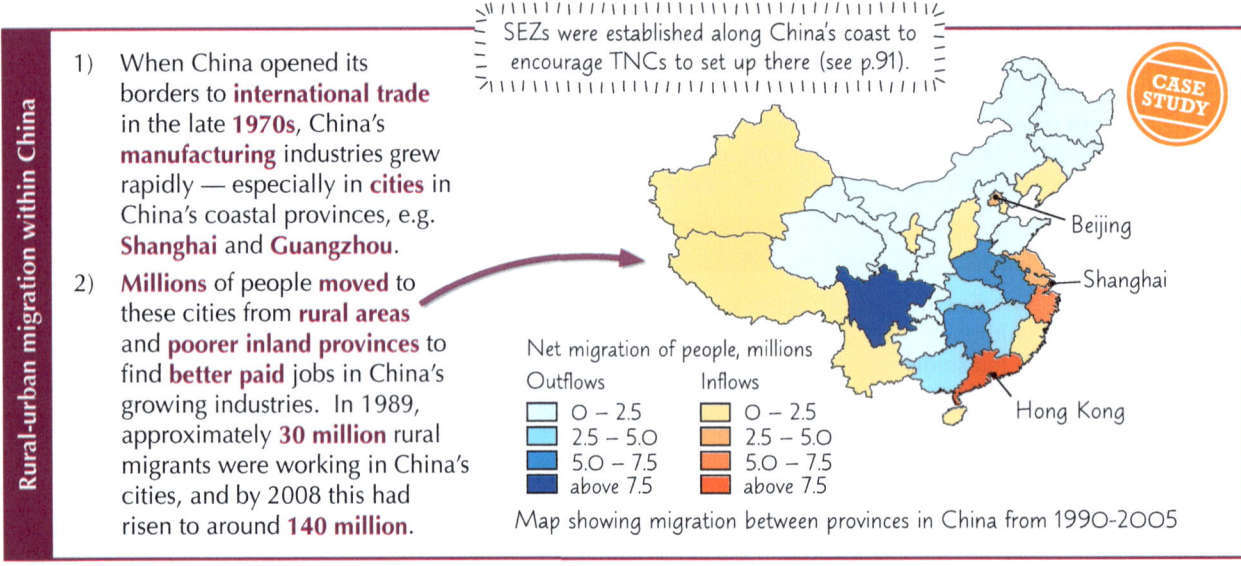

Rural-urban migration within China

1) When China opened its borders to **international trade** in the late **1970s**, China's **manufacturing** industries grew rapidly — especially in **cities** in China's coastal provinces, e.g. **Shanghai** and **Guangzhou**.

2) **Millions** of people **moved** to these cities from **rural areas** and **poorer inland provinces** to find **better paid** jobs in China's growing industries. In 1989, approximately **30 million** rural migrants were working in China's cities, and by 2008 this had risen to around **140 million**.

SEZs were established along China's coast to encourage TNCs to set up there (see p.91).

CASE STUDY

Beijing
Shanghai
Hong Kong

Net migration of people, millions

Outflows	Inflows
0 – 2.5	0 – 2.5
2.5 – 5.0	2.5 – 5.0
5.0 – 7.5	5.0 – 7.5
above 7.5	above 7.5

Map showing migration between provinces in China from 1990-2005

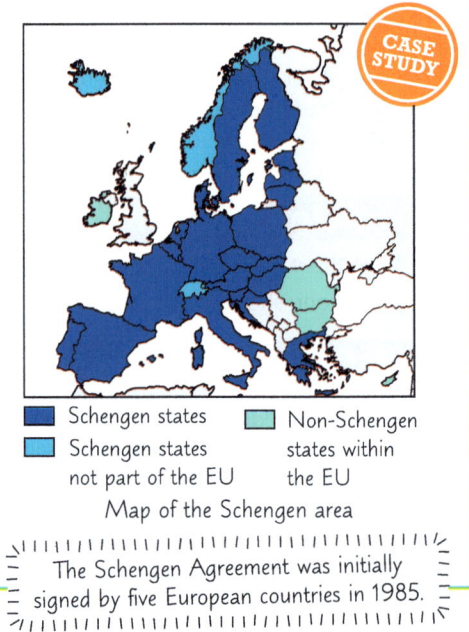

International migration between countries in the EU-Schengen area

1) **Political** and **economic** changes that have happened (as **part** of **globalisation**) include some countries **encouraging** the **free movement** of people **across international borders**. This has **increased** migration by making it **easier** for people to **live** and **work** in different countries.

2) In the **EU**, citizens of **member countries** have the **right** to **live**, **work** and **study** in other member countries without needing special **visas** or **permits**. The level of globalisation differs between countries. Overall, countries in Western Europe are **more globalised** than countries in Eastern Europe, and have **more skilled jobs** available and **higher wages**, which drives migration. For example, in 2020, nearly **200 000 Romanians** migrated to **Germany**.

3) Most EU countries are also members of the **Schengen Agreement**. This is an agreement between **27** European countries. Within the Schengen area, **anyone** can travel **between** countries without needing a **visa** or having to show a **passport**.

CASE STUDY

Schengen states	Non-Schengen states within the EU
Schengen states not part of the EU	

Map of the Schengen area

The Schengen Agreement was initially signed by five European countries in 1985.

Globalisation and Migration

A country's **policies** affect its **levels** of **international migration**

1) The **size** of a country's **migrant population** depends on how **involved** the country is in the **global economy**, **international trade** and what its **policies** are around **migration**. Some countries, e.g. **North Korea**, are **closed off** from international trade and have very **strict rules** around migration in and out of the country, so migrants make up a very **small proportion** of their population. Other countries, e.g. **Singapore**, are highly **integrated** in the global economy and **encourage** certain groups of people to migrate there, meaning that they have a **higher proportion** of migrants.

> International migrants are people who move to another country temporarily. Immigrants are people who move to another country permanently.

2) In 2020, there were approximately **281 million international migrants** across the world. This is equal to around **3.6%** of the **global population** living **outside** their **country of birth**.

3) However, the **proportion** of migrants within a country's population **varies** between countries. Some countries have a **large** migrant population, while others **take in** very **few** migrants. In 2020, the **USA** was the world's **biggest recipient** of migrants — it was home to **50.6 million** international migrants, around **15%** of its population. However, other countries, such as **Qatar**, have a **lot more**

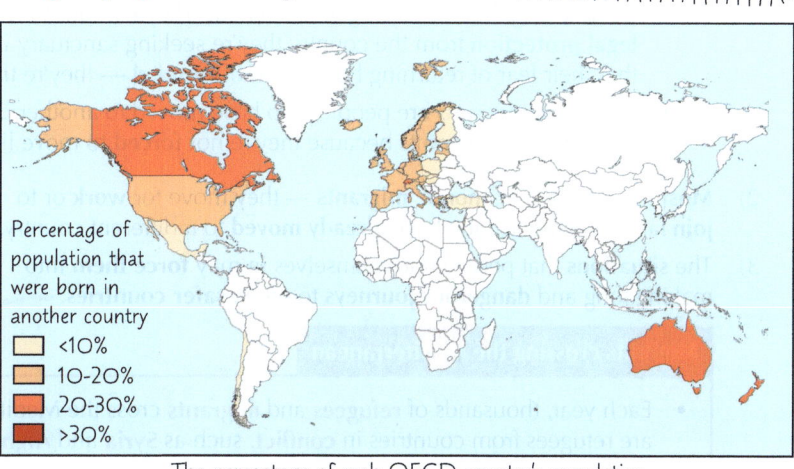

Percentage of population that were born in another country
- ☐ <10%
- ☐ 10-20%
- ☐ 20-30%
- ☐ >30%

The percentage of each OECD country's population born in a foreign country in 2018/2019

migrants **relative** to the size of their **population**. For example, Qatar has a population of almost **3 million** but received around **2.2 million migrants** in 2020, **73%** of its population.

Immigration policies in Japan

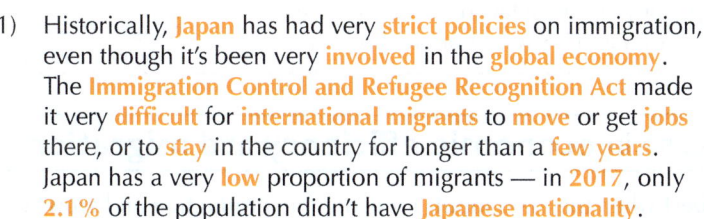

CASE STUDY

1) Historically, **Japan** has had very **strict policies** on immigration, even though it's been very **involved** in the **global economy**. The **Immigration Control and Refugee Recognition Act** made it very **difficult** for **international migrants** to **move** or get **jobs** there, or to **stay** in the country for longer than a **few years**. Japan has a very **low** proportion of migrants — in **2017**, only **2.1%** of the population didn't have **Japanese nationality**.

	Nationality
Japanese	97.9%
Chinese	0.6%
Korean	0.4%
Filipino	1.1%

2) However, in recent years, the Japanese government has become **concerned** about the country's **ageing population** and **labour shortages**. It's **loosened** some of the restrictions on immigration to allow **more people** of **working age**, especially **skilled workers** in certain sectors like **construction**, **farming** and **health care**, to enter the country.

Warm-Up Questions

Q1 Explain how globalisation has resulted in an increase in rural-urban migration.

Q2 Explain why some countries might relax their immigration policies.

PRACTICE QUESTIONS

Exam Question — A-Level

Q1 Explain how globalisation has contributed to migration. [8 marks]

Hopefully these two pages aren't riddled with push factors...

Remember, push factors are negative things that *PUSH* you away from somewhere and pull factors are positive things that *PULL* you towards somewhere else. Simple right? I know usually if it sounds too good to be true it is, but not in this case...

Migration — Causes and Policies

There are a number of reasons why someone might migrate from one place to another, including different policies.

People **migrate** for **different** reasons

1) There are different **circumstances** and sets of **push** and **pull** factors that can cause migration. These determine the **type** of migrant:

- **Refugees** are people who have been **forced** to **flee** their country and are **unable to return** because of **persecution**, **conflict** or changes to the **environment** (e.g. natural disasters).

- **Asylum seekers** are people who have fled their country, but have not yet had their application (to be **recognised as a refugee**) accepted. They can only receive formal **assistance** and **legal protection** from the country they're seeking sanctuary in once it's been demonstrated that their fear of returning home is **well-founded** — they're then granted **refugee status**.

- **Economic migrants** are people who have moved to another country to **work**. They're also known as '**voluntary**' migrants because they're not **forced** to move like refugees and asylum seekers.

2) **Most** migrants are **economic migrants** — they move for **work** or to **join** family members who have **already moved** to a different country.

3) The **situations** that people find themselves in may **force them** into making long and **dangerous journeys** to other, **safer countries**.

Migrants crossing the Mediterranean

CASE STUDY

- Each year, thousands of **refugees** and **migrants** cross the Mediterranean to Europe. Many of these are refugees from countries in **conflict**, such as **Syria** and **Afghanistan**, who are hoping to find **asylum** in **Europe**.

- In 2015, **more than 60%** of people who made the crossing were fleeing from **war-torn countries**. Other people are fleeing countries because they are experiencing **human rights violations** and risk the journey to live in a **safer society**.

- In 2021, over **123 000 refugees** and **migrants** arrived on Europe's Mediterranean coast from Turkey and North Africa. It's a very **dangerous route** — around **3200** people **died** or went **missing** trying to make the crossing that year. Since 2014, nearly **25 000 people** have **died** attempting to cross the Mediterranean.

- When refugees and migrants reach Europe they are often faced with **unhygienic conditions**, a lack of **sheltered accommodation** and **hostility** from authorities.

Refugees crossing the Mediterranean Sea

Different **policies** can cause an increase in **economic efficiency** and **migration**

1) Some **economists** believe that societies work **best** when the main **inputs** for economic **production** and **growth** (**goods**, **capital** and **labour**) aren't **restricted** by regulations, national **borders** or **government** interference. The belief is that this will cause businesses to generate wealth which will benefit all the country's citizens. The **idea** of this **free market society** is known as **neoliberalism**. Three key neoliberal policies are:

- **Free trade** — goods can be traded between countries without restrictions or tariffs

- **Deregulated financial markets** — laws and regulations on moving and investing money are removed

- **Open borders** — people can move between countries without needing border checks or visas

Casper was beginning to regret setting up a free trade agreement with Sasha.

2) Allowing the economy to work **freely** is beneficial because it can be **more efficient**, as the movement of **goods**, **money** and the **workforce** can respond to **supply** and **demand** as they're **needed**.

3) This means that **opening borders** to migration can have **economic benefits**. It allows people to **move** to places where their **labour** and **skills** are **needed**, helping **industries** to **grow**.

4) However, open borders are **controversial**. Some people argue that **opening borders** to migrants **undermines** the idea of **national sovereignty** — if **anyone** can cross a country's **border**, this raises the **question** of how **meaningful** the border is. If people of **all nationalities** and **cultures** can **cross borders** and live **anywhere**, this also challenges the idea of **national identity** and what it means to be from a **particular nation**.

5) In many countries, this has led to **debates** about **how far** borders should be **opened** to migration.

Migration — Causes and Policies

Environmental, economic and political events affect migration patterns

1) Events like **natural disasters**, economic **recessions** and **wars** can **displace** people or **encourage** them to move. These events **affect** where the main **source areas** (the places that migrants are mostly likely to **come from**) and main **destinations** are.

2) For example, during the 2000s, **Syria** was a major **destination** country for people displaced by **war** in **Iraq**. However, since the start of the **Syrian civil war** in 2011, over **6.8 million** Syrians have been forced to **flee** to other countries.

3) **Economic migrants** may be encouraged to move by the prospect of more **job opportunities** with higher salaries, **better schools** and **universities** and an improved **quality of life**.

4) The **distribution** of **source** and **destination** areas is **always changing**, and will continue to **change**, especially as **climate change** leads to more **environmental pressures** and **natural disasters** in some parts of the world.

Some policies encourage economic migration

1) As with the free movement of **goods** and **money**, some economists believe that the **unrestricted movement of labour** within and between **nation states** helps to **promote economic growth**. They argue that **freedom of movement** allows people to move to the places where their **skills** and **labour** are most in **demand**.

2) **Industries** that need to **attract** people from elsewhere can offer **higher wages**, encouraging people to **move** there. Economists believe that this is more **efficient** than governments trying to **control** everyone's **movement** and **allocate** labour where it's needed.

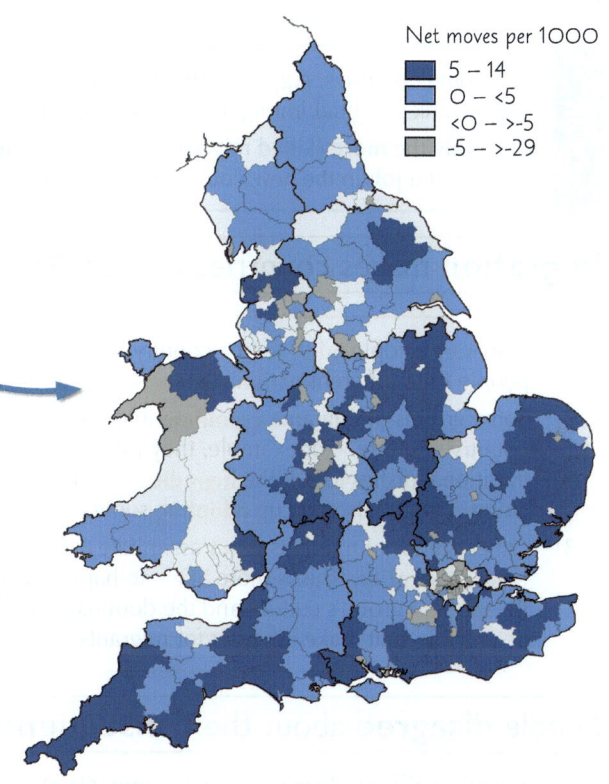

Net moves per 1000

Movement within the UK — CASE STUDY

- In some **countries** and regions, **governments** have **freedom of movement policies** to **encourage** economic migration. Within the UK, there aren't any **restrictions** on people moving between **different areas**.

- Around **1 million** people move a significant distance within England and Wales each **year**. Many of those moving are **young**, **skilled** people moving to **cities** for work — but this is **balanced** by an **outflow** of **older** people moving to **rural** areas.

3) Some global **regions** also encourage economic migration, e.g. the **EU** and the **Schengen Agreement** (see p.262) allow people to **move** between and **work** in different **European countries**. However, freedom of movement **doesn't apply** on a **global scale** — some people are **worried** that having **open borders** might **threaten** their **national sovereignty** (see p.250). For example, some people argued that the UK's membership of the EU threatened its national sovereignty and that leaving the EU would allow the UK to regain control of its future.

A map of England and Wales showing net movement of people between local authority areas in 2013

Warm-Up Questions

Q1 What's the difference between a refugee and an asylum seeker?

Q2 Explain how environmental change can affect migration patterns.

Exam Question — A-Level

Q1 Explain why some economists believe that the removal of restrictions on goods, capital and labour help a country's economy. [6 marks]

You're not the only one tempted to migrate from the desk to the sofa...

There are lots of reasons why people migrate, but fleeing your A-levels isn't a good one — get off that airline website and have another read of these pages. If you get a question on migration in the exam, adding some specific details is sure to impress.

Migration — Consequences

Migration can have a variety of impacts on the host country — and people have different opinions about those impacts...

Some people find it **easier** to **migrate** than others

Not **everyone** has the same **ability** to move between countries and this can affect the consequences of migration. Some groups of people are able to **get past** border controls and **cross** international borders more **easily** than others, depending on:

Nationality

1) There are different levels of border controls between countries.

2) Some countries have agreements allowing their citizens to travel between the countries without needing visas, meaning that their citizens have more opportunities to migrate.

3) The presence or absence of international border controls can determine how easy or difficult it is for people to migrate.

Income

1) Migration can be **expensive**, with costs including **travel tickets** and **visa fees**.

2) **Poorer** migrants, and those who aren't able to get visas, may resort to **illegal**, more **dangerous** routes with the assistance of **people smugglers**.

	Number of destinations
Japan	193
USA	186
South Africa	105
The Philippines	67
Haiti	49
Afghanistan	27

The number of destinations a citizen of different countries can travel to without obtaining visa in advance (Henley Passport Index, 2022)

Skills

1) Some countries, e.g. **Australia**, have immigration **policies** that use **points** to rank potential immigrants according to their **skills** and **abilities**.

2) Only the **more skilled** migrants are allowed **entry**. More skilled migrants might also be able to get a **job** in the new country **before** they arrive — this makes it **easier** for them to get a **visa**.

Migration makes countries more **culturally** and **ethnically diverse**

1) The **more** people **migrate** to a country, the **greater** the country's **mix** of people from different **nationalities**, **cultures** and **ethnic** groups.

2) Some groups of migrants **assimilate** into the **main culture** of their host country — they **adopt** the host country's culture. For example, they might **celebrate** the host country's main **festivals** or **adapt** their traditional dishes to suit the **ingredients** and **tastes** of the host country. In turn, people from the **host country** might also end up **adopting** some **elements** from the **migrant** group's culture, e.g. Wagamama.

3) Other groups of migrants **don't** assimilate as **much** — instead, they maintain their own **distinct cultural identity**. This is more **likely** to happen when there are **big differences** between the **migrant** group's culture and the dominant culture of the **host** country, e.g. **language** or **religion**, that make it harder for migrants to **adjust** to being part of a **new culture**.

People **disagree** about the **consequences** of **migration**

1) People have **different views** about the **impacts** of migration — some people feel that **overall**, migration makes a positive contribution to society, whereas others argue that migration should be more **limited**.

2) These different **perceptions** of migration can lead to political **tensions**. For example, within the EU:

Migration between EU states

CASE STUDY

- The EU allows the **free movement** of people between **member states**. In 2015, around **4%** of people **born in the EU** were living in a **different** country within the EU than where they were born.

- For example, **Germany** has the **highest** number of **EU migrants**. In 2019, nearly **600 000** people migrated from other EU member states to Germany — this accounted for **40% of total migration** to Germany.

- **Perceptions** of migration **vary** throughout EU member states. For example, in 2018, **62%** of **Swedish nationals** believe that migrants make their country **stronger**, compared to only **10%** of **Greeks**.

- In the **2016 Brexit Referendum**, some people voted for the UK to leave the EU as they were unhappy with the **free movement of people** within the EU.

In 2016, 51.9% of people that voted in the EU referendum in the UK chose to leave the EU. In January 2020, the UK officially left the EU.

Migration — Consequences

3) In the **USA**, there's a growing **debate** around migrant workers from **Mexico**, and whether the US-Mexico **border** should be more **open** or **closed**.

Migration between Mexico and the USA

- **Mexico** is one of the biggest **source** countries of migration to the **USA** — it's home to around **11 million** Mexican migrants, about a **quarter** of the total number of migrants in the USA.

- The USA is the most popular **destination** for migrants from Mexico, who include both **low-skilled workers** and more **highly skilled professionals**. Most Mexican migrants live in the states of **California** and **Texas**, which share a **border** with Mexico.

The US-Mexico border at San Diego, California (USA) (left) and Tijuana, Baja California (Mexico) (right)

4) **Perceptions** of **migration** from Mexico to the USA have been both **positive** and **negative**:

	Positive Perceptions	Negative Perceptions
Social	Most Mexican migrants move to the USA to **work**, which means that they pay **taxes** there. This helps to support social services like **health care**, **transport** and **education**.	Around **5 million** Mexican migrants in the USA are **undocumented** — they don't have **visas**, so they're in the USA **illegally**. This makes it **harder** for the US **government** to **regulate** migration.
Economic	Mexican migrants help to meet **demand** for **labour**, e.g. in the **agriculture** and **construction** industries. In 2011, their work contributed around **4%** of the USA's **GDP**.	Some Americans believe that Mexican migrants are **filling the jobs** that they would otherwise want.
Cultural	Lots of aspects of Mexican **culture**, e.g. Mexican **food** and **language**, have become **popular** in the USA. For example, **13%** of the US population speak **Spanish**.	Some Americans feel that Mexican migrants don't try to adopt US culture. They worry that this **threatens** the future of **American culture**.
Demographic	Mexican migrants tend to be **young**, so they contribute to the USA's **working age** population (the average age of Mexican migrants arriving to the USA was **31** in 2017). This is **good** in areas with an **ageing population** — younger migrants **support** the older population by **working** and paying **taxes**.	Some people **worry** that the increasing numbers of Mexican migrants will result in white Americans becoming a **minority** in the country. In 2018, around 50% of surveyed Americans said they view a white minority in the USA **negatively**.

5) Migration from Mexico has become a **politically charged topic** in the USA and often plays a role in general elections. For example, in 2016, **Donald Trump** was elected as **US president** following a presidential campaign that included **promising** to build a 2000-mile-long **wall** along the **Mexico-USA border** to reduce illegal immigration.

It's important to recognise that these views are just some people's perceptions, they're not facts.

Warm-Up Questions

Q1 How can open borders be seen as making the economy more efficient?

Q2 Give an example of a cultural, a demographic and a social perception of the impacts of migration.

Exam Question — A-Level

Q1 Evaluate the view that the economic benefits of international migration are outweighed by the political tension it causes. **[20 marks]**

Skipping over these pages will have serious consequences for the exam...

Migration can have both positive and negative impacts on the recipient country. It's important to note that people's views on migration can vary significantly, and the positive and negative impacts are viewed differently by different people.

Nation States

There's some new terms on this page, but don't panic — in most cases, a nation state is the same as a nation or country.

Nation states come in all **shapes** and **sizes**

1) A **nation state** is a **country** or **territory** that's **recognised** as an **independent**, **sovereign** nation by its **citizens** and **other countries**. The UK, Botswana, India, Venezuela, New Zealand — they're all nation states.

2) Nation states **vary** a lot — some nation states are made up of **citizens** who are very **similar** to each other, and are **united** by the same **language**, **ethnicity** and **culture**. Other nation states are much more **diverse**. The diversity of a nation state is shaped by factors like how migration has **changed over time** and how **isolated** the nation state is.

Diversity in Iceland

1) Iceland has a very **homogeneous** population — around **93%** of its population is ethnically **Icelandic**, mostly descended from **Viking** settlers. There's not much **cultural** or **religious** diversity — the dominant religion is **Christianity** (up to **70% of Icelanders** belong to the Lutheran Church), and around **96%** of the population speak **Icelandic**.

> **homogeneous**
> *similar or alike*

2) One **cause** of Iceland's **lack** of diversity is the country's **isolation**. As a **small island** in the middle of the **Atlantic**, it's relatively **difficult** for migrants to get to — this was especially the case before **planes**. Iceland also maintained its **cultural unity** by **encouraging** immigrants to **assimilate** — up until 1996, immigrants had to change their **name** to an **Icelandic** name. However, people applying for **citizenship** are still required to prove a **basic understanding** of Icelandic. This policy helps to maintain **cultural homogeneity** in Iceland.

Diversity in Singapore

1) Singapore is a small nation state between Malaysia and Indonesia. There's no single Singaporean ethnicity — the population is made up of people from Chinese, Malay, Indian and other ethnic groups. Each of these groups is also fairly diverse — they include speakers of several languages (e.g. English, Mandarin and Malay) and dialects and followers of different religions (e.g. Buddhism, Christianity and Islam).

2) This diversity is a result of Singapore's long history of migration and its central location on trade routes between India and China. Before Singapore became an independent nation in 1965, it was colonised by the UK. The growth of Singapore's economy (before and after gaining independence) encouraged migrants from countries surrounding Singapore to provide labour.

Photo of a multi-lingual street sign in Singapore

© Paul Brown / Alamy Stock Photo

> You need to learn both these case studies for the exam, so you can compare the two.

National **borders** are the result of **geography** and **history**

1) The **borders** of many nation states have **evolved** over time. They often follow **physical** barriers, such as **rivers** (e.g. the Rio Grande between the USA and Mexico) or **mountain ranges** (e.g. the Pyrenees mountains between Spain and France). Other borders are the outcome of **historical development** such as **conflicts** and **treaties** between different **groups** or **countries** as they **negotiated** over who had control of **territory** and **resources**.

2) When **European** countries **colonised** other parts of the world, they **divided** territories between themselves, drawing **new borders** around their **colonies**. Colonial **administrators** often drew borders as **straight lines** on maps. Many of these borders **ignored** the **reality** on the **ground** — e.g. **social differences** (such as ethnicities, religion or language) between groups of people living there or the terrain of the area.

3) This has contributed to **tensions** between different groups, leading people to **question** the **legitimacy** of the borders. In some places, people from the **same** ethnic group were **divided** by national borders, e.g. the land inhabited by **Kurdish** people was divided between Iraq, Iran, Syria, Turkey and Armenia:

The Iraq border

1) In 1916, the UK and France drew up plans, called the Sykes-Picot Agreement, to divide part of the Middle East between them. This boundary ran approximately along the modern boundary between Syria and Iraq.

2) The plans took little notice of ethnic or religious groupings, meaning that some groups, such as the Kurds, became minorities in several countries, and had little power. The UK also supported Sunni Muslims in the region, making them more powerful than other groups.

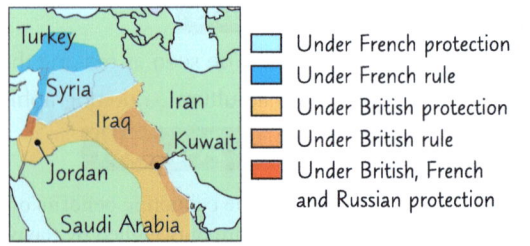

A map of the boundaries of the Sykes-Picot Agreement

Under French protection
Under French rule
Under British protection
Under British rule
Under British, French and Russian protection

3) Tension between different groups and disputes over boundaries have led to decades of conflict in the region, including calls for a Kurdish state and challenges to current national boundaries by the Islamic State group (ISIL). Such disputes undermine the legitimacy of national governments and can challenge countries' sovereignty.

Nation States

Some **nation states** and **national borders** are **contested**

Not all **nation states** are **recognised** by other countries, and countries don't always **agree** about where **borders** should be. This can lead to **tension** or even **conflict** between countries. It can also cause big **population movements**, as people **migrate** to **avoid** conflict or to move into a **different** nation state.

The Ukrainian and Russian border

1) From **1783 to 1954**, the **Crimean Peninsular** was part of **Russia**. In **1954**, Crimea became **part of Ukraine** and in **1991** Ukraine became **independent**.

2) In February 2014, Russia **annexed** the Crimean Peninsular in southern Ukraine (see p.226), and **conflict** between Russia and Ukraine **intensified**. Russian forces (along with pro-Russian Ukrainian separatists) took **control** of the **military bases** and the **Crimean parliament**.

3) **Armed conflict** between the two countries began in April 2014 in the Donbas region of eastern Ukraine.

4) In 2022, the conflict between Russia and Ukraine **escalated** when Russia launched a full invasion of Ukraine. This led to large **population movements** and at the end of 2022 **over 7.8 million** Ukrainian refugees were recorded throughout Europe.

A map showing the border between Ukraine and Russia

Conflict between China and Taiwan

1) **Taiwan** is an **island** off the coast of China in the western Pacific Ocean. During the **Chinese Civil War** (1945-1949), China's **government** was **defeated** by the **Chinese Communist Party**. In 1949, the government's forces **fled** to Taiwan and took **control** of the island. They **claimed** that they were **still** the **leaders** of China, and planned to **take back control** of **mainland** China. Meanwhile, China's new **Communist government** refused to **recognise** their **authority** — it still argues that Taiwan is actually **part** of **China** and isn't a **sovereign state**.

2) China has **reinforced** this message by **refusing** to have official **diplomatic relations** with countries that recognise Taiwan. **Fewer** than **20 countries** officially recognise Taiwan, and some **international organisations**, e.g. the **UN**, don't treat it as a sovereign nation. This makes it **difficult** for Taiwan to expand its **international influence**.

3) There's a **risk** of **conflict** between the two countries. China has fired **missiles** and carried out **military exercises** in the **sea** between mainland China and Taiwan. Tensions have **increased** as the **USA** (who doesn't recognise Taiwan as a country) has started making several visits to Taiwan. However, the USA still claims to be **against** an independent Taiwan.

Map of China and Taiwan

Warm-Up Questions

Q1 Name two factors that can shape a country's borders.

Q2 Explain why one country may have a different level of diversity to another.

Exam Question — A-Level

Q1 Explain how borders that don't consider ethnic or religious differences can lead to problems. [8 marks]

Learning also comes in all shapes and sizes...

Sadly, there's not one foolproof way of revising — I know that would make things a whole lot easier. Everyone learns differently and it's OK to go against the grain and do things a bit differently if that's what suits you. OK, inspirational talk over.

Nationalism

Nationalism has shaped the world as we know it, read on to find out why and learn some case studies along the way…

Nationalism in the 19th century led to colonisation and conflict

1) **Nationalism** is when people have a **strong sense of loyalty** towards their country. Nationalists often believe that **their nation** is **superior** to **other nations**.

2) During the **19th century**, nationalism became **popular** in European countries like **Britain**, **France** and **Germany**. These countries were fighting **each other** over national **borders** in Europe. This gave people a stronger **sense** of **national identity** — e.g. what it **meant** to be **British** rather than **French** or **German**.

3) In the 19th century, European countries were **expanding** their **empires** in other parts of the world. Their **nationalism** meant that they believed that they were **superior** to the countries they were **colonising**. This meant that they thought it was a **good thing** to impose their **authority**, **culture** and **language** on other nations.

4) European countries became increasingly **competitive** with each other, vying to be the **most powerful** country, with the most **colonies**, **resources** and **military power**. At the end of the **19th century**, European nations were **competing** for the **control** of **African countries**, during a period known as the 'Scramble for Africa'.

5) One of Britain's **aims** during the 'Scramble for Africa' was to maintain links to **India**:

> **India and the British Raj**
>
> 1) **Britain** began colonising parts of **India** around 1600. These colonised areas were controlled **indirectly**, through **trade agreements** made between **existing rulers** and the **East India Company**.
>
> 2) In 1857, **Indian soldiers** working for the East India Company **rebelled** against their **British officers**, setting off a wave of **violence** between Indian and British people. In **response**, the British government decided to bring India under its **direct control**. This period of direct control was known as the **British Raj**. Indian rulers had to swear an **oath** of **allegiance** to **Queen Victoria**. The Raj used **military force** to keep **control**, and **promoted** British **culture**, e.g. by setting up **English-language schools** with a euro-centric curriculum.
>
> 3) Many Indians were **unhappy** with British **colonialism**. Young Indian men who'd received a **British education** but had been **refused jobs** in areas like the **civil service** started to **discuss** the **idea** of India becoming an **independent nation**. In 1885, the **Indian National Congress** had its first meeting. It grew into a **nationalist movement**, calling for India to become **independent** from Britain and be governed by **Indians** instead. This involved the **boycotting** of British goods and schools as well as **uprisings** against British control.

Some people argue that by imposing British nationalism on people in India, the British government actually contributed to the rise of a new Indian nationalism.

CASE STUDY

After the Second World War, lots of new nation states gained independence…

1) By the end of the **Second World War** in **1945**, nationalist movements had emerged in many colonies. The **devastation** of the war left European countries **weak** and in **debt**. They put more **pressure** on their colonies to provide them with **cheap resources**, but were also **less able** to **control** what happened in the colonies.

2) In **1947**, after years of **negotiation** with the British, **India** and **Pakistan** became **independent**. In the next three decades, many other nations were **created** from former **colonies** as they became **independent** and empires **disintegrated**.

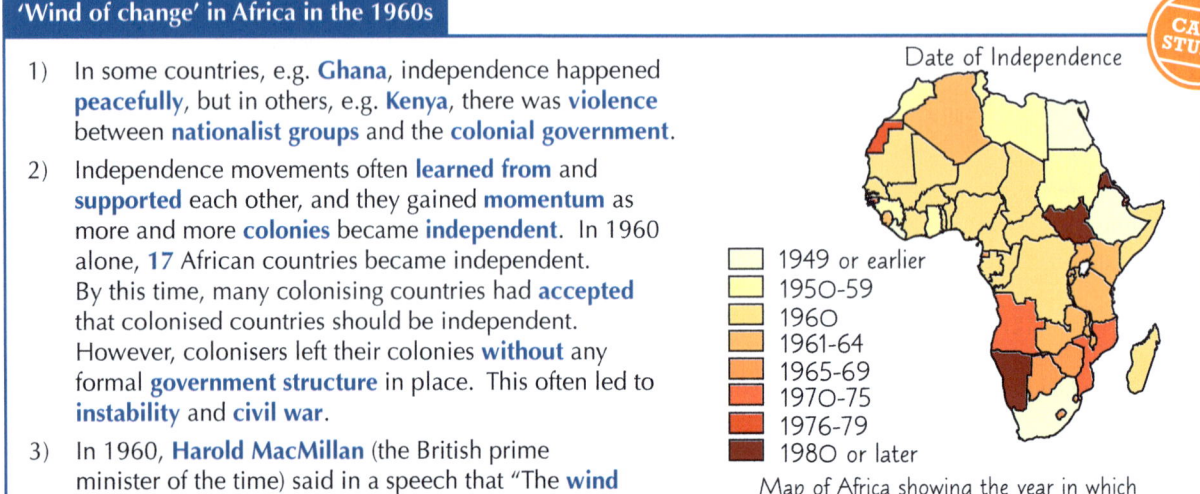

> **'Wind of change' in Africa in the 1960s**
>
> 1) In some countries, e.g. **Ghana**, independence happened **peacefully**, but in others, e.g. **Kenya**, there was **violence** between **nationalist groups** and the **colonial government**.
>
> 2) Independence movements often **learned from** and **supported** each other, and they gained **momentum** as more and more **colonies** became **independent**. In 1960 alone, **17** African countries became independent. By this time, many colonising countries had **accepted** that colonised countries should be independent. However, colonisers left their colonies **without** any formal **government structure** in place. This often led to **instability** and **civil war**.
>
> 3) In 1960, **Harold MacMillan** (the British prime minister of the time) said in a speech that "The **wind of change** is **blowing** through this **continent**".

Date of Independence

CASE STUDY

- ☐ 1949 or earlier
- ☐ 1950-59
- ☐ 1960
- ☐ 1961-64
- ☐ 1965-69
- ☐ 1970-75
- ☐ 1976-79
- ☐ 1980 or later

Map of Africa showing the year in which each country gained independence

Nationalism

...but in some countries this had **very negative** consequences

1) For many people in **former colonies**, independence was **exciting** — it was hoped that it would bring **peace** and **prosperity** to the new nations. However, this **wasn't always** the case — for some countries, the **transition** from **colony** to **independent country** was difficult.

2) The **borders** of many newly independent countries were shaped by **colonial interests**, rather than marking the territory of a **coherent**, **unified** nation. New nations were often made up of very **different social groups** (see p.268). Some colonisers actively **encouraged divisions** between these groups to make them **less likely** to form a **nationalist movement** for independence. This made it **hard** for people to agree **how** the new country should be **run** and **who** should **lead** it. In some cases, this led to **conflict**, causing huge **social**, **economic** and **environmental** problems.

Post-independence conflict in Sudan and South Sudan

CASE STUDY

1) In the 1890s, Sudan was colonised by the British. Between 1899-1955, Sudan was under British-Egyptian rule and in 1956 it became independent.

2) The new nation was made up of two very different areas. Northern Sudan has a dry climate, and most of its population are Muslim and speak Arabic. The south has a wetter climate and is more fertile, and it's home to a more diverse set of religious (e.g. Christians and Muslims) and ethnic groups (e.g. Dinka and Nuer). The south is also rich in resources, including 75% of Sudan's oil reserves. However, northern Sudan has relatively higher levels of human development than the south. In contrast, southern Sudan has the lowest Human Development Index in the world.

3) Before Sudan became independent, the southern areas of the country wanted to be given a degree of independence from the central government. These demands were refused.

4) Tensions between the north and south grew, leading to a civil war between 1955 and 1972. Years later, a violent conflict broke out between the government and the Sudan People's Liberation Army from 1983-2005.

5) Eventually, in 2011, South Sudan became independent from Sudan, but conflict has continued, especially around the control of oil reserves. This is because the infrastructure needed to transport this oil goes through Sudan. There has also been a series of devastating conflicts in Darfur, a region in western Sudan.

6) From 2013 to 2020, there was a civil war in South Sudan between different ethnic groups, who supported different leaders.

7) Conflicts in the region have had many negative consequences:

- **Social impact** — between 2013 and 2018, around 400 000 people were killed by the civil war in South Sudan, and nearly 4 million were displaced. NGOs reported many human rights abuses during the civil war, including civilian killings, sexual assault and looting.

- **Economic impact** — in 2013, South Sudan spent more than half its entire national budget on its military, and GDP per capita in 2018 was around 30% what it would have been if there had been no conflict. In Sudan, the conflict in Darfur cost nearly US$90 billion from 2003 to 2017. Spending on education and health care is low in both countries, further restricting economic development.

- **Environmental impact** — Sudan and South Sudan are vulnerable to desertification as a result of climate change, and this has been made worse by conflict. There's been severe degradation of the land around refugee camps, and the lack of political stability has meant that governments haven't focused on promoting sustainable development.

Map showing the location of Sudan and South Sudan, Africa

© John Robert / Shutterstock.com

Drying river bed, South Sudan

Desertification is when fertile land (such as shrubland and grassland) is transformed into desert. It is caused by a combination of drought conditions and overpopulation.

Nationalism

Migration from **former colonies** still **influences** nation states

Migration between **former colonies** and the **imperial core** (the **country that colonised them**) have shaped the **cultural heterogeneity** and **ethnic diversity** of nation states.

19th and 20th centuries

1) During the 19th and early 20th century, millions of people **left** European countries to **settle** in colonies like **Australia** and **South Africa**.

2) In the **latter** part of the 20th century, this pattern was **reversed** — large numbers of people moved to **Europe** from **formerly colonised countries** such as South Africa, India, Pakistan and the Caribbean.

3) After the **Second World War**, the UK **invited** people from its **colonies** and **former colonies** to come and help **rebuild** the UK (under the British Nationality Act of 1948).

4) Over the next 20 years, around **900 000 people** moved to the UK from **India**, **Pakistan**, countries in **Africa** and the **Caribbean**. This changed the UK's **ethnic composition**, making it more **diverse**.

5) For example, between 1956 and 1970 **London Transport** actively recruited around **6000 employees** from the Caribbean. Additionally, by 1960 nearly **40%** of NHS **junior doctors** were from India, Bangladesh, Pakistan and Sri Lanka.

6) Migration also **increased** the UK's **cultural heterogeneity** (diversity), as migrant communities introduced new **food**, **music**, **languages** and **art** to the UK. For example, curry became a popular dish in the UK as a result of the migration of people from countries such as India and Pakistan (see p.156).

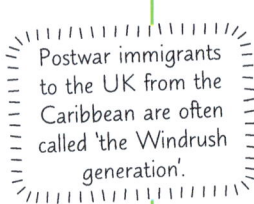
Postwar immigrants to the UK from the Caribbean are often called 'the Windrush generation'.

© Contraband Collection / Alamy Stock Photo

Photo of the Empire Windrush, which brought nearly 500 workers from the Caribbean to the UK in 1948.

Other former colonising countries have similar patterns — e.g. France has a significant migrant population from its former colonies in North Africa and West Africa.

Current Trends

1) People have **continued** to migrate to former **imperial core countries** from former colonies.

2) This is due to several factors, e.g. the countries' shared **history** and **language**, and the fact that **new** migrants can join their extended families and **existing** migrant **communities** from their home country.

3) In **2021**, over 33% of the migrants living in the UK were originally from **former colonies** like India, Pakistan, Kenya and Nigeria.

4) Some countries choose to **restrict immigration**. From the **early 1960s**, the UK passed laws that specifically limited immigration from **former colonies**. This changed the **pattern of migration** to the UK.

Country of birth	Number of people living in the UK in 2021
India	896 000
Pakistan	456 000
Kenya	144 000
Nigeria	312 000
Singapore	63 000

The number of people living in the UK who were born in former British Colonies.

Warm-Up Questions

Q1 What is nationalism?

Q2 Give an example of an environmental cost of a conflict in a post-colonial nation state.

Exam Question — A-Level

Q1 Explain how colonial migration patterns have shaped ethnic and cultural diversity in former imperial core countries.

[8 marks]

One thing I am passionate about is passing this exam...

Nationalism can lead to serious conflict. Learning about it, however, is a lot less risky. In fact, I'd actively advise you learn the last three pages so you stand the best chance of smashing the exam. It's daunting I know, but I believe in you...

Globalisation and Inequality

Globalisation has changed the way nation states work, and even led to the creation of new kinds of nation state.

Globalisation has led to the development of tax havens

FSI scores rank countries on their ability to hide money and the scale at which this happens. A higher score means a higher level of secrecy.

1) Globalisation has caused **big changes** to the **global economic system**, including the **removal** of **regulations** around **financial markets** and the **movement** of **money**.

2) Some countries have taken **advantage** of the **lack** of **regulation** by setting themselves up as **tax havens**. This usually means they charge **low** or no **taxes** on **financial dealings** — this is called a **low tax regime**.

3) This makes them very **attractive** to **people** or **companies** with a lot of **money**, who want to **avoid** paying **high rates** of tax on their **wealth**.

4) **Financial deregulation** (see p.241) has made it fairly **easy** for people to **keep** their money in a **tax haven**, even if the money is **made** in **another country**, because there are **fewer laws** in place.

5) Many **TNCs** base their **headquarters** in tax havens to **make the most** of this. Some **wealthy individuals** also choose to live as **expatriates** in tax havens, or **invest** their money in **banks** based there.

	Financial Secrecy Index (FSI) 2022
USA	1951
Switzerland	1167
Singapore	1167
Hong Kong	927
Luxembourg	804

The world's largest tax havens ranked by FSI scores.

6) Some of the world's tax havens are **small island states**, e.g. the **Bahamas**, or the **overseas territories** of bigger countries, e.g. the **Cayman Islands** (which are an overseas territory of the UK).

There's disagreement over whether tax havens should be allowed to exist

1) People and organisations **disagree** about whether to **allow** tax havens to **continue**. Tax havens have **pros** and **cons**:

Advantages
- Tax haven status can be a good way for small countries to make money, by charging fees for people to register their businesses there.
- Tax havens help TNCs (see p.93) make bigger profits. Some people argue that these can then be used to expand the company and employ more people, which can promote economic growth.

Disadvantages
- Governments lose taxes — up to $600 billion worth of corporate taxes are thought to be lost globally each year due to tax havens. These taxes could be spent on key services and human development.
- Secrecy around finances can allow illegal activities to take place, e.g. international drug trafficking.

2) **Governments** and **IGOs** mostly **accept** that tax havens are **part** of the **global economic system** — they're the **price** they pay for **deregulated financial markets** (see p.241). IGOs like the **EU** and the **OECD** (see p.110) have taken some **measures** to make TNCs more **accountable**, but their power over companies is limited. Governments **worry** that if they try to **force** companies to pay tax, **TNCs** will **move** somewhere else where they will be taxed less.

3) Some **NGOs**, e.g. **Oxfam**, have called for **stronger financial regulation** and an **end** to tax havens. They argue that by allowing companies and individuals to **avoid** paying tax, **wealthy** people become even **richer**, which also increases **global inequality**. Meanwhile, this **reduces** governments' income from taxes, so they have less money to spend on **health care** and **education** — this can increase **poverty**. As a result of this, most people have to pay **higher taxes** or **spending** has to be **cut**.

Global inequality is growing

1) Globalisation has **contributed** to **economic growth**. It's made it **easier** for **TNCs**, and the people who **own** them, to make large amounts of **money**. This has increased the wealth divide between the world's **richest** and **poorest**. In 2020, the richest **1%** of the **global population** owned nearly **46%** of the world's **wealth**.

2) Countries where income inequality is **rising** experiences **slower economic growth**, which affects the **global economy**. The main reason for this is that **unequal access** to **education** limits the skills that people from poorer backgrounds can acquire, which in turn **limits** their potential contribution both to the economy and to progress, e.g. in technology. Inequalities in **health care** can also **reduce productivity**, which limits economic growth.

3) There's rising inequality **between** countries — for example, the world's wealthiest people are **concentrated** in wealthy countries like the **USA** and **Japan**. There's also growing inequality **within** many countries — between different **social groups**, **men** and **women**, and **wealthy elites** and the **rest** of the population. These groups also experience **other kinds** of inequality, for example access to **education** and **human rights** (see p.244).

4) These growing inequalities could **threaten** the **stability** and **sustainability** of the global economic system. Inequality can increase **tensions** between **different social groups** and lead to **conflict** or **protests** against the **government**. This can weaken **social** and **political stability**.

5) **International organisations** are starting to **recognise** these problems. In **2020**, the **UN** called for countries to work towards **greater equality**, especially for **women** and **minority groups**.

Globalisation and Inequality

Ecuador has reduced inequality by following a different economic model

Some countries have **moved away** from the **neoliberal** economic model that's been promoted as part of **globalisation** (see p.88). Instead, they've adopted **alternative models** that aim to **reduce inequality** and **poverty** rather than increasing **economic growth**.

This is the moment Larry learned that the word 'Ecuador' is Spanish for 'Equator'.

An alternative economic model in Ecuador

1) Since 2007, Ecuador has brought in more **socialist** policies that have helped it to become a more **equal** society.

2) In the 1980s and 1990s, Ecuador was **unstable** and experienced **conflict** and **economic recessions**. It had large **international debts**.

3) In 2006, **Rafael Correa** was elected as Ecuador's **president**. He announced that his **new government** would put **Ecuadorian** people **first**, moving away from the approach of **earlier governments**. Correa believed they had focused **too much** on making **profits** from Ecuador's **oil reserves** and paying the country's **debts**.

A map showing the location of Ecuador within South America.

Gini Coefficient, 2003-2020, Ecuador

Since 2003, Ecuador's Gini coefficient has generally declined

4) In 2008, the new national **constitution** was approved, which included the idea of *buen vivir*. This idea comes from **indigenous** South American philosophy. It translates as 'good living', but it means **more** than that — e.g. that **everyone** should have a **good life** that doesn't **damage** the **planet**, and that the wellbeing of the **community** is more **important** than the success of one **individual**. This challenges **neoliberal ideas** (see p.241) about the importance of **individual choice** and **economic growth**.

5) Correa increased **public spending** on services and temporarily **raised taxes** in order to fund Ecuador's recovery from an earthquake. His government also **sued** a US-based **oil company** for causing **environmental damage** in the **Amazon**.

6) Ecuador's inequality decreased and **poverty** rates nearly **halved** between 2006 and 2016. This can also be seen by the **decrease** in the country's **Gini coefficient** (see p.105), which has shown an overall decline since 2003. However, Correa and his policies **weren't popular** with everyone. He was **criticised** by **indigenous groups** and **activists** for his **authoritarian** approach and for drilling for **oil** in a **remote** part of the **Amazon** rainforest.

7) Following the **COVID-19 pandemic**, Ecuador entered a **recession** which had a negative impact on poverty levels within the country. In 2021, a government plan was drawn up to try and improve the **socio-economic** situation by focusing on creating opportunities for its citizens.

Correa's presidency ended in 2017.

Warm-Up Questions

Q1 Give one advantage and one disadvantage of tax havens.

Q2 Explain one way that global inequality can affect global stability.

Exam Question — A-Level

Q1 Evaluate the view that 'the global economic system will need to change in order to become sustainable.' [20 marks]

Top Tip #236 for budding geographers: make Larry the lemur proud...

Globalisation and inequality is a complicated issue — there's a fair amount to take in on these two pages. You don't want a face like Larry's when you sit down to do the exam, so get stuck in, don't be scared and you'll be thriving...

United Nations

Since the Second World War, several organisations have been set up to sort out global issues, but the UN is the original...

The **United Nations** is a **global organisation**

1) The **United Nations** (**UN**) was formed in 1945 to establish a **peaceful** and **fair** world. It was the **first IGO** to be set up **after the Second World War**, and its importance in **global governance** has continued to grow.

2) The UN currently has **193 member countries** — the UN has a lot of **authority** because almost **every country in the world** is a member.

3) When countries join the UN, they have to sign the **United Nations Charter**. This sets out the basic **principles** of global governance and the **functions** of the UN. According the Charter, the UN's **aims** are:

- to maintain **global peace** and **security**.
- to develop **friendly relations** between nations.
- to use **cooperation** to solve international problems.
- to bring countries together to **settle** disputes.

Euan was tired of people assuming he could also solve global problems.

4) In practice, the UN tries to deal with **global problems**. It's made up of several agencies and organisations that address different issues:

① **Environmental** — the UN set up the United Nations Framework Convention on Climate Change in 1992 (see p.222), an international treaty to get member countries to tackle **climate change**.

② **Socio-economic** — the UN has several organisations that work on **reducing poverty** and **inequality** (e.g. the UN Development Programme) or supporting **vulnerable groups** (e.g. the UN Refugee Agency).

③ **Political** — the UN **Security Council** (see below) is responsible for maintaining **global peace and security** while the General Assembly acts as a 'parliament of nations' and **makes decisions** on a range of issues.

Members of the **UN Security Council** have **different** ideas about **geopolitics**

1) The **UN Security Council** has **15 member countries** — **5 permanent** ones and **10 temporary** members that **change** every **two years**. This ensures that each country is **represented** at some point.

2) The members of the Security Council, especially the **permanent** members, include countries with very different **political systems**, **allies** and **foreign policies** — e.g. **China** is an **authoritarian** state, whereas **France** is a **democracy**. This can lead to **disagreements** over **Security Council** decisions, e.g. whether to launch an **intervention** to protect **human rights**.

Permanent members of the UN Security Council:
- 🟦 China, France, Russia, UK, USA

Non-permanent members of the UN Security Council:
- 🟥 African countries
- 🟥 Asia-Pacific countries
- 🟧 Latin American countries
- 🟨 Western European countries (and Australia, Canada, New Zealand and Turkey)
- ⬜ Eastern European countries

3) The Security Council votes on **decisions** about how the UN and its member countries should **respond** to situations that threaten **global peace and security**, e.g. wars. Permanent members have **veto rights** — this means that they have the power to **block** a decision made by the rest of the Security Council. For example, in 2020 the USA vetoed a decision made on counter-terrorism by the UN security council.

The UN has intervened **directly** and **indirectly** to **protect human rights**

1) The UN tries to maintain **peace** and **security** and protect **human rights** by **intervening** in situations where these are **threatened**. The UN might apply **economic sanctions** or launch a **direct military intervention**:

- The UN and its member countries put **political** and **economic pressure** on the countries involved in the crisis.
- This can range from **writing letters** to leaders to applying **economic sanctions**, e.g. **trade embargoes** that restrict trade with a country.
- In very **serious** crises, the UN Security Council might choose to launch a **military intervention**, sending in the UN's **peacekeeping forces**.

2) Interventions by the UN have had a **mixed record** of success — in some cases, the intervention has **resolved** or **prevented** a crisis, but in others, intervention has been unsuccessful.

United Nations

UN intervention in the Democratic Republic of the Congo

1) In 1998, **conflict** broke out between rival **ethnic** and **political** groups in the **Democratic Republic of the Congo** (DRC). Each side was supported by **neighbouring countries**, e.g. Rwanda and Angola. Over **5 million** people lost their lives between 1998 and 2008.

2) The UN Security Council called for a **ceasefire** and **warned** other countries not to get involved in the conflict, although this warning was **ignored** by some countries. For example, Zimbabwe sent over **10 000 troops** into the DRC and held over 300 prisoners of war.

3) The UN Security Council set up a **peacekeeping mission** to help the different groups to reach an **agreement** and hold **democratic elections**. This involved deploying over 19 000 peacekeeping **troops** to end conflicts that continued in some regions of the DRC.

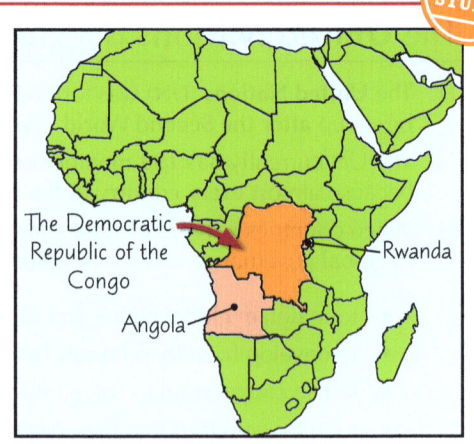

A map showing the location of the DRC, Africa

Photo of UN peacekeepers in the DRC

4) The mission managed to **resolve conflict** in some areas and elections were held for the first time in over 40 years. However, **violence** and serious **human rights abuses** continued in eastern DRC.

5) In 2013, the UN Security Council approved the creation of an '**Intervention Brigade**' to tackle the armed groups behind the violence. The aim was to strengthen the **peacekeeping mission** and **stabilise** the conflict in the region.

6) Progress has been **slow** — at the end of 2022, there were still around 12 800 UN soldiers in DRC and the **conflicts continued**.

Members of the UN have carried out their own interventions

1) The UN Security Council has strict **rules** about when an intervention (especially a direct military intervention) is **justified**, and member countries often **disagree** about the best way to resolve a situation. Some of the UN's member countries have acted **independently**, carrying out interventions in other countries **without** the Security Council's **approval**.

2) A **coalition** led by the USA and the UK intervened in **Afghanistan** in 2001 and in **Iraq** in 2003. They argued that this was part of a '**war on terror**'. By intervening in **unstable** or '**failed states**' like these, they argued that they would **prevent** a global rise in **terrorism** and help to maintain global stability (see p.256). **Russia** has also carried out its own interventions, e.g. providing military support to the **Syrian** government to fight rebel groups (since 2015).

3) None of these interventions were approved by the UN Security Council. This was due to **disagreements** about whether the interventions were **necessary** and **doubts** about whether they would **help** to protect human rights.

4) By acting without the Security Council's approval, these interventions have undermined **geopolitical relations** between countries, as the intervening countries are **ignoring** the principles of the UN Charter. Independent interventions have also threatened **global stability** — e.g., the intervention in Iraq led to a long conflict that **displaced** millions of people and contributed to **violence** in **neighbouring countries**.

Warm-Up Questions

Q1 When was the UN set up and why?

Q2 Give one way in which the differing geopolitical views of countries in the UN may affect the way the UN works.

Exam Question — A-Level

Q1 Explain why the UN and its member states may intervene in other countries. [8 marks]

UN-sure what do to next? First things first, pop the kettle on...

Before you dive in, my advice would be to get your fave brew on the go. Revision doesn't need to be stressful, in fact, it can actually be quite enjoyable once you get into it. UN-believable, I know? Sorry, no more UN puns from now on, I swear...

IGOs and Trade

There are three main IGOs that set the rules for global trade and finance — buckle in, there's lots to learn here...

Three IGOs play a big role in global trade — the IMF, World Bank and WTO

1) After the Second World War, the winning Allied nations set up **new economic institutions** to promote international **trade** and help to maintain the dominance of **capitalism**. The IGOs that grew out of these institutions still play an **important role** in the **global economic system**. The **three** main ones are:

> (1) **The World Trade Organisation (WTO)** was set up to **increase trade** and help **resolve trade disputes** between member countries. It sets **rules** about **how** countries should **trade with each other**.

> (2) **The International Monetary Fund (IMF)** monitors the global economy and **advises governments** on how they could **improve** their **economic situation**. It also gives **loans** to countries with economic problems.

> (3) **The World Bank** provides **loans** to **developing** countries to invest in areas like **health**, **education** and **infrastructure**.

2) All three IGOs are **based in the West** (the IMF and the World Bank are in the USA and the WTO is in Switzerland) and they're based on western ideas about what a **capitalist economic system** should look like. The **main ideas** are:

- **Free trade** — there should be as **few barriers** to trade as possible, **between and within** countries.
- **Privatisation** and **deregulation** — companies and resources are **managed more efficiently** when they're privately owned, and the economy is more efficient when it's **free** from **government interference**.
- **Democracy** — people should have the **freedom to choose**, from their **government** to **what they read**.

3) These IGOs use their **influence** to **manage** the **global economy**, e.g. by overseeing agreements between countries and lending money to countries to **prevent** an economic crisis. They also use this influence to **promote** their **western capitalist model**. For example, by establishing policies about trade and financial markets as well as offering countries loans that are **conditional** on adopting free trade policies.

IGO policies on trade and borrowing have helped developed countries

1) IGOs have helped **increase economic growth** in **developed** countries. Policies that encourage **international trade** and reduce **economic barriers** have helped these countries to export their goods and services, and to import cheaper goods from developing countries, boosting the **economic growth** of developed countries. It's estimated that the USA's membership of the WTO **increases** its **GDP** by around US $87 billion a year.

2) The IGOs set **borrowing rules**, e.g. the IMF only lends money to **mitigate** or **prevent economic crises**. This can benefit developed countries — e.g. in 1976, the UK government was approved a loan of $3.9 billion from the IMF to help it **avoid a financial crisis** (although the UK didn't end up withdrawing the full amount). Interest rates **fell** and the overall economic situation **improved** as a result of the loan.

3) Developed countries also have **more influence** in the running of the IGOs, so they have **more power** to make sure the rules work in their favour. The USA has a 17.4% share of voting rights in the IMF, so it has a lot of sway over decisions.

IGO policies on economic restructuring in developing countries have caused issues

1) IGOs have encouraged developing countries to adopt policies like **free trade**, **privatisation** and **deregulation**. They've done this by attaching **conditions** to their loans. These conditions take **two main** forms:

Structural Adjustment Programmes (SAPs)	Heavily Indebted Poor Countries (HIPC) initiative
The **IMF** or the **World Bank** agrees to provide a **loan** to a country, on the condition that it makes **big changes** to its **economic policy**. This can include reducing **government spending**, privatising **state-owned** industries and reducing regulations and other barriers to **international investment**.	The **IMF** and the **World Bank** offer **debt relief** (reducing the amount of money that countries have to pay them back) to **developing countries** that meet strict criteria. These criteria include adopting **economic reforms** and taking steps towards **reducing poverty**.

2) SAPs and HIPC policies have had **mixed results**. In some cases, SAPs have helped to promote **economic growth** and HIPC initiatives have freed up money for governments to spend on health care and education instead of paying back debts. However, in some countries, they've had **negative impacts** (see the next page). They've also been **criticised** for **undermining** countries' **economic sovereignty** — people argue that IGOs use their power and wealth to prevent countries from making their own economic decisions.

IGOs and Trade

Structural adjustment policies have caused problems in Jamaica

Jamaica's structural adjustment programme

1) Jamaica is a developing country in the Caribbean. In the 1970s and 1980s, it struggled with economic recessions and debt. This led to Jamaica accepting loans from the IMF and the World Bank with structural adjustment conditions.

2) One of the main conditions was adopting austerity measures — reducing government spending on public services like health care. This had a big impact on these services — e.g. the number of registered nurses fell by 60%.

3) In 1991, Jamaica agreed to more structural adjustment in exchange for more loans — this time, the focus was on deregulating Jamaica's economy, with policies like keeping wages low to encourage foreign investment.

4) Social and economic wellbeing have declined since these reforms were introduced — for example, in 1990, 97% of children completed primary school, whereas 85% of children did in 2019.

5) A large proportion of the Jamaican government's spending goes towards paying back foreign debts, limiting the amount it can spend on health care and education. Jamaica still has a high burden of debt, however, it doesn't qualify for an HIPC initiative because its GDP means it's considered an upper-middle-income country, and is therefore 'too rich' for this financial assistance.

CASE STUDY

Most countries are members of global IGOs, but there are regional trade groups too

1) **Almost all** countries are members of the IMF, World Bank and the WTO. This is because the **three IGOs** are so **influential** that countries find it **difficult** to **take part** in the **global economy** without becoming members.

The main exceptions are countries like North Korea, who are economically and politically closed off from the rest of the world.

2) Additionally, some countries are members of **regional trade blocs**, like **NAFTA**:

North American Free Trade Agreement (NAFTA)

1) **NAFTA** was a **treaty** between Canada, the USA and Mexico. It was signed in 1992 and committed the three countries to **removing** any **trade barriers** between them, e.g. tariffs and customs duties on imports and exports. This allowed corporations from Canada and the USA to take advantage of the lower labour costs in Mexico, and Mexico was able to import high-quality goods from Canada and the USA. NAFTA also involved agreements about how to resolve **trade disputes** between the three countries.

2) NAFTA has had **advantages** and **disadvantages** for its members — it increased trade and foreign investment between the countries, but it's also **created problems** in some sectors, e.g. family farmers in Mexico have struggled to compete with cheap agricultural imports from the USA. NAFTA was **replaced** by a new United States-Mexico-Canada (USMCA) agreement in 2020.

CASE STUDY

European Union

1) The EU is a common market area — most barriers to trade and the movement of people, goods and money between the member countries have been removed. Some member countries use a shared currency, the euro, to remove these barriers even further. EU members have shared policies on trade with countries outside the EU.

2) However, the EU isn't just a trade bloc — it's also developed its own political agreements and institutions, e.g. the European Parliament. EU members have joint policies or make joint decisions on issues like agriculture, climate change and foreign policy. The European Parliament is an example of how the EU has led to closer political unity.

Warm-Up Questions

Q1 What is an HIPC initiative?

Q2 Give an example of a regional trade bloc and describe how it works.

PRACTICE QUESTIONS

Exam Question — A-Level

Q1 Explain how policies made by intergovernmental organisations such as the WTO, IMF and the World Bank can be seen to benefit developed countries more than developing countries. [8 marks]

I rode my bike into a non-tariff barrier just last week...

Trade is mentioned an awful lot in this book. Get to grips with IGOs and trade and you'll be at least part of the way there...

IGOs and the Environment

Many environmental problems are global, so international organisations have been established to try to manage them.

IGOs have been created to tackle **environmental challenges**

1) Countries have set up **IGOs**, international **treaties** and **agreements** to deal with **global environmental problems** like pollution, biodiversity loss and climate change.

2) Some environmental IGOs and agreements have been **more successful** than others.

The Montreal Protocol **limited** the use of **CFCs**

The Montreal Protocol

1) The **ozone layer** is mostly found above the Earth's surface in the stratosphere (part of the atmosphere). It absorbs some of the sun's harmful **ultraviolet rays**, making it safer for life to live on Earth. In the 1970s, scientists realised that industrial chemicals, such as chlorofluorocarbons (**CFCs**), were damaging the ozone layer.

2) In 1987, the UN set up an **agreement** between countries to limit and eventually ban the production and use of CFCs and other ozone-damaging chemicals. This was called the **Montreal Protocol on Substances that Deplete the Ozone Layer**, and it was signed by **every country** in the world.

3) By 2010, the Montreal Protocol had succeeded in **phasing out 98%** of the ozone-damaging chemicals that were being emitted in 1990. The biggest area of ozone damage, above the Antarctic, has shown **signs of recovery**. However, from 2012 to 2017, there was an **increase** in emissions of **one type of CFC**, suggesting that some countries **broke the global ban** in this period. Global ozone levels are expected to **fully recover** by around **2050**, but the damage over **Antarctica** may **not** fully recover until around **2100**.

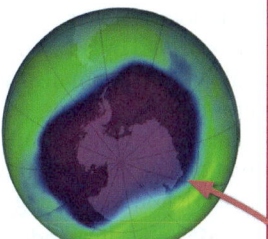

The ozone hole over Antarctica, 2006

Geoffrey just hadn't been able to get the same volume from his hairspray since the CFC ban.

The ozone hole is the region in purple over Antarctica. It's an area where there are extremely low levels of ozone in the atmosphere. The size of the ozone hole changes throughout the year due to changes in temperature and stratospheric circulation patterns.

CITES **stops** people **trading endangered** species

CITES

1) CITES is the **Convention on International Trade in Endangered Species of Wild Flora and Fauna**. It's an international agreement between governments (currently over 180) about trading wild plant and animal species as well as products made from them.

2) The convention, which came into force in 1975, covers around **6000 animal species** and **32 000 plant species**. These are sorted into three categories, depending on how high the **risk** of them being **traded internationally** is.

3) Trade involving the most **endangered** species, such as tigers, have the **strictest** rules. Species that are less at risk can be traded by some countries and in certain circumstances, as long as the trade is **monitored** and **regulated**. CITES holds training workshops in order to help countries to enforce the treaty and tackle illegal trade. CITES can be enforced through national legislation.

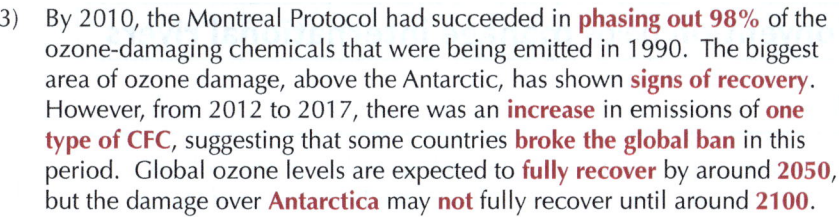

A vicuña (a previously endangered species)

4) CITES has helped to make trade more sustainable and **protect** some species. For example, the **vicuña** (an animal related to llamas and native to South America) was seriously endangered in the 1970s because of demand for its fur, but populations have now **recovered** — by 2020, populations had reached over 450 000.

5) However, CITES hasn't been completely successful — sanctions are **rarely imposed**, membership is **voluntary** and it can be **slow** to take action. For example, **pangolins** were designated as threatened with **extinction** in 2014 but weren't added to CITES **most endangered** list till 2017.

A pangolin (an endangered species)

Between 2010 and 2013, around 1 million pangolins were traded illegally.

IGOs and the Environment

The **UN Convention on the Law of the Sea** is designed to **protect oceans**

The United Nations Convention on the Law of the Sea

1) The **United Nations Convention on the Law of the Sea** (UNCLOS) is an international treaty that covers the governance of the sea, including **environmental management** and the use of **marine resources**. Under the convention, countries have a **responsibility** to protect the oceans and marine creatures. For example, countries have the right to fish, but only within sustainable limits. The convention ensures that all countries that sign up **cooperate** in **protecting** the marine environment.

Cod fishery, Northern Norway

2) UNCLOS provides a **basis in law** for using marine resources more sustainably, but it doesn't cover **everything** — some aspects of marine conservation are covered by **other organisations** and agreements, e.g. the International Whaling Commission is responsible for conservation of whales and for making whaling sustainable. Some people argue that to really protect the oceans, there needs to be an **overarching treaty** that covers **all aspects** of marine conservation and sustainability. UNCLOS is also **hard to enforce** — for example, some countries get around the restrictions on fishing by moving their catch onto other boats in international waters.

3) **Exclusive economic zones** were adopted as part of UNCLOS. These are areas where coastal nations have **exclusive rights** to the **resources** within 200 nautical miles of their shore.

The **Helsinki Water Convention** helps **manage international rivers**

The Helsinki Water Convention

1) Many countries **share** water resources, e.g. rivers that flow through several countries, or lakes that border on more than one country. In 1996, the **Convention on the Protection and Use of Transboundary Watercourses and International Lakes** (Water Convention) came into force after being set up in **Helsinki** by the UN to help countries manage their shared water resources in a sustainable way.

2) The **Water Convention** has helped countries to develop **agreements** about specific water resources, e.g. the Convention on the Protection of the Rhine, which involves five countries, plus the European Community. It's also supported lots of **monitoring** and **research** around water quality, e.g. the monitoring of chloride levels. However, not all countries have joined the Water Convention — as of 2022, 47 countries have ratified the convention.

The Rhine flowing next to Lorch, Germany

The Helsinki Water Convention is seen as an important tool for achieving the SDGs (see p.243).

The **Millennium Ecosystem Assessment** monitored the state of the **environment**

The Millennium Ecosystem Assessment

1) The UN set up the **Millennium Ecosystem Assessment** (MA) in 2001 to carry out a global **study** of the state of the environment and to monitor environmental change. The idea was that this would provide a **scientific basis** for tackling environmental problems and the impact they can have on **human wellbeing**. Over 1360 scientists and experts contributed to the MA, which was published in 2005.

2) The MA's findings were that most of the Earth's ecosystems are being **damaged** and **degraded**, and that human activities have caused huge biodiversity loss. These results were used to inform **environmental policy** and ongoing work on existing agreements like the **Convention on Biological Diversity**. However, some countries and organisations have ignored the results — the MA was a **summary** of the **global environmental situation**, but didn't have any political or legal power to drive change.

IGOs and the Environment

The Antarctic Treaty System **protects Antarctica** from **exploitation**

1) Antarctica **isn't owned** by any one country, although **several countries** have made **territorial claims** to it. International agreements have been set up to try to make sure that it's governed peacefully and used for **scientific research** rather than exploited for its resources.

There are a total of 14 articles in the Antarctic Treaty.

2) The **Antarctic Treaty (1959)** is an agreement about how to sustainably manage Antarctica's ecosystems. It has now been signed by 55 countries. The rules laid out in the treaty include:

- Antarctica should only be used for **peaceful reasons** — no army bases or weapons are allowed on Antarctica.
- Countries should **cooperate** on **scientific research** in Antarctica by sharing plans, researchers and results.
- Antarctica should **remain** in the **global commons** — individual countries can't make any claims to it.

3) Overall the Antarctic Treaty is viewed as a **success**. It ended territorial claims being made and Antarctica has remained **largely unexploited**, while scientists working there have **collaborated**, increasing knowledge about the Antarctic environment. Under the 1959 treaty, all bases and equipment in Antarctica can be **inspected** at any time, with different countries taking responsibility for carrying out inspections. However, inspections **don't** occur very **often**.

Emperor penguins in Antarctica

4) The **Protocol on Environmental Protection to the Antarctic Treaty** was signed in **1991** and added to the Antarctic Treaty. It focuses on protecting Antarctica's fragile **environment**. This **banned all mining** in Antarctica. It also set rules to help **protect** Antarctic **plants** and **animals**, **regulate waste disposal** and **prevent pollution**. Under the 1991 protocol, an Environmental Impact Assessment (EIA) is required for any new activities.

5) However, there's **no system** to ensure all countries abide by the rules. If there are **disputes** between countries then they are encouraged to negotiate. Otherwise, disputes can be taken to the **International Court of Justice**. The countries involved must reach a **consensus** over all decisions regarding Antarctica. Tackling problems can therefore be **slow** and **difficult** — e.g. between 2012 and 2016, plans for Antarctic marine reserves repeatedly failed because of opposition from Russia and Ukraine. The **number of countries** involved in the treaty and their different **priorities** can make **decisions** around further environmental protection **difficult**.

These claims were all made before the Antarctic Treaty was signed in 1959.

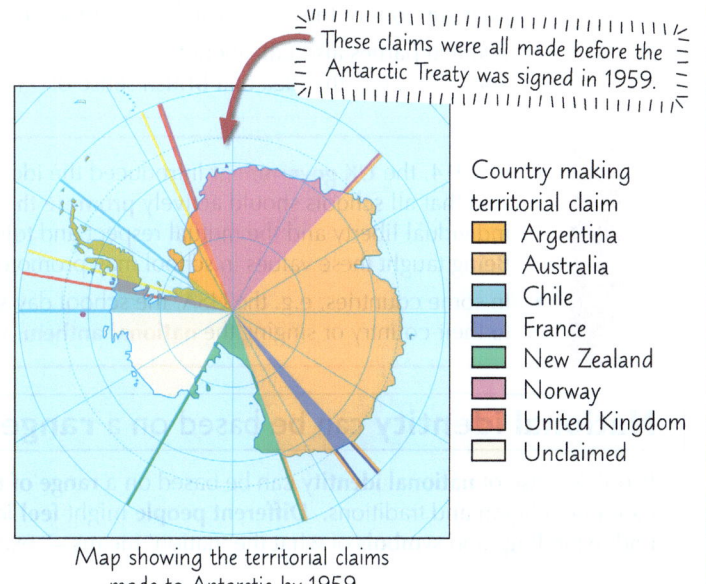

Country making territorial claim
- Argentina
- Australia
- Chile
- France
- New Zealand
- Norway
- United Kingdom
- Unclaimed

Map showing the territorial claims made to Antarctic by 1959

6) Although the treaty has largely been viewed as a success, there are serious **threats** to the Antarctic that the treaty can't deal with, e.g. **climate change** is causing **warming** in the area, which can **increase melt rates** and encourage the spread of **non-native plant species**.

Warm-Up Questions

Q1 What is the Montreal Protocol?

Q2 Explain how the Antarctic Treaty System works.

PRACTICE QUESTIONS

Exam Question — A-Level

Q1 Explain how IGOs attempt to manage environmental problems affecting the oceans. [6 marks]

Wailing — one response to exploitation in Antarctica...

It's snow joke — Antarctica's a cool customer, but it's also a key case study for the exam, so make sure you're clued up on the Antarctic Treaty System and how it protects Antarctica from exploitation. That'll protect you from slipping up in the exam.

National Identity

Nationalism is important in culture and politics — but it means very different things to different people.

Education, sport and politics help to reinforce nationalism

1) **Nationalism** helped to shape the world as we know it (see p.270), and it continues to be important in many ways, from **everyday life** to **debates** around **culture** and **politics**. It's usually based on two main ideas:

1 **Identity** — a sense of **belonging** to a particular nation or group.

2 **Loyalty** — a feeling of **allegiance** to your nation, which can also involve a feeling of dislike towards people or countries that are seen to be threatening it.

2) **Social activities**, **institutions** and ways of **talking** about a nation can **reinforce** people's ideas about **nationalism** along with their national **identity** and **loyalty**:

French sports fans

Sport
- People support their **national team** in international **sporting competitions**, e.g. the football World Cup or the Olympics. Winning against teams or athletes from other countries can reinforce people's **pride** in their nation.
- For example, after the 2012 London Olympics, a survey of what makes people feel proud to be British found that they were prouder of the **British Olympic team** than the **royal family**.

Politics
- Some **politicians** and political parties try to win **support** from **voters** by emphasising their **loyalty** to the nation. They often do this by talking about the importance of **institutions** that are important to national identity, e.g. in the UK, politicians stress their support for the **NHS**, which is an important institution to many people.
- Political parties also talk about their **commitment** to national **values** and **ideals**, e.g. the idea of a British sense of '**fair play**'.

A British sense of 'fair play' refers to following the rules and everyone being treated equally.

Education
- In 2014, the **UK government** introduced the idea of 'British Values' into the education system and said that all schools should actively **promote these values**. They are democracy, the rule of law, individual liberty and the mutual respect and tolerance of those of different faiths and beliefs. Being taught these values in school may promote a **sense of national identity** among students.
- In some countries, e.g. the USA, the school day starts with children swearing a **pledge of allegiance** to their country or singing the **national anthem**, which can reinforce their feeling of **national loyalty**.

National identity can be based on a range of ideas and emblems

People's sense of **national identity** can be based on a **range of things** — food, language, clothing, religion and traditions. **Different people** might **feel loyalty** to different **understandings** and **symbols** of what the 'nation' means — examples of this include:

- **LEGAL SYSTEMS** — some people believe that their nation's legal institutions and processes are **better or fairer** than other countries', e.g. some British people have claimed that **Britain's justice system** is the best in the world — one argument for this is the use of a jury.

- **METHODS OF GOVERNANCE** — some people are proud of their **nation's system of government**, e.g. many people celebrate the USA's parliamentary democracy. The United States Congress consists of two institutions: the **House of Representatives** and the **Senate**. In order for laws to be passed, both the House of Representatives and the Senate must be in **agreement** — strengthening the democratic process in the USA.

The Capitol (meeting place of the US Congress), Washington D.C.

- **NATIONAL CHARACTER** — there are **stereotypes** about the **personality traits** of people from **different nations**, e.g. that Germans tend to be hardworking and efficient, Italians tend to be passionate and British people tend to have a 'stiff upper lip' (they're able to put up with hardship without complaining). Although these are **stereotypical**, some people might feel like they **identify** with their **national character**.

The use of stereotypes and generalisations should be avoided. Although they can be positive, they can also cause offence.

National Identity

National identity and the English countryside

1) England's landscape is very **varied**, but some people have a particular idea of the **English countryside** that's tied up with their idea of **England** and what it means to be English. People's notion of the English countryside is often an image of a **green**, **peaceful** landscape — this has been celebrated and reinforced through art, music and literature, e.g. William Blake's poem *Jerusalem* refers to 'England's green and pleasant land'.

2) The landscape that people picture may **no longer exist**, or may never have really existed, but the **image** of it is **powerful**. It's often linked to **nostalgia** for a different period in English history, before industrialisation and urbanisation — and for some people, a time before immigration made England more multicultural.

3) **Specific landscapes** in England are connected to the **national identity** of many people — e.g. the **White Cliffs of Dover** are often used as a symbol of Britain as an independent **island nation** as well as being linked to **defence** and viewed as '**home**'. These landscapes were particularly used by the government during the **Second World War** to encourage **patriotic** feelings.

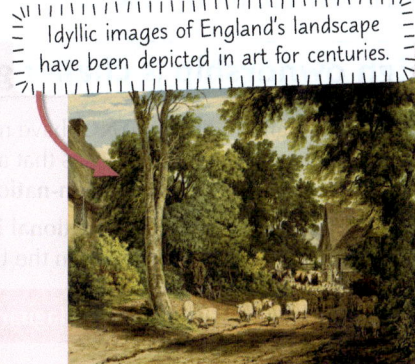

Idyllic images of England's landscape have been depicted in art for centuries.

CASE STUDY

'Driving home the flock', painted in 1812 by Robert Hills

Chiltern Hills, Berkshire

White Cliffs of Dover, Kent

Painting of Stonehenge, Wiltshire

National identity and loyalty can be **complex** in a **multinational world**

1) Most countries are **multinational**, home to people from lots of **countries** and different **ethnic groups**. **Globalisation** has increased diversity in many countries by **encouraging** a greater **exchange** of people, culture and ideas between countries.

2) An increasingly multinational world has **challenged** ideas about national **identity** and **loyalty**. For example, immigrants might support the sports teams from their home country as well as their host country. People from ethnic minorities may decide to **retain** the **language**, **customs**, **food** and **traditions** of their country of origin, or they may **combine elements** of the culture of their home country and host country.

3) This can lead to **tensions**, as some people believe that you can only be loyal to one nation — e.g. they may view someone who supports two football teams as not being a 'proper fan'. Some people also believe that immigrants should embrace the values of their host nation — e.g. they should speak the language of their host country. However, it can also lead to **new understandings** of national identity that are based around **diversity**.

Warm-Up Questions

Q1 Give two examples of factors that might inform someone's sense of national identity.

Q2 Explain how globalisation has challenged ideas about national identity.

PRACTICE QUESTIONS

Exam Question — A-Level

Q1 Explain how landscape can shape people's ideas of national identity and loyalty. [8 marks]

Queuing — tell us you're in the UK without telling us you're in the UK...

Is there anything more British than a good old-fashioned queue? Queuing seems to have taken off at the start of the 19th century as a result of mass urbanisation and industrialisation. It also sums up the idea of British 'fair play'. Interesting eh?

Challenges to National Identity

In a globalised world, foreign ownership of businesses and property is raising questions about what national identity means.

Foreign ownership is challenging ideas about national identity

1) **Globalisation** and the rise of **TNCs** have made it **easier** for people to **invest** in businesses, property and land in **other countries**. This means that an increasing number of **businesses**, and an increasing amount of **property** and **land**, are under **non-national ownership** — they're owned by **foreign nationals**.

2) This has **challenged** ideas about **national identity** — e.g. what it means for something to be 'Made in Britain' if it's made in the UK by a foreign-owned company.

Foreign-owned UK companies — Jaguar Land Rover

CASE STUDY

1) Many people see **Jaguar** and **Land Rover** as symbols of British **expertise** in car **engineering** and **manufacturing**.

2) Both Jaguar and Land Rover have been owned by several different companies — Land Rover's previous owners include British Leyland, German-owned BMW and US-owned Ford (who owned Jaguar Cars at the time). In 2008, **Tata Motors Ltd**, a company based in **India**, bought both brands and created a new company, Jaguar Land Rover. Tata Motors Ltd moved a lot of the car manufacturing overseas, for example to India, China and Slovakia, but some of the manufacturing still happens in the UK — for example, Range Rover cars are made in **Solihull**.

Entrance to the Land Rover factory in Solihull, prior to the creation of Jaguar Land Rover

3) Some people are **unhappy** about these brands coming into foreign ownership — they feel that British brands should be British-owned. Other people point out that foreign ownership can bring **economic benefits**. For example, both Jaguar and Land Rover were facing **financial problems** when they were bought by Tata Motors Ltd. **Investment** from Tata Motors Ltd helped to keep the companies going, and Jaguar Land Rover employs around 30 000 people in the UK. However, people also worry if Tata Motors Ltd **move** more of its **manufacturing** to other countries, it will lead to **job losses** in the UK.

The 'Leaping Jaguar' bonnet mascot on a classic Jaguar car

4) These changes in ownership have made the idea of British products and the label '**Made in Britain**' more **complicated**. Jaguar Land Rover continues to use its British **identity** in **advertising** — for example, adverts for Jaguar cars often use famous British actors to emphasise their **British identity**. However, some people question how British the brands really are now. Some Jaguar and Land Rover cars may be manufactured in the UK, but they might be assembled using parts made elsewhere, and they're ultimately owned by an Indian company.

Foreign-owned London property

CASE STUDY

1) An increasing amount of **property** and **land** in the **UK**, especially in **London**, is being bought by **foreign companies** and individuals. Wealthy people and companies buy **expensive properties** in exclusive parts of London like **Kensington** and **Mayfair** because it's a **good investment** — for example, they can make money by renting buildings out. Buying property in other countries can also be a way of **avoiding** paying **tax** in their own country.

2) **Qatar**, a small and very wealthy country in the Middle East, has been one of the biggest investors in London property. The biggest landowner in London is a company that's jointly owned by a Qatari firm. Some of London's most **iconic buildings**, e.g. the **Shard**, the **Queen Elizabeth Olympic Park** and **Harrods**, are owned (or majority-owned) by the Qatar Investment Authority.

The Shard, London

3) **Russia** is another big **investor** in **London property**, e.g. there are over 200 Russian-owned properties in Westminster. Following Russia's invasion of Ukraine in 2022, the UK government **imposed sanctions** on Russian property owners with links to Vladimir Putin.

4) These trends have led to **concerns** about rising house prices, increased pressure on London's housing market, and tax avoidance where properties are bought by companies registered in tax havens (see p.273). They've also raised questions about **national identity** — many of London's famous **landmarks** are no longer **British-owned**, so some people may see them as **less British**.

London Borough	Number of properties owned by foreign companies (2022)
City of Westminster	10 704
Kensington and Chelsea	5607
Camden	2302
Tower Hamlets	2218
Wandsworth	1830

Challenges to National Identity

Big corporations have helped to promote US culture and capitalism

1) **Globalisation** has contributed to a process called '**westernisation**'. This refers to the spread of Western culture, especially US culture, around the world (including people's values, e.g. democracy and capitalism, and languages, e.g. the dominance of English).

2) One way this has happened is through **Western corporations** expanding to other countries. **Entertainment** companies promote Western **culture** and **lifestyles** through media like films, music, literature and TV (e.g. *Disney*®). **Retail** corporations sell Western products, e.g. food, clothes and electronics. They encourage people to buy things they've seen, e.g. in Western entertainment, so that they can adopt similar lifestyles.

3) Entertainment and retail corporations have also promoted a **positive view** of **capitalism** and **consumer culture**. This has big advantages for the corporations, as more people looking to buy and consume their products means bigger profits.

4) Some people have **criticised** westernisation — they argue that **traditional culture** and **values** are being **lost**. This can challenge **national identity**, as the things that people associate with their nation (e.g. language, traditional crafts or non-capitalist values) are **replaced** with Western culture. However, other people have argued that local cultures take on and adapt these Western cultures through the process of **glocalisation** (see p.93).

'Westernisation' is a form of soft power. Turn to p.212 to learn more about it.

Molly couldn't quite make up her mind about US culture.

McDonald's®

1) Since the first McDonald's restaurant opened in the **USA** in 1940, it has grown to become one of the **most recognised brands** in the world.

2) McDonald's currently has over **38 000 restaurants** in over **100 different countries** worldwide.

3) Other **Western corporations** which have gained **global recognition** include Netflix, NIKE and Apple.

4) Globalisation has enabled these companies to **expand** across the world, and is an example of **westernisation**.

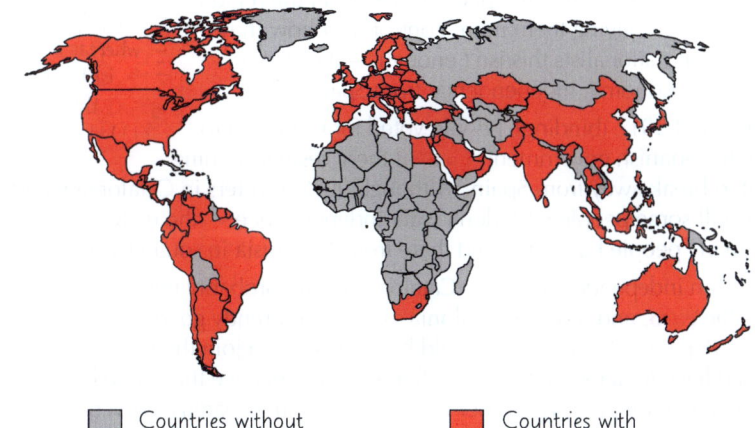

McDonald's in Yichang, China

© Imaginechina Limited / Alamy Stock Photo

☐ Countries without McDonald's restaurants

■ Countries with McDonald's restaurants

Map showing the countries where McDonald's restaurants are currently located, as of 2022.

Warm-Up Questions

Q1 What is westernisation?

Q2 Explain one way that foreign ownership can challenge national identity.

Exam Question — A-Level

Q1 Explain why the concept of 'Made in Britain' is becoming increasingly complex. [8 marks]

Know your jaguar from your leopard to avoid cat-astrophe...

Just in case you're interested (which I'm sure you are), although they may look very similar, jaguars have spots which often have a dot in the middle of them. We don't want a case of mistaken identity on our hands now do we...

Division Within Nations

Groups of people within nations aren't always united — this has consequences for the nation and further afield.

In some countries, **nationalist movements** are trying to make **new nation states**

1) Many nation states are made up of different **ethnic groups**, people speaking different **languages**, or areas that were **separate countries** or **territories** in the past. In some countries, these different groups or regions have formed their own **nationalist movements** and **campaigned** to form new, smaller states, **independent** from the main nation state.

2) Although these nationalist and separatist movements want to become independent, in many cases they still want to **continue** to be members of bigger **trading groups** like the EU, so that they can keep hold of the benefits of having trade and political links with other countries.

Catalonia and independence from Spain

Around 16% of Spain's population live in Catalonia.

CASE STUDY

1) **Catalonia** is a region in northeast **Spain**. Historically, it was part of the Kingdom of Aragon before becoming part of Spain. It has a distinct **culture** and its own **language**, Catalan. Catalan groups have been campaigning for independence from Spain since the 17th Century.

2) Catalonia is a **wealthy** part of Spain, producing around 19% of its GDP and over 25% of its foreign exports. Catalan nationalists believe that Catalonia contributes more to the Spanish economy than it gets in return. These frustrations grew after the 2008 **financial crisis**, as Spain struggled to repay its international debts.

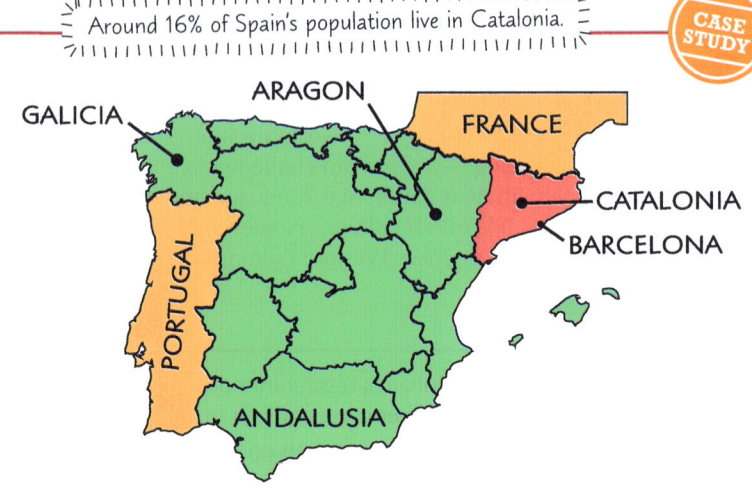

Map showing location of Catalonia in Spain.

'Binding' means it is legally enforceable, whereas 'non-binding' isn't enforceable through law.

3) Catalonia's political status has caused a lot of tension with Spain. In 1979, all the regions in Spain became '**autonomous communities**', giving them more control over how they are governed. However, for some Catalan nationalists this isn't enough — they want Catalonia to have complete political and economic independence from Spain.

4) In **referendums** in 2014 (non-binding) and 2017 (binding), people voted for Catalonia to become an independent nation, but the Spanish government **rejected** these results, arguing that Catalonia doesn't have the right to break away from Spain. Although 90% of voters in Catalonia voted for independence in 2017, there are still some people who don't want independence — the region has many links to the rest of Spain, e.g. many people have migrated there from Andalusia in southern Spain.

5) Some people argue that to **succeed** as an independent country, Catalonia would probably need to stay part of the **EU** — its **currency** is the **euro**, and most of Catalonia's **exports** currently go to EU countries. However, as a **newly independent country**, it would have to **apply** to join the EU, and the EU's member countries would have to agree to it joining. Some people believe this could be **tricky** — Spain might be reluctant to let Catalonia join, as might other EU countries with separatist movements.

Globalisation has caused **political tensions** in **emerging** countries

1) The **impacts** of **globalisation** have been **uneven**. The **BRICS countries** (see p.215) and other **emerging nations** have experienced some of the **biggest changes** as a result of globalisation, as manufacturing and other industries have relocated there from developed countries.

2) These changes have had both **costs** and **benefits**. In emerging countries like **China** and **India**, foreign investment and economic growth have helped to create a new **middle-class** of wealthy consumers with more **disposable income** — they add more money to the wider economy. However, not everyone has felt these benefits. For example, in China, **rural inland provinces** haven't seen much economic improvement, leading to migration away from these areas in search of better jobs. Emigration can cause **further economic decline** in these areas. People may also feel exploited if, for example, their **land** has been **claimed** for a **development project** or if they're working for **low pay** for a TNC.

3) These inequalities can cause **political tensions** — people might **lose trust** in the government and **publicly criticise** them or **protest** against their policies. Some **BRICS countries** have become more **politically divided**, e.g. in Brazil, a general election in 2022 resulted in 51% of voters voting for left-wing Lula da Silva and the remaining 49% of voters voting for the far-right Bolsonaro. In 2023, **thousands** of Bolsonaro's supporters stormed the Brazilian Congress in **violent protest** at the result.

Division Within Nations

National identity can be weaker in 'failed states'

1) A country's **government** can influence whether people have a **strong sense** of **national identity**. Some **authoritarian** states, e.g. North Korea, try to enforce a particular national identity on their citizens, whereas countries with a **weak** government might have a **weak** sense of national identity, especially if there are also a lot of **divisions** between social groups.

2) **'Failed states'** are countries where the government is no longer able to **assert authority** within the country and has lost control over its **national boundaries**. This leads to a **weaker national identity** within that country. States often fail in situations where there is **conflict** or huge **inequality** between different social groups. This can happen where the combination of a **weak government** and **globalisation** leads to big **differences** between a powerful, **wealthy elite** (along with foreign investors and corporations) and the **wider population**.

Weakened national identity in Somalia

1) Various people have argued that **Somalia** is a **'failed state'**. A dictatorial government was overthrown in 1991, and this was followed by decades of **conflict**. Two **disputed** territories — Somaliland and Puntland — split off in the 1990s and are governed **independently**, though not recognised internationally. There is ongoing fighting between government forces and clan-based opposition groups, which has led to the **displacement** of people, civilian **casualties** and human rights **abuses**.

Hargeisa, Somaliland

Berbera City, Somaliland

2) Somalia is one of **poorest** countries in the world, with around **70%** of people living in poverty. Only about 60% of Somalis have access to clean water, and about 10% to sanitation. There is marked **inequality** within the country. For example, while the majority of people **subsist** by farming or small-scale trade, wealthy Somali businessmen and politicians can afford to **invest** in **property** and **international trade**, making large profits and enjoying a **relatively high** standard of living compared to the rest of the population. This **widens** the gap between the rich and poor even more, which leads to greater levels of **inequality** within the country.

3) Somalia attracts some **FDI**, with around US $460 million of investment in 2020. However, further FDI has been **limited** by ongoing conflict, weak governance, high levels of corruption and a lack of infrastructure. In some cases, TNCs have **paid** government officials or other groups for preferential treatment (e.g. protection or lower taxes), financing further conflict. There's also **division** over where the **profits** from exploiting Somalia's resources should go — to TNCs, to the government, or to local people.

4) These problems have undermined people's **trust** in the government to protect them from **conflict** or provide **basic services**. Poverty and inequality have also increased recruitment to armed factions, increasing division and challenging the idea of a coherent Somali **national identity**.

CASE STUDY

Warm-Up Questions

Q1 Give an example of a nationalist movement that has tried to create an independent state.

Q2 Give an example of a 'failed state'. How has inequality weakened the state?

PRACTICE QUESTIONS

Exam Question — A-Level

Q1 Explain how globalisation has led to political tensions in emerging countries. [8 marks]

Now I have your undivided attention...

You may or may not be relieved to hear that you've made it to the end of the main content of the book unscathed! It was a long old slog I know, but give yourself a pat on the back for making it this far. It's no easy feat that's for sure...

Independent Investigation & Fieldwork

Ahh, it's time to don the wellies, grab a clipboard and venture out into the real world. Sadly, it's not just for fun — once you're done investigating you've got to write up your findings. Don't panic though — these pages have lots of handy hints.

You have to complete an **independent investigation**

1) You must carry out a **independent investigation** based on part of the **geographical** content of your A-Level course, then produce a **report** on it.

2) The report needs to be **3000-4000 words** long and there are **70 marks** available (20% of your A-Level).

3) Your report needs a good **structure** — to get full marks you need to present a **clear** and **well-argued case**.

You need to have a **clear title** and **research question**

1) When you're doing fieldwork and research you will not get very far without a **title**. This might be written as a **research question** or an **issue** to investigate.

2) A research question is **what you want to find out**, e.g. 'Do coastal defences at Happisburgh affect the rate of erosion of the coastline?'. It should be closely related to the content you have studied for A-Level.

Herman's hypothesis that large grey rabbits could play miniature pianos was very specific (and seemed to be correct).

3) Before you write your introduction, you should research relevant **literature sources** that are related to your **research question** or **issue**.

4) Your reading will help you to write a **hypothesis**. A hypothesis is a **specific testable statement** such as 'Coastal defences at Happisburgh don't decrease the rate of erosion of the coastline downdrift'.

5) Your hypothesis should be '**developed**'. This just means that it must be **specific**. For example, the hypothesis above is better than 'Coastal defences do affect erosion'.

6) In your introduction, you should write about what you have **read** and what is already **known** about your topic. Summarise your findings in your report to show you understand the **geographical theory** related to your hypothesis.

You need to **collect data**

1) You'll **collect data** when you're doing your **fieldwork** and when **researching** your independent investigation.

2) There are two types of data — **qualitative** and **quantitative**:

- **Qualitative** data is **descriptive** — it might be in words or images, so you can't easily use it in calculations, e.g. interviews with residents.

- **Quantitative** data is **numerical** — it can be measured and used in calculations, e.g. pedestrian/traffic counts.

3) You'll have to collect **primary** data, and you might also use **secondary** data:

- **Primary** data is data you **collect yourself** (i.e. the data you get from your **fieldwork**).

- **Secondary** data is data **someone else** has **collected** (i.e. the data you get from your **research**).

4) There are lots of **secondary data sources** that you could use, including:

- **Images** — e.g. historical and present-day images can help to show how an area has changed over time.

- **Factual text** — e.g. articles about a place or processes that you are investigating.

- **Creative material** — e.g. stories or songs about real places can provide information about how a place is perceived or what it was like in the past.

- **Spatial data** — information with a location, e.g. maps (see p.118) and GIS.

- **Crowd-sourced data** — information that has been contributed by members of a community, e.g. online. One way it can be used is after a natural disaster, when people on the ground contribute information about who needs help, the extent of damage to infrastructure and so on.

- **Big data** — very large datasets that require powerful computers to analyse. They are often created from logs of digital actions, e.g. social media posts, transactions completed, journeys made.

5) In your report, you need to describe how you collected your data. This includes things like what type of **equipment** you used (e.g. a velocity meter), **how you did it** (e.g. you conducted a questionnaire containing 10 questions) or **what source** it came from (e.g. 2011 census data).

6) You need to **justify** the techniques you used to collect your data — e.g. why the **methods** you used were appropriate, **why** you took observations **when** you did, and **why** you chose particular **sampling techniques** (see next page).

7) You also have to **critically examine** any data you use — this means pointing out any **limitations** of the **data**, e.g. whether it could be **biased** due to the collection method, or if it might not be **representative** of the whole population.

Independent Investigation & Fieldwork

You need to **select** your **sites carefully**

1) When you're investigating a large area, e.g. a city, river or coast, you **can't study** the whole thing, so you have to **select sites** to investigate instead.

2) Selecting sites can be **tricky** though. You need places that are **easy to get to** (e.g. places with **footpath access**) and **not too far** from a **parking place** (if you have got **heavy equipment** to carry you do not want to be walking for miles). But you also need sites that are a **good representation** of all the things you want to study, e.g. if you're studying how **characteristics** of a city change in different areas, it's no good selecting three sites in the **city centre** — this would be a **biased sample**. You need to select sites in **different** locations, from the **city centre** to the **rural-urban fringe**.

3) To make sure your sites are **representative**, you might want to use one of these sampling strategies:

> *Don't forget to do a risk assessment — it might affect which sites you can use.*

1 Random Sampling

E.g. using a **random number table** to find the distance of a site from the city centre.
As long as you're using a big enough sample, this should **remove** any **bias** because each possible sample site has the same **probability** of being chosen. However, there is a **risk** that **large areas** of the survey area might not be chosen if the random samples **happen** to fall in such a way that a large area is not sampled.

2 Systematic Sampling

This involves selecting sites in a **regular**, **structured way**, e.g. **every 2 km** along a coastline, or **every third shop** on the high street. Doing it this way means you should be able to **cover** the **whole area** in an unbiased way.

3 Stratified Sampling

This is when **different parts** of the study area are **identified** and **sampled separately** in **proportion** to their size or importance in the study area as a **whole**. For example, in an urban independent investigation, you might identify **different neighbourhoods** and interview a **different number of people** in each in proportion to their population size. This **reduces** the likelihood that some areas could be **under-represented** in your investigation.

You also need to **process** and **analyse** your data...

1) Once you've collected your results, you'll need to **process** the data into a form that you can **use**, e.g. **collating** responses to questionnaires and adding them to a **spreadsheet**.

2) You'll then need to **present** your data, e.g. in a **graph** (see p.295) or on a **map** (see p.296-297). To get top marks, you have to use a **range** of presentation techniques, so don't just stick to the same old bar graph every time.

3) You'll also need to **analyse** your data, e.g. by using **statistics** (see p.298-301), to find any **patterns**.

4) You can then **interpret** your results to work out what they **show** in relation to your **original research question**.

...and use it to draw **conclusions**

Draw conclusions about your data

A conclusion is a **summary** of what you found out in relation to the **original research question**. It should include:

- A **summary** of what your results show.

- An **answer** to the question you're investigating (does your data agree with your **hypothesis**?), and an **explanation** of why that is the answer.

- An explanation of how your conclusion fits into **wider geographical knowledge**.

Evaluate your data and methods

Evaluation is about **assessing** what went **well** and what could have been **improved** in your investigation.

- Identify any **problems** with any part of the investigation, e.g. problems with sampling or inability to access big data sources.

- Describe how **accurate** the results are and any problems with the **methods** used that might have affected the results.

- Comment on the **validity** of your conclusion — could any problems you encountered have **affected** your conclusion?

- Consider any **ethical issues** that your fieldwork might raise, e.g. whether people were free to opt out of the research.

- Think about how your results could be used by **other people** or in **further investigations**.

Exam Skills

Answering Questions

Doing well in your exams isn't only about revision. It's also about learning how to answer questions properly.

Make sure you read the question properly

It's very easy to **misread** a question and **waste time** writing about the wrong thing. To avoid this, try to follow these simple tips:

1) Figure out if it's a **case study question** — if the question wording includes 'using named examples' or 'with reference to a **place you have studied**' you need to include a case study.

2) Underline the **command words**. Some of these words are listed in the table on the right.

3) Underline the **key words**. These are the words that tell you what the question is about.

4) For **essay** questions, **re-read the question** a couple of times **when you're answering it**, just to make sure you're still **sticking** to what the question is asking you to do.

5) For **all** questions, **re-read the question** and **your answer** when you've **finished**, just to check that your answer really does address all parts of the question being asked. A **common mistake** is to **miss a bit out** — like when questions say 'use data from the graph in your answer' or 'use evidence from the map'.

Command Word	Means write about...
Analyse	what the information **means**
Assess	the **advantages** and **disadvantages** OR the **arguments** for and **against**
Evaluate	
Compare	the **similarities** AND **differences**
Explain	**why** it's like that (i.e. give reasons)
Suggest	
To what extent	**both** sides of the argument AND **your opinion**
Study figure	**evidence** from the figure in relation to the **main ideas** in the **question**. You should **quote data** or information in the figure to illustrate your points.

Work out your structure before you start

For any **longer answers**, it's a good idea to write a **plan**. This is to make sure that you:

- structure your ideas in a **logical** order.
- **answer the question** and don't deviate from it.
- give a **well-balanced response** if the question requires it.
- use and choose **case studies** wisely.

Clearly **label** your **plan** and your **answer**, so the examiner knows which is which.

Q1 Assess whether modifying the event or modifying the losses is most effective in managing a hazard.

PLAN

1) Intro — define modifying event / losses

2) Modifying event, e.g. earthquake resistant buildings / land-use zoning / soft engineering

3) Modifying losses, e.g. emergency aid / overseas aid / insurance

4) Conclusion — modifying event is most effective — reduces impact of event rather than dealing with consequences of event.

ANSWER

There are various ways to manage hazards, including the modification of the hazard event itself or the modification of the losses after the event...

Include relevant geographical terms

Use the **proper geographical terms** for things, e.g. 'migration' instead of 'the movement of people'. Make sure you also know how to **spell** those terms, especially if they're tricky. It might be a good idea to **include** these in your **plan**, and to check for these when you **re-read your answer** too.

Aisha hoped the yellow flour would bring her good luck in her spelling test.

Don't forget all the usual rules either

1) Your answer should be **legible**. Use the **correct grammar** and **double-check** that everything is **spelt correctly**. Unfortunately, you won't get many marks if the examiner can't read your answer.

2) Use **diagrams** where they're appropriate. Drawing a **diagram** may be quicker than describing the same thing in words. Make sure you give the diagram lots of space. Use a pencil first but quickly go over it in pen so the examiner can definitely read it. Make sure you **annotate the diagram** if it's not obvious what it shows.

3) If you're **running out of time** at the end of the exam, **don't panic** — just write what you can as **bullet points**. You'll still get some marks for doing this providing what you've written is correct.

Answering Questions

Answers to case study questions need plenty of relevant details

1) Your **case study** needs to include things like **place names**, **dates**, **statistics** and **names of organisation or companies**.

2) Both the **case study** and its **details** need to be **relevant**. The examiner **doesn't** want you to just **write everything** you know about a place or situation. They want you to **apply** what you know to a **specific question**. For example, it's no good including the exact number of people displaced by a volcanic eruption when the question is about the formation of volcanoes.

Here's an example answer to a case study question

> Q1 With reference to a <u>named example</u>, <u>explain</u> how <u>international migration</u> has resulted in a <u>distinct built environment</u> in an <u>urban</u> place. *[6 marks]*
>
> Named urban place: *London*
>
> PLAN
>
> 1) *Describe the forms of international migration that have taken place in London*
>
> 2) *Places of worship — explain why these are more diverse — data from Southall (Sikh community)*
>
> 3) *Retail space and restaurants — e.g. Brick Lane*
>
> 4) *House / apartment sizes — wealthy migrants in London, e.g. Kensington*
>
> ANSWER
>
> *The British Nationality Act (1948) gave people from crown colonies, British overseas territories and Commonwealth countries British nationality and citizenship. This led to many international migrant groups moving to the UK from former British colonies in the decades following the end of the Second World War. The majority of these migrants settled in cities, with London being the most popular choice. These migrants often lived together in clusters for several reasons — e.g. working in similar environments, to join family members who had already migrated or because of hostility.*
>
> *The clustering together of migrants results in services that are made to meet their particular needs. Migrant groups have a wide range of religious views and beliefs. As a result, places of worship are built to accommodate these religious beliefs. For example, due to the large Indian and Pakistani community in Southall in London, there are several Gurdwara temples which serve the Sikhs living in the area. Many parts of London also have clothing shops, food shops and restaurants which specialise in goods that migrants are familiar with. For example, Brick Lane in East London has a large number of curry houses, reflecting the Bangladeshi community that lives there. These are examples of how international migration has resulted in distinct urban environments because it has changed these urban areas to meet the needs of international migrants.*
>
> *In central London, where land prices are high, some argue it would make economic sense to reduce the size of new build apartments to make them more affordable to Londoners. However, the migration of the global elite to London boroughs such as Kensington is allowing developers to build larger apartments and sell them at a higher price. For example, estate agents estimate that 7% of London's properties valued at £1 million or more are Russian-owned. Developments targeting this lucrative market are creating a distinct built environment by meeting the demands and needs of these wealthy migrants.*

A six-mark question **doesn't** necessarily need a **plan** — one is included here to show how this answer has been **approached**.

It's a good idea to think carefully about which **urban place** you will choose **before** you start writing your answer. Make sure you know enough **specific details** so that you don't just write a generic answer.

If you choose a large and diverse urban place, like London, make sure you mention **particular areas or districts** within it.

Using **key facts** and **figures** links the answer to the urban place mentioned in the answer, e.g. the number of Sikhs the Gurdwara temples serve and the percentage of London properties that are Russian-owned.

In a six-mark answer, you **don't need** a conclusion as the answer should be relatively **concise** and to the **point**.

- You'll know your answer is **detailed enough** with **case study** materials if you try to **substitute** the name of **another city** in the place of London and the **answer doesn't work**.

- If you can put the name of another urban place (e.g. Liverpool, Birmingham or Manchester) at the top of your plan and the answer still works, then it's **too generic** and you haven't used enough case study details.

Answering Questions

You might get a **question based** on a **map**...

In the exam, you'll get some questions where you will have to **interpret** a **resource** and use it to answer the question fully.

Q2   Study Figure 2b. <u>Explain</u> the <u>role</u> of <u>prevailing wind</u> in creating <u>distinctive coastal landforms</u>. *[6 marks]*

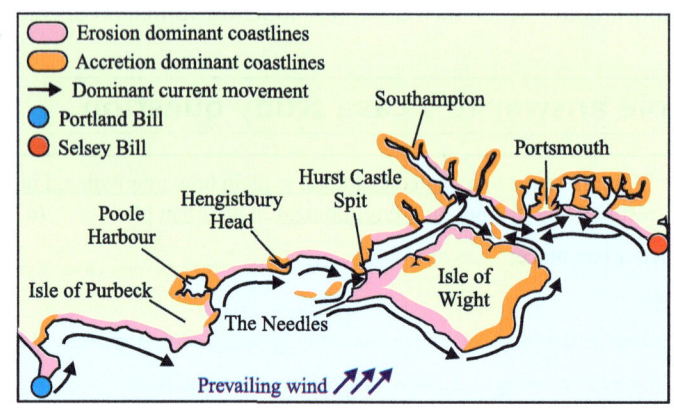

Figure 2b: The sediment cell between Portland Bill and Selsey Bill

According to the map, this sediment cell has a south-westerly prevailing wind. This means that most sediment is moving from west to east. This sediment will be transported by the dominant currents labelled by the arrows on the map. On the sections of coastline that face the prevailing wind directly, such as the Isle of Purbeck in the map, erosive landforms will form. This is because the waves will have more energy, and therefore be more destructive. This causes erosion, which wears away the rock facing the wind to leave behind landforms such as headlands (e.g. Hengistbury Head) and stacks (e.g. The Needles). In areas that are sheltered from the prevailing wind and currents (e.g. Poole Harbour), depositional landforms will form. This is because the weaker wind causes waves to lose their energy, so they can no longer carry their sediment. They can deposit this sediment to form landforms such as spits, like Hurst Castle Spit. This area is protected from the SW currents by the Isle of Wight.

Use the key in the figure as a sort of **checklist** to make sure you've **covered everything** you need to write about. In this case, you should mention **erosion**, **deposition** and the dominant **current movement**.

Make sure you use the correct **compass directions** if you're **describing** something you can **see in a map**.

Include **specific names** of **places** and **features**.

...or a **photo**...

Q3   Study Figure 3b. <u>Explain</u> how <u>landforms</u> are created in <u>periglacial landscapes</u>. *[6 marks]*

Figure 3b: A periglacial landscape in Northern Canada

Periglacial landscapes contain features such as ice wedges, as shown in Figure 3b. As temperatures drop in the winter, frost contraction occurs. This is where the ground shrinks and the permafrost cracks. In spring, the active layer thaws and meltwater seeps into these cracks. The lower ground temperature caused by the permafrost layer means the water refreezes in the cracks. Some ice wedges in Figure 3b are larger than others. This occurs when the cracks reopen in subsequent years and are filled in by water, which then refreezes and widens the ice wedge. Figure 3b also shows a pingo. Pingos form when water freezes underground, forming a core of ice which pushes the ground above it upwards and creates a dome like the one visible in Figure 3b.

Use key **geographical terminology** to **explain** the **formation** of a **feature**, e.g. frost contraction and active layer.

You should **directly refer** to the **figure** in your answer.

Answering Questions

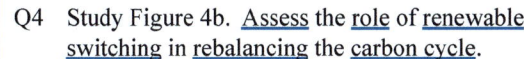

...or a **graph**...

Q4 Study Figure 4b. <u>Assess</u> the <u>role</u> of <u>renewable switching</u> in <u>rebalancing</u> the <u>carbon cycle</u>.

[12 marks]

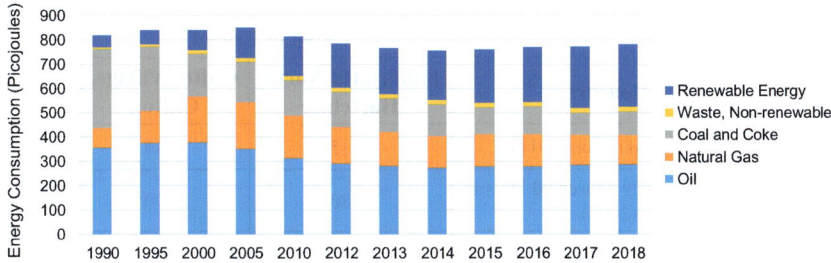

Figure 4b: Denmark's energy consumption by source between 1990 and 2018

<u>PLAN</u>

1) Define renewable switching + rebalancing

2) How renewable switching can help rebalance the carbon cycle
    — compare fossil-fuel and renewable consumption

3) How renewable switching is not the only solution
    — reducing consumption in total — quote from figure

4) Other strategies can be used to rebalance the carbon cycle — e.g. afforestation

5) Conclusion — renewable switching is an important part of rebalancing
    the carbon cycle but other factors also play a role

<u>ANSWER</u>

Renewable switching is the process where consumers change their consumption of energy from one mostly made up of fossil fuels (such as coal, natural gas and oil) to one mostly made from renewable sources (such as wind and solar power). Industries or national governments can lead the process of renewable switching. This would allow more energy that a country receives to be from renewable sources.

Rebalancing the carbon cycle means reducing the amount of carbon dioxide emissions that get released into the atmosphere. Fewer emissions mean that oceans and plant life may be able to reduce atmospheric carbon, through photosynthesis and carbon sequestration, at a more sustainable pace.

Renewable switching can help to rebalance the carbon cycle as a higher proportion of a country's energy mix will come from sources that release less carbon dioxide or none at all. In Denmark, roughly a third of consumed energy was produced from renewable sources in 2018, compared to less than 10% in 1990. This means that in this 28-year period, the country has significantly decreased its carbon footprint in relation to energy production.

However, fossil fuel use still accounted for more than half of the country's total energy production (500 out of 780 picojoules) in 2018. This means carbon is still being released into the atmosphere and it will take more work before Denmark becomes a carbon neutral energy consumer. Therefore, the role of carbon switching is important, but it is not the only strategy that a country can implement to rebalance the carbon cycle. Reducing overall energy consumption is also important. In Denmark, this occurred most noticeably between 2005 and 2014 but started to level off between 2015-2018.

As well as reducing overall consumption of energy, other strategies such as afforestation can also be implemented to rebalance the carbon cycle. It is important to recognise that the balance of the carbon cycle relies on reducing the release of carbon into the atmosphere and removing carbon from it too. In time, trees planted as part of afforestation schemes will become important carbon stores and these could (if implemented on a global scale) play an important role in rebalancing the carbon cycle. However, afforestation is unlikely to be able to rebalance the carbon cycle as effectively as renewable switching.

In conclusion, renewable switching has an important role to play in rebalancing the carbon cycle, but it is not the only strategy to consider. For example, reducing energy consumption and afforestation can also play an important role. While switching to renewable energy will make an immediate impact on the carbon released, rebalancing the carbon cycle means reversing the damaging effects of fossil fuel combustion that has happened over the last 250 years. Success in rebalancing the carbon cycle will therefore rely on a mix of different strategies being used in different countries.

Use **examples** of different **energy sources** to illustrate your point.

Use **data** quoted **directly** from the **source**.

Concise **definitions** at the **start** of your answer mean you don't have to **break up** the main body of your response with **explanations**.

Use the **data** to show **general trends** and **patterns**.

Make sure you clearly **answer the question** in the **conclusion** and actually **assess** the **main idea**.

Answering Questions

Q5 Study Figure 5b, which shows some social and economic indicators for two areas of Bristol. <u>Assess</u> the reasons for <u>economic</u> and <u>social inequalities</u> in <u>cities</u>. *[12 marks]*

Measure	Area of Bristol	
	Fulford Rd North, Hartcliffe	**Clifton Village, Clifton**
Total population (2011)	1515	1449
Average weekly household income (2008)	£480	£800
% Unemployed (2011)	8.1	1.8
% 'not good' general health (2011)	23.6	7.8
Police recorded burglaries (2016)	21	16
% people aged 16 or over with no qualifications (2011)	41.7	2.7
% households with >1 person per bedroom (2011)	35.0	19.0

Figure 5b: Table showing information about two areas of Bristol

When **defining** a word that's in **common use**, make sure you **define** it in your answer within a **geographical context**.

PLAN

1) Define social and economic inequalities

2) Economic reasons
   — low incomes lead to low investment
   — spiral of decline

3) Social reasons
   — poor housing linked to health
   — poor education linked to income

ANSWER

Inequality is the uneven distribution of resources or services. It can take the form of economic inequalities, such as differing job opportunities, or social inequalities, such as access to healthcare or education.

Refer **directly** to **data** you see in the **figure**.

Figure 5b shows that Hartcliffe and Clifton had very different employment statistics in 2011. Hartcliffe had more than four times as many unemployed people as Clifton. This contributes to lower incomes and lower tax revenue in this area, meaning there is less money available for improving services and developing the area, widening social inequalities. A lack of employment opportunities and investment can cause an area to enter a 'spiral of decline', where it becomes unattractive to companies, so employment opportunities are reduced further. Once an area begins to decline, people on higher incomes will often move out of the area whilst people on lower incomes are left behind, further widening inequalities within a city.

Try to **link** different aspects of the **data** together.

Social inequalities, such as variations in health, may be due to the level of care people are able to access or other factors such as poor quality housing. People on lower incomes are more likely to live in poor quality, overcrowded housing — this is shown in Figure 5b, where lower household incomes in Hartcliffe correlate with a higher percentage of households with more than one person to a bedroom. Poor quality housing and overcrowding can affect physical health, for example through increased spread of respiratory infections, as well as negatively affecting mental health. This may help account for the higher proportion of people in 'not good' health in Hartcliffe than in Carlton. Education is another indicator of social inequality, and it can also contribute to economic inequality. Figure 5b shows that around fifteen times as many adults have no qualifications in Hartcliffe compared to Clifton. This may partly account for the inequality in income, as people with little education are less able to secure well-paid jobs.

Make sure you **assess** the **causes of inequality** as you **write** about them.

In conclusion, there are a number of reasons why economic and social inequalities might exist within a city, including low incomes and poor living conditions. Inequalities in one area are likely to contribute to inequalities in other areas, leading to a spiral of decline that, over time, widens inequalities between areas.

Graph and Map Skills

As sure as death and taxes, there'll be graphs and maps in your exam. So make sure you know all the different types...

There are **loads** of **different types** of **graphs** and **maps**

1) There are some types of graphs and maps that you'll have come across before. These include **line graphs**, **bar charts**, **pie charts**, **scatter graphs**, **atlas maps** and **sketch maps**.

2) Some graphs and maps are **trickier than others** and these pages will help you interpret the tougher ones.

3) When you're **interpreting** graphs and maps, you need to remember to **read** the **scale** or **key really carefully**. Some graphs may also have **two axes** (such as climate graphs) that will require you to look at them closely.

4) If you have to read from a graph, **draw working lines** on to help you get an accurate figure.

Triangular graphs show **percentages** split into **three categories**

1) To read a triangular graph, start by **finding the point** you want on the graph.

2) **Follow** the **line** that goes **down** from the **point** to the **lowest end** of the **scale** and record the percentage.

3) Then, **turn the graph around** so that the next axis is at the **bottom**, follow the line down to the lower end of the scale and record that percentage.

4) Do the same for the **third axis**.

5) Check that the three readings add up to **100%**.

6) The graph on the right shows the age distribution of three populations. There are **three age groups** so a triangular graph can be used. **Each point** represents **one population**.

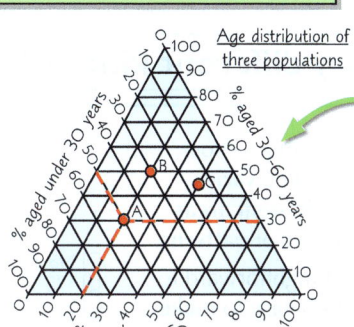

Age distribution of three populations

Population A
Under 30 — 50%
30 – 60 — 30%
Over 60 — 20%

Population B
Under 30 — 30%
30 – 60 — 50%
Over 60 — 20%

Population C
Under 30 — 15%
30 – 60 — 45%
Over 60 — 40%

On this scale, the lowest end is on the **left**, so to find the percentage you follow the line down and towards the left of the scale.

Dispersion diagrams show the **frequency** of data

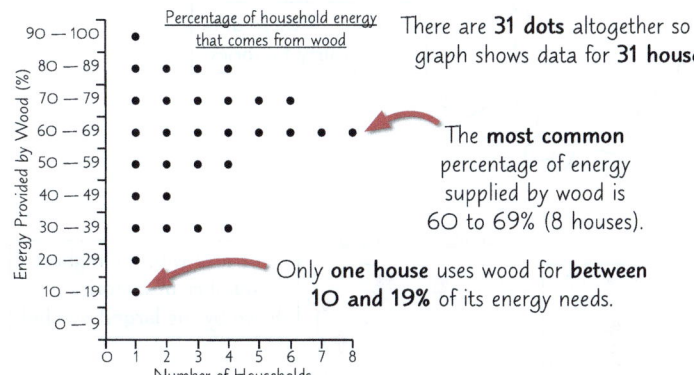

Percentage of household energy that comes from wood

There are **31 dots** altogether so the graph shows data for **31 houses**.

The **most common** percentage of energy supplied by wood is 60 to 69% (8 houses).

Only **one house** uses wood for **between 10 and 19%** of its energy needs.

1) Dispersion diagrams are a cross between a **tally chart** and a **bar chart**.

2) The **range** of **data that's measured** goes on one axis. **Frequency** goes on the other axis.

3) **Each dot** represents **one** piece of **information**. The **more dots** there are in a particular category, the **more frequently** that event has happened.

4) The dispersion diagram on the left shows the **percentage** of **household energy** that comes from **wood** for **houses** in a **particular village**.

Logarithmic scales are used when the **data range** is **large**

1) The **intervals** on logarithmic scales are **not fixed amounts**. Instead, the **intervals** get **increasingly larger** at the **top end** of the scale (such as 10, 20, 40, 80).

2) This lets you fit a **very wide range** of **data** onto one **axis** without having to draw an enormous graph.

3) The graph on the right uses a **logarithmic scale** on the **vertical axis** to show how the world's population changed between 1950 and 2000.

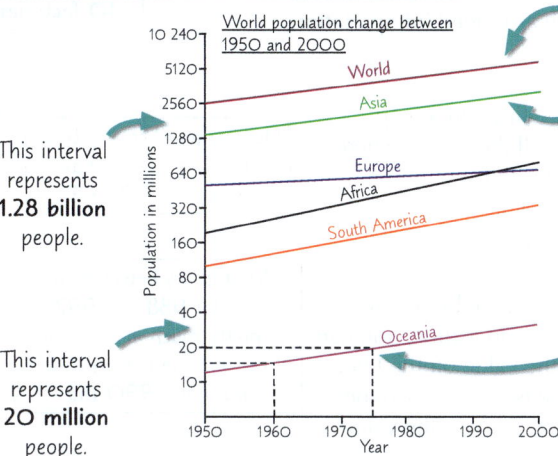

World population change between 1950 and 2000

This interval represents **1.28 billion** people.

This interval represents **20 million** people.

Be careful, it looks like the world's population isn't much bigger than Asia's but that's only because there are **big jumps** at this end of the scale.

Graphs with log scales are **really tricky** to read. It's OK if your working line hits a label on the log axis (e.g. there were 20 million people in Oceania in 1975), but if it doesn't it's easiest to **give a range** (e.g. it was between 10 and 20 million in 1960).

Graph and Map Skills

Kite diagrams are one way of presenting data

Kite diagrams

1) Kite diagrams show the **frequency**, **density** or **distribution** of something over a **specific distance**. For example, how abundant different animal or plant species are in a specific area.

2) These diagrams are usually used to **compare** the **distribution** of something, e.g. plant species, in relation to each other, rather than show the actual number of individuals.

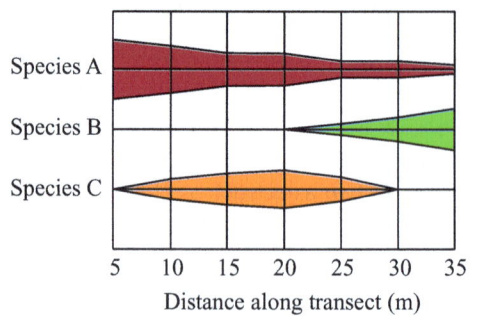

transect
a path or line along which the distribution of species is recorded

Jake hoped his exam skills would take off better than this kite.

There are many different kinds of maps...

Choropleth maps

1) Choropleth maps show how something **varies** between **different areas** using **colours** or **patterns**.

2) They're straightforward to read but it's easy to **make mistakes** with them as the patterns can be very similar.

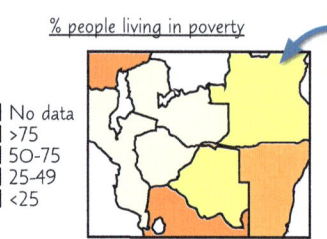

% people living in poverty

Between 25 and 49% of people in this region are living in poverty.

The maps in exams often use cross-hatched lines and dots.

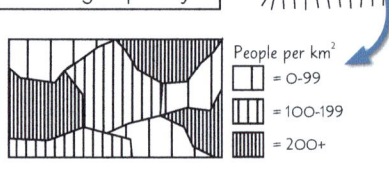

Dot maps

1) Dot maps use **identical dots** to show how something is **distributed** across an **area**.

2) The **key** shows the **quantity** each dot represents.

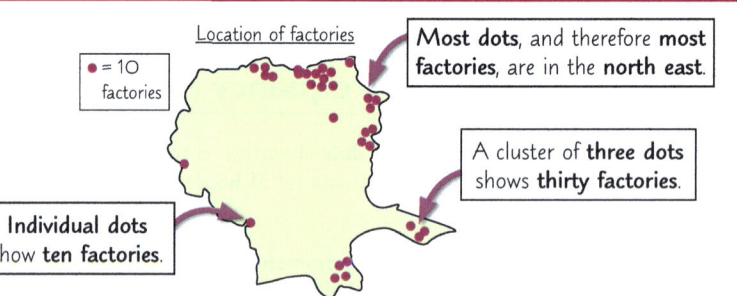

Location of factories

● = 10 factories

Most dots, and therefore **most factories**, are in the **north east**.

A cluster of **three dots** shows **thirty factories**.

Individual dots show **ten factories**.

Proportional symbol maps

1) Proportional symbol maps use symbols of **different sizes** to represent **different quantities**.

2) A **key** shows the quantity each symbol represents. The **bigger** the **symbol**, the **larger** the **amount**.

3) The symbols might be **circles**, **squares**, **semi-circles** or **bars**. They might even be pie charts or other forms of data which will also need interpretation.

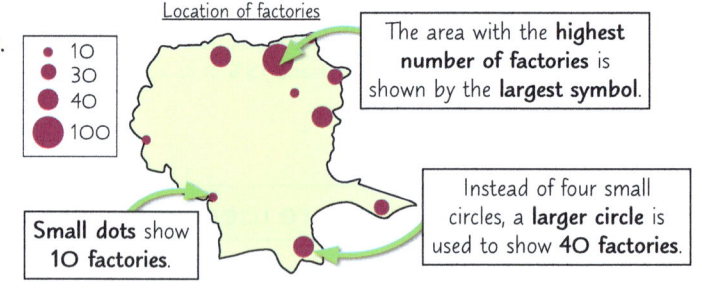

Location of factories

● 10
● 30
● 40
● 100

The area with the **highest number of factories** is shown by the **largest symbol**.

Instead of four small circles, a **larger circle** is used to show **40 factories**.

Small dots show **10 factories**.

Isoline maps

1) **Isolines** are lines on a map **linking** up all the **places** where something is the **same**. For example, isolines on **weather maps** show places that have the **same air pressure**.

2) If the place you're being asked about lies **on** an isoline, you can just **read** the value off the line. If the place is **between** isolines you should **estimate** the value.

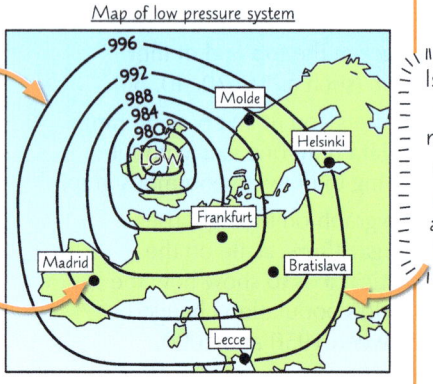

Map of low pressure system

Helsinki and Lecce both lie **on** this line so both have a pressure of **996 mb**.

Madrid lies **between** the lines for **988** and **992**. It's pretty much in the middle of the lines, so has a pressure of roughly **990 mb**.

Isolines on weather maps that show air pressure are called 'isobars'.

Graph and Map Skills

Flow lines and trip lines show movement...

1) **Flow line maps** have **arrows** on them, which can be **different sizes**, to show how many things **move** (or are moved) from **one place to another**.

2) The flow line map on the right shows the movement of people **into** and **out of** a **region**. The width of the arrows shows **how many** people are moving.

3) **Trip line maps** have straight lines showing the **origin**, **destination** and **direction** of movements, but they **don't** show the **volume** of movement.

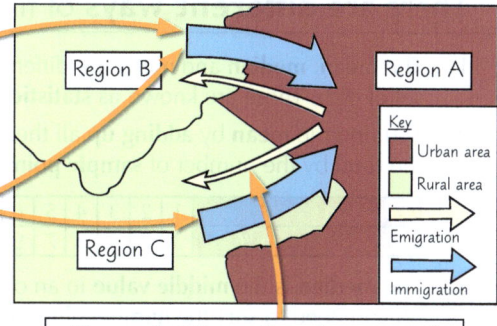

Some of the flows of people to and from Region A

The **largest flows** of people are **to Region A**, as these are the **largest arrows**.

Roughly the same number of people are **immigrating to Region A** from Regions B and C. This is shown by the arrows, which are the **same size**.

The **smallest flows** of people are **out of Region A**, as these are the **smallest arrows**.

...and so do desire lines

1) **Desire line maps** have **straight lines** that show **journeys** between two locations, but they **don't follow roads** or **railway lines**.

2) They're used to show **how far** a population has **travelled** to get to a **place**, e.g. a shop or a town centre, and **where** it's **come from**.

3) In the map on the right, the **number of lines** represents the **number of journeys**. The **width** of the lines can also be changed to show the **volume** of the movements.

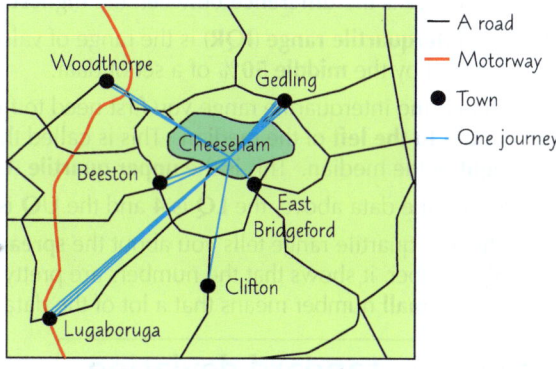

Desire Lines showing journeys to Cheeseham

— A road
— Motorway
● Town
— One journey

Ordnance Survey maps show detailed information of all areas

1) Ordnance Survey® (OS®) maps use lots of **symbols**. It's a good idea to **learn** the most common ones.

2) You can find places on OS maps using **grid references**.

3) **Four-figure grid references** direct you to a 1 km × 1 km **square** on the map, e.g. for **1534** go **across** to the number **15** (the **eastings** value) and then **up** to the number **34** (the **northings** value). This grid reference refers to the **square above** and to the **right** of the point 1534.

4) **Six-figure grid references** are more precise and can direct you to a more **exact spot** (a 100 m × 100 m square). E.g. for 15**5**341 the eastings value is 155, so go across to 15 again and then a further **5 "tenths"** across the square. For the northings value of 341 go up to 34 and a further **1 "tenth"** of that square. The spot you're looking for is where the easting and northing values **cross**.

5) Every map has a **scale** so that you can work out the **distance between points**. If the scale is **1:25 000**, it means that every **1 cm** on the map represents **25 000 cm** (250 m) in real life.

6) **Altitude** (height above sea level) is shown on OS maps using a type of isoline called **contour lines**. The **closer together** the contour lines are, the **steeper the gradient** is. Sometimes, the altitude of specific **spot heights** is also given.

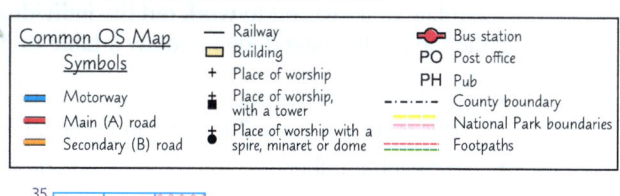

Common OS Map Symbols

— Railway
▭ Building
+ Place of worship
⛪ Place of worship, with a tower
⛪ Place of worship with a spire, minaret or dome
— Motorway
— Main (A) road
— Secondary (B) road
🚌 Bus station
PO Post office
PH Pub
-·-·- County boundary
▤ National Park boundaries
▦ Footpaths

Grid reference: 1534

Grid reference: 155341

Getting your northings and eastings mixed up can cause havoc — Chaz thought he was heading to the bus station.

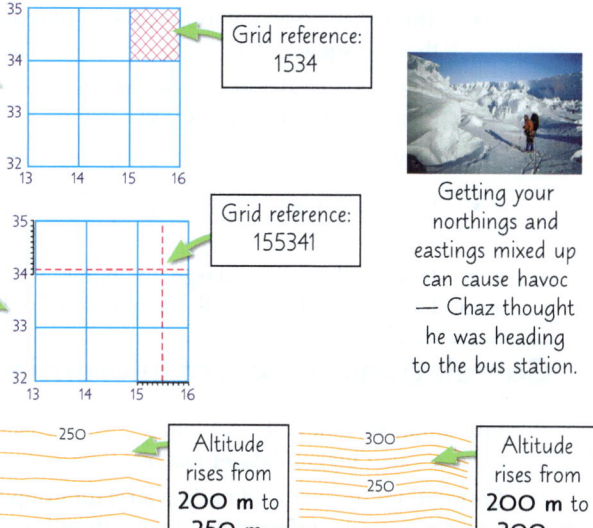

Altitude rises from **200 m to 250 m**.

Altitude rises from **200 m to 300 m**.

The contour lines on the right are closer together and show a **steeper slope** (there's a **greater increase in height** over the **same distance**).

Data and Statistical Skills

As if knowing about lots of weird graphs and maps wasn't enough, you also need to be pretty familiar with statistics...

There are **different ways** of finding the **average** value of a set of data

1) The **mean**, **median** and **mode** are different ways of finding the **average** value of a set of data. These are known as **statistics of central tendency**.

2) You find the **mean** by **adding up** all the numbers in a set of data, then **dividing** the **total** by the number of **sample points**, **n**. Take a look at the data in this table:

n = 11, so you divide the data by 11.

Location	1	2	3	4	5	6	7	8	9	10	11
Temperature in °C	3	7	4	3	7	9	9	5	5	7	6

$$\frac{3+7+4+3+7+9+9+5+5+7+6}{11} = 5.9\,°C$$

3) The **median** is the **middle value** in an ordered set of data. So you need to **sort the numbers into order**, then work out which one is in the middle. The median for this data is **6**.

3 3 4 5 5 ⑥ 7 7 7 9 9

4) The **mode** is the **most common value** in a set of data. Here the mode is **7**.

3 3 4 5 5 6 ⑦ ⑦ ⑦ 9 9

The **interquartile range** is a **measure of dispersion**...

1) The **range** is the **difference** between the **highest** and the **lowest** values.

2) The **interquartile range** (**IQR**) is the range of values covered by the **middle 50%** of a set of data.

      LQ   Median   UQ
3 3 ④ 5 5 ⑥ 7 7 ⑦ 9 9
            IQR

3) To find the interquartile range you first need to find the median of the values **to the left** of the median. This is called the **lower quartile** (**LQ**). Next, find the median of the values **to the right** of the median. This is the **upper quartile** (**UQ**). Then you just **subtract** the **LQ from the UQ** to give you the **IQR**.

4) So, for the data above, the **LQ** is **4** and the **UQ** is **7**, and the interquartile range is UQ – LQ = 7 – 4 = **3 °C**.

5) The interquartile range tells you about the **spread** of data **around** the **median**. If it's a **big** number, it shows that the numbers are pretty **spread out**. And yep, you've guessed it — a **small** number means that a lot of the data is pretty **close** to the **median**.

...and so is **standard deviation**

1) The **standard deviation** is a bit trickier to calculate than the IQR, but it's often a **more reliable** measure of dispersion (spread). The symbol for it is *σ*.

The formula is $\sigma = \sqrt{\dfrac{\sum(x - \bar{x})^2}{n}}$

2) To calculate it, it's easiest to **work out** the **individual bits** in the formula **first**, e.g. the mean. It's a good idea to **draw** a **table** to help you. Below is a simple example for the set: 5, 9, 10, 11, 14.

- For these numbers, the **mean** is (5 + 9 + 10 + 11 + 14) ÷ 5 = **9.8**. This is shown in the 2nd column in the table.

- For each number, **calculate** $x - \bar{x}$ (3rd column in the table).

- Then **square** each of those values (4th column) — remember that the square of a **negative number** is always **positive**.

- Then **add up** all the squared numbers you've just worked out — this will give you $\sum(x - \bar{x})^2$.

- Now just **divide** your total by **n**, then take the **square root**.

- In this example, n = 5, so $\sigma = \sqrt{\dfrac{42.8}{5}} = $ **2.93** (2 d.p.)

x	$\bar{x}$	$x - \bar{x}$	$(x - \bar{x})^2$
5	9.8	−4.8	23.04
9	9.8	−0.8	0.64
10	9.8	0.2	0.04
11	9.8	1.2	1.44
14	9.8	4.2	17.64
		$\sum$	42.8

3) If the standard deviation is **large**, the numbers in the set of data are **spread out** around the **mean**. If it's **small**, the numbers are **bunched** closely around the mean.

Standard deviation can be represented by σ or s.

Correlations are part of **inferential statistics**...

1) **Inferential statistics** allow an analyst to understand geographical ideas about a **bigger dataset** just by looking at a **smaller sample**.

2) **Correlations** are a way of finding a **relationship** between two factors. It can be **inferred** that the **relationship** found within the **bigger dataset** will be the **same** as the one in the **smaller sample**.

3) Correlations can be found by drawing a **scatter graph** of the results.

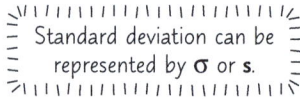

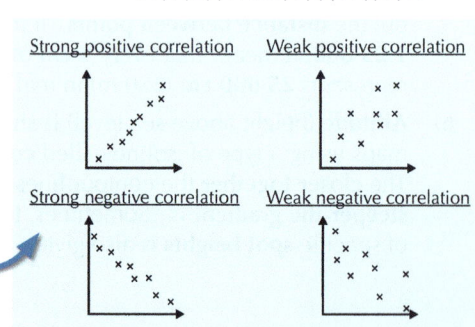

Strong positive correlation   Weak positive correlation

Strong negative correlation   Weak negative correlation

Data and Statistical Skills

...and work with **lines of best fit**

1) If you need to draw a **line (or curve)** of **best fit** on a **scatter graph**, draw the line through or as near to as many points as possible, ignoring any **anomalous** (different to what is expected) **results**.

2) The **same number of points** should sit **below and above** the line.

3) You have to be very careful when drawing conclusions from scatter graph data. This is because a **correlation** between two variables doesn't always mean that a change in one variable **causes a change in the other**. For example, the correlation could be due to **chance** or there could be another variable (or variables) having an effect.

Make sure you know how to find **Spearman's Rank correlation coefficient**

The **Spearman's Rank correlation coefficient** is a test to find out whether two sets of numbers are actually **correlated**. The example below uses the test to see if **GDP per capita** ($) and **life expectancy** (in years) are correlated.

1) The first step is to give a **rank** to each number in both sets of data. The **highest** number is given rank **1**, the second highest is given rank 2... you get the idea.

2) Then you **calculate 'd'**, the **difference** between the ranks for each item, e.g. if the ranks for Country F are 4 and 6, the difference is 2.

3) Next, you **square 'd'** and **add up** the **d²** values to give $\sum d^2$, which you use in the formula below.

4) Finally you need to work out the **Spearman's Rank correlation coefficient** (known as r_s).

The formula is: $r_s = 1 - \dfrac{6\sum d^2}{n^3 - n}$

Country	GDP per capita ($)	GDP rank	Life expectancy	Life expec. rank	d	d²
A	14 000	5	72	5	0	0
B	19 000	4	71	6	2	4
C	9000	9	67	8	1	1
D	6000	11	61	11	0	0
E	21 000	3	75	3	0	0
F	13 000	6	74	4	2	4
G	22 000	2	76	2	0	0
H	35 000	1	78	1	0	0
I	5000	12	60	12	0	0
J	7000	10	65	9	1	1
K	11 000	8	64	10	2	4
L	12 000	7	69	7	0	0

$\sum d^2$ = 14

5) So for the example above, $\sum d^2 = 14$ and n = 12. So $r_s = 1 - \dfrac{6\times14}{12^3 - 12} = 1 - \dfrac{84}{1716} = 1 - 0.05 = \textbf{0.95}$

6) The number you get is always **between –1 and +1**.

7) A **positive number** means the variables are **positively correlated** — as one variable **increases** so does the **other**. The **closer** the number is to 1, the **stronger** the correlation.

8) A **negative number** means that the two sets of variables are **negatively correlated** — as one variable **increases** the other **decreases**. The **closer** the number is to –1, the **stronger** the correlation.

9) If the coefficient is **0**, or near 0, there probably isn't much of a relationship between the figures.

10) The value of r_s in the example above was **0.95**, which is **close to 1**, so there's a **strong positive correlation** between the **data** for GDP per capita and life expectancy.

Take a look at the scatter graphs at the bottom of the previous page to see what these correlations look like.

You have to **check** the **correlation** is **significant** though

1) A **Spearman's Rank correlation coefficient** might tell you that **two sets of numbers** are **correlated**. But you need to check whether there is a genuine **link** between the **two quantities** you're looking at.

2) You can check whether it's evidence for a genuine link by looking at the **probability** that a correlation would happen by **chance**. If there's a 5% (or higher) probability that a correlation is because of chance, then it's **not significant** evidence for a link. If there's a **0.1% or less** chance, then it's **very significant** evidence for a link.

3) To test whether the value of r_s is evidence for a relationship between GDP per capita and life expectancy, you'll need a **graph** like the one on the right, or a **table** of critical values. You'll also need to know the **degrees of freedom** (in the example above this is just n – 2, so 12 – 2 = 10). Since r_s = **0.95**, you can use the graph to find that this correlation has a **less than 0.1%** probability of being due to chance. This means you have **very significant** evidence for a **relationship** between GDP per capita and life expectancy.

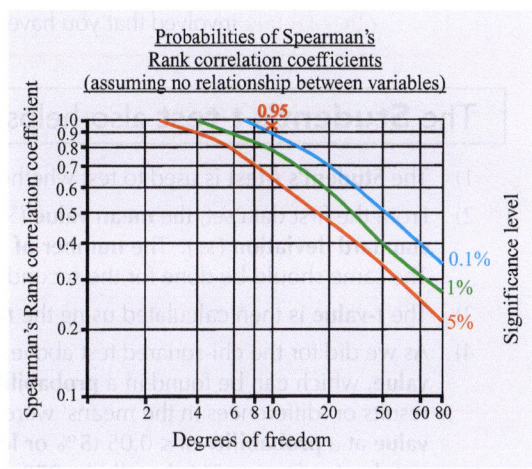

Probabilities of Spearman's Rank correlation coefficients (assuming no relationship between variables)

Data and Statistical Skills

The chi-squared (χ^2) test tells you whether two variables are linked

1) Imagine a student wants to find out whether the **number of wells** in three areas of Africa is related to **how much rainfall** these areas receive each year. They would need to come up with a **hypothesis** and a **null hypothesis** to do this.

2) A **hypothesis** is your theory that there's a link between two variables, e.g. the **number of wells** (variable one) and the **amount of rainfall** the area receives (variable two). The **null hypothesis** is always that the two variables are **not linked**.

3) You can use the chi-squared (χ^2) test to check for this link.

4) First, use the null hypothesis to **predict** a **result** — this is called the **expected result**. For example, there are 144 wells in total. If there was **no link** between rainfall and number of wells, you'd **expect** there to be an equal number of wells in each area, i.e. **48** ($144 \div 3 = 48$).

	Area A (lower rainfall)	Area B (medium rainfall)	Area C (higher rainfall)	
Expected	48	48	48	
Observed	64	49	31	
O – E	16	1	–17	
(O – E)²	256	1	289	
$\frac{(O - E)^2}{E}$	5.33	0.02	6.02	**11.37**

5) Next, the experiment is carried out and the **actual result** is recorded — this is called the **observed result**. The **observed** result in this case is the actual number of wells in each area.

6) The chi-squared (χ^2) **test** is then carried out and the **outcome** either supports the **null hypothesis** or allows you to **reject** it.

Put your observed values (O) and expected values (E) into **this equation** to work out chi squared (χ^2) one step at a time:

1) First calculate O – E for each area, e.g. $64 - 48 = 16$.

2) Then, square each of the resulting numbers, e.g. $16^2 = 256$.

3) Next, divide each of these figures by the expected result, e.g. $256 \div 48 = 5.33$.

$$\chi^2 = \Sigma \frac{(O - E)^2}{E}$$

4) Finally, add the numbers for Areas A, B and C together to get χ^2. $5.33 + 0.02 + 6.02 = 11.37$, so $\chi^2 = 11.37$.

7) The χ^2 value shows whether there is a **significant difference** between your observed and expected results. If there **is** a significant difference, this suggests your two variables are **linked** (and you can **reject** the null hypothesis).

Compare your result to the critical value

1) To find out if there is a significant difference between your observed and expected results, you need to compare your χ^2 value to a critical value. You also need to compare your t-value to a critical value if you want to find out if there is a significant difference between the means of your two data sets.

2) The critical value is the value that corresponds to a 0.05 (5%) level of probability that the difference between your data sets is due to chance.

3) If your χ^2 or t-value is smaller than the critical value, then there is no significant difference. This means your variables are independent and you accept your null hypothesis.

If you're not given the critical value, you might have to look it up in a table.

4) If your χ^2 or t-value is larger than the critical value, then there is a significant difference and something other than chance is causing the difference. This suggests that there is a link and the null hypothesis is rejected.

5) For this example, the critical value for the data above is 5.99. The χ^2 value of 11.37 is bigger than 5.99, so there is a significant difference between the observed and expected results and the null hypothesis is rejected.

6) Be careful with your conclusions though — your χ^2 value is evidence supporting your hypothesis that there is a link between the number of wells and the amount of rainfall. However, it doesn't prove this link — there could be other factors involved that you haven't considered in your investigation. This is true for any investigation you do.

The Student's t-test also helps you compare two sets of data

1) The **Student's t-test** is used to test whether there's a **significant difference** between the **means** of two data sets.

2) From the first data set, the **mean value** ($\overline{x_1}$) should be calculated along with the **standard deviation** (s_1). The **number of values** in the set should also be noted (n_1). The same should be done for the second set of data ($\overline{x_2}$, s_2 and n_2).

3) The **t-value** is then calculated using the **t-test formula** on the right.

$$t = \frac{|\overline{x_1} - \overline{x_2}|}{\sqrt{\frac{(s_1)^2}{n_1} + \frac{(s_2)^2}{n_2}}}$$

4) As we did for the chi-squared test above, the calculated value for t is **compared** to a **critical value**, which can be found in a **probability table**. This helps you decide how likely it is that the results or 'differences in the means' were due to chance. If the **t-value** is greater than the **critical value** at a **probability** of ≤ 0.05 (**5% or less**), then you can be **95% confident** that the difference is **significant** and not due to chance. This is called a **95% confidence limit**. This is good enough to **reject** the **null hypothesis**.

Data and Statistical Skills

A **Lorenz curve** is a way of showing the **diversity** within a **data set**

1) Lorenz curves compare an **expected** or **normal** distribution (where the data is symmetrical either side of the mean) with the actual **observed** data.

2) In this example, the **annual income** of a population has been **compared** to investigate **economic income inequality**. In an equal society, every person is expected to own an **equal** amount of the **wealth**. This means that in percentages, 40% of people should own 40% of the wealth, 60% should own 60% etc.

3) This expected distribution is expressed as a **straight line** on the graph.

4) In order to plot the actual observed data, the **raw data** is placed in size order, from largest to smallest and given a **ranking** from 1 to 5.

Annual Income Category	Expected number of people in category	Observed number of people in category
up to £25 000	40	96
£25 000 to £30 000	40	54
£30 000 to £50 000	40	32
£50 000 to £60 000	40	8
£60 000 to £75 000	40	10

5) In this example, 200 people were sampled in total, so as there are five categories, the expected results would be that 40 people would occupy each category. The percentage of the total is calculated for each value and a **cumulative total** is also put into a table to give the results below.

Ranking	Expected number of people in category	% of total number of people	Cumulative % total	Observed number of people in category	% of total number of people	Cumulative % total
1	40	20	20	96	48	48
2	40	20	40	54	27	75
3	40	20	60	32	16	91
4	40	20	80	10	5	96
5	40	20	100	8	4	100

The cumulative % columns are plotted against the rank to create a straight line for the expected data and a curve for the observed data.

6) This actual data is then expressed as a curve. The **size** of the curve (i.e. how much it **bends away** from the straight line) is a measure of **how similar** the observed data is from the expected data. They're commonly used to show **levels of inequality** — the larger the curve, the more unequal something may be.

7) This level of inequality can be measured by finding the **Gini coefficient**. This is the **area** between the expected data line and the observed data curve (area **A** on the graph to the right), **divided** by the **total area** above the expected data line (area **A + B**).

The Gini coefficient is a number between 0 and 1, where 0 represents complete equality and 1 represents complete inequality.

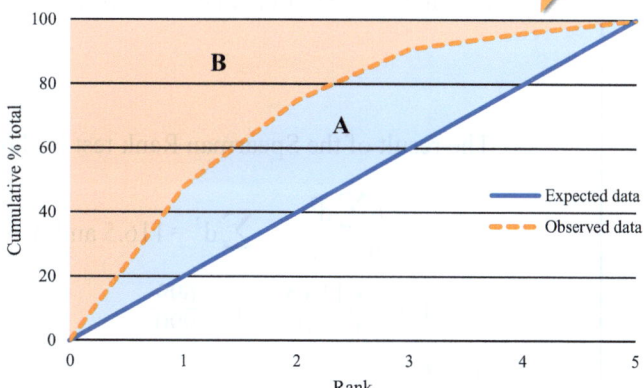

A Lorenz curve showing the expected data against the observed data for the number of people in each annual income category

Anomalies are **not always** what they **seem**

1) When **analysing tables** of data or data in a graph, a value that lies **outside** the main body of the data is often called an **anomaly**. Sometimes, these pieces of data are **ignored**. This is because they might represent **random occurrences** rather than be part of a pattern of events. This means that considering them part of a data set can be **problematic** as they **might not fit** with **scientific knowledge**.

2) In geography, anomalies are often dealt with differently. Providing the data has been collected fairly, this data can be **really important**. It **might not fit** with the **general pattern**, but it's a good idea to start with the 'anomalous' data and try to think of the **geographical reasons** why the data is like this. In geography, it's important to remember that the real world is hugely complex — it's likely that things won't always appear to 'fit'.

3) If the data has been **carefully considered** and no reason can be found for it not fitting the main body of data, only then should it be considered **anomalous** and treated **differently** to the **rest of the data**.

Exam Practice

Well, that's all the Geography topics done and dusted. Now there's only one way to find out how much you've learnt — here are some practice questions to get you started.

Make sure you answer the questions from Topics 1, 3, 5, 6 and 7 — they're compulsory topics. The other questions you answer will depend on which topics you've studied.

---

1   Tectonic Hazards and Processes
Answer all questions on this topic.

1.1 Explain how volcanic eruptions differ at destructive and constructive plate margins.

(4 marks)

1.2 **Table 1** shows a partly completed Spearman's Rank correlation between the Volcanic Explosivity Index (VEI) and the number of fatalities.

Table 1

Volcanic eruption, country, year	VEI	Rank	Fatalities	Rank	d	d²
Mount Merapi, Indonesia, 2010	4	8	353	9	1	1
Kelud, Indonesia, 2014	4	8	7	4	4	16
Mount Ontake, Japan, 2014	3	5.5	63	8	2.5	
Anak Krakatoa, Indonesia, 2018	3	5.5	430	10	4.5	20.25
Stromboli, Italy, 2019	2	2.5	1	1	1.5	2.25
Whakaari / White Island, New Zealand, 2019	2	2.5	21	5	2.5	6.25
Taal, Philippines, 2020	4	8	39	7	1	1
Mount Nyiragongo, DRC, 2021	2	2.5	32		3.5	12.25
Popocatépetl, Mexico, 2022	2	2.5	1	1	1.5	2.25
Hunga Tonga-Hunga Ha'apai, Tonga, 2022	6	10	5	3	7	49
					Σd²	116.5

The result of the Spearman Rank test is shown below.

$$r_s = 1 - \frac{6\sum d^2}{n^3 - n} \qquad \sum d^2 = 116.5 \text{ and } n = 10$$

$$r_s = 1 - \frac{6 \times 116.5}{10^3 - 10} = 1 - \frac{699}{990}$$

$$r_s = 0.29$$

Critical Values of Spearman's Rank Correlation Test

	Significance Level	
n	0.05	0.01
10	0.564	0.745

(i) Complete **Table 1** by filling in the two missing cells.

(2 marks)

(ii) Explain what this result shows about the relationship between VEI and fatalities.

(3 marks)

1.3 Assess the advantages and disadvantages of modifying the vulnerability of a community when managing hazards.

(12 marks)

Exam Practice

2A Glaciated Landscapes and Change
Only answer the questions for 2A if you've studied this topic.
If you answer questions from 2A, don't answer those in 2B.

2A.1 Explain how the process of nivation alters glacial landscapes.

(4 marks)

2A.2 **Figure 1** and **Figure 2** show information about a glacier in the Tien Shan mountains, Kyrgyzstan.
Figure 1 shows the change in the elevation of the glacier between 2003 and 2012.
Figure 2 shows the glacier's mass balance (the difference between ice melt and accumulation) between 2003/4 and 2013/14.

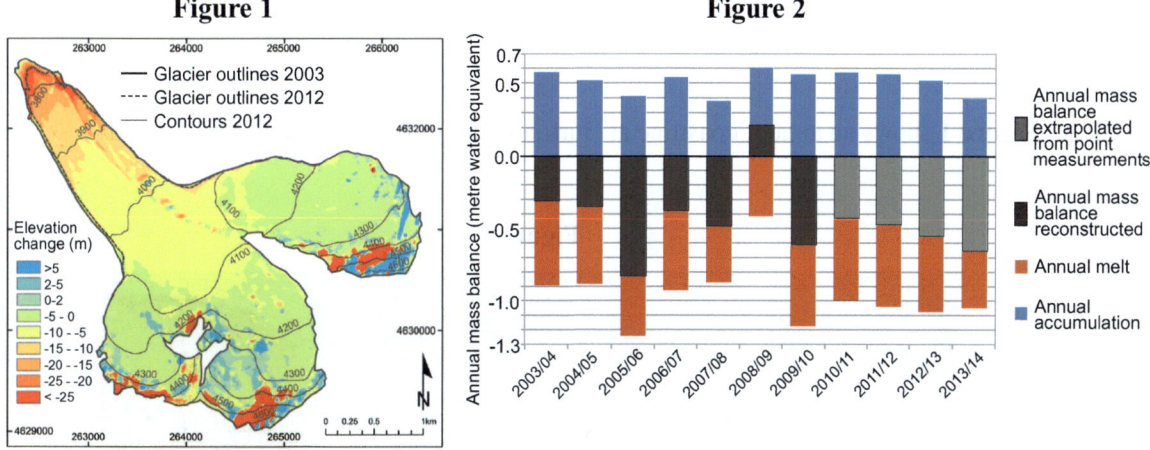

Explain the relationship between the data shown in **Figure 1** and **Figure 2**.

(6 marks)

2A.3 **Figure 3** shows the total number of days of surface melting on the Greenland Ice Sheet in the 2019 summer season. **Figure 4** shows the difference between the total number of melt days in 2019 and the mean number of melt days from 1981-2010.
Figure 5 shows the difference in air temperature in 2019 from the 1981-2010 mean.

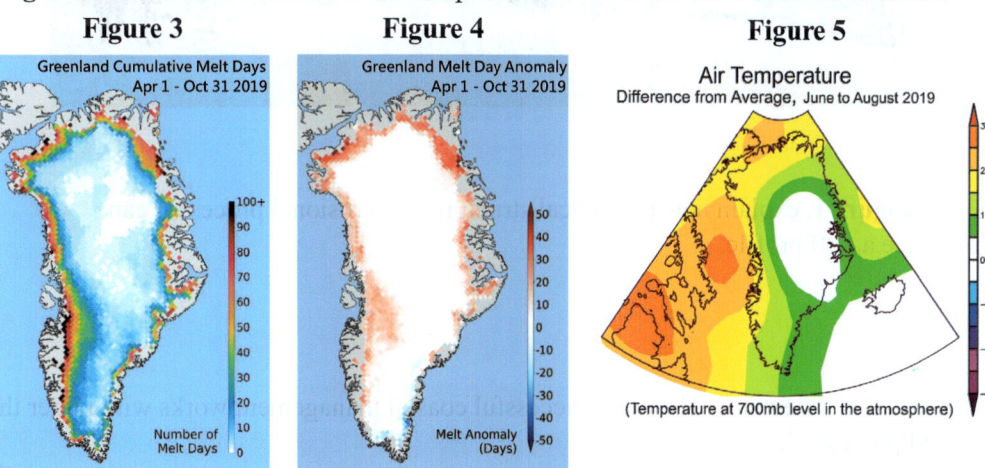

Using **Figures 3-5**, explain the likely impacts of climate change on glacial landscapes such as Greenland.

(6 marks)

2A.4 Evaluate the relative importance of erosional and depositional processes in shaping glacial landscapes.

(20 marks)

Exam Practice

2B   **Coastal Landscapes and Change**
Only answer the questions for 2B if you've studied this topic.
If you answer questions from 2B, don't answer those in 2A.

2B.1   Explain how the major landforms associated with emergent coastlines are formed.

(4 marks)

2B.2   Explain how the physical factors of a coastline can impact its local flood risk.

(6 marks)

2B.3   **Figure 1** shows an area of coastline at Burton Cliffs in Dorset, southwest England.

Figure 1

Using **Figure 1**, explain how geological structure and erosional processes can influence a cliff profile.

(6 marks)

2B.4   Evaluate this statement: 'The most successful coastal management works with rather than against coastal processes'.

(20 marks)

Exam Practice

3 **Globalisation**

Answer all questions on this topic.

3.1 Explain how changes to flows of goods have contributed to the process of globalisation.

(4 marks)

3.2 Explain how Foreign Direct Investment (FDI) has contributed to globalisation.

(4 marks)

3.3 **Figure 1** shows the location of a smartphone TNC's operations throughout the world.

Figure 1

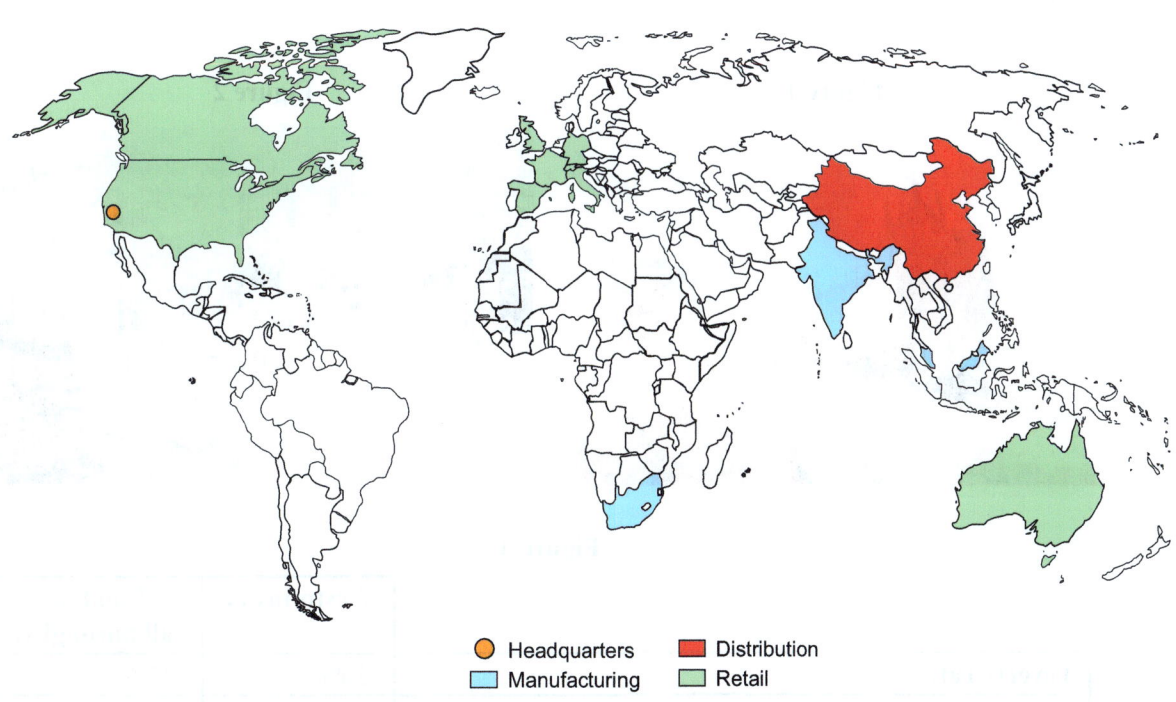

Using **Figure 1**, explain how TNCs have taken advantage of economic liberalisation.

(8 marks)

3.4 Evaluate this statement: 'Interdependence between countries only benefits the wealthiest and most powerful countries.'

(20 marks)

Exam Practice

4A   **Regenerating Places**
Only answer the questions for 4A if you've studied this topic.
If you answer questions from 4A, don't answer those in 4B.

4A.1   Explain how physical factors can determine changes to the characteristics
and function of a place.

         (4 marks)

4A.2   Suggest why the lived experiences of different groups in a particular place may differ.

         (6 marks)

4A.3   **Figure 1** is a painting of Westminster in London. **Figure 2** is a photo of Westminster.
Figure 3 shows data for Westminster and for London as a whole.

Figure 1

Figure 2

Figure 3

	Westminster	London (all boroughs)
Poverty rate	36%	27%
Ratio of pay to house price	20.4	13.7
Average rent paid as a proportion of pay	64%	46%
Houses rented from council/Housing Association	31%	23%
Unemployment rate	8%	5%
Degree of overcrowding	4.8%	7.5%
Access to open space	42%	52%

The data in this table has been taken from 2010-2022.

Using **Figures 1**, **2** and **3** and your own knowledge, explain why different groups might have
different perceptions of the need for regeneration in Westminster.

         (6 marks)

4A.4   Evaluate the view that rural areas need completely different rebranding strategies
to urban areas.

         (20 marks)

Exam Practice

4B Diverse Places

Only answer the questions for 4B if you've studied this topic.
If you answer questions from 4B, don't answer those in 4A.

4B.1 Explain one opportunity and one challenge created by international migration to rural areas.

(4 marks)

4B.2 **Figure 1** shows data from the 2011 census about the ethnicity of Devon's population. **Figure 2** is an excerpt from an article by the writer Louisa Adjoa Parker, in which she talks about her experiences of growing up as a mixed-race child in Devon in the 1980s.

Figure 1

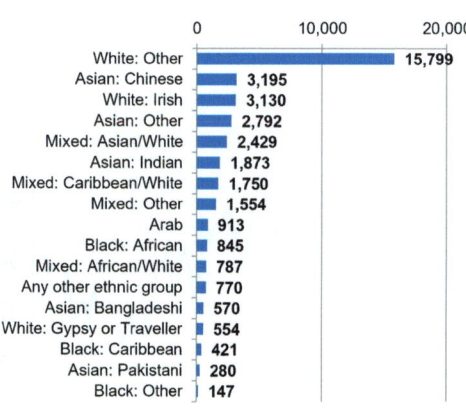

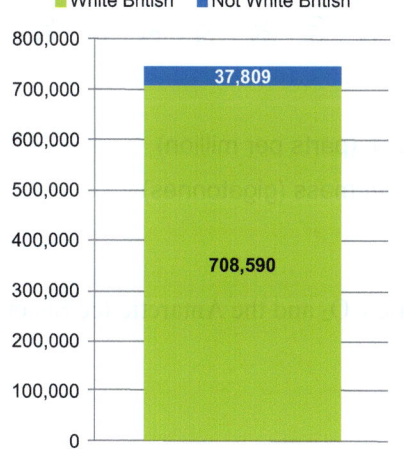

Figure 2

In 1985 I moved to Paignton, Devon, with my white, English mum, brother and sister. My parents had recently split up, and Mum wanted to be near my English grandparents. The move was like going back in time, and the early '70s was not a place I wanted to go back to. Huntingdon, with its flat fenlands and burning stubble, had been slightly multicultural – we had a Japanese friend, and another friend whose mum was mixed, like us. Devon was white, whiter than white. I don't think I even set eyes on another brown face (apart from those of my siblings) for a couple of years. We were the only ones for miles around.

Although I hadn't wanted to move, south Devon, with its red earth, green hills and beaches was at least familiar to me, as we'd been visiting the grandparents for years. I quickly set about trying to befriend the natives. The girl next door informed me that her dad hadn't wanted blacks living next door. She was nice, but the boys racially abused us. My brother later told me they sometimes beat him up.

Using **Figure 1**, **Figure 2** and your own knowledge, assess how useful quantitative sources are in conveying the character of a place.

(12 marks)

4B.3 Evaluate the view that people's perceptions of place are based on their lived experiences, and therefore local people are unlikely to be affected by media representations.

(20 marks)

Exam Practice

5   **The Water Cycle and Water Insecurity**
Answer all questions on this topic.

5.1   Explain how changing land use can increase flood risk.

(4 marks)

5.2   **Figure 1** shows the mass balance of the Antarctic Ice Sheet and mean
global atmospheric levels of carbon dioxide between 2002 and 2017.

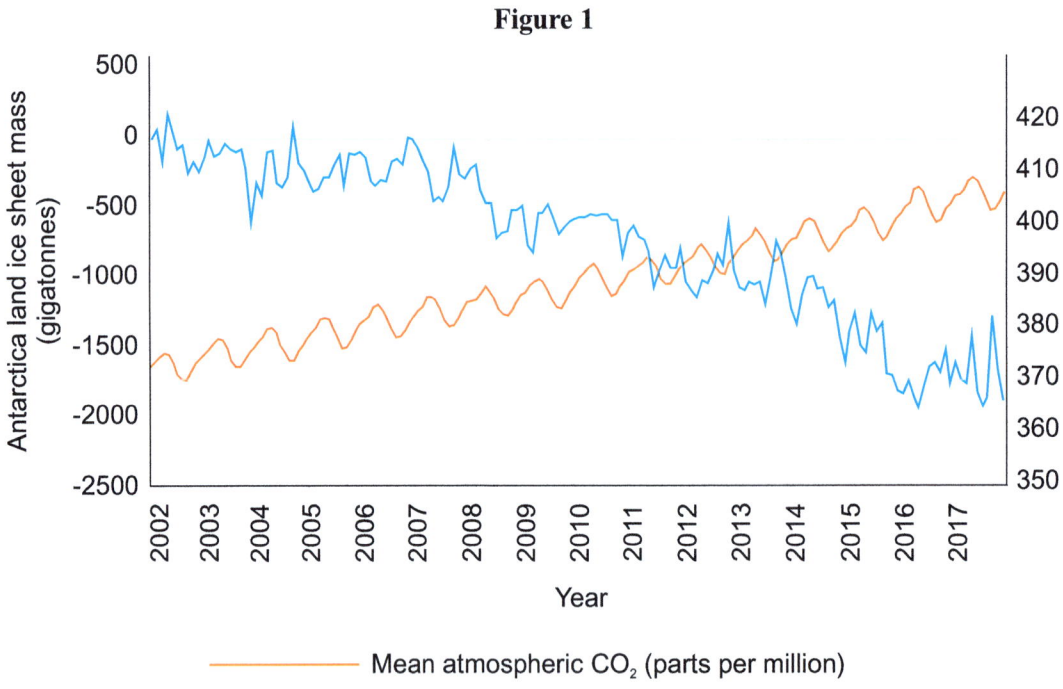

Figure 1

——— Mean atmospheric CO₂ (parts per million)
——— Antarctic land ice sheet mass (gigatonnes)

(i) Explain the relationship between mean atmospheric CO$_2$ and the Antarctic Ice Sheet mass.

(4 marks)

(ii) Suggest how this relationship would affect the water cycle.

(6 marks)

5.3   Evaluate the view that human activities are the main cause of water insecurity.

(20 marks)

Exam Practice

6   The Carbon Cycle and Energy Security
Answer all questions on this topic.

6.1 Explain how carbon is moved between different stores in the carbon cycle.

(4 marks)

6.2 **Figure 1** shows the number and energy production capacity of land and sea wind turbines in Denmark between 1977 and 2017.

Figure 1

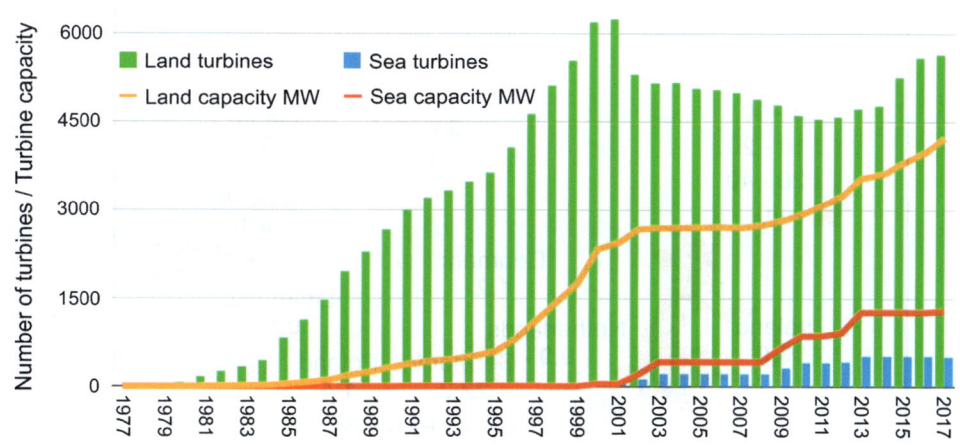

(i) Describe what the data in **Figure 1** shows.

(3 marks)

(ii) Explain why the use of renewable energy might vary between developed countries.

(6 marks)

(iii) For an energy source you have studied, explain the impacts of its development.

(8 marks)

6.3 Evaluate the view that achieving energy security relies on reducing energy consumption rather than increasing energy supply.

(20 marks)

Exam Practice

7   Superpowers
Answer all questions on this topic.

7.1 Explain why a world with a unipolar pattern of power is likely to be unstable in the long term.

(4 marks)

7.2 Study **Figure 1**, which shows how much Chinese FDI EU countries received in 2000-2018.

Figure 1

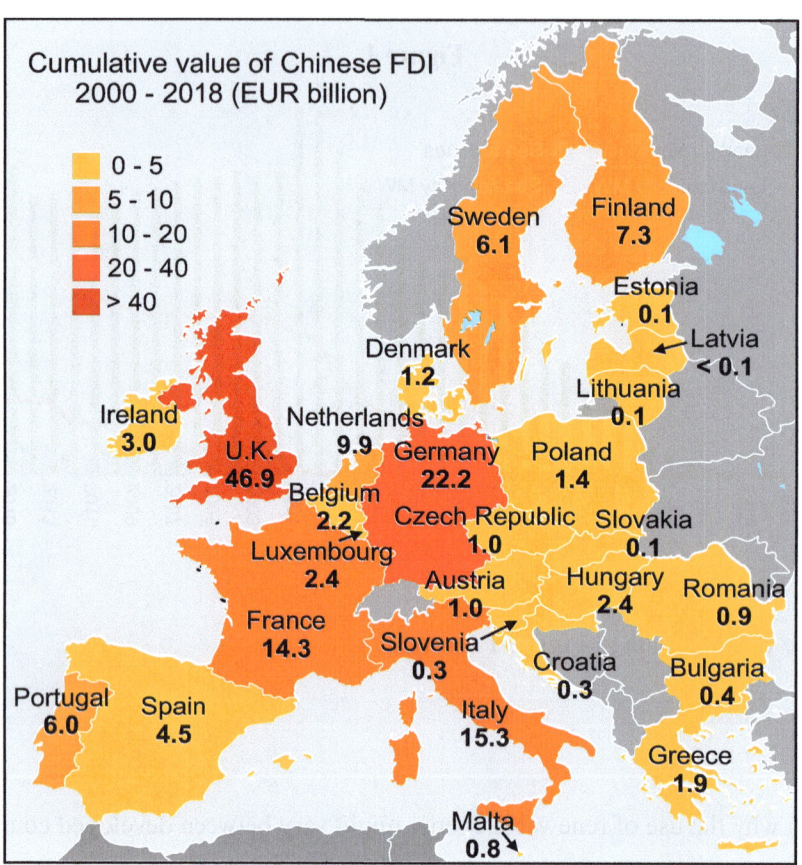

Using **Figure 1** and your own knowledge, assess the view that China is a superpower.

(12 marks)

7.3 Explain why tensions can develop over ownership of disputed areas, such as the Arctic.

(8 marks)

Exam Practice

8A   **Health, Human Rights and Intervention**
Only answer the questions from 8A if you've studied this topic.
If you answer questions from 8A, don't answer those in 8B.

8A.1   Explain why some countries have **not** signed the Universal Declaration of Human Rights.

(4 marks)

8A.2   **Table 1** shows the Gini coefficient (GC) and the Gender Inequality Index (GII) for selected countries. **Figure 1** shows this data as a graph.

Table 1

Country	Gini coefficient (GC), 2019	Gender Inequality Index (GII), 2021
Belgium	27.2	0.048
Bulgaria	40.3	0.210
Cyprus	31.2	0.123
El Salvador	38.8	0.376
Greece	33.1	0.119
Honduras	48.2	0.431
Sweden	29.3	0.023
Zimbabwe	50.3	0.532

Figure 1

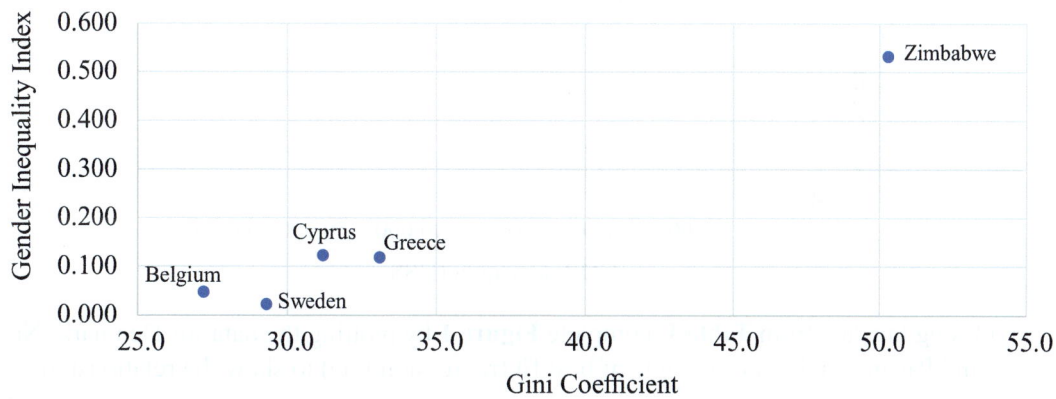

(i) Using the data from **Table 1**, complete **Figure 1** by plotting the data for Bulgaria, El Salvador and Honduras and drawing a line of best fit (regression line) to show the relationship.

(2 marks)

(ii) Suggest reasons for the relationship between the Gini coefficient and Gender Inequality Index.

(6 marks)

8A.3   Explain why the Millennium Development Goals were replaced with the Sustainable Development Goals.

(8 marks)

8A.4   Evaluate the view that inequality is the main cause of spatial variations in health and life expectancy.

(20 marks)

Exam Practice

8B Migration, Identity and Sovereignty

Only answer the questions from **8B** if you've studied this topic.
If you answer questions from **8B**, don't answer those in **8A**.

8B.1 **Table 1** shows GDP per capita and the net migration rate for selected countries.
Figure 1 shows this data as a graph.

Table 1

Country	GDP per capita (US $), 2019	Net migration rate (‰ people), 2019
Argentina	9963.7	0.1
Denmark	59,593.0	4.0
Haiti	1324.8	-3.1
Japan	40,458.0	1.5
Namibia	5126.2	-3.5
Panama	15,826.1	2.5
South Africa	6688.8	0.4
UK	42,747.1	3.3

Figure 1

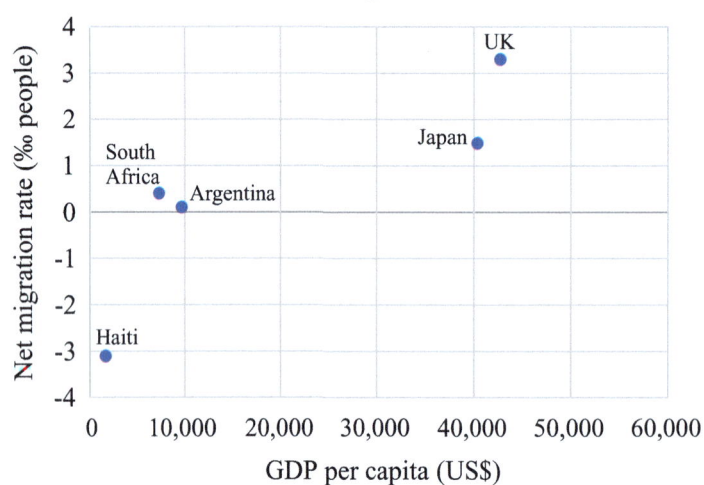

(i) Using the data from **Table 1**, complete **Figure 1** by plotting the data for Denmark, Namibia and Panama and drawing a line of best fit (regression line) to show the relationship.

(2 marks)

(ii) Suggest reasons for the relationship between GDP per capita and net migration rate.

(6 marks)

8B.2 Explain why international migration is easier for some people than for others.

(8 marks)

8B.3 Evaluate the extent to which globalisation has weakened the idea of national identity.

(20 marks)

Exam Practice

9 **Synoptic Question Practice**
Answer all questions. Use the resources provided and your own knowledge and understanding from across your course.

Figure 1

Water stress per country, 2019

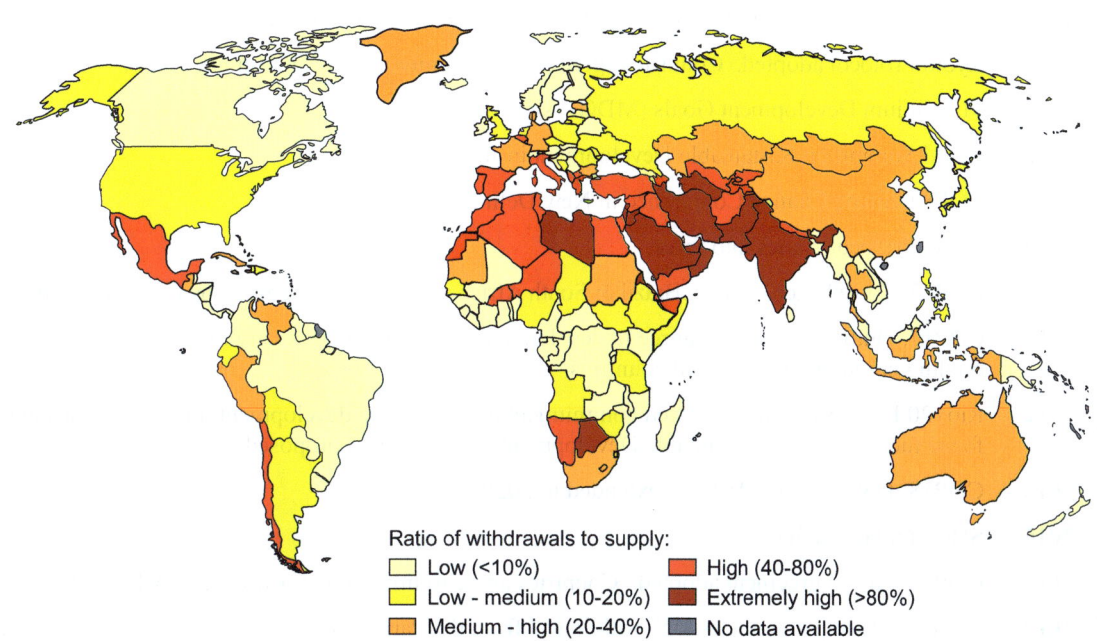

Ratio of withdrawals to supply:
Low (<10%)
Low - medium (10-20%)
Medium - high (20-40%)
High (40-80%)
Extremely high (>80%)
No data available

Figure 2

Global Water Consumption, 1900 - 2025

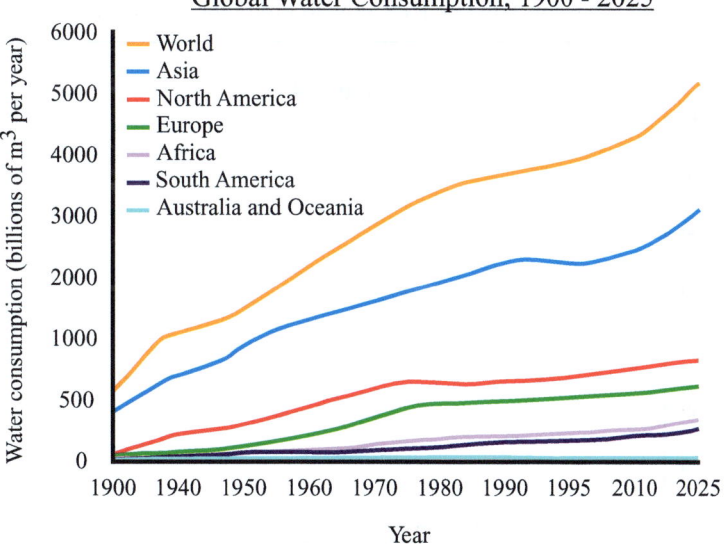

9.1 Explain why climate change can cause water shortages.

(4 marks)

9.2 Suggest reasons for the data shown in **Figure 1** and **Figure 2**.

(6 marks)

9.3 Explain why **Figure 2** may not give a reliable picture of water stress in different regions.

(4 marks)

Exam Practice

Figure 3

Timeline of climate change commitments

1987 — Montreal Protocol adopted. Aimed to protect the ozone layer by phasing out the use of CFCs.

1988 — Intergovernmental Panel on Climate Change (IPCC) formed.

1992 — Earth Summit in Rio de Janeiro. UNFCC signed. Developed countries agree to reduce emissions.

1997 — Kyoto Protocol adopted. Countries agree to reduce emissions by an average of 5%.

2000 — Millennium Development Goals (MDGs) signed.

2002 — World Summit on Sustainable Development in Johannesburg.

2002 — COP8 Climate Change Conference in New Delhi.

2005 — Kyoto Protocol comes into force.

2009 — COP15, Copenhagen. China, Brazil & South Africa agree to reduce emissions for the first time.

2011 — COP17, Durban. Countries agree to develop a new global climate treaty by 2015, with legally binding commitments from all countries.

2012 — Rio+20 Earth Summit. Renewed commitment to sustainable development, including phasing out fossil fuel subsidies. Sustainable Development Goals (SDGs) proposed.

2012 — COP18, Doha. Kyoto Protocol extended to 2020.

2015 — SDGs replace MDGs.

2015 — COP21, Paris Agreement adopted. Countries agree to limit global warming to below 2°C.

2017 — President Trump announces U.S. withdrawal from the Paris Agreement.

2020 — President Xi Jinping of China pledges to be carbon neutral by 2060.

2021 — Under President Biden, the U.S. rejoins the Paris Agreement.

2021 — COP26, Glasgow. Glasgow Climate Pact aims to limit global warming to 1.5°C.

Figure 4

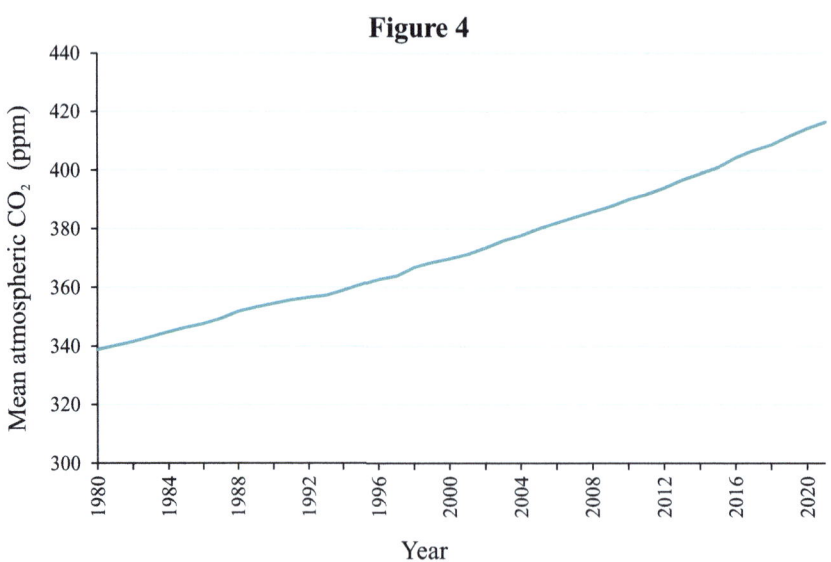

9.4 Study **Figure 3** and **Figure 4**. Evaluate the view that attempts to mitigate climate change have been too little and too late.

(18 marks)

9.5 Using all the resources, evaluate the view that climate change is the strongest factor in changing patterns of geopolitical power.

(24 marks)

Answers

Topic 1 — Tectonic Processes and Hazards

Page 8 — Plate Tectonics Theory

1   One mark for each valid point, up to a maximum of 4 marks.
E.g. One feature of destructive plate margins is ocean
trenches, which are formed when oceanic crust is forced to
subduct under the continental crust *[1 mark]*. This is because
the oceanic crust is denser than the continental crust so is
more easily forced downwards *[1 mark]*. A second feature
is volcanoes, which can form on the surface of both types of
crust either side of the margin *[1 mark]*. This is because the
newly formed magma (made when the subducted oceanic
crust melts) is less dense than the crust above it, so it forces its
way to the surface at any weak spot in the crust *[1 mark]*.

Page 11 — Earthquakes and Tsunamis

1   Maximum of 6 marks available. This question is level marked.
HINTS:
  • Start by briefly outlining that the map shows that earthquakes tend
    to happen along all three types of plate boundary, though more
    frequently at destructive plate boundaries.
  • Then, explain that this is because there is a high chance of plates
    jolting as the pressure is released when one plate is subducted
    beneath another.
  • Next, outline that there is more activity around the Pacific ocean as
    there are many smaller plates making up a larger system of tectonic
    movement.
  • Then, explain how very little earthquake activity occurs in the
    middle of tectonic plates as there are no areas of widespread plate
    movement.

Page 14 — Volcanoes

1   Maximum of 6 marks available. This question is level marked.
HINTS:
  • Start by briefly stating that volcanic island arcs are chains of hotspot
    volcanoes, which form at mantle plumes.
  • Then, explain how a mantle plume creates a hotspot volcano. Make
    sure you explain that heat from the mantle plume melts part of the
    crust above, which creates large volumes of magma.
  • Next, explain how island arcs are formed, e.g. that include the plume
    stays in one place while the plate moves over it. You could include
    an example of an island arc, such as the Hawaiian islands.

Page 16 — Natural Disasters

1   One mark for each valid point, up to a maximum of 4 marks.
E.g. The pressure and release model shows that a disaster
comes not just from the nature of the hazard itself, the
intensity of which is difficult to predict *[1 mark]*, but also
from the conditions found in the country *[1 mark]*. A number
of economic, political, and social structures within a country,
(such as their ability to invest in public services or their
amount of international debt) can contribute to their level of
vulnerability to the hazard. This means that the root causes
of vulnerability can create a pressure *[1 mark]* which makes
a disaster more likely to happen in the event of a hazard
occurring. The more ways in which a country is vulnerable,
and the greater the magnitude of that vulnerability, the greater
the risk of disaster for that population *[1 mark]*.

Page 18 — Vulnerability to Tectonic Hazards

1   Maximum of 12 marks available. This question is level marked.
HINTS:
  • Start by briefly outlining the connection between a location (with
    reference to rural areas especially) and the potential for isolation
    of a community. Outline some of the different ways a community
    might be isolated such as physically, through lack of communication
    networks, economically etc.
  • Then, explain the connection between communication problems and
    the impact that may have on hazard management. Explain how
    poor communication networks make some locations more vulnerable
    to hazards.
  • Next, discuss why rescue teams may find access to some areas more
    challenging than access to others. Relate this to hazard vulnerability.
  • Complete the answer by comparing the relative income levels of rural
    communities with urban communities and the impact this may have
    on secondary factors such as health care, education and housing
    quality in some countries.
  • Make sure you include references to specific case study information
    to support the points you make above.
  • Write a conclusion that assesses the link between the severity of the
    impacts of a tectonic hazard and the level of economic development
    within a country. This should be based on your personal opinion but
    should also draw on the evidence already provided in your answer.

Page 20 — Hazard Profiles

1   Maximum of 12 marks available. This question is level marked.
HINTS:
  • Start by briefly explaining what a hazard profile is and what
    information it aims to show. You should mention the six different
    variables that hazards are measured against.
  • Then, explain how each of the variables might link to impacts. For
    example, you should mention that the greater the magnitude of the
    hazard, the more likely it is to result in fatalities and injuries. Make
    sure you discuss the potential for impacts rather than vulnerability.
  • Next, explain the problems with using a hazard profile as a
    predictive tool, giving examples of hazards that don't fit with a
    certain profile.
  • You should also explain how the variables may be linked, for example
    the magnitude of a tsunami is directly linked to the areal extent of
    its impact.
  • Make sure you include references to specific case study information
    to support the points you make above.
  • Write a conclusion that assesses the extent to which hazard profiles
    give an indication of the potential impacts. This should be based on
    your personal opinion but should also draw on the evidence already
    provided in your answer.

Answers

Page 24 — Tectonic Hazards — Contrasting Impacts

1 Maximum of 12 marks available. This question is level marked.
HINTS:
- Start by outlining the nature of tectonic hazards and the kind of impacts (social, economic and environmental) that can be experienced as a result of their occurrence.
- Then, make suggestions and explain the links between economic development and the impacts of a hazard. For example, you could explore the idea that developed countries have more money to be able to prepare for hazards, monitor them and mitigate their impacts than developing countries. Use clear examples to illustrate your point.
- Next, you should then write a counter-argument by highlighting how other factors can shape the impact of tectonic hazards, such as location or time of day. Use clear examples to illustrate your point. Make a judgement about how other factors shape the impacts of a tectonic hazard.
- Write a conclusion that assesses whether the impacts of a tectonic hazard are shaped by economic development. This should be based on your personal opinion but should also draw on the evidence already provided in your answer.

Page 25 — Tectonic Disaster Trends

1 One mark for each valid point, up to a maximum of 4 marks. E.g. Since 1960, the number of people affected by tectonic hazards has increased *[1 mark]*. This is partly due to the increase in global population, which has caused more people to live in more high-risk and densely populated areas *[1 mark]*. The economic impact of natural hazards has also increased *[1 mark]*. This is because as a country develops economically and becomes wealthier, they can afford to invest in more expensive infrastructure. However, this infrastructure becomes more costly to rebuild in the event of a tectonic hazard causing a longer economic recovery *[1 mark]*.

Page 26 — Tectonic Mega-Disasters

1 Maximum of 12 marks available. This question is level marked.
HINTS:
- Start by defining a mega-disaster and giving an example of one to illustrate your definition.
- Then, explain reasons why the world is likely to see more mega-disasters, e.g. the increasing interconnectedness of the world means that a disaster in one country has more knock-on effects in other countries. Use clear examples to illustrate your points, such as the eruption of Eyjafjallajökull. Make a judgement about how likely the world is to see an increasing number of mega-disasters in this century.
- Next, explain the reasons why we are unlikely to see an increasing number of mega-disasters in this century, e.g. increased interconnectedness may allow countries to be able to learn from each other's mistakes, which may eventually reduce the knock-on effects of a disaster. Use clear examples to illustrate your points. Make a judgement about how the world is unlikely to see an increasing number of mega-disasters in this century.
- Write a conclusion that assesses whether the world is likely to see more mega-disasters. This should be based on your personal opinion but should also draw on the evidence already provided in your answer.

Page 27 — Multiple-Hazard Zones

1 One mark for each valid point, up to a maximum of 4 marks. E.g. The Philippines is a multiple-hazard zone because it experiences different types of hazards, including tectonic hazards such as earthquakes (Luzon 1990) and volcanoes (Mount Pinatubo 1991) and hydrometeorological hazards such as typhoons (Haiyan 2013) *[1 mark]*. The Philippines is close to a destructive boundary and multiple fault lines run though the centre of the country making it more vulnerable to tectonic hazards *[1 mark]*. The Philippines also sits in the far west of the Pacific Ocean and so receives a large number of typhoons *[1 mark]*. Hazards can also cause secondary hazards. Earthquakes in the region have caused tsunamis (Moro Gulf 1976) and landslides are common in the aftermath of typhoons. These further define the Philippines as a multiple-hazard zone *[1 mark]*.

Page 29 — Managing Tectonic Hazards — Theory

1 Maximum of 12 marks available. This question is level marked.
HINTS:
- Start by describing the Hazard Management Cycle and its four different stages.
- Then, state the importance of these stages and how the successful implementation of each one affects the success of the following stage. Explain why one of the stages (such as preparedness) may be seen as more important than the others.
- Next, state some of the factors that can affect how successful a transition from one stage of the cycle to the next might be. Explain how the type of hazard, level of global interdependence experienced by the country, economic development level and living in a multiple hazard zone may affect the success of each stage of the model.
- Write a conclusion that assesses how important the different stages of the model are, bearing in mind the large number of factors that affect their success. This should be based on your personal opinion but should also draw on the evidence already provided in your answer.

Page 31 — Managing Tectonic Hazards

1 Maximum of 12 marks available. This question is level marked.
HINTS:
- Start by describing what is meant by 'modifying a hazard' and give some examples of ways in which this management strategy is done. Then do the same for 'modifying hazard losses'.
- Then, in separate paragraphs, state the advantages and disadvantages of trying to modify the hazard as a form of management. Be sure to refer to how a country's economic development will impact its ability to manage a tectonic hazard. Remember to use real-life examples of tectonic hazard management where possible to illustrate your ideas.
- Next, repeat the process for modifying the losses associated with a hazard.
- Write a conclusion that assesses whether modifying the event or modifying the losses is the more effective option in tectonic hazard management. This should be based on your personal opinion but should also draw on the evidence already provided in your answer.

Answers

Topic 2: Option 2A — Glaciated Landscapes and Change

Page 33 — Climate Change and Glaciation

1   Maximum of 8 marks available. This question is level marked.
    HINTS:
    • Start by briefly outlining what a short time period is in the context of climate change. You should also make reference to some of the short-term climate change episodes that have happened in relative recent history, such as the Little Ice Age.
    • Then, explain how solar output can affect climate. For example, you could mention how the level of sunspot activity coincides with global temperature changes. Use clear examples to illustrate your point.
    • Then, explain how volcanic eruptions can impact climate. Mention should be made of both ash clouds and their impact on sunlight penetration and of sulphur dioxide and how it behaves in the upper atmosphere. Again, be sure to link your points to the impact these factors will have on the climate.

Page 35 — Ice Cover on Earth

1   Maximum of 6 marks available. This question is level marked.
    HINTS:
    • Start by briefly stating what the cryosphere is and what is meant by a global system.
    • Then, explain the role of the cryosphere in the hydrological cycle. The discussion should focus on meltwater and how this affects climate (and especially precipitation levels). An explanation of how the cryosphere is a significant water store is also needed.
    • Next, explain the role of the cryosphere in climate systems. For example, you could explain how the albedo effect affects global temperatures.
    • Complete the answer by explaining the role of permafrost and how this affects global systems. For example, you could mention how it reduces CO_2 in the atmosphere.

Page 38 — Glaciers

1   Maximum of 8 marks available. This question is level marked.
    HINTS:
    • Start by defining accumulation and ablation. Then, explain that the relationship between these two processes in a glacial system over a year is the glacial mass balance.
    • Then, state how the difference between accumulation and ablation tells us whether the glacial budget is positive or negative. Explain what either scenario means for the amount of ice held in the glacier. You could give examples of scenarios that would lead to a positive and a negative regime.
    • Then, explain how glaciers can gain mass through accumulation by snowfall, avalanches and wind deposition. Go on to explain how glaciers can lose mass through ablation by sublimation, calving and wind. Give reasons why glaciers may retreat or grow in different budget scenarios.

Page 41 — Glacial Movement

1   Maximum of 6 marks available. This question is level marked.
    HINTS:
    • Start by defining pressure melting point and explain an example of the conditions that are needed for it to raise or lower.
    • Then, explain why a change in pressure melting point affects the speed at which a glacier moves. You should be sure to make the connection between pressure, basal temperature and the amount of meltwater that a glacier might sit on.
    • Then, explain how meltwater affects speed of movement. You might like to make an additional reference to the notion of positive feedback and how greater levels of meltwater will cause heat generation through friction and subsequently create even more levels of meltwater.

Page 44 — Glacial Erosion — Processes and Landforms

1   Maximum of 12 marks available. This question is level marked.
    HINTS:
    • In the introduction, briefly outline what a corrie is.
    • Then, explain the role that plucking, frost shattering and freeze-thaw weathering have on shaping the back wall of a corrie. Explain how the two processes work and how they shape the corrie. Make a judgement about how significant these processes are in shaping the corrie.
    • Next, explain the role of abrasion and how it shapes the corrie base, as well as the corrie lip. You could mention that the effect abrasion has on a corrie depends on how much rock material is produced by other erosive processes, such as plucking. Make a judgement about how significant abrasion is in shaping the corrie.
    • Write a conclusion that assesses the importance of different erosional processes in the formation (or shaping) of a corrie. This should be based on your personal opinion but should also draw on the evidence already provided in your answer.

Page 47 — Glacial Deposition — Processes and Landforms

1   One mark for each valid point, up to a maximum of 4 marks. E.g. Moraines are landforms that are created from unsorted till deposited by glaciers. There are different types of moraines *[1 mark]*. Lateral moraine forms when till is deposited at the sides of glaciers after freeze-thaw weathering erodes the side of the valley *[1 mark]*. Medial moraine is formed when two lateral moraines have joined together *[1 mark]*. Terminal moraine is formed when deposited till builds up at the snout of the glacier. Recessional moraine forms in a similar way to terminal moraine, but occurs when a glacier temporarily retreats or advances *[1 mark]*.

Page 49 — Fluvioglacial Processes and Landforms

1   Maximum of 6 marks available. This question is level marked.
    HINTS:
    • Start by briefly stating how meltwater can flow through a glacial system and that meltwater carries and deposits material. Make sure you reference the sorting of material and the effect this has on the size and shape of the feature that's formed.
    • Then, explain how meltwater in contact with the ice can lead to the formation of features such as eskers, kames and kame terraces.
    • Next, explain how meltwater can form proglacial landforms such as meltwater channels, sandurs, kettle holes and proglacial lakes.

Answers

Page 53 — Periglacial Processes and Landforms

1 One mark for each valid point, up to a maximum of 4 marks. E.g. Very low temperatures in winter makes the ground contract and cracks to form in the permafrost *[1 mark]*. The active layer thaws during spring and this meltwater seeps into the cracks *[1 mark]*. As the permafrost layer is still frozen, this meltwater freezes once in the cracks, which is what forms ice wedges *[1 mark]*. Ice wedges can grow each year as the ground contracts and reopens existing cracks, which fill with water and freeze again, causing further expansion of the ice wedge *[1 mark]*.

Page 55 — Reconstructing Glacial Dynamics

1 Maximum of 8 marks available. This question is level marked.
HINTS:
- Start by briefly outlining that different glacial landforms are found throughout glacial landscapes.
- Then, explain how the direction a glacier moved in can be determined by studying certain glacial landforms. For example, identifying the rock type of an erratic can tell you where the erratic originally came from and therefore the direction the glacier moved in.
- Next, explain how lateral moraines can reveal the specific path a glacier took, by mirroring its path. Then go on to say that recessional and terminal moraines show the extent of a glacier's snout.
- Complete your answer by explaining how the orientation of drumlins can reveal how fast a glacier was moving. Outline how dividing the length of the longest axis by the width of the widest point gives you the elongation ratio of the drumlin. The higher the elongation ratio, the faster the glacier was moving.

Page 57 — Value of Glaciated Landscapes

1 Maximum of 6 marks available. This question is level marked.
HINTS:
- Start by outlining the natural resources that can be found in glacial areas. For example, you could mention mineral resources, forests and water.
- Then, explain how each of these resources can be valuable to the economy of a glaciated area. You could explain how meltwater may be used to create hydroelectric power or how mining of mineral resources may take place. Refer to specific examples where possible.
- Next, you could explain how forestry can add economic value to a glaciated landscape. For example, forestry plantations in Canada and northern Europe contribute to the economy by supplying timber and paper.

Page 60 — Threats to Glaciated Landscapes

1 Maximum of 20 marks available. This question is level marked.
HINTS:
- Start by briefly outlining that glaciers are threatened in different ways by human activity and natural hazards.
- Then, explain how humans use glacial areas and how human activity threatens glacial areas. For example, you could talk about how the construction of HEP changes the course of a river, which can have an effect on sediment levels. You could explain the negative effect that urbanisation in glacial areas can have. For example, urbanisation can create small urban heat islands, which can increase surrounding temperatures and cause settlements to subside if they're built on areas of permafrost.
- Next, explain how natural hazards threaten glacial areas. For example, you could explain how avalanches can threaten glacial upland landscapes and how they are a risk to human life. You could go on to explain how avalanches, if triggered by a volcanic reaction, can mix with soil to form a lahar. Explain how natural hazards, such as avalanches and glacial outburst floods, also threaten glacial landscapes. For example, they can move glacial deposits and create new erosional features.
- Then, explain the threat that humans outside of glacial areas can have. You could explain how tourism threatens glacial areas. For example, tourism in Antarctica has threatened wildlife habitats and caused stress in penguin colonies, which has negatively impacted their ability to breed and reduced biodiversity in the area.
- Write about the effect that climate change can have on glacier mass balance. You could explain how a reduction in meltwater from glaciers could cause an increase in the concentration of pollution and threaten water quality.
- Write a conclusion that evaluates whether or not humans that live in glacial areas are the greatest threat to glacial areas. This should be based on your personal opinion and also draw on the evidence already provided in your answer.

Page 63 — Managing Glaciated Landscapes

1 Maximum of 20 marks available. This question is level marked.
HINTS:
- Start by explaining that the management of glacial environments sits on a spectrum between total protection and total exploitation, with various stages in between.
- Then, discuss the ways in which management strategies can and have been successful. This should refer clearly to case studies of glacial landscapes, e.g. Yosemite National Park. Make a judgement about the extent of success and whether such strategies are sustainable in the long term.
- Next, discuss some of the barriers to successfully managing glacial areas and give examples. Make a judgement about how these barriers might change over time and whether new barriers to management of glacial areas will play a role in the future.
- Finally, you could write about how different glacial landscapes have unique qualities and different needs and so there may not be one management strategy that is overall better than any other. You could also reference the idea that all management strategies have to be decided on in the context of a changing climate, which can affect how likely they are to be successful.
- Write a conclusion that evaluates whether successful management of different glacial areas is possible. This should be based on your personal opinion and also draw on the evidence already provided in your answer.

Answers

Topic 2: Option 2B — Coastal Landscapes and Change

Page 66 — The Coastal System

1 Maximum of 6 marks available. This question is level marked.
HINTS:
- Start by outlining what constructive waves are and what they look like. Go on to explain this appearance by describing how water moves underneath the surface of the sea.
- Then, describe the frequency of constructive waves. Outline that they have a low frequency of around 6 to 8 waves per minute.
- You should then compare the relative strength of the swash and backwash for the constructive wave and explain what this does to the shore.
- Complete your answer by describing how the swash of a constructive wave is able to carry material up the beach, depositing it at a higher point.

Page 69 — Coastal Landscapes and Geology

1 Maximum of 6 marks available. This question is level marked.
HINTS:
- Start by explaining what the term permeability means, giving examples of different rocks with different permeabilities.
- Explain how impermeable rocks create greater levels of surface run off, which can further erode rocks. Describe the impact that this might have on recession rates.
- Then, explain how permeable rocks might be weakened by pressure that the water exerts within the rock. Describe the impact that this might have on recession rates.
- Finally, link permeability to mass movement and explain how water can lead to slumping and landslips.

Page 71 — Coastal Weathering and Mass Movement

1 Maximum of 8 marks available. This question is level marked.
HINTS:
- Start by outlining what the different types of weathering are and briefly describe the main differences between them.
- Then, explain how mechanical weathering can create opportunities for biological weathering to take place. For example, you could refer to how seeds may fall into cracks formed by the mechanical weathering and how the plants that grow can penetrate the crack further.
- Then, explain how the slow break down of rock chemically can mean it is more vulnerable to mechanical weathering. For example, you could refer to cracks within a rock face may develop more easily if they are weakened by oxidation.
- Complete your answer by explaining how all three types of weathering are likely to happen simultaneously (making mass movement and cliff recession more likely) and there's no specific order to weathering as a process.

Page 73 — Coastal Erosion — Processes and Landforms

1 Maximum of 6 marks available. This question is level marked.
HINTS:
- Start by defining what a wave-cut platform is and its typical characteristics.
- Then, explain the role of weathering and wave erosion in creating a notch in the cliff face. Go on to highlight how further erosive processes (e.g. abrasion) increase the size of the notch, until it eventually forms a cave.
- Next, explain that over time the rock above the cave may collapse as there isn't anything to support it and it becomes unstable.
- Complete your answer by explaining how further weathering and erosion will cause a new notch to form in the cliff face. This process will repeat until eventually a wave-cut platform is formed.

Page 75 — Coastal Erosion and People

1 Maximum of 20 marks available. This question is level marked.
HINTS:
- Start by defining coastal recession and explaining how this can impact the coastal landscape.
- Then, discuss how the economic costs outweigh the social costs. For example, you could mention the economic impact coastal recession has on different players, such as residents, businesses and local authorities. Use examples to illustrate your ideas wherever possible.
- Next, discuss how the social costs of coastal recession can be very important too. For example, you could discuss the way coastal recession can break down communities and the impact it may have on people's physical and mental wellbeing. Again, remember to use examples as part of your argument.
- Write a conclusion that evaluates the extent that the economic costs of coastal recession outweigh the social costs. This should be based on your personal opinion and also draw on the evidence already provided in your answer.

Page 78 — Coastal Deposition — Processes and Landforms

1 One mark for each valid point, up to a maximum of 4 marks. E.g. Once a coastal depositional feature, such as a spit, has created an area of sheltered water, plants that are tolerant to salt (e.g. eelgrass) will start to colonise the muddy sediments that develop there *[1 mark]*. The roots of the plant bind silt together to form dense mud *[1 mark]*. They do this by trapping the sediment as the seawater moves around the mudflats *[1 mark]*. The plants allow the mud to grow upwards, meaning it rises above the high tide level and over time begins to dry out *[1 mark]*.

Page 80 — Sea Level Change

1 Maximum of 6 marks available. This question is level marked.
HINTS:
- Start by briefly defining climate change and outline how sea level can change through eustatic processes.
- Then, explain the impact rising global temperatures will have on the cryosphere and how this links to changes in sea level.
- Next, explain the impact rising global temperatures will have on the volume of water held in ocean basins and large bodies of water. You could mention the thermal expansion of water.
- Complete your answer by explaining how climate change is not the only cause of sea level change (e.g. tectonic processes can also cause sea level change).

Answers

Page 82 — Coastal Flooding

1   Maximum of 8 marks available. This question is level marked.
    HINTS:
    • Start by outlining what it means to be vulnerable to coastal flooding,
      then explain some of the reasons that might make coastal flooding
      occur (such as higher sea levels or storm conditions).
    • Then, discuss why land at a low elevation is more vulnerable to
      coastal flooding. You could refer to a country, such as Bangladesh
      or Tuvalu, to illustrate your points.
    • Then, discuss why the economy of some places makes them more
      vulnerable. You could refer to how a place's ability to protect itself
      and build defences affects its vulnerability to coastal flooding.
    • Finally, the answer should discuss the local topography (besides
      elevation) and the effect this might have on where water is able to
      flow to and where it might pool and cause flooding.

Page 84 — Coastal Management
        — Hard and Soft Engineering

1   Maximum of 12 marks available. This question is level
    marked.
    HINTS:
    • Start by defining both soft engineering and sustainable coastal
      management.
    • Then, explain how soft engineering defences can be sustainable. For
      example, you could mention that soft engineering strategies generally
      try to work with and adapt to natural coastal processes. Examples
      of soft engineering that demonstrate this (such as dune stabilisation)
      should be explored.
    • Then, highlight how soft engineering may not always be sustainable.
      For example, you could mention that strategies such as cliff
      reprofiling and how the expense and disruption this can cause people
      makes it unsustainable.
    • Write a conclusion that assesses the view that soft engineering
      defences are a sustainable form of coastal management. This should
      be based on your personal opinion and also draw on the evidence
      already provided in your answer.

Page 87 — Integrated Coastal Management

1   Maximum of 20 marks available. This question is level
    marked.
    HINTS:
    • Start by outlining the idea of coastal management and what this
      might include. Then, define what economic sustainability means.
    • Then, discuss how coastal management aims to create economic
      sustainability. You could refer to techniques such as cost-benefit
      analysis and how this is used to calculate the most feasible outcome.
      You should refer to examples and case studies in your answer.
    • Next, discuss how coastal management must also consider other
      factors, such as its impact on the environment and communities.
      For example, you could refer to the role an Environmental Impact
      Assessment has in decision making. You should refer to examples
      and case studies in your answer.
    • Then, you could outline ICZM and briefly explain how it can be
      beneficial to the sustainability of coastal management, but also how
      it can create conflict between different groups of people.
    • It should be noted that although economic sustainability may not be
      the overall goal of coastal management, the affordability of different
      coastal management schemes is often a deciding factor.
    • Write a conclusion that evaluates the view that coasts are managed
      primarily for economic sustainability. Your conclusion should be
      based on your personal opinion and also draw on the evidence
      already provided in your answer.

Topic 3 — Globalisation

Page 89 — Introduction to Globalisation

1   One mark for each valid point, up to a maximum of 4 marks.
    E.g. Technological innovations and advancements in transport
    during the nineteenth and twentieth centuries have allowed
    trade to take place over much wider distances and at greater
    speed than previously possible *[1 mark]*. For example, at
    the start of the nineteenth century, goods that needed to be
    transported overseas would have been transported via slow
    sailing ships *[1 mark]*. Today these goods can be moved by
    shipping container or by jet aircraft, which is far quicker and
    allows more direct transportation to places without a coastline
    [1 mark]. Travelling the same distance is now possible in
    much less time and therefore distant places feel closer and the
    world seems smaller *[1 mark]*.

Page 91 — Globalisation — Key Players

1   Maximum of 6 marks available. This question is level marked.
    HINTS:
    • Start by briefly outlining the differences between the broad economic
      policies that governments may follow (such as protectionism versus
      free market liberalism). It may be possible to suggest whether such
      policies encourage or discourage the spread of globalisation.
    • Then, explain specific actions and policies that national governments
      engage with that open them up to further globalisation. E.g.
      you could mention privatisation of services (with reference to an
      example) and the incentives governments provide in order to attract
      international industries (such as SEZs).
    • Then, define the term 'trade bloc' and discuss the extent to which
      national governments can decide to be part of trade blocs. You
      could then explain the impact the growth of trade blocs can have on
      the spread of globalisation. Here you could also refer to an example
      of a trade bloc and the impact it might have had on globalisation.

Page 92 — Extent of Globalisation

1   Maximum of 12 marks available. This question is level marked.
    HINTS:
    • Start by stating how different countries experience differing levels of
      globalisation.
    • Then, examine some of the different factors that can contribute to
      globalisation taking place. You should refer to specific information
      and data about different countries to support your ideas. You should
      assess the relative strengths of these factors as you discuss them.
    • Next, explore some of the different factors that can hinder
      globalisation. These might be political, social, economic or even
      related to physical geography. This paragraph should also assess the
      relative strengths of each factor.
    • Write a conclusion that gives an overall assessment of the different
      factors that can affect the extent of globalisation. This should
      be based on your personal opinion but should also draw on the
      evidence already provided in your answer.

Answers

Page 93 — Trans-National Corporations

1 Maximum of 12 marks available. This question is level marked.
HINTS:
- Start by defining a TNC and describe what you think is meant by the role of TNCs in globalisation.
- Then, make the point that TNCs have a significant role to play in driving globalisation. This could make reference to the nature and extent of TNC supply chains, what is traded between countries, as well as how TNCs make best use of outsourcing and offshoring in the way they operate. Your answer should make reference to specific examples of TNCs and how they operate.
- Then, make a point which highlights the power of TNCs in deciding where to invest FDI and create new markets. You could point out that TNCs' goals are usually to maximise profits regardless of where they actually end up operating.
- Then, highlight how other bodies and factors have a role to play in driving globalisation. Mention the role national governments and trade blocs (such as the EU) have in the process. You could also reference how international organisations, such as the WTO and IMF, might be seen to encourage the spread of globalisation.
- Write a conclusion that evaluates the role that TNCs have in driving globalisation. This should be based on your personal opinion but should also draw on the evidence already provided in your answer.

Page 96 — Global Winners and Losers

1 Maximum of 12 marks available. This question is level marked.
HINTS:
- Start by defining global shift and briefly outline how this has come about.
- Then, state some of the ways in which the global shift has been beneficial for people and places. You should refer to specific information and data about different countries to support your ideas. You should also be sure to make references to social as well as economic advantages and disadvantages. You should assess the relative strengths of these different benefits as you discuss them.
- Next, state some of the ways in which the global shift has had a negative impact on people and places. Remember to highlight how different groups of people may feel these impacts differently and again, assess the relative strengths of these impacts as you discuss them. This paragraph should also include references to specific case study information as well as covering both social and environmental impacts.
- Write a conclusion that gives an overall assessment of the impacts of the global shift. This should be based on your personal opinion but should also draw on the evidence already provided in your answer.

Page 99 — Migration and Urban Growth

1 Maximum of 12 marks available. This question is level marked.
HINTS:
- Start by defining international migration and briefly outlining some of the ways it may have an impact on the source country (such as economically, socially and environmentally).
- Then, begin to discuss negative aspects of international migration in more detail, making sure you remain focused on the source country. It will be possible to develop your points to include more than one idea such as linking the departure of working age people with pressure in essential public services. You should try to bring in real world examples to illustrate your ideas where you can.
- In order to make a full assessment, you will also need to comment on the positive impacts international migration has on the source country as well as concepts which are both positive and negative in nature, or which might affect different people in the source country in different ways.
- Write a conclusion that gives an overall assessment of whether international migration is for the most part a negative experience for the source country or not. This should be based on your personal opinion but should also draw upon the evidence already provided in your answer.

Page 103 — Global Culture

1 Maximum of 12 marks available. This question is level marked.
HINTS:
- Start by defining global culture and explain why a global culture is likely to be western in nature.
- Then, explain in detail the ways in which a global culture benefits people. Remember to think about, social, environmental and economic benefits and to use specific case studies to illustrate the points you make.
- Next, explain in detail how there may be costs that come with the spread of global culture. Again, make reference to different types of costs and examples to support your points.
- As part of your discussion of costs and benefits, discuss the idea that some groups of people are likely to benefit from the spread of global culture while at the same time, and in the same location, other people will feel the negative consequences of it.
- Write a conclusion that gives an overall assessment of whether the spread of western global culture has more costs than benefits. This should be based on your personal opinion but should also draw on the evidence already provided in your answer.

Page 105 — Globalisation and Development

1 Maximum of 6 marks available. This question is level marked.
HINTS:
- Start by discussing the economic indicators of development and include examples of these indicators. Explain how they show what a country is like. Outline the type of data that can come from single indicators.
- Then, discuss the social indicators of development (including examples). Compare the nature of the data that social indicators provide with data gathered through economic indicators. Discuss the advantages of using composite indicators and index scores.
- Briefly discuss environmental indicators of development (including examples) and discuss what kind of image they might provide of a country. Compare the quality of the information environmental indicators can provide with the quality of social and economic indicators.

Answers

Page 108 — Attitudes to Globalisation

1   Maximum of 12 marks available. This question is level marked.
    HINTS:
    - Start by outlining the spread of globalisation and give some examples of it in practice. Then, briefly discuss why there has been criticism of it.
    - Then, describe the first attempt (trade protectionism / resource nationalism / migration policies / censorship) you wish to discuss. Explain how the spread of globalisation has created a need for this measure and how the measure controls that spread. You should use case studies and examples wherever possible to illustrate your ideas.
    - Follow the same paragraph structure for at least one other attempt you wish to discuss. Remember that in order to show assessment, you must say how successful the measure (trade protectionism / resource nationalism / migration policies / censorship) was at controlling the spread of globalisation.
    - Write a conclusion that gives a summary of which measure, if any, has been most successful at controlling the spread of globalisation. This should be based on your personal opinion but should also draw on the evidence already provided in your answer.

Page 110 — Responses to Globalisation

1   Maximum of 6 marks available. This question is level marked.
    HINTS:
    - Start by defining what is meant by ethical consumption and give some examples of what this might look like. Outline the role of international workers in the global supply chain (that ends in developed countries). Suggest why they may be experiencing negative impacts of globalisation.
    - Then, describe a positive impact of ethical consumption and carefully explain how it can benefit international workers. Remember to be clear on what aspect of their lives it affects and to use brief examples to illustrate your ideas.
    - Continue this structure for a second positive impact of ethical consumption. Ideas you might like to explore include boycotting certain brands, buying locally produced goods and buying Fairtrade items.

Topic 4: Option 4A — Regenerating Places

Page 113 — Places — Economic Variations

1   Maximum of 6 marks available. This question is level marked.
    HINTS:
    - Start by briefly outlining the circumstances that can cause people to have low average incomes.
    - Then, explain the link between a low income and a poor diet, giving examples of circumstances in which people may not be able to afford to eat healthily. You may want to mention different types of ill health.
    - Finally, you could mention the reduced ability of low earners to afford higher-quality housing and the impact this can have on their health.

Page 116 — Places — Changing Characteristics

1   One mark for each valid point, up to a maximum of 4 marks. E.g. As the function of a place changes, it may bring with it changing economic opportunities. For example, moving from an industrial to a retail-based economy increases the need for a wider variety of workers *[1 mark]*. These opportunities may require new forms of labour and the place may become attractive to migrants *[1 mark]*. Migrants may come from other parts of the UK or from overseas, possibly changing the ethnic composition of the place *[1 mark]*. A more ethnically diverse population often indicates that a place is experiencing an increase in economic success brought about by a change in function *[1 mark]*.

Page 117 — Places — Influences

1   Maximum of 20 marks available. This question is level marked.
    HINTS:
    - Start by defining what is meant by global connections and give some examples.
    - Then, explain how places might be affected by global influences. These influences can be generic in nature, but reference should also be made to specific places you've studied. Discuss the specific roles that people and organisations have in influencing places.
    - Next, discuss some of the different global influences evident in your chosen case study. Make sure you say how these have influenced the place.
    - Then, briefly explain how places might be affected by other factors, such as national policies. You should make it clear where national policies interact with global influences. You could relate your places to past and present influences to show changes over time.
    - Write a conclusion that evaluates the extent to which a place can be influenced by global connections. This should be based on your personal opinion but should also draw on the evidence already provided in your answer.

Answers

Page 120 — Places — Representations and Identity

1 Maximum of 6 marks available. This question is level marked.
HINTS:
- Start by defining personal identity as well as briefly outlining why someone might migrate to a new place.
- Then, explain the concept of cultural enhancement through migration. Discuss how merging cultures can change how someone views themselves in relation to others. Examples can be used from the places you've studied to briefly illustrate this.
- Next, explain how migration can lead to tensions between people in relation to cultural erosion and how it can affect identity.
- Finally, discuss the link between migration and the growing inequality in some areas. Be sure to make it clear throughout your answer that changes to identity can come about through both real and imagined changes.

Page 123 — Inequalities of Place

1 Maximum of 20 marks available. This question is level marked.
HINTS:
- Start by defining the types of economic model in question. In this context, cumulative causation and the spiral of decline should be discussed.
- Then, evaluate the extent to which cumulative causation can explain why some areas are perceived as economic successes. References should be made to examples of places and case studies to help you to illustrate your points.
- Next, evaluate the extent to which the spiral of decline is representative of how different places come into periods of economic decline. This should highlight aspects of the model that would be true in all cases (such as lower consumer spending having a knock-on effect on small businesses) and the parts of the model which do not always occur (e.g. emigration of the most socially mobile)
- Write a conclusion that evaluates the extent to which different models explain the relative success of a place. You might also like to comment on whether models apply to all countries or to all time frames. Your conclusion should be based on your personal opinion but should also draw on the evidence already provided in your answer.

Page 126 — Experiences of Place

1 Maximum of 6 marks available. This question is level marked.
HINTS:
- Start by explaining the term voter turnout and then look at the different groups of people that might need to be considered in the answer.
- Then, discuss the difference between the voting patterns of younger and older people and how this might reflect their experience of a place. Relate this to youthful and ageing population structures.
- Next, explain how level of education can affect how people engage with a place and how this might affect whether they vote or not. Link this idea to the different socio-economic groupings that might be found within a settlement or area.
- Finally, explain how voting can differ between people of different ethnic groups. Explain how this relates to experiences of a place and links to the different socio-economic groups in some areas.

Page 128 — The Need for Regeneration

1 Maximum of 6 marks available. This question is level marked.
HINTS:
- Start by stating which place you plan to use as your example, and briefly outline the sources of information that can be used to assess the need for regeneration.
- Then, explain why the statistical analysis of data might not give the full picture regarding the need for regeneration. Reference should be made to the fact that data goes out of date and that numerical data doesn't show how individuals might feel about a place. You could give examples of statistics from your chosen place and comment on what they show about that place's need for regeneration.
- Next, explain why media depictions of a place are also unreliable. Comments should include the idea of subjectivity and bias towards certain ideas depending on the storyline or agenda of the writer or presenter. You could give examples of how your chosen place is depicted in the media and why this may not be accurate.

Page 132 — Regenerating Place — Government Policies

1 Maximum of 6 marks available. This question is level marked.
HINTS:
- Start by stating the planning policies you will be discussing and what is meant by economic regeneration.
- Then, explain how planning policies can benefit regeneration. You could refer to examples of policies such as 'planning gain' and how it can expand the scope of any development undertaken by private investors. The short and long term benefits of this could be discussed.
- Next, explain how planning policies can have a negative effect on regeneration. Green Belt policies and the restrictions that can come about through designated areas of land (such as national parks) could be discussed.
- Conclude your answer with a brief comment on how all planning policy can affect the regeneration and redevelopment of an area, and how this can both positively and negatively affect the outcomes.

Page 135 — Rebranding Places

1 Maximum of 12 marks available. This question is level marked.
HINTS:
- Start by defining rebranding and briefly describing how it's used by both rural and urban areas.
- Then, explain how rebranding could be seen to be a marketing tool for tourism. Explain the link between tourism and economic development in a region and how the perceived 'image' of a place is central to the number of visitors it might receive.
- Next, explain how rebranding can be about local people as well as those who wish to invest in the area from outside the place. Explain the impact this may have on a place. Remember to use examples of real places to illustrate your ideas.
- Write a conclusion that assesses the extent to which rebranding is only about attracting tourists. You could make a clear link between tourism, local spending, external investment and how they all affect each other. Your conclusion should be based on your personal opinion but should also draw on the evidence already provided in your answer.

Answers

Page 140 — Measuring the Success of Regeneration

1   Maximum of 20 marks available. This question is level marked.
HINTS:
   • Start by defining the term lived experience and how a successful regeneration project might be measured.
   • Then, discuss the view that local people with lived experiences are valuable evaluators. You could write about the purpose of regeneration and who it's for, as well as examples of local community groups you've studied who have long-standing lived experiences of your chosen place.
   • Next, evaluate the extent to which a wider and more varied selection of viewpoints should be considered. Highlight the benefits of different perspectives from different stakeholders in regeneration, such as the national government, which may be able to plan and see the wider picture of regeneration across the UK as a whole. Make clear references to real life regeneration projects to illustrate your ideas.
   • Write a conclusion that evaluates the view that a regeneration project's success should only be measured from the viewpoint of people with lived experiences of the place. Be sure to recognise the different forms of lived experience that people might have had. Your conclusion should be based on your personal opinion but should also draw on the evidence already provided in your answer.

Topic 4: Option 4B — Diverse Places

Page 144 — Places — Changing Populations

1   Maximum of 6 marks available. This question is level marked.
HINTS:
   • Start by defining the demographic characteristics of a place and give some examples.
   • Then, discuss the typical characteristics of a rural place and explain why certain demographic groups tend to live in those areas.
   • Then, explain the same for urban places. Take care not to repeat the same reasoning you used in your previous points, and include historical examples such as international migration to the UK from Commonwealth countries.

Page 145 — Places — Influences

1   Maximum of 12 marks available. This question is level marked.
HINTS:
   • Start by defining global and national influences.
   • Then, explain how places might be affected by global influences. You should refer to specific places such as Liverpool, Lerwick or the places you have studied. Discuss specific players and assess the role they have had in influencing places.
   • Next, explain how places might be affected by national influences. You should refer to the same places that you mentioned earlier in your answer. Discuss specific players and assess the role they have had in influencing places.
   • Write a conclusion that states whether places are impacted more by global or national influences. This should be based on your personal opinion but should also draw on the evidence already provided in your answer.

Page 148 — Places — Representations and Identity

1   Maximum of 6 marks available. This question is level marked.
HINTS:
   • Start by defining personal identity and briefly outline why someone might migrate to a new place.
   • Then, explain the concept of cultural enhancement through migration. Discuss how merging cultures can change how someone views themselves in relation to others. Examples can be used from the places you have studied to briefly illustrate this.
   • Next, explain how migration can lead to tensions between people in relation to cultural erosion and how this can impact a person's identity. Again, use specific examples to illustrate how this affects people.
   • Finally, discuss the link between migration and the growing inequality in some areas and how this can change people's sense of identity. You might want to make it clear throughout your answer that changes to identity can come about through both real and imagined changes brought about through migration.

Page 150 — Perceptions of Urban Places

1   Maximum of 6 marks available. This question is level marked.
HINTS:
   • Start by defining a life-cycle stage and give some examples.
   • Then, explain what young people who have just left education are looking for from an urban place. Explain how those values affect their perception of urban places.
   • Next, explain why people with families of different ages may perceive an urban place differently. Be sure to differentiate between inner-city spaces and suburban places.
   • Finally, discuss how retired and elderly people may view an urban place differently to people at other life-cycle stages. Link this to what they might value most in life and whether the city is able to provide those things.

Answers

Page 152 — Perceptions of Rural Places

1 Maximum of 20 marks available. This question is level marked.
HINTS:
- Start by defining lived experience and the rural idyll.
- Then, discuss the view that the rural idyll is an accurate image of life in rural places. Make sure you discuss the aspects of your chosen rural place that are stereotypically 'rural' to outside observers.
- Next, discuss the view that the lived experience of your chosen place is different to the idyllic image. Highlight the challenges your rural area faces and why these contradict the image people may have of the place. You should make clear references to things like service provision, isolation and leisure facilities.
- Write a conclusion that evaluates the extent to which a lived rural experience is similar to the 'rural idyll'. Be sure to recognise the different forms of lived experience that can be considered, such as the experiences of contrasting groups of people within rural areas.
- Your conclusion should be based on your personal opinion but should also draw on the evidence already provided in your answer.

Page 153 — Perceptions of Living Space

1 Maximum of 6 marks available. This question is level marked.
HINTS:
- Start by defining perception of place.
- Then, explain why the statistical analysis of data might not give the full picture of what a place is really like. Reference should be made to the fact that data is often out of date and that numerical data does not comment on how individuals might feel about a place. You could give examples of statistics from your chosen place and comment on whether they reinforce a person's perceptions.
- Next, explain why media depictions of a place are equally unreliable in assessing what a place is like. You could give examples of how your chosen place is depicted in the media and whether this validates or contradicts the lived experience of place.
- Finally, you may want to comment on and give examples of a situation where someone's perception of place matches the lived experience.

Page 156 — Cultural Diversity in the UK

1 One mark for each valid point, up to a maximum of 4 marks. E.g. As migrants from different ethnic and nationality groups move into an urban area, they might require specific places of worship, which leads to new buildings being built to cater for their needs *[1 mark]*. Shops and businesses might be set up by entrepreneurial migrants who wish to serve the migrant community with food and clothes that reflect their home environments *[1 mark]*. Restaurants and cafes may serve food from the migrant's home region to allow migrants to enjoy familiar social spaces *[1 mark]* and street signage may be written in both the local language and the language used by migrants in their homes *[1 mark]*.

Page 158 — Changing Places

1 Maximum of 20 marks available. This question is level marked.
HINTS:
- Start by defining social exclusion and ethnic tension.
- Then, explain how the built environment can create real and perceived feelings of social exclusion and ethnic tension. Some of your points can be generic in nature, but you should also refer to specific places that you have studied. Make sure you clearly define the exact elements of the built environment that have created social exclusion and ethnic tension.
- Next, explain how other factors have a role to play in creating feelings of social exclusion and ethnic tension. Reference should be made to political policies and the histories that places have in relation to migration. You may want to discuss how these factors have changed over time.
- Finally, discuss how the built environment may create greater levels of social cohesion. You could give examples of places where this has happened and what changes were made to the built environment.
- Write a conclusion that evaluates the view that social exclusion and ethnic tension are caused by the built environment. This should be based on your personal opinion but should also draw on the evidence already provided in your answer.

Page 160 — Managing Cultural Issues

1 Maximum of 6 marks available. This question is level marked.
HINTS:
- Start by defining cultural assimilation and give some real-world examples.
- Then, explain how data (e.g. crime data in relation to hate crimes and racist abuse) can be used to show different attitudes to different cultures and how well a host community might embrace diversity. Explain how comparisons in this data should be made between areas of different sizes to show relative change.
- Next, explain why voter turnout and political engagement is a sign of assimilation. Be sure to highlight the connection between voter turnout and a community's feelings of investment in a place.
- Finally, discuss how mapping segregation can show changes in the degree of assimilation. Link this to the idea of change over time and how second and third generation migrants may feel more socially integrated than their parents.

Page 161 — Managing Change in Urban Areas

1 One mark for each valid point, up to a maximum of 4 marks. E.g. Community groups, such as those connected with places of worship, can become a driving force for increasing levels of community integration and cohesion *[1 mark]*. By opening their doors and sharing their festivals with a wider community, they can remove any perceptions of being insular *[1 mark]*. Local groups may have a stronger feel for the problems and issues a particular community faces, which puts them in a stronger position to address those issues directly *[1 mark]*. Local residents' associations can bid for funding and lobby local authorities for change *[1 mark]*. Community groups can also set up local consultations that ensure that everybody's voice is heard *[1 mark]*.

Answers

Page 162 — Managing Change in Rural Areas

1 Maximum of 6 marks available. This question is level marked.
HINTS:
- Start by defining the different types of landowners such as residents, farmers and industry owners.
- Then, discuss the role of each in turn. Be sure to link the landowners' roles to things such as employment, provision of rural services and environmental protection.
- Make sure your answer briefly states the actions that landowners can take and explains how these link to change in the rural space. It would be a good idea to try to link your ideas to both protecting rural areas from unwanted change as well as actively promoting beneficial change.

Topic 5 — The Water Cycle and Water Insecurity

Page 164 — The Hydrological Cycle

1 Maximum of 6 marks available. This question is level marked.
HINTS:
- Start by defining the term non-renewable water. You could also make a reference to residence time.
- Next, explain why more non-renewable water is being used. For example, you could mention how water demand is increasing so new sources are needing to be explored to meet demand.
- Finally, explain how this is reducing the overall amount of non-renewable water in the hydrological cycle. For example, you could mention how there is no effective way of recharging non-renewable water reserves, so once it has gone, we will not be able to use it again in the same way.

Page 167 — Drainage Basins

1 Maximum of 12 marks available. This question is level marked.
HINTS:
- Start by introducing the idea that both physical and human factors can cause variations in the drainage basin cycle. You could also explain that the drainage basin is an open, local hydrological cycle within the global hydrological cycle.
- Then, explain some of the physical factors. You could mention vegetation, differences in soil structure, the shape and size of the drainage basin etc. Remember to explain how each factor influences the drainage basin cycle, e.g. its processes. Make a judgement about how important these factors are in influencing the drainage basin cycle.
- Next, explain some of the human factors. For example, you could mention changing land usage, deforestation and water abstraction on the water cycle. Remember to explain how each factors influences the drainage basin cycle. Make a judgement about how important these factors are in influencing the drainage basin cycle.
- Write a conclusion that assesses the relative importance of physical and human factors in influencing the drainage basin cycle. This should be based on your personal opinion but should also draw on the evidence already provided in your answer.

Page 170 — Water Budgets and River Discharge

1 Maximum of 8 marks available. This question is level marked.
HINTS:
- Start by defining river discharge and describing the ways in which it may vary. Make sure to refer to the river's regime too.
- Then, explain how different physical factors can affect discharge patterns. For example, you could mention basin size and shape, soil, geology, vegetation etc. You should refer to two different scenarios in your answer — one where the discharge is high and another where it's low.
- Then, explain how human factors can have an effect on discharge patterns. For example, you could mention urbanisation and changing land use. As with the physical factors, show how they can increase and decrease river discharge in two different river scenarios.

Answers

Page 172 — Drought
1   Maximum of 6 marks available. This question is level marked.
HINTS:
- Start by defining what is meant by ecological resilience and how it might apply to a wetland area.
- Next, explain how a drought might affect a wetland area. Discuss the impact it would have on the landscape, the vegetation and the animals found in a wetland.
- Then, explain how a wetland would suffer in both the short and long term and how well different elements of the ecosystem would be able to recover from a drought.
- Finally, discuss how the size and location of a wetland might affect its ecological resilience as well as how severe the drought was.

Page 175 — Flooding
1   Maximum of 20 marks available. This question is level marked.
HINTS:
- Start by defining what a flood is and briefly describe the ways in which both human and physical factors make floods more damaging.
- Then, assess the extent to which humans can make these impacts worse. For example, you could mention river management schemes and changing land use. Remember to use case studies and examples to illustrate your ideas. Make a judgement about how much these factors can make the damage caused by flooding worse.
- Next, argue the view that flooding is a purely natural process. For example, you could mention how monsoonal rains, snowmelt in upland glacial areas and other physical factors can all influence the damage from flooding. Make sure you recognise that some types of flooding (in terms of their cause) will produce hugely destructive floods, regardless of the level of human action. Remember to use case studies and examples to illustrate your ideas. Make a judgement about how these factors influence the damage caused by flooding.
- Write a conclusion that evaluates the view that damage caused by flooding is made worse by human actions. You need to make a judgement about whether or not human actions are the most significant factor. This should be based on your personal opinion but should also draw on the evidence already provided in your answer.

Page 177 — Climate Change and the Hydrological Cycle
1   Maximum of 6 marks available. This question is level marked.
HINTS:
- Start by briefly outlining the potential impacts of climate change, e.g. temperature and precipitation change.
- Next, explain how changes in temperature can affect water stores that might be used to supply communities with water. For example, you could mention how rising temperatures will melt glacial snow, which will initially increase nearby water supplies, but will eventually diminish when there is no more snow to melt. Make sure you link your answer back to water security.
- Then, discuss how climate change will have an impact on glacial stores, the amount of meltwater these produce and how this will affect water stores used by communities.
- Finally, explain how climate change will affect precipitation. For example, you could mention how wet areas are expected to get wetter and dry areas are expected to get drier, further. Make sure you link your answer back to water security.

Page 180 — Water Insecurity — Causes
1   Maximum of 8 marks available. This question is level marked.
HINTS:
- Start by defining what water scarcity is and how it might apply to different countries. You should mention the connection between economic water scarcity and physical water scarcity.
- Next, explain how different human factors can create water scarcity. You should discuss issues around pollution and hydroelectric power.
- Then, explain how over-abstraction can lead to water scarcity. You could refer to an example, such as the Aral Sea.
- Make sure that your answer explains how human actions contribute to, rather than directly cause water scarcity and show how it is usually a combination of many factors that create water scarcity.

Page 182 — Water Insecurity — Consequences
1   Maximum of 8 marks available. This question is level marked.
HINTS:
- Start by defining the term water security and give some examples of conditions that might make a place water insecure. Briefly outline the link between water security and economic development.
- Then, discuss how a secure water supply is an important part of driving energy supplies and how this is critical for the economic development of many industries.
- Next, discuss how poor water security can affect education levels and school attendance. Make sure you include the point that links literacy levels to further points of economic development.
- Then, discuss the link between water security, water cleanliness and health. Make sure this is then linked to economic development and explain in detail how a healthy population is a more productive one. You should also explain how a country's healthcare budget might be better spent on care if water cleanliness was not a problem.

Page 186 — Managing Water Supply
1   Maximum of 12 marks available. This question is level marked.
HINTS:
- Start by briefly discussing why there is a need to manage water supplies in different countries.
- Then, evaluate the effectiveness of water conservation as a management strategy. Discuss both the advantages and disadvantages of conservation schemes. You should mention specific examples and case studies to highlight your points.
- Next, evaluate the effectiveness of hard engineering schemes such as water transfer schemes and desalination plants, again mentioning both advantages and disadvantages of each. You should mention specific examples and case studies to highlight your points.
- Make sure you recognise that the scale of management is an important consideration as well as where the management will take place.
- Write a conclusion that assesses the view that water conservation is a more effective strategy to manage water supplies. You need to make a judgement about whether water conservation is a more effective strategy than other approaches at managing water supply. This should be based on your opinion and also draw on the evidence already provided in your answer.

Answers

Topic 6 — The Carbon Cycle and Energy Security

Page 190 — The Carbon Cycle

1   Maximum of 6 marks available. This question is level marked.
    HINTS:
    • Start by briefly defining what is meant by a geological process.
    • Then, explain how volcanic and tectonic activity can release carbon dioxide into the atmosphere. Discuss the details of the process of outgassing and what affects its intensity.
    • Next, explain how rocks can be weathered by chemicals. Acid rain should be clearly defined and the process that causes it should be explored.
    • Finally, make it clear that volcanic activity and chemical weathering occur at different time scales and deal with different volumes of carbon.

Page 193 — The Role of the Carbon Cycle

1   Maximum of 6 marks available. This question is level marked.
    HINTS:
    • Start by briefly defining what is meant by soil health. Discuss organic matter and the different types of ecosystem or habitat.
    • Explain how carbon is added to the soil and include how these inputs will affect soil health. Explain the factors that will affect these processes and therefore impact soil health.
    • Explain how carbon is removed from the soil and include how these outputs will affect soil health. Explain the factors that will affect these processes and therefore impact soil health.

Page 196 — Energy Security

1   Maximum of 20 marks available. This question is level marked.
    HINTS:
    • Start by defining the term energy security and give an example of how a country might know if it is energy secure.
    • Then, discuss the view that national governments have a key role to play in managing energy security. Explain the different scales that national governments work at — both in managing their domestic supply and demand and managing their wider role in international politics. Make sure you clearly evaluate their role. Additionally, include a case study of a country/countries you have studied that have successfully managed their energy security issues.
    • Next, argue how other players have a role to play in energy security and its management, including consumers, TNCs and large organisations such as OPEC. Remember to reference case studies or examples you have studied to illustrate your ideas.
    • Write a conclusion that evaluates the view that national governments have the most important role in managing energy security. Make sure to recognise that governments can directly and indirectly influence energy security. Also, be clear in your conclusion that national governments of some countries will have a bigger role than in other countries. Your conclusion should be based on your personal opinion and also draw on the evidence already provided in your answer.

Page 200 — Fossil Fuels

1   Maximum of 8 marks available. This question is level marked.
    HINTS:
    • Start by defining what an unconventional fossil fuel is and give some examples.
    • Then, explain why the demand for fossil fuels is growing. This should include references to population changes and the growth of emerging economies. Use key terms to explore the concept of finite resources and how this affects the pricing of fuels on the market.
    • Then, explain how advancements in technology have made the extraction of unconventional fossil fuels more feasible. Make sure you use examples to illustrate these points. You might also want to mention the advancement of engineering in energy pathways and the ways this has made the transfer of unconventional fossil fuels more viable.

Page 203 — Fossil Fuel Alternatives

1   Maximum of 6 marks available. This question is level marked.
    HINTS:
    • Start by defining what a biofuel is and give some examples of crops that can be grown to produce biofuels.
    • Then, explain the impact growing biofuels might have on the environment and how this makes it a potentially unsustainable fuel option. You could mention deforestation and the increased use of fertilisers and pesticides as well as the potential damage these chemicals can do to water courses.
    • Next, explain the impact growing biofuels might have socially. You could mention the use of land that could be used to produce food crops and the investment that might be needed in the design and manufacture of vehicles in order to make them biofuel compatible. If applicable, add in an example of a location you have studied that uses biofuels.

Page 205 — Human Impacts on the Carbon Cycle

1   Maximum of 6 marks available. This question is level marked.
    HINTS:
    • Start by defining ocean acidification and ecosystem services.
    • Then, explain the impact ocean acidification will have on marine plant life and phytoplankton. Make it clear how this has an impact on the wider health of the ocean, e.g. through food webs and fish stocks (for human consumption).
    • Next, explain the link between ocean acidification and coral bleaching. Explain the knock-on impact this can have in some areas that rely on tourism (e.g. reduced demand for boat trips and scuba diving).

Page 207 — Changes to the Carbon Cycle — Impacts

1   Maximum of 12 marks available. This question is level marked.
    HINTS:
    • Start by outlining what an environmental Kuznets Curve shows and briefly discuss the three stages.
    • Then, weigh up how good a representation the curve is of how countries view and treat their environment. Use some named examples of places that accurately represent different sections of the curve. Link the theory clearly to both the carbon and water cycles.
    • Next, assess how the curve is inaccurate. Explain how different scales of different problems make the stages difficult to predict. For example, a country may be developed but still be suffering from environmental degradation due to the legacy of industrialisation.
    • Write a conclusion that summarises the extent to which a standard environmental Kuznets Curve holds true for different countries or regions. Make sure to include how it can vary from country to country. This should be based on your personal opinion but should also draw on the evidence already provided in your answer.

Answers

Page 210 — Future Changes to the Carbon Cycle

1 Maximum of 20 marks available. This question is level marked.
<u>HINTS</u>:
- Start by defining the terms mitigation and adaptation and give a brief example of each.
- Then, explore how adaptation strategies play an important role in responding to future climate change. Evaluate the potential impact of adaptation strategies on people, the economy and on the wider environment. Mention specific strategies and any examples of places where these strategies are being used. You should also evaluate the likelihood of the strategy being a success. Ensure that you leave space to discuss the limitations of adaptation as a sole approach to climate change.
- Next, explore how mitigation strategies play an important role in managing future climate change. As with the adaptations, evaluate the potential impact of mitigation strategies on people, the economy and on the wider environment. Mention specific strategies and any examples of places where these strategies are being used. You should also evaluate the likelihood of the strategy being a success. Your answer should estimate the scale of impact and disruption — whether the strategies are designed for local, national or international management.
- Write a conclusion that evaluates the view that mitigation is a better response to climate change than adaptation. Make sure you recognise that a combination of both sets of strategies is likely to see greater engagement from the general public and have greater success. Your conclusion should be based on your personal opinion and also draw on the evidence already provided in your answer.

Topic 7 — Superpowers

Page 212 — Introduction to Superpowers

1 One mark for each valid point, up to a maximum of 4 marks. E.g. Hard and soft power is seen as a spectrum because some forms of influence can be seen as harder or softer depending on how they are used *[1 mark]*. Hard power is when a country uses its power by force. This is often through military intervention (representing an extreme use of hard power) or through economic sanctions (representing a softer approach) *[1 mark]*. Soft power is the use of power through attractive policies or ideologies *[1 mark]*. With soft power, countries are likely to be influenced voluntarily *[1 mark]*.

Page 214 — Patterns of Power

1 Maximum of 12 marks available. This question is level marked.
<u>HINTS</u>:
- Start by outlining the key differences between direct and indirect control. Briefly introduce the main examples you plan to use in your answer.
- Then, explain how direct control can create and stimulate further economic power. Make reference to examples, such as the British Empire, to illustrate your points. Make a judgement about how significant direct control is in creating economic power.
- Next, explain how indirect control continues to create economic power for powerful countries (through neocolonialism). Examples should be used where possible. Make a judgement about how significant indirect control is in creating economic power.
- Write a conclusion that assesses the extent to which direct and indirect control contributes to economic power. You also need to make a judgement about which type of control is most significant. For example, indirect control in many circumstances is not possible without there first being a period of direct control. This judgement should be based on your personal opinion but should also draw on the evidence provided in your answer.

Page 217 — Emerging Powers

1 Maximum of 12 marks available. This question is level marked.
<u>HINTS</u>:
- Start by outlining what a development theory should aim to do. Briefly introduce the names of the different development theories.
- Then, explain how the Modernisation Theory models patterns of world power. Critique the theory by showing how different countries and powers fit (or do not fit) the model. Make a judgement about how useful the modernisation theory is at explaining patterns of power.
- Next, explain how the dependency theory models patterns of world power. Again, critique the theory and compare the Dependency Theory with the Modernisation Theory. Make a judgement about how useful dependency theory is in explaining patterns of power.
- Finally, explain how the World Systems Theory models patterns of world power. Highlight the advantages and disadvantages of this theory in comparison with the other two mentioned. Give a judgement about how useful the world systems theory is in explaining patterns of power.
- Write a conclusion that assesses the extent to which any development theory is accurate in explaining patterns of world power. You also need to make a judgement about which theory (if any) is most useful. This should be based on your personal opinion but should also draw on the evidence already provided in your answer.

Answers

Page 219 — Influence of Superpowers

1 One mark for each valid point, up to a maximum of 4 marks.
E.g. Free market capitalism is a type of economic system
based on the private ownership of industries and wealth
[1 mark]. Goods and services are traded competitively
on a free market where price is based on supply and
demand *[1 mark]*. On the other hand, in centrally planned
economies, all industry and profits are owned by the state, as
well as land and property *[1 mark]*. There is no competition
between industries and the state controls the supply of
products, so prices for goods and services remain stable
[1 mark].

Page 222 — Global Governance

1 Maximum of 12 marks available. This question is level marked.
HINTS:
- Start by outlining what an intergovernmental organisation is and
  give some examples of IGOs working at different scales.
- Then, explain how IGOs can support and strengthen superpowers.
  For example, you could make reference to the power that
  superpowers have within the structure of IGOs and how there is
  potential for members to make decisions that benefit themselves.
  Make a judgement about the extent that IGOs might strengthen
  the superpower status of a county.
- Next, explain how membership of IGOs might disadvantage the
  most powerful superpowers. Highlight examples such as climate
  change conferences and agreements that have produced plans that
  challenge the power of a country. Make a judgement about the
  extent that IGOs might weaken the superpower status of a county.
- Write a conclusion that assesses the extent to which the decisions
  made by IGOs may strengthen the influence of superpowers. This
  should be based on your personal opinion but should also draw on
  the evidence already provided in your answer.

Page 224 — Superpowers
— Environmental Impacts

1 Maximum of 12 marks available. This question is level marked.
HINTS:
- Start by outlining the types of environmental issues that would be
  classed as global, for example climate change and carbon emissions.
  Briefly introduce the country or group of countries you are going to
  be assessing.
- Then, explain how specific countries reacted to the various climate
  change agreements and the impacts the targets had on both
  different countries and on carbon emissions. Quote data from
  particular countries to support your answer. Make a judgement
  about how well the country / group of countries addressed the issue
  through this strategy.
- Next, explain the role that your country or group of countries had
  in regional schemes (such as the EU emissions trading scheme)
  and how this had an impact on the global issue of climate change.
  Critique the scheme and again, quote data and specific information
  from the countries involved to support your answer. Make a
  judgement about how well the country or group of countries
  addressed the issue through this strategy.
- Write a conclusion that assesses how a country or group of
  countries attempted to address the global environmental issue. This
  should be based on your personal opinion but should also draw on
  the evidence already provided in your answer.

Page 227 — Conflict Over Resources

1 One mark for each valid point, up to a maximum of 4 marks.
E.g. Disputes can arise from overlapping claims based on
the size of each country's Exclusive Economic Zone (EEZ)
[1 mark]. An EEZ is the marine area 200 nautical miles off
the coastline of any country and is included in the territory of
the country in question *[1 mark]*. Controlling these areas can
provide countries with valuable natural resources and military
strongholds, and can grow their sphere of influence *[1 mark]*.
Disputes occur when countries are not able to agree on
how to share an area that is considered to be economically,
politically and militarily important *[1 mark]*.

Page 231 — Superpowers
— International Relations

1 Maximum of 12 marks available. This question is level marked.
HINTS:
- Start by outlining the tensions found in the Middle East and the
  countries that are involved.
- Then, explain the economic reasons why the Middle East is
  experiencing tension and conflicts. Use clear examples to illustrate
  your ideas, such as the role of widespread economic decline across
  the region which led to the Arab Spring. Make a judgement about
  the significance of these economic reasons in the context of the wider
  tensions in the Middle East.
- Next, explain the political reasons why the Middle East is
  experiencing tension and conflicts. Again, use clear examples to
  illustrate your ideas. You could mention how new and relatively
  inexperienced systems of governance can create weaknesses in
  leadership. Make a judgement about the significance of these
  political reasons in the context of the wider tensions in the
  Middle East.
- Finally, explain the cultural reasons why the Middle East is
  experiencing tension and conflicts. Use clear examples to illustrate
  your ideas such as how one religious group may feel that they are
  being undermined by the rise of another group. Make a judgement
  about the significance of these cultural reasons in the context of the
  wider tensions in the Middle East.
- Write a conclusion that assesses the relative importance of economic,
  political and cultural reasons for tension in the Middle East. Make
  sure you show how your examples are linked to one another. Your
  conclusion should be based on your personal opinion but should also
  draw on the evidence already provided in your answer.

Answers

Page 233 — Challenges for Superpowers

1 Maximum of 12 marks available. This question is level marked.
HINTS:
- Start by outlining what is meant by 'economic sustainability'. Briefly state the level of spending that different countries might be using to further themselves in their military and space exploration.
- Then, explain the ways that funding for the military may be sustainable (or unsustainable) for a country. Use clear examples to illustrate your ideas. Highlight how that level of sustainability may affect their superpower status. Make a judgement about the degree of economic sustainability that is possible.
- Next, explain the ways that funding for space exploration may be sustainable (or unsustainable) for a country. Use examples to illustrate your ideas. Highlight how that level of sustainability may affect their superpower status. Make a judgement about the degree of economic sustainability that is possible.
- Finally, suggest that the economic sustainability of things can change over time. It is worth highlighting that having an awareness of past events can help countries to better prepare for the future. However, predicting long-term economic sustainability requires an ability to accurately forecast future world events, which is impossible.
- Write a conclusion that assesses the extent to which spending on military advancement and space exploration is economically sustainable for superpowers. Your conclusion should be based on your personal opinion but should also draw on the evidence already provided in your answer.

Topic 8: Option 8A — Health, Human Rights and Intervention

Page 235 — Introduction to Development

1 One mark for each valid point, up to a maximum of 4 marks. E.g. The USA has a much higher GDP and HDI than Costa Rica and Nepal, but a lower HPI score *[1 mark]*. This could be because economic growth often causes environmental damage, which would give it a big ecological footprint *[1 mark]*. Costa Rica has a much lower GDP than the USA but it has the highest HPI score of the three countries *[1 mark]*. This suggests that Costa Rica might have higher levels of wellbeing than the other countries, even though it's not as wealthy as the USA *[1 mark]*.

Page 237 — Measures of Development

1 Maximum of 20 marks available. This question is level marked.
HINTS:
- Start by briefly outlining what is meant by human development and the different factors that affect it.
- Then, go on to discuss the ways that education is fundamental to economic development. E.g. by investing in education, countries can develop a more skilled, knowledgeable workforce who can get better-paid jobs and help the economy grow. You should also discuss how education can help to promote human rights and gender equality, and how this is a key step in development.
- Next, you could discuss countries where universal access to education is not seen as a key development goal. E.g. in some cultures, men are expected to earn a living while women look after the home, so educating females is seen as less important in terms of driving development.
- Then, discuss some of the other factors that could be seen as key factors in development, including GDP and health. You could discuss how, in some countries, increasing GDP may be necessary in order to increase educational opportunities.
- You could bring in relevant examples from specific countries you have studied.
- Write a conclusion that sums up your points and evaluates the view that education is the most important factor in increasing human development. This should be based on your personal opinion but should also draw on the evidence already provided in your answer.

Page 239 — Development and Health

1 Maximum of 8 marks available. This question is level marked.
HINTS:
- Start by briefly describing how health and life expectancy vary between and within countries in the developed world.
- Then, explain how differences in lifestyle affect the types of diseases that occur in a country, e.g. an unhealthy diet can increase the chances of developing heart disease.
- Next, you could explain how access to good quality medical treatment varies between developed countries with free or affordable health care and those where you have to pay for health care. E.g. health care in the UK is free whereas in the USA patients have to pay for their medical care.
- Then, you could describe how deprivation affects health and life expectancy in the developed world, e.g. inequalities in the UK mean that people in wealthier areas often have longer life expectancies than people in deprived areas.
- Finally, explain how deprivation also influences lifestyle and access to health care, reinforcing the impact they have on health and life expectancy in the developed world.

Answers

Page 243 — Increasing Development — Role of Government

1   Maximum of 20 marks available. This question is level marked.
    HINTS:
    - You could start by briefly describing the approaches to development taken by IGOs.
    - Then, you could discuss ways in which IGOs have been effective at deciding the goals of development. You could use the example of the Millennium Development Goals and the progress they made in reducing global poverty.
    - Next, you could discuss cases where IGOs have promoted particular development goals, but this has caused problems for development, e.g. Structural Adjustment Programmes promoted by the IMF in South America.
    - Write a conclusion that evaluates the view that IGOs have been effective in deciding development goals. This should be based on your personal opinion but should also draw on the evidence already provided in your answer.

Page 245 — Human Rights

1   Maximum of 6 marks available. This question is level marked.
    HINTS:
    - Start by briefly describing what the Geneva Convention is and what it does for human rights.
    - Next, you could give examples of human rights that are listed in the Geneva Convention and explain how having an agreement of these rights helps to protect civilians during wars. For example, actions such as terrorism, torture, taking hostages and sexual assault are banned under the Geneva Convention.
    - Complete your answer by explaining how the ICC was established to enforce the Geneva Convention and briefly describe how it's been used to prosecute war criminals to protect the human rights of civilians.

Page 247 — Human Rights — International Variability

1   Maximum of 20 marks available. This question is level marked.
    HINTS:
    - Start by briefly describing what human rights are, and how they are often presented as a priority in international forums.
    - Then, you could explain how governments of developing countries justify this position, e.g. by arguing that meeting citizens' basic needs is more important than human rights, or that economic development will lay the foundation for promoting human rights later on.
    - Next, you could explain the ways in which human rights and economic development can be contradictory. Remember to use case studies and examples to illustrate your ideas. For example, you could talk about coal mining in Chhattisgarh, and how local people's right to their land and culture has been compromised by policies to promote national economic growth.
    - Write a conclusion that evaluates the view that governments prioritising economic development over human rights is justifiable. This should be based on your personal opinion but should also draw on the evidence already provided in your answer.

Page 249 — Human Rights — National Variability

1   Maximum of 8 marks available. This question is level marked.
    HINTS:
    - Start by outlining how variations in access to human rights are often closely related to variations in levels of health and education.
    - Then, you could explain the idea of systemic discrimination and how groups that are given fewer rights by the government are often also marginalised in terms of government funding and support. E.g. you could mention that in 2019 the number of Indigenous American and Alaska Native students who graduated from high school was 74% compared to 86% of all ethnic groups in the USA.
    - Next, you could explain how the lack of rights, such as freedom of speech, makes it harder for oppressed groups to improve their situation. E.g. you could mention that Indigenous populations are more likely to be arrested or killed by the police compared to white people.

Page 251 — Defending Human Rights

1   Maximum of 8 marks available. This question is level marked.
    HINTS:
    - Start by outlining the idea that different types of intervention are favoured by different groups for various reasons.
    - Then, suggest which types of intervention are usually promoted by different groups. Make sure you give reasons why each group would prefer that intervention. For example, human rights NGOs often seek to avoid any human rights abuses, so they tend to favour peaceful interventions such as petitions to put pressure on oppressive governments.
    - Next, you could explain how superpowers are more likely to favour military interventions in order to stop human rights abuses. For example, the USA argued that the US-led intervention in Iraq was justified in order to remove Saddam Hussein's oppressive regime.

Page 254 — Increasing Development — Aid

1   Maximum of 20 marks available. This question is level marked.
    HINTS:
    - Start by explaining what development aid is and outline some of the forms it takes. For example, emergency aid from NGOs, loans from IGOs or bilateral aid from governments and IMF loans.
    - Then, discuss ways that development aid can cause problems, such as dependency and corruption. Refer to case studies and examples to illustrate your points, for example, you could mention how some of the money raised in response to the 2010 earthquake in Haiti was spent on the salaries and accommodation of NGO workers.
    - Next, discuss ways that development aid has helped people by increasing development or promoting human rights. Refer to case studies and examples to illustrate your points, for example, progress in reducing deaths from malaria or towards greater gender equality.
    - Write a conclusion that evaluates the view that development aid harms more people than it helps. This should be based on your personal opinion but should also draw on evidence already provided in your answer.

Answers

Page 256 — Increasing Development — Military Intervention

1 Maximum of 8 marks available. This question is level marked.
HINTS:
- Start by explaining how some governments see themselves as advocates of human rights on the international stage and have used human rights to justify military interventions.
- Then, discuss how this has been undermined by their actions. For example, by using military interventions for ulterior motives, providing military aid to countries with poor human rights records or by carrying out torture. You could draw on examples you have studied in your answer, e.g. the USA-led intervention in Iraq in 2003.
- Next, provide a counter-argument suggesting why a country's stance on human rights is not undermined by its actions. For example, you could refer to progress in human rights as a result of military or development interventions.

Page 258 — Increasing Development — Measuring Success

1 One mark for each valid point, up to a maximum of 4 marks. E.g. Freedom of expression is people's right to express their beliefs without government censorship *[1 mark]*. It allows people to criticise the government and hold it accountable, which is central to a successful democracy *[1 mark]*. It also allows people to debate and discuss different ideas, helping them to make a more informed decision about who to vote for in democratic elections *[1 mark]*. Freedom of expression also prevents one group (e.g. followers of a particular religion or political ideology) from dominating or repressing the beliefs of others, which is also key to democracy *[1 mark]*.

Page 261 — Success of Aid and Military Intervention

1 Maximum of 20 marks available. This question is level marked.
HINTS:
- Start by briefly outlining some of the ways in which the impact of aid on human rights and development can be measured. You could write about changes in indicators like life expectancy and education levels, as well as improvements in the upholding of human rights such as free speech.
- Next, discuss the view that aid can fail to boost human development or human rights. You could discuss the problems that aid can create, such as dependency, as well as issues around corruption and lack of organisation, which can mean that aid does not reach the people who need it. You should draw on examples you have studied, such as the aid efforts in Haiti following the 2010 earthquake, where the government and NGOs failed to spend much of the aid provided on helping affected people.
- Then, examine the successes that aid can bring, such as improving infrastructure, reducing the impact of diseases such as malaria, and promoting human rights such as freedom of speech. Again, you should bring in examples to support your points. You could use the example of Ebola, which has been largely brought under control with the help of international aid.
- You could also discuss the fact that, if poorly managed, aid money can be spent on development projects that boost GDP but damage the environment, which has knock-on impacts on human rights.
- Write a conclusion that sums up your points and evaluates the view that aid brings more failure than success. This should be based on your personal opinion but should also draw on evidence already provided in your answer.

Topic 8: Option 8B — Migration, Identity and Sovereignty

Page 263 — Globalisation and Migration

1 Maximum of 8 marks available. This question is level marked.
HINTS:
- Start by briefly describing how globalisation has made it easier for people to travel and communicate worldwide which has led to a shift in migration patterns.
- Next, you could explain how globalisation has changed labour demands, creating more jobs in some areas. This has affected push and pull factors, e.g. encouraging people to migrate to both within and between different countries in search of jobs. You should draw on an example that you have studied, e.g. China, to support your answer.
- Then, you could explain how globalisation has led some countries to encourage international migration by opening their borders. Remember to use case studies and examples to illustrate your ideas. For example, you could look at the EU-Schengen area, and give evidence of how globalisation has increased international migration.

Page 265 — Migration — Causes and Policies

1 Maximum of 6 marks available. This question is level marked.
HINTS:
- Start by briefly outlining what is meant by the removal of restrictions on goods, capital and labour.
- Next, explain how removing these restrictions may increase economic efficiency, e.g. it can enable the economy to work more freely by being able to respond to the supply and demand of goods, capital and labour. Open borders encourages migration between countries to where people's skills are needed most. This can help businesses to grow and therefore promote economic growth.
- Then, mention that some economists believe that the increased profit for businesses from the removal of restrictions on goods, capital and labour will 'trickle down' to the rest of society, further boosting the economy.

Page 267 — Migration — Consequences

1 Maximum of 20 marks available. This question is level marked.
HINTS:
- Start by briefly describing what international migration is, how it has increased due to globalisation. You could also introduce the idea that there are debates about its consequences.
- Then, you could outline the economic benefits of international migration. For example, you could explain how economic theory promotes the free movement of labour. Remember to use case studies and examples to illustrate your ideas. For example, you could look at how Mexican migrant labour contributes to the USA's economy.
- Next, you could explain how international migration can lead to political tensions and what the consequences of these tensions are. For example, you could look at how migration from Mexico to the USA has become a key argument in general elections.
- Write a conclusion that evaluates the view that the economic benefits of international migration are outweighed by the political tension it causes. This should be based on your personal opinion but should also draw on the evidence already provided in your answer.

Answers

Page 269 — Nation States

1   Maximum of 8 marks available. This question is level marked.
HINTS:
  • Start by introducing the different ways that borders are created. For example, by physical geography, history and colonisation.
  • Next, you could explain how this resulted in borders that brought together groups with ethnic or religious differences, and in others split some ethnic groups in places. You should draw on an example that you've studied, e.g. the area inhabited by Kurdish people was divided between Iraq, Iran, Syria, Turkey and Armenia.
  • Then, explain how when borders don't consider ethnic and religious differences, problems such as tensions and conflicts between different social groups can arise. You should draw on an example that you've studied, e.g. Iraq, to support your answer.

Page 272 — Nationalism

1   Maximum of 8 marks available. This question is level marked.
HINTS:
  • Start by briefly describing the main migration patterns of the colonial period, between the colonies and a former imperial core country such as the UK. For example, white settlers migrating from the UK to the colonies, and later trends of people from former colonies migrating to the UK.
  • Next, explain how this affected the ethnic diversity of the UK, e.g. by making it more heterogeneous. You should draw on an example that you've studied, e.g. the Windrush generation in the UK.
  • Then, explain how cultural diversity tends to increase as a result of ethnic diversity because immigrants bring their own culture to their new country, e.g. curry was introduced to the UK by immigrants from former colonies.
  • Finally, explain how these historic patterns continue to influence migration and diversity, e.g. mentioning that many new migrants to the UK come from the UK's former colonies, and that in 2021 over 33% of migrants living in the UK came from former colonies.

Page 274 — Globalisation and Inequality

1   Maximum of 20 marks available. This question is level marked.
HINTS:
  • Start by briefly explaining the current global economic system and neoliberal ideas such as deregulation and free trade.
  • Then, explain some of the flaws with this system, e.g. the removal of regulations around the movement of money have allowed some tax havens to develop. Some NGOs argue that, by helping wealthy individuals and companies to avoid paying tax, tax havens allow rich people to become richer and therefore increased global inequality.
  • Next, discuss how these flaws could make the current global economic system unsustainable.
  • You should also discuss alternatives to the current system, and explain how they may be more sustainable. For example, you could discuss the concept of 'buen vivir' in Ecuador, and explain how it helped to reduce poverty and inequality in the country. You could suggest how it might be applied at a global scale, and what impacts this might have on the economy.
  • Finally, you should also look at arguments against the view that current economic system needs to change. You could discuss some of the strengths of the current economic system, e.g. you could explain how IGOs work to prevent economic crises, and how they provide loans such as SAPs and HIPCs to developing countries, which can help to reduce poverty.
  • Write a conclusion that gives an overall assessment of the extent to which you agree with the statement. This should be based on your personal opinion but should also draw on the evidence already provided in your answer.

Page 276 — United Nations

1   Maximum of 8 marks available. This question is level marked.
HINTS:
  • Start by briefly outlining what the UN is and its main aims.
  • Next, explain some of the reasons why the UN may intervene in a country, e.g. you could mention maintaining security and protecting human rights.
  • Then, you could give a brief example of a time when the UN has intervened in a country and why intervention was thought to be necessary, e.g. attempts to resolve conflict and allow democratic elections in the DRC.
  • Finally, explain why some UN member states may independently intervene in other countries. You could use the examples of the USA and UK intervention in Afghanistan and Iraq to illustrate this.

Page 278 — IGOs and Trade

1   Maximum of 8 marks available. This question is level marked.
HINTS:
  • Start by briefly outlining some of the policies made by IGOs.
  • Then, explain how different policies can benefit developed countries more than developing countries. You could discuss trade liberalisation and global borrowing rules.
  • Next, you could mention that developed countries have a disproportionate voting share in IGOs, so they are able to influence the rules and make sure they work in their favour.
  • Finally, you should explain policies targeted at developing countries and discuss some of the ways that they could be seen to have failed. For example, SAPs and HIPCs have been criticised for forcing countries to fit a western, capitalist model, which may not suit all developing countries.

Page 281 — IGOs and the Environment

1   Maximum of 6 marks available. This question is level marked.
HINTS:
  • Start by outlining some of the major environmental challenges affecting the world's oceans. For example, you could mention overfishing.
  • Next, explain how IGOs have been created to manage these problems. For example, you could describe the UN Convention on the Law of the Sea (UNCLOS) and how it set up exclusive economic zones which gives coastal nations exclusive rights to resources.
  • Then, you could mention the ways in which IGOs have addressed the environmental issues faced by the oceans, e.g. under UNCLOS countries are only allowed to fish within sustainable limits — this helps to ensure that countries can access the resources they need without overexploitation.

Page 283 — National Identity

1   Maximum of 8 marks available. This question is level marked.
HINTS:
  • Start by explaining what is meant by national identity and loyalty, and how they can be influenced by a range of factors, including landscape.
  • Then, discuss ways in which ideas about landscape can shape national identity and loyalty. For example, you could explain how people may have idealised the English countryside as green and peaceful, and how this may form part of their perception of what it means to be English.
  • Then, you could explain how certain landscapes may be especially meaningful for some people. For example, you could discuss how, for some people, the White Cliffs of Dover symbolise the UK's identity as an island nation, and may inform their national identity, e.g. they may see the UK as strong and independent.

Answers

Page 285 — Challenges to National Identity

1 Maximum of 8 marks available. This question is level marked.
HINTS:
- Start by briefly outlining how globalisation has led to increased foreign ownership of companies across the world, including in the UK.
- Next, outline the ways in which this challenges the idea of 'British' companies or 'Made in Britain', since many British brands and companies are now owned by non-nationals. Draw on an example you have studied, e.g. Jaguar Land Rover, to support your answer.
- Then, you could explain how the ownership of British brands by non-nationals adds to the complexity of the idea of 'Made in Britain'. For example, Jaguar Land Rover is owned by an Indian company, and much of their manufacturing happens overseas. However, some manufacturing still happens in the UK, and they are marketed as a British brand, for example by using British actors in their adverts.

Page 287 — Division Within Nations

1 Maximum of 8 marks available. This question is level marked.
HINTS:
- Start by explaining that globalisation has led to significant change in emerging countries, which has created both costs and benefits.
- Next, you could give examples of benefits, e.g. the creation of a wealthy middle-class who are able to contribute more money to the economy and costs, e.g. poverty in rural areas.
- Then, describe how these costs and benefits are unevenly distributed between different regions and social groups. You should draw on an example that you've studied, e.g. the regional variation in economic growth in China resulting from people migrating from rural inland provinces in search of jobs, to support your answer.
- Finally, explain how these inequalities have contributed to political tensions. You should use evidence from an example that you've studied, e.g. division in Somalia over where profits gained from Somalia's resources should go.

Exam Practice

Page 302 — Tectonic Hazards and Processes

1.1 One mark for each valid point, up to a maximum of 4 marks. E.g. Volcanic eruptions at destructive plate margins are often more explosive than those at constructive plate margins *[1 mark]*. Andesitic and rhyolitic lavas at destructive plate margins, are cooler and have a higher viscosity than the basaltic lava at constructive plate margins *[1 mark]*. This means that the lava at destructive margins flows less freely and cools more quickly, which can cause blockages in the volcano's vents *[1 mark]*. This leads to a build-up of pressure and can cause more violent eruptions than at constructive plate margins. This is because hotter, basaltic lava at constructive margins flows more easily and over longer distances, so is less likely to block the volcano's vents *[1 mark]*.

1.2 i) One mark for each valid point, up to a maximum of 2 marks.
Mount Ontake: $d^2 = 6.25$ *[1 mark]*
Mount Nyiragongo: Rank = 6 *[1 mark]*

ii) One mark for each valid point, up to a maximum of 3 marks. This result shows that there is a positive correlation between VEI and fatalities, so as VEI increases so do fatalities *[1 mark]*. However, the figure of 0.29 is below both the 99% and 95% significance level, so it is only a weak correlation *[1 mark]*. This means that the number of fatalities is not strongly correlated with the magnitude of the eruption *[1 mark]*.

1.3 Maximum of 12 marks available. This question is level marked.
HINTS:
- Start by briefly explaining what is meant by modifying the vulnerability of a community to a hazard and outline an example.
- Then, explain the advantages of modifying the vulnerability of a community to a hazard. For example, you could mention that monitoring systems are able to help predict when and where a hazard may occur, which can help communities to be more prepared and less vulnerable to a hazard event.
- Next, explain the disadvantages of modifying the vulnerability of a community to a hazard. For example, you could mention that evacuations can be very expensive to carry out and so communities must be certain that a hazard event is likely to happen, which may cause delays.
- You could also discuss the fact that modifying the vulnerability of a community to a hazard often cannot prevent hazards or fully mitigate their impacts.
- Write a conclusion that assesses the advantages and disadvantages of modifying the vulnerability of a community when managing hazards. This should be based on your personal opinion but should also draw on the evidence already provided in your answer.

Page 303 — Glaciated Landscapes and Change

2A.1 One mark for each valid point, up to a maximum of 4 marks. E.g. Nivation is a set of processes that deepen hollows in a sloped landscape by the repeated freezing and thawing of snow in a hollow when temperatures fluctuate around 0 °C *[1 mark]*. When the snow in the hollow freezes, it expands, which breaks up the rock at the base of the hollow by frost shattering *[1 mark]*. When the snow melts, the meltwater carries debris away from the hollow. When snow fills the hollow again, the process repeats and the hollow becomes deeper *[1 mark]*. The slope may also collapse due to erosion and waterlogging with meltwater, making the hollow deeper and wider *[1 mark]*.

Answers

2A.2 Maximum of 6 marks available. This question is level marked.
HINTS:
- Start by identifying the relationship between the two figures. You should mention that the negative mass balance shown in Figure 2 has led to the shrinking of the glacier shown in Figure 1.
- Then, describe how the data shows this relationship. For example, you could mention that Figure 1 shows that the glacier's size generally decreased between 2003 and 2012, and that Figure 2 shows a negative mass balance every year between 2003/04 and 2013/14, with the exception of 2008/09. Make sure you use exact figures in your answer.
- Next, explain the relationship. For example, you could explain how the negative mass balance shown in Figure 2 means that there is more ablation than accumulation, which causes the glacier to shrink and retreat. This shrinkage is demonstrated in Figure 1 by the glaciers diminished size and decreased elevation between 2003 and 2012.

2A.3 Maximum of 6 marks available. This question is level marked.
HINTS:
- Start by briefly describing what the figures show and how this is linked to climate change.
- Then, you could explain how it is likely that climate change will cause glaciers to retreat. Temperatures in the Arctic are expected to increase, which is supported by the above-average air temperatures over Greenland shown in Figure 5. Higher air temperatures are likely to contribute to the higher number of melt days seen in Figure 4 and surface melting of the Greenland Ice Sheet in Figure 3.
- Next, you could mention how climate change can cause permafrost in glacial landscapes to melt. You could also explain how this causes a positive feedback loop.

2A.4 Maximum of 20 marks available. This question is level marked.
HINTS:
- Start by briefly outlining the differences between erosional and depositional processes, and the landforms that they create.
- Then, you should discuss the importance of erosional processes in shaping glacial landscapes. For example, you could mention the power of erosion and how it creates different landforms in the landscape, many of which are large-scale.
- Next, you should discuss the importance of depositional processes in shaping glacial landscapes. For example, you could mention that depositional processes and landforms are dependent on the erosional processes supplying material.
- Next, you could discuss how the processes and landforms that dominate a landscape vary between places and over time. For example, erosional processes may dominate during glacials, whereas depositional processes may have more impact on the landscape during interglacials.
- Write a conclusion that evaluates the relative importance of erosional and depositional processes in shaping glacial landscapes. This should be based on your personal opinion but should also draw on the evidence already provided in your answer.

Page 304 — Coastal Landscapes and Change

2B.1 One mark for each valid point, up to a maximum of 4 marks. E.g. The drop in sea level relative to the coast that forms emergent coastlines leads to the formation of various landforms *[1 mark]*. Raised beaches form when the fall in sea level lowers the high tide mark, meaning the beaches remain exposed and become vegetated *[1 mark]*. This also leads to the exposure of wave-cut platforms, which are left raised above their former level *[1 mark]*. As the sea level lowers, the cliffs along the coastline above the raised beaches are no longer eroded by the sea, which allows them to become vegetated *[1 mark]*.

2B.2 Maximum of 6 marks available. This question is level marked.
HINTS:
- Start by outlining the impact of elevation on the risk of flooding. For example, low-lying areas are more at risk, particularly during storm events.
- Then, explain subsidence, especially in deltas, e.g. some areas are subsiding due to an increase in sediment being washed out to sea by increased river flow. This lowers the land level, making it more at risk of localised flooding.
- Next, explain how the presence of vegetation in coastal areas can decrease flood risk. For example, mangrove forests reduce the height and power of oncoming waves, reducing the flood risk to the areas behind them.

2B.3 Maximum of 6 marks available. This question is level marked.
HINTS:
- Start by briefly outlining that a cliff profile is influenced by a combination of dip, rock resistance and the energy of the coastline.
- Then, explain how geological structure influences cliff profiles. You should discuss how the dip of the rock strata has influenced the way in which erosion has taken place and the resulting cliff profile. For example, the rock strata dip horizontally in Figure 1, which would leave areas of softer rock exposed. This softer rock would be easily eroded, forming a jagged cliff profile with wave-cut notches.
- Next, explain how erosional processes can influence cliff profile. For example, you could explain how erosion has caused a wave-cut notch to form in Figure 1, leaving an overhang above. You could then explain that continued erosion may deepen the notch until the overhang collapses, resulting in a more vertical profile and may cause a rockfall, as seen in the background of Figure 1.

Answers

2B.4 Maximum of 20 marks available. This question is level marked.
HINTS:
- Start by outlining the main approaches to coastal management: hard and soft engineering, sustainable management and integrated coastal zone management, and the extent to which each of these works 'with' or 'against' coastal processes.
- Then, you could discuss the costs and benefits of management strategies that work 'with' coastal processes. For example, soft engineering strategies reduce the damage of coastal processes rather than preventing them, e.g. through restoring natural barriers like dunes. These have positive impacts like restoring habitats, but can also present challenges, such as not being suitable in developed areas.
- Next, discuss the costs and benefits of coastal management strategies that work 'against' coastal processes, such as hard engineering. For example, hard engineering strategies like building sea walls or offshore breakwaters can be effective at preventing flooding, but are often very expensive and can have negative impacts on the local environment and other coastlines.
- You could refer to case studies and examples to illustrate your points, e.g. you could discuss different coastal management schemes along the Happisburgh coastline in Norfolk and evaluate their success.
- Write a conclusion that evaluates the extent to which you agree that the most successful coastal management strategies work with coastal processes. This should be based on your personal opinion but should also draw on the evidence already provided in your answer.

Page 305 — Globalisation

3.1 One mark for each valid point, up to a maximum of 4 marks. E.g. The manufacture of many goods has shifted from more developed countries to less developed countries ('global shift') *[1 mark]*. Lower labour costs in less developed countries have encouraged companies to relocate the production side of their businesses overseas *[1 mark]*. This means that many goods are now produced in one country, then exported to a different country to be sold *[1 mark]*. As a result of these changes, international trade in manufactured goods has increased, making the world more interconnected *[1 mark]*.

3.2 One mark for each valid point, up to a maximum of 4 marks. E.g. FDI is money invested by individuals or companies from one country in another country *[1 mark]*. Most commonly, it moves from more developed to less developed countries, creating links between them and increasing globalisation *[1 mark]*. For example, companies may build factories overseas, meaning the investing country depends on the host country to produce goods, while the host country relies on the investor for jobs and economic growth *[1 mark]*. The investor may also invest in infrastructure, e.g. roads and ports, in the host country, opening the host country up to more investment and trade *[1 mark]*.

3.3 Maximum of 8 marks available. This question is level marked.
HINTS:
- Start by briefly explaining what is meant by economic liberalisation.
- Then, you could mention that TNCs can take advantage of economic liberalisation through offshoring and outsourcing. For example, you could mention that the TNC in Figure 1 has moved its manufacturing and distribution elements to countries that provide cheaper labour and lower running costs, such as China and India. This helps TNCs to grow more rapidly and generate larger profits.
- Next, explain that TNCs often keep their headquarters and retail operations in more developed countries. For example, the TNC in Figure 1 has kept its headquarters in the USA. This is so the TNC can charge higher prices for the product and wages are often higher.
- Then, explain how economic liberalisation is allowing the TNC in Figure 1 to develop a more global presence. For example, it has locations throughout the developed and developing world. This increases the international recognition of the TNC, which increases the market for their product.

3.4 Maximum of 20 marks available. This question is level marked.
HINTS:
- Start by briefly explaining what is meant by interdependence, and how globalisation has increased interdependence between countries.
- Then, look at the ways in which interdependence between countries can benefit the wealthiest and most powerful countries. For example, wealthy countries tend to attract economic migrants seeking better-paid work. Migrants then add to the labour force and contribute to the economy, e.g. through taxes.
- Next, you could look at ways in which interdependence can benefit all countries, not just wealthy and powerful ones, for example by improving security and decreasing the chance of war.
- Then, look at how interdependence may cause challenges for less wealthy and powerful countries. For example, TNCs operating in less developed countries may cause pollution and land degradation, while sending most of their profits back to their home country. Less wealthy countries may also be disadvantaged by the emigration of well-educated, working-age people.
- You could also discuss how interdependence can bring disadvantages to the wealthiest and most powerful countries. For example, large numbers of migrants might put pressure on healthcare and education systems.
- Write a conclusion that evaluates the view that interdependence only benefits the wealthiest and most powerful countries. This should be based on your personal opinion but should also draw on the evidence already provided in your answer.

Page 306 — Regenerating Places

4A.1 One mark for each valid point, up to a maximum of 4 marks. E.g. Physical factors can limit the spread of an area, for example mountains could prevent a city from expanding outwards, increasing its population density *[1 mark]*. Physical factors such as climate and soil type determine the crops that can be grown in an area, so if one of these changes, the function and characteristics of the area will change too *[1 mark]*. For example, if the climate of a cold area warms, it may become possible to grow a greater range of fruit there, changing the employment structure and economy by making it more dependent on agriculture *[1 mark]*. Finally, natural processes, such as coastal erosion, can cause damage to buildings and infrastructure, potentially leading to affected areas being abandoned *[1 mark]*.

Answers

4A.2 Maximum of 6 marks available. This question is level marked.
HINTS:
- Start by briefly explaining what the term 'lived experience' means.
- Then, outline some of the factors that might affect lived experience, such as length of residence, age, ethnicity and level of deprivation.
- Next, outline why different groups may have different lived experiences based on these factors. For example, people whose family have lived in an area for generations may feel a strong connection to that place, whereas recent immigrants may feel disconnected from the place where they live. Make sure you do this for multiple factors.

4A.3 Maximum of 6 marks available. This question is level marked.
HINTS:
- Start by outlining the different groups who may have views on Westminster regeneration. For example, residents, workers and tourists.
- Then, explain how some groups may see the need for regeneration in Westminister. For example, Figure 3 shows that only 42% of people in Westminster have access to open space, while nearly 5% live in overcrowded conditions. You could explain that this means that residents may be more likely to feel that Westminster would benefit from regeneration to increase green space and improve housing stock.
- Next, explain how other groups may not see the need for regeneration in Westminster. For example, you could use Figures 1 and 2 to show that if tourists only experience the more vibrant and affluent side of Westminster (portrayed in Figures 1 and 2), they are likely to conclude that there is little need for regeneration.

4A.4 Maximum of 20 marks available. This question is level marked.
HINTS:
- Start by briefly outlining what rebranding is and why different groups may aim to rebrand a place.
- Then, explain how rebranding strategies used in rural areas are likely to differ from those used in urban areas. For example, you could talk about diversification of farms and promoting outdoor activities in rural areas, versus emphasising industrial heritage or building a reputation as a creative hub in cities.
- Next, you could also discuss ways in which rebranding strategies may be similar in both rural and urban areas. For example, rebranding often focuses on emphasising a particular historical or literary connection, which is specific to the area rather than being a feature of how rural or urban an area is. You could also mention that accessible rural areas may have more in common with cities than they do with remote rural areas, so the rebranding strategies used may be similar to those used in urban areas. You could also mention that, irrespective of area, some aspects of rebranding remain the same.
- You should refer to case studies and examples of places you have studied to illustrate your points, e.g. Glasgow and Brontë country.
- Write a conclusion that evaluates the view that rural areas need completely different rebranding strategies to urban areas. This should be based on your personal opinion but should also draw on the evidence already provided in your answer.

Page 307 — Diverse Places

4B.1 One mark for each valid point, up to a maximum of 4 marks.
E.g. International migrants moving to rural areas can fill vacancies that local people are unable or unwilling to fill *[1 mark]*, which can reduce unemployment levels and allow the rural area to fulfil its economic potential *[1 mark]*. However, large numbers of immigrants can also increase demand for housing, schools and health care *[1 mark]*. These services may be limited in rural areas, so increased demand can restrict access for local people *[1 mark]*.

4B.2 Maximum of 12 marks available. This question is level marked.
HINTS:
- Start by briefly outlining what quantitative sources are, and what they can show and give examples of some quantitative sources (e.g. the census).
- Then, discuss how quantitative data can be useful at providing accurate, detailed information about the characteristics of a place. You could use Figure 1 to support your point. For example, Figure 1 uses census data to show that Devon's population is ethnically homogeneous.
- Next, discuss the limitations of quantitative sources in conveying the character of a place. You could mention the inability of quantitative sources to convey people's views and attitudes, which contribute to the character of a place. While quantitative data is objective, it can be used subjectively and selected specifically to support a particular point of view.
- Then, you should mention that Figure 2 is a qualitative source which shows the character of a place. However, one limitation of Figure 2 is that it is only from one person's perspective.
- Write a conclusion that assesses how useful quantitative sources are in conveying the character of a place. This should be based on your personal opinion but should also draw on the evidence already provided in your answer.

4B.3 Maximum of 20 marks available. This question is level marked.
HINTS:
- You could start by briefly explaining what is meant by perception of place, and the different factors that might influence it.
- Then, discuss how perception of place can be shaped by lived experience. You could consider how a local person's lived experience and therefore their perception of place may vary depending on things like their age, gender, ethnicity and how long they have lived in a place, as well as the aspects of place that are most important to them.
- Next, consider how media representations of place may affect people's perceptions of that place, from both an insider and outsider perspective. You could also consider how far people's responses to media representations may be based on whether those representations back up or contradict their own views. You could mention that the form of media may influence their perception of place, for example, whether it is factual or fictional.
- Make sure you include examples from the places you have studied to support your points, e.g. in the 1980s, Liverpool was often portrayed in the news as dangerous following the Toxteth riots. However, people's lived experience often determined whether or not they believed the media's portrayal of Liverpool.
- Write a conclusion that evaluates the view that people's perceptions of place are based on their lived experiences and not on media representations. This should be based on your personal opinion but should also draw on the evidence already provided in your answer.

Page 308 — The Water Cycle and Water Insecurity

5.1 One mark for each valid point, up to a maximum of 4 marks.
E.g. As populations grow and urbanisation increases, floodplains are being built on more frequently *[1 mark]*. This means the floodplain is covered by an impermeable layer, which stops or slows infiltration, increasing flood risk *[1 mark]*. Removing vegetation for agriculture reduces interception, exposes the soil and makes it more likely to erode *[1 mark]*. Eroded soil carried into a river can decrease its carrying capacity, so it's more likely to flood *[1 mark]*.

Answers

5.2 (i) One mark for each valid point, up to a maximum of 4 marks. E.g. Figure 1 shows a negative relationship between atmospheric CO_2 and the Antarctic Ice Sheet *[1 mark]*. This means that as atmospheric CO_2 increased between 2002 and 2017, the Antarctic Ice Sheet generally decreased over the same period. This also shows the strength of the relationship between atmospheric CO_2 and the Antarctic Ice Sheet *[1 mark]*. An increase in CO_2 in the atmosphere will increase global temperatures, which causes a positive feedback loop that means more ice melts *[1 mark]*. This is demonstrated in Figure 1 by short-term increases in ice sheet mass tending to coincide with decreases in CO_2. Both lines also fluctuate annually, with short-term increases in ice sheet mass tending to coincide with decreases in CO_2 *[1 mark]*.

(ii) Maximum of 6 marks available.
This question is level marked.
HINTS:
- Start by briefly outlining some of the ways that melting ice would affect the water cycle — for example, reduced storage in the cryosphere and increased flows to the oceans.
- Then, you could explain some of the ways that melting ice would affect the size of water stores within the water cycle. For example, an increase in glacial meltwater will reduce the amount of water stored in the cryosphere. Meanwhile, the amount of water entering and being stored in the oceans will increase, which will cause sea levels to rise.
- Next, you could explain some of the ways that melting ice would affect the rates of flow between stores within the water cycle. For example, an increase in ice melting will increase the amount of glacial meltwater flowing into rivers. You could then mention that the rate of flow between the cryosphere and the oceans will decrease over time as the amount of glacial meltwater will reduce as glaciers shrink.

5.3 Maximum of 20 marks available. This question is level marked.
HINTS:
- Start by outlining what is meant by water insecurity, and explain that it varies from place to place.
- Then, explain some of the human factors that may cause water insecurity. You could discuss global trends, including population growth and increased demand for water caused by economic development, as well as smaller-scale causes such as over-abstraction and pollution.
- Next, explain how physical factors may cause water insecurity. For example, you could discuss the variability in rainfall, temperature and the general access to natural stores such as lakes or rivers.
- Then, give an opinion about the relative importance of human versus physical factors in causing water insecurity. For example, you could mention that human activities are driving climate change, which changes patterns of rainfall and increases water stress in some areas, as well as contributing to sea level rise.
- Finally, you could also comment on ways in which human activity is reducing water insecurity, e.g. by water transfer schemes, water conservation and desalination.
- Write a conclusion that evaluates the view that human activities are the main cause of water insecurity. This should be based on your personal opinion but should also draw on the evidence already provided in your answer.

Page 309 — The Carbon Cycle and Energy Security

6.1 One mark for each valid point, up to a maximum of 4 marks. E.g. The flow of carbon between different carbon stores is called a carbon flux *[1 mark]*. Carbon fluxes happen over a range of time and spatial scales *[1 mark]*. For example, respiration and photosynthesis are carbon fluxes that take place at plant scale and happen relatively quickly *[1 mark]*. At a larger scale, a range of carbon fluxes can happen. For example, the sequestration of carbon may take place, which can take millions of years *[1 mark]*.

6.2 (i) One mark for each valid point, up to a maximum of 3 marks. E.g. Denmark's onshore wind energy capacity increased from just above zero to around 4000 MW between 1987 and 2017 *[1 mark]*. Over the same period, the number of land turbines nearly quadrupled *[1 mark]*. Sea turbines came into use around 2001, and between 2001 and 2017 their number and capacity remained relatively low when compared to land turbines *[1 mark]*.

(ii) Maximum of 6 marks available.
This question is level marked.
HINTS:
- Start by outlining the idea that there are many factors that may influence a country's renewable energy use, and that these factors vary between countries.
- Then, discuss how some countries have a natural advantage in certain renewable energy sources. For example, Norway is mountainous with many natural lakes, making it ideal for hydroelectric power, which may not be a viable option in a flat country or one with few lakes.
- Next, you could also mention how the role of governments can influence the use of renewable energy. For example, some governments may prioritise the development of renewable energy sources, whereas others may still rely heavily on fossil fuels.
- You could also consider how some local people may be opposed to certain types of renewable energy and protest against their development, e.g. wind turbines.

(iii) Maximum of 8 marks available.
This question is level marked.
HINTS:
- Start by introducing the energy resource that you have chosen and outline its location and requirements for development.
- Then, explain the positive social, economic and environmental impacts of exploiting your chosen resource. E.g. biofuels just rely on growing crops, so they can be produced in most countries and are generally cheaper and less environmentally damaging than fossil fuels.
- Next, you could write about the negative environmental, social and economic impacts of producing your chosen resource. E.g. the removal of vegetation reduces biodiversity, plus the crops replacing them are often applied with fertilizers and pesticides that further reduce species as well as pollute water resources. Loss of pasture and land for food crops may reduce food security, especially for poorer people.
- You could refer to a specific case study in your answer, e.g. biofuels in Brazil, which produces cheaper fuels and in 2005 employed over half a million people.

Answers

6.3 Maximum of 20 marks available. This question is level marked.
<u>HINTS:</u>
* Start by explaining what is meant by energy security, and why either increased supply or reduced demand (or both) may be needed in the future.
* Then, discuss how decreasing energy consumption can be done, and how this may improve energy security. For example, by increasing the energy efficiency of homes and cars and by discouraging people from flying.
* Next, discuss the challenges of decreasing energy consumption. For example, an increasing global population and increasing economic development results in increased energy demand, and the initial cost of increasing energy efficiency is high.
* Then, discuss how increasing energy supply can be done, and how this may improve energy security. For example, increasing the use of renewable energy can increase energy supply. Unlike fossil fuels, renewable energy sources will not run out, which makes them a more viable long-term option for achieving energy security.
* Next, discuss the challenges of increasing energy supply. For example, exploiting unconventional reserves using methods such as hydraulic fracturing (fracking) to extract natural gas from shale can increase energy supply. This increases energy security, but it can cause major environmental issues, such as groundwater contamination and air pollution, as well as contributing to climate change.
* Write a conclusion that evaluates the view that people should focus on reducing energy consumption rather than attempting to increase the supply of energy resources. This should be based on your personal opinion but should also draw on the evidence already provided in your answer.

Page 310 — Superpowers

7.1 One mark for each valid point, up to a maximum of 4 marks. E.g. A unipolar world is where one superpower dominates the world, for example Imperial Britain *[1 mark]*. This is unlikely to be stable in the long-term, as it is hard for one superpower to maintain control for very long *[1 mark]*. Other countries are likely to exploit any areas of weakness *[1 mark]*, to try and become more powerful, causing conflict *[1 mark]*.

7.2 Maximum of 12 marks available. This question is level marked.
<u>HINTS:</u>
* Start by describing what a superpower is. You could also briefly outline China's economic and political history, including how its 'Open Door Policy' has opened it to global trade and investment, helping its economy and global influence to grow rapidly.
* Then, explain how China can be seen as a superpower. For example, you could use Figure 1 to explain that investing in other countries is a way to gain power and influence, as countries that depend on that FDI would have their economy damaged if the investing country withdrew that money. You could also mention that China has a large, well-educated workforce and a large military.
* Next, explain how China may not be seen as a superpower. For example, you could mention that the One Child Policy has created an ageing population, which leads to a higher dependency ratio. You could also mention China's major air and water pollution, and its heavy reliance on importing raw materials, which makes it dependent on other countries and limits its own power.
* Write a conclusion that assesses the view that China is a superpower. This should be based on your personal opinion but should also draw on the evidence already provided in your answer.

7.3 Maximum of 8 marks available. This question is level marked.
<u>HINTS:</u>
* Start by explaining why some areas such as the Arctic are disputed.
* Then, explain how tensions can develop over territory. For example, you could say that some countries have overlapping Economic Exclusive Zones, that multiple countries may try to claim a piece of unclaimed territory, and that passage through waters can also cause conflict.
* Next, explain that tension may develop because of access to physical resources. For example, you could mention that the Arctic has large resources of oil and gas, which makes countries eager to claim territory where these resources are found.
* Finally, explain that tension may develop due to conflicts of interest over the use of an area. For example, you could mention that some countries want to exploit the Arctic's resources, whereas others want to conserve it as a sensitive ecosystem.

Page 311 — Health, Human Rights and Intervention

8A.1 One mark for each valid point, up to a maximum of 4 marks. E.g. Some countries believe that the UDHR is biased towards western values *[1 mark]* and signing up to it means that these values will be imposed on them *[1 mark]*. For example, Saudi Arabia chose to abstain because they felt some articles went against Islamic teachings *[1 mark]*, such as the right to freedom of religion *[1 mark]*.

8A.2 (i) One mark for each valid point up to a maximum of 2 marks. Accurately plotting the data for Bulgaria, El Salvador and Honduras *[1 mark]* and drawing a line of best fit (see below) *[1 mark]*.

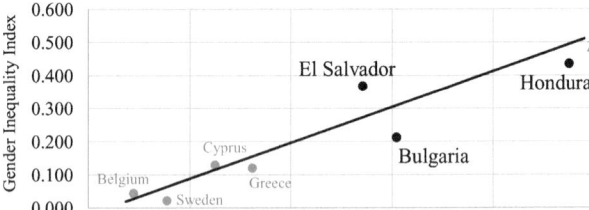

(ii) Maximum of 6 marks available. This question is level marked.
<u>HINTS:</u>
* Start by stating that the Gini Coefficient (GC) shows wealth inequality in a nation while the Gender Inequality Index (GII) measures the human development costs of gender inequality.
* Then, briefly outline the relationship between the GC and the GII shown on the graph. For example, you could mention that there is a positive relationship between GC and GII, so as wealth inequality increases gender inequality also increases.
* Next, you should suggest reasons for the relationship. For example, you could say that as countries become more developed, access to education is likely to become more universal, which should create opportunities for more people. This is likely to reduce wealth inequality as well as aspects of gender inequality, such as female secondary education and female representation in government.

Answers

8A.3 Maximum of 8 marks available. This question is level marked.
HINTS:
- Start by outlining what the Millennium Development Goals (MDGs) were.
- Then, explain why it was necessary to replace the MDGs. For example, you could mention that the progress towards achieving the MDGs was mixed. The overall number of people living in poverty halved, but this was inconsistent around the world, with poverty in Sub-Saharan Africa actually increasing. This suggested that the goals needed to be changed to ensure all regions were supported.
- Next, explain what the Sustainable Development Goals are and what they aim to do. For example, you could say that they aim to resolve some of the issues associated with the MDGs by being more ambitious and covering a wider range of issues.
- You could also explain that the SDGs are more focused on achieving development without environmental damage than the MDGs, for example they aim to tackle climate change and protect ecosystems.
- Overall, the SDGs were devised to build on the progress made towards the MDGs while resolving some of the issues associated with them.

8A.4 Maximum of 20 marks available. This question is level marked.
HINTS:
- Start by briefly outlining how health and life expectancy vary globally. For example, life expectancy tends to be significantly higher and health better in developed countries compared to developing countries.
- Then, explain how this links to inequality. E.g. in developed countries, most people have access to basic needs, such as clean water and sanitation, whereas many people in developing countries do not. This increases the risk of infectious diseases like cholera.
- Next, you could also discuss inequalities in access to health care between countries and the impact this is likely to have on health.
- Then, discuss inequalities in health and life expectancy within countries. E.g. you could write about regional variations in the UK, where people in wealthier areas tend to live longer and have better health than those from deprived areas.
- Finally, you could also discuss factors other than inequality that impact on health and life expectancy. For example, you could argue that many non-communicable diseases such as heart disease are more common in developed countries, partly as a result of lifestyle choices such as an unhealthy diet.
- Write a conclusion that sums up your points and assesses how far inequality is the main cause of spatial variations in health. This should be based on your personal opinion but should also draw on the evidence already provided in your answer.

Page 312 — Migration, Identity and Sovereignty

8B.1 (i) One mark for each valid point up to a maximum of 2 marks. Accurately plotting the data for Denmark, Namibia and Panama *[1 mark]* and drawing a line of best fit (see below) *[1 mark]*.

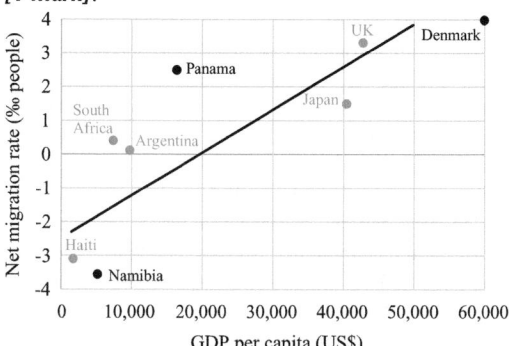

(ii) Maximum of 6 marks available.
This question is level marked.
HINTS:
- Start by explaining that GDP per capita is a measure of wealth while the net migration rate (NMR) measures the difference between the number of people moving into and out of a country.
- Then, outline the relationship between GDP per capita and NMR. For example, you could say that there is a positive relationship between GDP and NMR, so as GDP increases, more people tend to move into the country than move out of the country.
- Next, you should suggest reasons for this relationship. For example, you could say that countries with higher GDPs typically have more better-paid jobs, which is a pull factor for the more developed countries and a push factor from more developing countries.
- Finally, you could also discuss the fact that there may be other factors that affect both GDP and NMR. For example, countries affected by natural disasters, like Haiti, are likely to have a lower GDP due to reduced productivity and the cost of rebuilding. This may cause more people to leave a country, and reduce the number of people moving to a country.

8B.2 Maximum of 8 marks available. This question is level marked.
HINTS:
- Start by defining international migration and briefly outline some of the factors that affect people's ability to migrate.
- Then, explain that nationality affects people's ability to migrate. For example, some countries allow citizens of some other countries to enter without a visa, meaning migration is easier. Residents of Japan can travel to 193 countries without a visa, compared to 27 countries for residents of Afghanistan.
- Next, explain how skill level can affect people's ability to migrate. For example, you could say that countries, like Australia, have immigration policies that use points to rank applicants by their skills.
- Finally, explain how income can affect people's ability to migrate. For example, you could say that migration can be very expensive, which means it is easier for people with higher incomes.

Answers

8B.3 Maximum of 20 marks available. This question is level marked.
HINTS:
- Start by outlining what is meant by national identity and how some argue that globalisation has weakened it.
- Then, discuss some of the ways in which globalisation may have weakened national identity. This might include increased levels of immigration leading to increased ethnic diversity in many countries; global governance, which may override national laws; the spread of foreign brands and media causing blurring of national identity; and foreign ownership of businesses and property.
- For each point you make, discuss any ways it could be seen to have weakened national identity, as well as ways it may have strengthened national identity. E.g. The spread of western values and culture has led to the 'Westernisation' of some countries, with people seeking goods such as clothes and food associated with particular western cultures. This can erode traditional cultures, which may weaken national identity, e.g. people may choose to eat at McDonald's® rather than a local restaurant. However, some elements of western culture have been adapted to fit with local cultures, so people may come to incorporate this hybrid of cultures into their national identity.
- Next, you could talk about some of the ways in which national identity can be strengthened. For example, you could mention that global sporting events can increase national pride and strengthen national identity. Make sure you refer to case studies to support your answer. For example, you could mention that after the 2012 London Olympics, a survey revealed that the British were prouder of the British Olympic team than the royal family.
- Finally, you could also make the point that national identity can change over time, e.g. new values and customs can become part of a country's national identity, rather than just weakening it.
- Write a conclusion that evaluates the view that globalisation has weakened national identity. This should be based on your personal opinion but should also draw on the evidence already provided in your answer.

Pages 313-314 — Synoptic Question Practice

9.1 One mark for each valid point, up to a maximum of 4 marks. E.g. Climate change leads to higher atmospheric temperatures in some areas, which cause higher rates of evapotranspiration *[1 mark]*. Higher temperatures mean more water vapour can be stored in the atmosphere, so less falls as precipitation *[1 mark]*. This leads to lower soil moisture levels, which means there is less water available for evaporation, further reducing precipitation *[1 mark]*. Reduced precipitation coupled with reduced groundwater supplies causes water shortages *[1 mark]*.

9.2 Maximum of 6 marks available. This question is level marked.
HINTS:
- Start by describing the overall patterns in the figures.
- Then, suggest reasons for the patterns you have described, e.g. The large increase in water consumption in Asia suggests that an increase in withdrawal, rather than a decrease in supply, is primarily responsible for water stress. Factors such as rapid population growth and economic development are likely to have driven the increase in consumption.
- Next, you could discuss anomalies in the data and suggest reasons for them, e.g. despite Australia's water consumption not increasing significantly, the ratio of withdrawal to supply is 'Medium' there, suggesting that some places experience water scarcity because they are naturally dry.

9.3 One mark for each valid point, up to a maximum of 4 marks. E.g. Figure 2 only shows overall water consumption in each region, not water supply *[1 mark]*. If water supply is plentiful, regions that consume a lot of water may still not experience scarcity, e.g. Brazil *[1 mark]*. Figure 2 also divides the world into regions like 'Asia' and 'North America', but it gives no indication of variations within those regions *[1 mark]*. Therefore, a country in an area of high consumption may itself have low consumption and little water stress, but this will not be shown in Figure 2 *[1 mark]*.

9.4 Maximum of 18 marks available. This question is level marked.
HINTS:
- Start by outlining how long attempts to mitigate climate change have been going on for and some of the steps taken. You should use evidence from Figure 3 as well as your own knowledge.
- Then, discuss ways in which governments have been slow in making changes. E.g. you could comment that as far back as 1992, developed countries agreed to limit their emissions, yet Figure 4 shows that atmospheric CO_2 has continued to increase steadily since then.
- Next, discuss the fact that, in recent years, countries have committed to more ambitious targets. However, Figure 4 suggests that countries have struggled to meet their previous commitments, so major changes are needed if they are to be kept this time.
- Then, you could comment that, although CO_2 concentrations have continued to rise, it is unclear how high they would be if steps had not been taken to limit emissions, so such attempts may still have had a significant effect.
- Finally, you could also mention that scientists don't know whether trends which are caused by climate change such as melting sea ice and biodiversity loss can be reversed. In this sense, you could argue that it is too late.
- Write a conclusion that evaluates the view that steps to tackle climate change have come too little and too late. This should be based on your personal opinion but should also draw on the evidence already provided in your answer.

9.5 Maximum of 24 marks available. This question is level marked.
HINTS:
- Start by outlining the ways in which patterns of geopolitical power can change. You could also mention that there are various factors that affect how powerful a country is, such as its economy, size, military strength, etc.
- Then, outline how climate change may impact different aspects of a country's power. You could consider direct impacts, e.g. decreased precipitation may increase water stress, as shown in Figure 1, which will affect the ability of a country to grow food. This may negatively affect the economy, reducing the country's influence.
- Next, you could discuss how wealthy countries are better equipped to deal with the effects of climate change. As these countries already tend to have more power, this may mean that current patterns of power become more pronounced.
- Then, you could also comment on political leadership on tackling climate change, and how this may change patterns of power in the future. E.g. as climate change becomes increasingly important, governments seen to be tackling it successfully may gain political influence. If China is able to honour its pledge to become carbon neutral by 2060 (Figure 3) it will be seen as a leader in averting climate catastrophe.
- Finally, you should also consider factors other than climate change that can affect patterns of geopolitical power. For example, natural resources and military power can have significant effects on the balance of power — you could use recent events in Russia and Ukraine to illustrate this.
- Write a conclusion that evaluates the view that climate change is the strongest factor in changing patterns of geopolitical power. This should be based on your personal opinion but should also draw on the evidence already provided in your answer.

Acknowledgements

Pagan Island photo on p.7: © *NASA's Earth Observatory/NASA. https://earthobservatory.nasa.gov/images/77503/pagan-island-northern-marianas*

Himalayas, Manili image on p.7 by little byte of luck, 2006. https://www.flickr.com/photos/littlebyteofluck/394129303/in/photostream/.
Licensed under CC BY 2.0 https://creativecommons.org/licenses/by/2.0/

Japan earthquake tsunami image on p.11 by Chainless Soul, 2011. https://www.flickr.com/photos/ophelia_jane_julia/5534414080.
Licensed under CC BY-SA 2.0. https://creativecommons.org/licenses/by-sa/2.0/.

Caldera Toba image on p.13 by Yudhapohan, via Wikimedia Commons.
Licensed under CC BY-SA 4.0 https://creativecommons.org/licenses/by-sa/4.0

Mauna Kea from the lava fields of Mauna Loa on p.13 by pedrik, 2016. https://www.flickr.com/photos/pedrik/31642083920.
Licensed under CC BY 2.0. https://creativecommons.org/licenses/by/2.0/.

Data on p.17 from The World Risk Report 2021. https://weltrisikobericht.de/weltrisikobericht-2022-e/

GDP Data on p.17: The World Bank: GDP per capita, PPP (current international $) International. Source: International Comparison Program, World Bank |
World Development Indicators database, World Bank | Eurostat-OECD PPP Programme.
Licensed under CC-BY-4.0 https://creativecommons.org/licenses/by/4.0/

Mount St Helens photo on p.23: © *NASA's Earth Observatory/NASA.*
https://earthobservatory.nasa.gov/images/92469/an-astronauts-view-of-mount-st-helens

Aerial view of Mount Nyiragongo image on p.24 by MONUSCO/Neil Wetmore, 2014. https://www.flickr.com/photos/monusco/16067098841/.
Licensed under CC BY-SA 2.0 https://creativecommons.org/licenses/by/2.0/.

Trends in disaster data on p.25: EM-DAT, CRED / UCLouvain, Brussels, Belgium – www.emdat.be

Balapitiya mangroves image on p.26: Shankar S., 2011. https://www.flickr.com/photos/shankaronline/7567865130.
Licensed under CC BY 2.0. https://creativecommons.org/licenses/by/2.0/.

Image of Eyjafjallajökull on p.26 by Bjarki Sigursveinsson, 2010. https://www.flickr.com/photos/bjarkis/4530958802/in/photostream.
Licensed under CC BY-SA 2.0. https://creativecommons.org/licenses/by-sa/2.0/

Destruction of Typhoon Haiyan image on p.27 by Henry Donati/ DFID — UK Department for International Development.
Licensed under CC BY 2.0. https://creativecommons.org/licenses/by/2.0/.

Park Model on p.29: © *Chris Park*

*Construction of an 18 meter high embankment image on p.30 by Chubu Electric Power Co., 2012. https://www.flickr.com/photos/iaea_*
imagebank/8506930602/. Licensed under CC BY-SA 3.0. https://creativecommons.org/licenses/by-sa/3.0/deed

Base isolation dampers image on p.30 by Marshelec, 2011.
Licensed under CC BY-SA 3.0, https://creativecommons.org/licenses/by-sa/3.0/deed.en

Students practice earthquake drill image on p.31: Department of Foreign Affairs, 2013.
https://www.flickr.com/photos/dfataustralianaid/10727359493.
Licensed under CC BY 2.0. https://creativecommons.org/licenses/by/2.0/.

Map on pages 32 & 50: N. Ray and J.M. Adams, Internet Archaology, 2001.
http://intarch.ac.uk/journal/issue11/2/map/download_page_js.htm.
Licensed for re-use under the Creative Commons Attribution-Share Alike 3.0 Generic Licence. https://creativecommons.org/licenses/by/3.0/

Image of sunspots on p.33: Courtesy NASA/JPL-Caltech.

Image of the Larsen B ice shelf on p.34: MODIS, NASA's Earth Observatory

Image of Lower Curtis Glacier Mt Shuksan on p.34: brew books, 2006. https://www.flickr.com/photos/brewbooks/214250503/in/photolist-qP9AFz-
qPiHHM-jW6ki-jW6sx-jW6dc-pSiHdY-hHcnR-xRQtMR-xRQtc2-jW65p-jXx8N-jW6XY-jW6Sc-jXxca-jW6FN-jW6Mz-hvD3V-TnEtTX-Skd3MZ. Licensed under
CC BY-SA 2.0. https://creativecommons.org/licenses/by-sa/2.0/

Data on p.37: Courtesy of www.imbie.org

Glacier Length graph on p.37: Leclercq, P. W., Oerlemans, J., Basagic, H. J., Bushueva, I., Cook, A. J., and Le Bris, R.:
A data set of worldwide glacier length fluctuations, The Cryosphere, 8, 659–672, https://doi.org/10.5194/tc-8-659-2014, 2014.
Licensed under CC BY 3.0. https://creativecommons.org/licenses/by/3.0/

Image of recessional moraine on p.46: Concentric recessional moraines in the end of Valle del Silencio by Gagea, 2010.
Licenced under CC BY-SA 3.0 https://creativecommons.org/licenses/by-sa/3.0/

Textbook Erratic image on p.47: © *Val Vannet. https://www.geograph.org.uk/photo/142387.*
Licensed under CC BY-SA 2.0. https://creativecommons.org/licenses/by-sa/2.0/.

Image of slump on p.50: Slump D, Boris Radosavljevic, 2013. https://www.flickr.com/photos/139918543@N06/24531601650/.
Licensed under CC BY 2.0. https://creativecommons.org/licenses/by/2.0/.

Image of solifluction lobes on p.51: View of the high tundra from Skookum Pass. By Dr John Cloud.
National Oceanic and Atmospheric Administration / Department of Commerce.

Image on p.52 of ice wedge, seen from top, Spitzbergen/Svalbard, Hannes Grobe, 2007.
Licensed under CC BY 2.5, via Wikimedia Commons. https://creativecommons.org/licenses/by-sa/2.5/deed.en.

Image of Halley VI Antarctic Research Station on p.56: Hugh Broughton Architects, 2012.
Licensed under CC BY-SA 4.0, via Wikimedia Commons. https://creativecommons.org/licenses/by-sa/4.0/deed.en

HEP data on p.56: AMAP, 2012. Arctic Climate Issues 2011: Changes in Arctic Snow, Water, Ice and Permafrost. SWIPA 2011 Overview Report. Arctic
Monitoring and Assessment Programme (AMAP), Oslo. xi + 97pp

Avalanche fatality data on p.58: CAIC

Image of signage in Himalayas on p.59: Sarangkot, Nepal by Bijay Chaurasia, 2017.
Licensed under CC BY-SA 4.0 via Wikimedia Commons. https://creativecommons.org/licenses/by-sa/4.0/

Glacial lakes image on p.60: © *NASA.*

Data on p.60: Indian Space Research Organization (ISRO)

Data on p.60: The World Bank: Employment in agriculture (% of total employment) (modeled ILO estimate) - Nepal. Source:
International Labour Organization. "ILO modelled estimates database" ILOSTAT. Accessed January 2021. ilostat.ilo.org/data.
Licensed under CC BY 4.0.https://creativecommons.org/licenses/by/4.0/.

Acknowledgements

Image on p.61 of SHellNo Paddle in Seattle, Greenpeace, 2015. https://www.flickr.com/photos/backbone_campaign/33976978586.
Licensed under CC BY 2.0 https://creativecommons.org/licenses/by/2.0/

Map on p.66: Data from the Department for Environment Food & Rural Affairs © Crown copyright licensed under the Open Government Licence v3.0.
http://www.nationalarchives.gov.uk/doc/open-government-licence/version/3/

Image on p.71 of Cliffs above Redcliff Point © Nigel Mykura. https://www.geograph.org.uk/photo/906984.
Licensed under CC BY-SA 2.0. https://creativecommons.org/licenses/by-sa/2.0/

Image on p.71 of Slumping Cliffs at Naish Park © Mike Smith. https://www.geograph.org.uk/photo/2709863.
Licensed under CC BY-SA 2.0. https://creativecommons.org/licenses/by-sa/2.0/.

View to Lannacombe Bay on p.72: © Philip Halling. https://www.geograph.org.uk/photo/3861309.
Licensed under CC BY-SA 2.0. https://creativecommons.org/licenses/by-sa/2.0/

Dredging at Whitby Harbour on p.74: © James T M Towill. https://www.geograph.org.uk/photo/5470446.
Licensed under CC BY-SA 2.0. https://creativecommons.org/licenses/by-sa/2.0/

Landsat imagery on p.74 courtesy of NASA Goddard Space Flight Center and U.S. Geological Survey

Page 75 contains public sector information licensed under the Open Government Licence v3.0.
https://www.nationalarchives.gov.uk/doc/open-government-licence/version/3/.

Image on p.75: Cliffs with Former Coastguard Cottages, Birling Gap, Sussex. © Christine Matthews.
https://www.geograph.org.uk/photo/4128805. Licensed under CC BY-SA 2.0. https://creativecommons.org/licenses/by-sa/2.0/

Image on p.75: Brighstone Holiday Centre abandoned chalets. Editor5807.
Licensed under CC BY 3.0 via Wikimedia Commons. https://creativecommons.org/licenses/by/3.0/deed.en

Image on p.76: Cliffs near Port O' Warren © Ann Cook. https://www.geograph.org.uk/photo/1629366.
Licensed under CC BY-SA 2.0. https://creativecommons.org/licenses/by-sa/2.0/

Image on p.78: Forvie: mobile sand dunes. © Martyn Gorman. https://www.geograph.org.uk/photo/1399339.
Licensed under CC BY-SA 2.0. https://creativecommons.org/licenses/by-sa/2.0/

Image on p.78: Saltmarsh south of Ashlett Creek. © Jim Champion. https://www.geograph.org.uk/photo/331679.
Licensed under CC BY-SA 2.0. https://creativecommons.org/licenses/by-sa/2.0/

Graph of Sea Level Rise on p.80: Sweet, W.V., B.D. Hamlington, R.E. Kopp, C.P. Weaver, P.L. Barnard, D. Bekaert, W. Brooks, M. Craghan, G. Dusek, T. Frederikse, G. Garner, A.S. Genz, J.P. Krasting, E. Larour, D. Marcy, J.J. Marra, J. Obeysekera, M. Osler, M. Pendleton, D. Roman, L. Schmied, W. Veatch, K.D. White, and C. Zuzak, 2022: Global and Regional Sea Level Rise Scenarios for the United States: Updated Mean Projections and Extreme Water Level Probabilities Along U.S. Coastlines. NOAA Technical Report NOS 01. National Oceanic and Atmospheric Administration, National Ocean Service, Silver Spring, MD, 111 pp.

Image on p.81: Floating Agricultural Field by Nazmulhuqrussell. Licensed under CC BY 3.0 via Wikimedia Commons.
https://creativecommons.org/licenses/by/3.0/deed.en

Image on p.87 of Cyclone Shelter, Helena Wright, 2013. https://www.flickr.com/photos/climatescope/15265520285.
Licensed under CC BY 2.0. https://creativecommons.org/licenses/by/2.0/

Train icon on p.89 by Yask from Noun Project. https://thenounproject.com/icon/train-9595/.
Licensed under CC BY 3.0. https://creativecommons.org/licenses/by/3.0/.

Ship icon on p.89 by icon 54, from Noun Project. https://thenounproject.com/browse/icons/term/Ship.
Licensed under CC BY 3.0. https://creativecommons.org/licenses/by/3.0/.

Plane icon on p.89 by trang5000, from Noun Project. https://thenounproject.com/browse/icons/term/plane.
Licensed under CC BY 3.0. https://creativecommons.org/licenses/by/3.0/.

Container icon on p.89 by Rockicon, from Noun Project. https://thenounproject.com/browse/icons/term/container.
Licensed under CC BY 3.0. https://creativecommons.org/licenses/by/3.0/.

Data on p.91: The World Bank: Foreign direct investment, net inflows (BoP, current US$) — China. Source: International Monetary Fund,
Balance of Payments database, supplemented by data from the United Nations Conference on Trade and Development and official national sources.
Licensed under CC BY-4.0. https://creativecommons.org/licenses/by/4.0/

Data on p.91: The World Bank: Manufacturing, value added (% of GDP) — China. Source: World Bank national accounts data, and OECD National
Accounts data files. Licensed under CC BY 4.0. https://creativecommons.org/licenses/by/4.0/

Dell Technologies, Dell, and other trademarks are trademarks of Dell Inc. or its subsidiaries

Data on p.92 from "Global Cities: Divergent Prospects and New Imperatives in the Global Recovery. 2021 Global Cities Report".
Copyright A.T. Kearney, 2021. All rights reserved. Used with permission.

Data on p.92 from the KOF Globalisation Index. https://kof.ethz.ch/globalisation.

Photo on p.94 of glue works, Steve Jurvetson, 2005. https://www.flickr.com/photos/44124348109@N01/52580259/.
Licensed under CC BY 2.0. https://creativecommons.org/licenses/by/2.0/.

Data on p.94: The World Bank: Foreign direct investment, net inflows (BoP, current US$) — East Asia & Pacific, North America, Latin America & Caribbean,
Middle East & North Africa, European Union. Source: International Monetary Fund, Balance of Payments database, supplemented by data from the United
Nations Conference on Trade and Development and official national sources. Licensed under CC BY-4.0. https://creativecommons.org/licenses/by/4.0/

Data on p.94: World Bank and the Development Research Center of the State Council, the People's Republic of China. 2022. Four
Decades of Poverty Reduction in China: Drivers, Insights for the World, and the Way Ahead. Washington, DC: World Bank.
doi:10.1596/978-1-4648-1877-6. Licensed under CC BY 3.0 IGO. http://creativecommons.org/licenses/by/3.0/igo.

Data on p.95: The World Bank: Educational attainment, at least Bachelor's or equivalent, population 25+, total (%) (cumulative) — India.
Source: UNESCO Institute for Statistics (UIS). UIS.Stat Bulk Data Download Service. Accessed October 24, 2022. apiportal.uis.unesco.org/bdds.
Licensed under CC BY-4.0. https://creativecommons.org/licenses/by/4.0/.

Data on p.95: The World Bank: Poverty Headcount ratio at $2.15 a day (2017 PP) (% of population) — China, World. Source: World Bank, Poverty
and Inequality Platform. Data are based on primary household survey data obtained from government statistical agencies and World Bank country
departments. Data for high-income economies are mostly from the Luxembourg Income Study database. For more information and methodology, please
see pip.worldbank.org. Licensed under CC BY 4.0. https://creativecommons.org/licenses/by/4.0/

Image on p.95: wanted, Paul Keller, 2005. https://www.flickr.com/photos/paulk/66164294.
Licensed under CC BY 2.0. https://creativecommons.org/licenses/by/2.0/.

Acknowledgements

Image on p.96 of Dharavi, M M, 2010. https://www.flickr.com/photos/43423301@N07/5842973175/.
Licensed under CC BY-SA 2.0. https://creativecommons.org/licenses/by-sa/2.0/.

Image on p.96 of Derelict factories in Detroit, LHOON, 2007. https://www.flickr.com/photos/lhoon/2289113074.
Licensed under CC BY-SA 2.0. https://creativecommons.org/licenses/by-sa/2.0/.

Data on p.97 from the UN 2018 Revision of World Urbanization Prospects.
https://www.un.org/development/desa/en/news/population/2018-revision-of-world-urbanization-prospects.html

Data on p.97 from the UN The Worlds Cities in 2018 Data Booklet.
https://www.un.org/en/development/desa/population/publications/pdf/urbanization/the_worlds_cities_in_2018_data_booklet.pdf

Data on p.98 from the UN World Urbanization Prospects. https://population.un.org/wup/

London property data on p.98: Transparency International, https://www.transparency.org.uk/.
Licensed under CC BY-ND 4.0. https://creativecommons.org/licenses/by-nd/4.0/ .

Image on p.101 of Japanese lanterns in souvenir shop, Tokyo, Japan, Maksym Kozlenko.
Licensed under CC BY-SA 4.0 via Wikimedia Commons. https://creativecommons.org/licenses/by-sa/4.0/deed.en.

Data on p.101 reproduced from The Global Health Observatory, Prevalence of obesity among adults, BMI>=30
(crude estimate) (%) Copyright 2022. https://www.who.int/data/gho/data/indicators/indicator-details/GHO/
prevalence-of-obesity-among-adults-bmi-=-30-(crude-estimate)-(-) accessed November 2022.

Image on p.101 of Paralympic games, Leandro Neumann Ciuffo, 2016. https://www.flickr.com/photos/leandrociuffo/29073210394/in/
photolist-Li6Nyb-Mjqmpn-YJDxpp-MjqkF8-XteMkf-P5Z7Pm-YvY3sn-MwYbtL-MwYa5J-LZSig9-LKxoXW-LKxvms-MzEkJM-YvWtkn-MwYeCS-
LKxkFq-Mg345y-Mjou68-YsooNd-MoyyJh-P8NWBD-Moyyg3-XvhUtV-Lb7Sxd-XuBrgB-Yu12aG-dQ1kYJ-Y94p9U-P8NX4F-LZQQRs-
XtbcSj-YsqofL-XtagHh-dQ1m5j-MjoY3R-7cjXVx-XtaquW-M1gxV4-dPUJn4-yPPk1M-MrGxWv-MrGziZ-MjoZ7e-Mjqk5Z-Y8oZfs-SfBnnX-
Mjqjyt-SrQoUM-Ra8nAQ-Mjqj76. Licensed under CC BY 2.0. https://creativecommons.org/licenses/by/2.0/.

Image on p.102: The Voice Afrique francophone, Jean Mari Robert, 2018.
Licensed under CC BY-SA 4.0 via Wikimedia Commons. https://creativecommons.org/licenses/by-sa/4.0/deed.en

Image on p.102: Chicken Tikka Kebab, Sumit Sarai, 2016.
Licensed under CC BY-SA 4.0 via Wikimedia Commons. https://creativecommons.org/licenses/by-sa/4.0/deed.en

Image on p.102: A typical sing-sing scene of Papua New Guinea, a gathering of different tribes singing and dancing with
their unique make-up and traditional costumes, Jialiang Gao, 2008. Licensed under CC BY-SA 3.0 via Wikimedia Commons.
https://creativecommons.org/licenses/by-sa/3.0/deed.en.

Data on p.105: The World Bank: GDP (Current US$) — South Asia. Source: World Bank national accounts data, and OECD National Accounts data files.
Licensed under CC BY 4.0. https://creativecommons.org/licenses/by/4.0/

Data on p.105: The World Bank: GDP (Current US$) — South Asia, Europe & Central Asia. Source: World Bank national accounts
data, and OECD National Accounts data files. Licensed under CC BY 4.0. https://creativecommons.org/licenses/by/4.0/

Data on p.105: Piketty, Thomas, Li Yang, and Gabriel Zucman. 2019. "Capital Accumulation, Private Property,
and Rising Inequality in China, 1978-2015." American Economic Review, 109 (7): 2469-96. Copyright
American Economic Association; reproduced with permission of the American Economic Review.

Page 106 contains public sector information licensed under the Open Government Licence v3.0.
https://www.nationalarchives.gov.uk/doc/open-government-licence/version/3/.

2021 Census data on p.106 provided by the United States Census Bureau.

Image on p.106: Meeting 1er mai 2012 Front National, Blandine le Cain, 2012. https://www.flickr.com/photos/68651617@N07/7421301940.
Licensed under CC BY 2.0. https://creativecommons.org/licenses/by/2.0/.

Press Freedom Index data on p.107: © Reporters Without Borders. rsf.org

Ecological footprint data on p.109: York University Ecological Footprint Initiative & Global Footprint Network. National Footprint and Biocapacity
Accounts, 2022 edition. Downloaded 8th December 2022 from https://data.footprintnetwork.org.

Data on p.110 from OECD (2022), "Waste: Municipal waste", OECD Environment Statistics (database),
https://doi.org/10.1787/data-00601-en

Maps on p.111 contains National Statistics data © Crown copyright and database right 2019 & OS data © Crown copyright and database right 2019.
Used under the terms of the Open Government licence http://www.nationalarchives.gov.uk/doc/open-government-licence/version/3/

Page 112 contains ONS data www.ons.gov.uk licensed under the Open Government Licence v3.0
http://www.nationalarchives.gov.uk/doc/open-government-licence/version/3/

Data on p.112: © Crown copyright. Data supplied by National Records of Scotland.
Licensed under the Open Government Licence v3.0. http://www.nationalarchives.gov.uk/doc/open-government-licence/version/3/

Page 113 contains ONS data www.ons.gov.uk licensed under the Open Government Licence v3.0
http://www.nationalarchives.gov.uk/doc/open-government-licence/version/3/

Data on p.113: © Crown copyright. Data supplied by National Records of Scotland licensed under the Open Government Licence v3.0
http://www.nationalarchives.gov.uk/doc/open-government-licence/version/3/

Nigeria and Hong Kong GDP Data on p.113 — GDP per capita, PPP (current international $) — Hong Kong SAR, China, Nigeria. Source: International
Comparison Program, World Bank | World Development Indicators database, World Bank | Eurostat-OECD PPP Programme.
Licensed under CC BY 4.0. https://creativecommons.org/licenses/by/4.0/

Data on pages 114 & 144: ONS data www.ons.gov.uk licensed under the Open Government Licence v3.0
http://www.nationalarchives.gov.uk/doc/open-government-licence/version/3/

Data on pages 114 & 144: © Crown copyright. Data supplied by National Records of Scotland licensed under the Open Government Licence v3.0.
http://www.nationalarchives.gov.uk/doc/open-government-licence/version/3/

Map on p.115 showing the Metropolitan Green Belt and other green belts of England by Hellerick
https://commons.wikimedia.org/wiki/File:The_Metropolitan_Green_Belt_among_the_green_belts_of_England.svg.
Licensed under CC BY 3.0. https://creativecommons.org/licenses/by/3.0/deed.en via Wikimedia Commons.

Pages 116 & 146 contain ONS data www.ons.gov.uk licensed under the Open Government Licence v3.0
http://www.nationalarchives.gov.uk/doc/open-government-licence/version/3/

Acknowledgements

Page 116 contains Scottish Government data https://www.gov.scot/
Licensed under the Open Government Licence v3.0. http://www.nationalarchives.gov.uk/doc/open-government-licence/version/3/

Data on p.117: © Crown copyright. Data supplied by National Records of Scotland licensed under the Open Government licence v3.0
http://www.nationalarchives.gov.uk/doc/open-government-licence/version/3/

Page 117 contains Scottish Government data https://www.gov.scot/ licensed under the Open Government Licence v3.0
http://www.nationalarchives.gov.uk/doc/open-government-licence/version/3/

Fishing Boats, Hay's Dock image on p.118: © Andy Waddington https://www.geograph.org.uk/photo/5827668.
Licensed for reuse under CC BY-SA 2.0. https://creativecommons.org/licenses/by-sa/2.0/

Lerwick survey data on p.118: ekosgen, Survey of Young People 2015.

Image of Jarl's Squad Lerwick 2019 on p.120 by Malvara, 2019. https://commons.wikimedia.org/wiki/File:Jarl%27s_Squad_Lerwick_2019_(03).jpg.
Licensed for reuse under CC BY-SA 4.0 via Wikimedia Commons. https://creativecommons.org/licenses/by-sa/4.0/deed.en

San Francisco data in table on p.122 Source: US Census Bureau

Detroit data on page 123 Source: US Census Bureau

Page 124 Contains UK Parliament Data licensed under the Open Parliament Licence
https://www.parliament.uk/site-information/copyright-parliament/open-parliament-licence/

Page 124 Contains Open Government Information used under the terms of the open government licence
v3.0. https://www.nationalarchives.gov.uk/doc/open-government-licence/version/3/

Page 125 contains UK Parliament Data licensed under the Open Parliament Licence
https://www.parliament.uk/site-information/copyright-parliament/open-parliament-licence/

Image on p.125 of The new Students' Union © Copyright Alan Murray-Rust. https://www.geograph.org.uk/photo/3788201.
Licensed under CC BY-SA 2.0. https://creativecommons.org/licenses/by-sa/2.0/

Image on p.127 of coast path leading N to St Agnes Head © Copyright Colin Park https://www.geograph.org.uk/photo/5246293.
Licensed under CC BY-SA 2.0. https://creativecommons.org/licenses/by-sa/2.0/

Page 128 contains ONS data www.ons.gov.uk licensed under the Open Government Licence v3.0
http://www.nationalarchives.gov.uk/doc/open-government-licence/version/3/

Data on p.128: Crown copyright. Data supplied by National Records of Scotland licensed under the Open Government Licence v3.0
http://www.nationalarchives.gov.uk/doc/open-government-licence/version/3/

Page 128 contains Scottish Government data https://www.gov.scot/ licensed under the Open Government Licence v3.0
http://www.nationalarchives.gov.uk/doc/open-government-licence/version/3/

Page 129 contains Parliamentary information licensed under the Open Parliament Licence.
https://www.parliament.uk/site-information/copyright-parliament/open-parliament-licence/

Page 129 contains ONS data www.ons.gov.uk licensed under the Open Government Licence v3.0
http://www.nationalarchives.gov.uk/doc/open-government-licence/version/3/

Overseas ownership & buyers data on p.130: © Crown Copyright, licensed under the Open Government Licence v3.0
http://www.nationalarchives.gov.uk/doc/open-government-licence/version/3/

Image on p.131 of Liverpool School of Tropical Medicine, John Bradley.
Licensed under CC BY 3.0 via Wikimedia Commons. https://creativecommons.org/licenses/by/3.0/deed.en

Image on p.133 of barges at the east of Canning Dock in the Port of Liverpool, England. Public Domain via Wikimedia Commons

Image on p.133: Canning Half-tide Dock, Liverpool © Margaret. https://www.geograph.org.uk/photo/290814.
Licensed under CC BY-SA 2.0. https://creativecommons.org/licenses/by-sa/2.0/.

Image on p.134: Dowhill Farm shop and Restaurant © Kenneth Allen. https://www.geograph.org.uk/photo/250502.
Licensed under CC BY-SA 2.0. https://creativecommons.org/licenses/by-sa/2.0/.

Image on p.134 of Bay of Scousburgh, Shetland Islands, Scotland, Mustang Joe.
https://flickr.com/photos/63234672@N04/29618445318. Public Domain

Image of Haworth Main Street on p.135 by Jacqui Sadler, 2003. https://www.geograph.org.uk/photo/415579.
Licensed under CC BY-SA 2.0. https://creativecommons.org/licenses/by-sa/2.0/

Data on p.137: © Crown Copyright, licensed under the Open Government Licence v3.0
http://www.nationalarchives.gov.uk/doc/open-government-licence/version/3/

Image on p.137: Salford Quays by cliffajw https://www.flickr.com/photos/71337903@N00/255959082/.
Licensed under CC BY 2.0. https://creativecommons.org/licenses/by/2.0/

Image of Weedon on p.138: © Stephen McKay https://www.geograph.org.uk/photo/5264386
Licensed under CC BY 2.0. https://creativecommons.org/licenses/by/2.0/

Photo of Town hall at the Diamond on p.139 by Andreas F. Borchert https://commons.wikimedia.org/wiki/File:Coleraine_The_Diamond_2014_09_13.jpg.
Licensed under CC BY-SA 4.0 via Wikimedia Commons. https://creativecommons.org/licenses/by-sa/4.0/

Image on p.139 of NIMBY notice in Battenhall © Andrew Darge. https://www.geograph.org.uk/photo/1911227.
Licensed for reuse under CC BY-SA 2.0. https://creativecommons.org/licenses/by-sa/2.0/

Rural Development Programme for Northern Ireland data on p.139 © Crown Copyright, licensed under the Open Government Licence v3.0
http://www.nationalarchives.gov.uk/doc/open-government-licence/version/3/

Image of The Wilkins Building, University College London on p.141 by DAVID ILIFF https://commons.wikimedia.org/wiki/File:Wilkins_Building_2,_UCL,_
London_-_Diliff_(cropped).jpg. Licensed under CC BY-SA 3.0 via Wikimedia Commons. https://creativecommons.org/licenses/by-sa/3.0/deed.en

Page 141 contains ONS data www.ons.gov.uk licensed under the Open Government Licence v3.0
http://www.nationalarchives.gov.uk/doc/open-government-licence/version/3/

Page 143 contains ONS data www.ons.gov.uk licensed under the Open Government Licence v3.0
http://www.nationalarchives.gov.uk/doc/open-government-licence/version/3/

Data on p.143: © Crown copyright. Data supplied by National Records of Scotland licensed under the Open Government Licence v3.0
http://www.nationalarchives.gov.uk/doc/open-government-licence/version/3/

Acknowledgements

Image on p.146: The Hay Wain by John Constable

Scottish independence rally image on p.147 by Azerifactory https://commons.wikimedia.org/wiki/File:Scottish_independence_rally_2018_Largs.jpg.
Licensed under CC BY-SA 4.0 via Wikimedia Commons. https://creativecommons.org/licenses/by-sa/4.0/

Page 149 contains ONS data www.ons.gov.uk licensed under the Open Government Licence v3.0
http://www.nationalarchives.gov.uk/doc/open-government-licence/version/3/

Image on p.151 of English countryside — Shrewton by Giuseppe Milo, 2016. https://www.flickr.com/photos/giuseppemilo/29085750495.
Licensed under CC BY 2.0 https://creativecommons.org/licenses/by/2.0/

Image on p.151 of George and Dragon Inn, Burpham by Andrew Bowden, 2010. https://www.flickr.com/photos/bods/4779200011.
Licensed under CC BY 2.0 https://creativecommons.org/licenses/by/2.0/

Grade II listed thatched cottage image on p.151 by Peter K Burian. https://commons.wikimedia.org/wiki/File:The_thatch_cottage_stretton_8250_xlo.jpg.
Licenced under CC BY-SA 4.0 via Wikimedia Commons. https://creativecommons.org/licenses/by-sa/4.0/deed.en

Bank day, Baltasound image on p.151 © Mike Pennington https://www.geograph.org.uk/photo/1557489.
Licenced under CC BY-SA 2.0 https://creativecommons.org/licenses/by-sa/2.0/

Image on p.152 Far From the Madding Crowd © Colin Smith https://www.geograph.org.uk/photo/2925863.
Licenced under CC BY-SA 2.0 https://creativecommons.org/licenses/by-sa/2.0/

Page 153 contains ONS data www.ons.gov.uk licensed under the Open Government Licence v3.0
http://www.nationalarchives.gov.uk/doc/open-government-licence/version/3/

Data on p.153: © Crown copyright. Data supplied by National Records of Scotland licensed under the Open Government Licence v3.0
http://www.nationalarchives.gov.uk/doc/open-government-licence/version/3/

Page 153 contains Scottish Government data https://www.gov.scot/ licensed under the Open Government Licence v3.0
http://www.nationalarchives.gov.uk/doc/open-government-licence/version/3/

Page 154 contains ONS data www.ons.gov.uk licensed under the Open Government Licence v3.0
http://www.nationalarchives.gov.uk/doc/open-government-licence/version/3/

Pages 155 & 156 contain ONS data www.ons.gov.uk licensed under the Open Government Licence v3.0
http://www.nationalarchives.gov.uk/doc/open-government-licence/version/3/

Image on p.156 of Himalaya Palace Cinema, Southall © Danny P Robinson https://www.geograph.org.uk/photo/173961.
Licensed under CC BY-SA 2.0. https://creativecommons.org/licenses/by-sa/2.0/

Image of Community meal on p.157 by Harisingh https://commons.wikimedia.org/wiki/File:Langar.jpg.
Licenced under CC BY-SA 3.0 via Wikimedia Commons. https://creativecommons.org/licenses/by-sa/3.0/deed.en

Page 158 contains Scottish Government data https://www.gov.scot/ licensed under the Open Government Licence v3.0
http://www.nationalarchives.gov.uk/doc/open-government-licence/version/3/

Pages 158 &159 contain ONS data www.ons.gov.uk licensed under the Open Government Licence v3.0
http://www.nationalarchives.gov.uk/doc/open-government-licence/version/3/.

Data on p.159: © Crown copyright. Data supplied by National Records of Scotland licensed under the Open Government Licence v3.0
http://www.nationalarchives.gov.uk/doc/open-government-licence/version/3/

Page 159 contains ONS & gov.wales data licensed under the Open Government Licence v3.0
http://www.nationalarchives.gov.uk/doc/open-government-licence/version/3/

Page 160 contains ONS data www.ons.gov.uk licensed under the Open Government Licence v3.0
http://www.nationalarchives.gov.uk/doc/open-government-licence/version/3/

Love Leicester Hate Racism image on p.160: © Ashley Dace https://www.geograph.org.uk/photo/3093336.
Licensed under CC BY-SA 2.0. https://creativecommons.org/licenses/by-sa/2.0/

Page 161 contains ONS data www.ons.gov.uk licensed under the Open Government Licence v3.0
http://www.nationalarchives.gov.uk/doc/open-government-licence/version/3/

Rural Enterprise data on P.162: © Crown Copyright, licensed under the Open Government Licence v3.0
http://www.nationalarchives.gov.uk/doc/open-government-licence/version/3/

Data on p.163 from UCAR Center for Science Education used under the terms of the Attribution 4.0 International Licence (CC BY 4.0)
https://creativecommons.org/licenses/by/4.0/

Page 163: Source: Igor Shiklomanov's chapter "World fresh water resources" in Peter H. Gleick (editor), 1993,
Water in Crisis: A Guide to the World's Fresh Water Resources (Oxford University Press, New York).

Page 169: Yukon data on graph from Surface Water data for USA: USGS Surface-Water Monthly Statistics. Data compiled by the U.S. Geological Survey

Page 169: Indus data on graph from Current World Environment journal http://www.cwejournal.org/pdf/vol9no3/vol9_no3_670-685.pdf. Used under
the terms of the Attribution 4.0 International Licence (CC BY 4.0) https://creativecommons.org/licenses/by/4.0/

Page 169: Amazon data on Graph from Sea Surface Salinity Observations from Space with the SMOS Satellite: A New Means to Monitor the Marine
Branch of the Water Cycle. Reul et al. 2014

Image on p.173: TRO Emergency Rescue Team carries a woman to safety by trokilinochchi, 2008.
https://www.flickr.com/photos/29593080@N08/3065561529. Licensed under CC BY 2.0. https://creativecommons.org/licenses/by/2.0/

Image of monsoon floods in Sri Lanka on p.173 by trokilinochchi via Wikimedia Commons, licensed under CC BY 2.0.
https://creativecommons.org/licenses/by/2.0/

Image on p.175: Flood defences at Frankwell car park, Shrewsbury. https://www.geograph.org.uk/photo/310794.
© David Gruar and licensed for reuse under CC BY-SA 2.0. https://creativecommons.org/licenses/by-sa/2.0/

Precipitation map on p.176: NOAA Geophysical Fluid Dynamics Laboratory. Edited.

Map of global water scarcity on p.178: © 2014 World Resources Institute.
Licenced under the Creative Commons Attribution 3.0 License. http://creativecommons.org/licenses/by/3.0/

Image on p.183: Irrigation Canal taking off from the dam, Nvvchar, 2011. https://commons.wikimedia.org/wiki/File:Irrigation_Canal_taking_off_from_the_
dam.jpg. Licensed under CC BY-SA 3.0 via Wikimedia Commons. https://creativecommons.org/licenses/by-sa/3.0/deed.en

Image on p.182: Ethiopia0002, Oxfam East Africa, 2011. https://www.flickr.com/photos/oxfameastafrica/5933226731/
Licensed under CC BY 2.0. https://creativecommons.org/licenses/by/2.0/

Acknowledgements

Image on p.183: Irrigation Canal taking off from the dam, Nvvchar, 2011. https://commons.wikimedia.org/wiki/File:Irrigation_Canal_taking_off_from_the_dam.jpg. Licensed under CC BY-SA 3.0 via Wikimedia Commons. https://creativecommons.org/licenses/by-sa/3.0/deed.en

Image on p.184 of South-North Water Transfer Project Central route starting point by Nsbdgc. https://commons.wikimedia.org/wiki/File:South%E2%80%93North_Water_Transfer_Project_Central_route_starting_point_taocha.jpg. Licensed under CC BY-SA 4.0 via Wikimedia Commons. https://creativecommons.org/licenses/by-sa/4.0/deed.en

Image on p.185: NEWater Bottle NDP 2014, Hz. Tiang, 2014. https://commons.wikimedia.org/wiki/File:NEWater_Bottle_NDP_2014.jpeg. Licensed under CC BY-SA 4.0 via Wikimedia Commons. https://creativecommons.org/licenses/by-sa/4.0/deed.en.

Image on p.188: Great Falls of Tinkers Creek, Tim Evanson, 2018. https://www.flickr.com/photos/timevanson/36416393475/. Licensed under CC BY-SA 2.0. https://creativecommons.org/licenses/by-sa/2.0/.

Image on p.188: Limestone outcrops, Great Asby Scar © Richard Webb. geograph.org.uk/p/5871552. Licensed for reuse under CC BY-SA 2.0. https://creativecommons.org/licenses/by-sa/2.0/

Soil carbon content data on p.192: Annual Review of Ecology, Evolution, and Systematics. Robert B. Jackson, Kate Lajtha, Susan E. Crow, Gustaf Hugelius, Marc G. Kramer, and Gervasio Pineiro8, Volume 48:419-445 (C) 2017; permission conveyed through Copyright Clearance Center, Inc.

Energy mix data on p.195: IEA, 2022, Europe data explorer, All rights reserved.

Energy consumption map on p.195: © BP Statistical Review of World Energy 2016.

USA energy consumption data on p.196: IEA, 2022, United States data explorer, All rights reserved.

France energy consumption data on p.196: IEA, 2022, Europe data explorer, All rights reserved.

Fossil fuel consumption data on p.197: BP Statistical Review of World Energy.

Image on p.198 of Coal Train, Dumfries. © Stephen McKay. https://www.geograph.org.uk/photo/380947. Licensed for reuse under CC BY-SA 2.0. https://creativecommons.org/licenses/by-sa/2.0/

Image on p.198 of Panama Canal Gatun Locks, Stan Shebs, 2022. https://commons.wikimedia.org/wiki/File:Panama_Canal_Gatun_Locks.jpg. Licensed under CC BY-SA 3.0 via Wikimedia Commons. https://creativecommons.org/licenses/by-sa/3.0/deed.en

Map of the major existing and proposed russian natural gas transportation pipelines in Europe on p.199: Samuel Bailey, 2009. https://commons.wikimedia.org/wiki/File:Major_russian_gas_pipelines_to_europe.png. Licensed under CC BY 3.0 via Wikimedia Commons. https://creativecommons.org/licenses/by/3.0/deed.en.

Data on p.199 based on IEA data from IEA 2022 World Energy Balances Highlights, www.iea.org/statistics, All rights reserved; as modified by CGP Ltd, UK.

Image on p.200: Tar sands, Alberta. Howl Arts Collective, 2008. https://www.flickr.com/photos/67952496@N05/6544064931. Licensed under CC BY 2.0. https://creativecommons.org/licenses/by/2.0/

Image on p.200 of Marcellus Shale Gas Drilling Tower 1, Ruhrfisch, 2009. https://commons.wikimedia.org/wiki/File:Marcellus_Shale_Gas_Drilling_Tower_1_crop.jpg. Licensed under CC BY 3.0, via Wikimedia Commons. https://creativecommons.org/licenses/by-sa/3.0/deed.en

Fossil fuel data on p.202 based on IEA data from IEA 2022 World Energy Balances Highlights, www.iea.org/statistics, All rights reserved; as modified by CGP Ltd, UK.

Image on p.202: Image on p.202: Badging used to identify the Brazlilian Honda City flex-fuel. Mariordo, 2012. https://commons.wikimedia.org/wiki/File:Honda_City_flex_fuel_09_2012_BSB_4397.JPG. Licensed under CC BY 3.0 via Wikimedia Commons. https://creativecommons.org/licenses/by-sa/3.0/deed.en

Data on p.203: © Crown copyright. Licensed under Open Government Licence v3.0. https://www.nationalarchives.gov.uk/doc/open-government-licence/version/3/

Image on p.205: Keppel bleaching, Acropora, 2011. https://en.wikipedia.org/wiki/File:Keppelbleaching.jpg. Licensed under CC BY-3.0 via Wikimedia Commons. https://creativecommons.org/licenses/by/3.0/deed.en

Data on p.206: The World Bank: Forest area (% of land area) — Low Income, High Income, World. Source: Food and Agriculture Organization. Licensed under CC BY 4.0. https://creativecommons.org/licenses/by/4.0/

Figure SPM.8 (a) on p.210 from IPCC, 2021: Summary for Policymakers. In: Climate Change 2021: The Physical Science Basis. Contribution of Working Group I to the Sixth Assessment Report of the Intergovernmental Panel on Climate Change [Masson-Delmotte, V., P. Zhai, A. Pirani, S.L. Connors, C. Péan, S. Berger, N. Caud, Y. Chen, L. Goldfarb, M.I. Gomis, M. Huang, K. Leitzell, E. Lonnoy, J.B.R. Matthews, T.K. Maycock, T. Waterfield, O. Yelekçi, R. Yu, and B. Zhou (eds.)]. Cambridge University Press, Cambridge, United Kingdom and New York, NY, USA, pp. 3–32, doi:10.1017/9781009157896.001.

Data on p.211 from the SIPRI database, https://www.sipri.org/databases/milex

GDP information on p.215: World bank national accounts data and OECD National Accounts data files. Licensed under CC BY 4.0. https://creativecommons.org/licenses/by/4.0/.

Apple is a trademark of Apple Inc., registered in the U.S. and other countries

Data on p.223: Global Carbon Project (2019). Supplemental data of Global Carbon Budget 2019 (Version 1.0) [Data set]. Global Carbon Project. https://doi.org/10.18160/gcp-2019 Licensed under CC BY 4.0. https://creativecommons.org/licenses/by/4.0/

Data on p.227: OECD/EUIPO (2021), Global Trade in Fakes: A Worrying Threat, Illicit Trade, OECD Publishing, Paris.

Chinese FDI 2020 Data on p.228: CEIC Data

Image on p.229: Zambia, by BlueSalo, 2008. https://commons.wikimedia.org/wiki/File:Zambia_3.JPG. Licensed under CC BY-SA 3.0 via Wikimedia Commons <https://creativecommons.org/licenses/by-sa/3.0>

GDP information on p.229: World bank national accounts data and OECD National Accounts data files. Licensed under CC BY 4.0. https://creativecommons.org/licenses/by/4.0/.

Image on p.229: Hong Kong Protests, Studio Incendo, 2019. https://flickr.com/photos/29418416@N08/48618372858. Licensed under CC BY 2.0. https://creativecommons.org/licenses/by/2.0/

EU Population data on p.232: Population structure and ageing © European Union, 1995-2022. Licensed under CC BY 4.0 https://creativecommons.org/licenses/by/4.0/

Data on p.232 from the SIPRI database, https://www.sipri.org/databases/milex.

Page 232 contains Parliamentary information licensed under the Open Parliament Licence. https://www.parliament.uk/site-information/copyright-parliament/open-parliament-licence/

Acknowledgements

Data on p.233: The World Bank: Fertility rate, total (births per woman). Source: (1) United Nations Population Division. World Population Prospects: 2019 Revision. (2) Census reports and other statistical publications from national statistical offices, (3) Eurostat: Demographic Statistics, (4) United Nations Statistical Division. Population and Vital Statistics Reprot (various years), (5) U.S. Census Bureau: International Database, and (6) Secretariat of the Pacific Community: Statistics and Demography Programme. Licensed under the terms of the following licence: CC BY 4.0. https://creativecommons.org/licenses/by/4.0/.

Data on p.234: The World Bank: GDP (current US$) — United States, United Kingdom, China, Brazil, Nigeria. Source: World Bank national accounts data, and OECD National Accounts data files. Licensed under the terms of the following licence: CC BY 4.0. https://creativecommons.org/licenses/by/4.0/

Data on p.234: The World Bank: GDP (current US$) — World, Vanuatu. Source: World Bank national accounts data, and OECD National Accounts data files. Licensed under the terms of the following licence: CC BY 4.0. https://creativecommons.org/licenses/by/4.0/

Data on p.235: The World Bank: Poverty headcount ratio at national poverty lines (% of population) — Bolivia. Source: World Bank, Poverty and Inequality Platform. Data are compiled from official government sources or are computed by World Bank staff using national (i.e. country-specific) poverty lines. Licensed under the terms of the following licence: CC BY 4.0. https://creativecommons.org/licenses/by/4.0/.

Data on p.235: The World Bank: GDP (current US$) — Bolivia. Source: World Bank national accounts data, and OECD National Accounts data files. Licensed under the terms of the following licence: CC BY 4.0. https://creativecommons.org/licenses/by/4.0/.

Data on p.235: The World Bank: GDP (current US$) — United States, Costa Rica, Nepal. Source: World Bank national accounts data, and OECD National Accounts data files. Licensed under the terms of the following licence: CC BY-4.0. https://creativecommons.org/licenses/by/4.0/.

Human Development Index data on p.235: Human Development Insight, United Nations Development Programme, 2023. Licensed under CC BY 3.0 IGO. https://creativecommons.org/licenses/by/3.0/igo/

Maps on p.236: Free to use, gapminder.org. Licensed under CC BY 4.0. https://creativecommons.org/licenses/by/4.0/.

Data on p.238: The World Bank: Life expectancy at birth, total (years) — United Kingdom, Zambia. Source: (1) United Nations Population Division. World Population Prospects: 2022 Revision, or derived from male and female life expectancy at birth from sources such as: (2) Census reports and other statistical publications from national statistical offices, (3) Eurostat: Demographic Statistics, (4) United Nations Statistical Division. Population and Vital Statistics Reprot (various years), (5) U.S. Census Bureau: International Database, and (6) Secretariat of the Pacific Community: Statistics and Demography Programme. Licensed under the terms of the following licence: CC BY 4.0. https://creativecommons.org/licenses/by/4.0/.

Data on p. 238: The World Bank: Maternal mortality ratio (modeled estimate, per 100,000 live births) - United Kingdom, Zambia. Source: WHO, UNICEF, UNFPA, World Bank Group, and the United Nations Population Division. Trends in Maternal Mortality: 2000 to 2017. Geneva, World Health Organization, 2019. Licensed under the terms of the following licence: CC BY 4.0. https://creativecommons.org/licenses/by/4.0/.

Data on p.238: The World Bank: Maternal mortality ratio (modeled estimate, per 100,000 live births) — United Kingdom, Zambia. Source: WHO, UNICEF, UNFPA, World Bank Group, and the United Nations Population Division. Trends in Maternal Mortality: 2000 to 2017. Geneva, World Health Organization, 2019. Licensed under the erms of the following licence: CC BY 4.0. https://creativecommons.org/licenses/by/4.0/.

Breast cancer mortality rate map on p.238 reprinted from 'Estimated age-standardized mortality rates (World) in 2020, breast, females, all ages' © International Agency for Research on Cancer 2023

Disease rate of tuberculosis in the population map on p.239: Reproduced with permission from (World : Estimated TB incidence rates, 2017), Geneva, World Health Organization (WHO), (2018) (http://gamapserver.who.int/mapLibrary/Files/Maps/Global_TB_incidence_2017.png, accessed 10 February 2023). WHO does not endorse any specific companies, products or services.

Life expectancy data on p.239 licensed under the Open Government Licence v3.0. https://www.nationalarchives.gov.uk/doc/open-government-licence/version/3/.

Aboriginal life expectancy data on p.239: Australian Bureau of Statistics. Used under the terms of CC BY 4.0 https://creativecommons.org/licenses/by/4.0/

Aboriginal peoples depression data on p.239: Smiling Minds

Data on p.240: The World Bank: Government expenditure on education, total (% of GDP) — Turkmenistan, Sweden, United Kingdom. Source: UNESCO Institute for Statistics (UIS). UIS.Stat Bulk Data Download Service. Accessed October 24, 2022. apiportal.uis.unesco.org/bdds. Licensed under the terms of the following licence: CC BY 4.0. https://creativecommons.org/licenses/by/4.0/.

Millennium Development Goal data on p.242 from The Millennium Development Goals Report 2015 © 2015 United Nations. Used with the permission of the United Nations.

Millennium Development Goal table on p.242 from The Millennium Development Goals 2015 Progress Chart © 2015 United Nations. Used with the permission of the United Nations.

Age of democracy data on p.246: Bastian Herre and Max Roser, Our World in Data, 2013. Licensed under CC BY 4.0. https://creativecommons.org/licenses/by/4.0/

Corruption Risk Map on p.247: Global Risk Profile Sàrl, https://risk-indexes.com/corruption-map/. Used with permission.

Diabetes data on p.248: National Center for Health Statistics. National Vital Statistics System, Mortality

Infant mortality rate data on p.248: National Center for Health Statistics. National Vital Statistics System, Linked birth/infant death file

Graduation data on p.248: National Center for Education Statistics. (2022). Public High School Graduation Rates. Condition of Education. U.S. Department of Education, Institute of Education Sciences. Retrieved [02.02.23], from https://nces.ed.gov/programs/coe/indicator/coi.

Gender pay gap data on p.248: Office for National Statistics (ONS), released 26 October 2022, ONS website, statistical bulletin, Gender pay gap in the UK: 2022. Licensed under the Open Government Licence v3.0. https://www.nationalarchives.gov.uk/doc/open-government-licence/version/3/

Data on p.249: © Commonwealth of Australia, Department of the Prime Minister and Cabinet, Closing the Gap Report 2020. Licensed under CC BY 4.0. https://creativecommons.org/licenses/by/4.0/

Data on p.251: Vine, D. et al. (2021, August 19). Creating Refugees: Displacement Caused by the United States' Post-9/11 Wars. Costs of War, Watson Institute, Brown University.

Oxfam financial data on p.252: Oxfam GB Annual Report 2020/21 is adapted by the publisher with the permission of Oxfam, Oxfam House, John Smith Drive, Cowley, Oxford OX4 2JY UK www.oxfam.org.uk. Oxfam does not necessarily endorse any text or activities that accompany the materials, nor has it approved the adapted text.

Gender Inequality Index Data on p.253: Human Development Reports, 2023. Licensed under CC BY 3.0 IGO. https://creativecommons.org/licenses/by/3.0/igo/

Acknowledgements

Data on p.253: The World Bank: Proportion of seats held by women in national parliaments (%). Source: Inter-Parliamentary Union (IPU) (ipu.org). For the year of 1998, the data is as of August 10, 1998.
Licensed under the terms of the following licence: CC BY 4.0. https://creativecommons.org/licenses/by/4.0/.

Arms exporters data on p.255: SIPRI database, 2020

Data on p.258: The World Bank: GDP (current US$) — Vietnam. Source: World Bank national accounts data, and OECD National Accounts data files. Licensed under the terms of the following licence: https://creativecommons.org/licenses/by/4.0/.

Vietnam poverty data on p.258: The World Bank: Poverty headcount ratio at $2.15 a day (2017 PPP) (% of population) — Vietnam. Source: World Bank, Poverty and Inequality Platform. Data are based on primary household survey data obtained from government statistical agencies and World Bank country departments. Data for high-income economies are mostly from the Luxembourg Income Study database. For more information and methodology, please see pip.worldbank.org. Licensed under CC BY-4.0. https://creativecommons.org/licenses/by/4.0/

Data on p.258: The World Bank: Life expectancy at birth, total (years) — Vietnam. Source: (1) United Nations Population Division. World Population Prospects: 2022 Revision, or derived from male and female life expectancy at birth from sources such as: (2) Census reports and other statistical publications from national statistical offices, (3) Eurostat: Demographic Statistics, (4) United Nations Statistical Division. Population and Vital Statistics Reprot (various years), (5) U.S. Census Bureau: International Database, and (6) Secretariat of the Pacific Community: Statistics and Demography Programme. Licensed under the terms of the following licence: https://creativecommons.org/licenses/by/4.0/.

Image on p.259: President Clinton and Secretary Clinton Pose for a Photo With Workers at Caracol Industrial Park. Copyright Kendra Helmer/USAID

Data on p.259: OECD/DAC 2023 (OECD DAC Aid at a glance — Zambia), https://public.tableau.com/views/OECDDACAidataglancebyrecipient_new/ Recipients?:embed=y&:display_count=yes&:showTabs=y&:toolbar=no?&:showVizHome=no

Data on p.259: OECD/DAC 2023 (OECD DAC Aid at a glance — Mali), https://public.tableau.com/views/OECDDACAidataglancebyrecipient_new/ Recipients?:embed=y&:display_count=yes&:showTabs=y&:toolbar=no?&:showVizHome=no

Gini index data on p.259: The World Bank: Gini Index — Mali. Source: World Bank, Poverty and Inequality Platform. Data are based on primary household survey data obtained from government statistical agencies and World Bank country departments. Data for high-income economies are mostly from the Luxembourg Income Study database. For more information and methodology, please see pip.worldbank.org. Licensed under CC BY-4.0. https://creativecommons.org/licenses/by/4.0/

Data on p.260: Vine, D. et al. (2021, August 19). Creating Refugees: Displacement Caused by the United States' Post-9/11 Wars. Costs of War, Watson Institute, Brown University

Map on p.262 showing migration between provinces in China http://faculty.washington.edu/kwchan/, used by permission

Data on p.262: National Bureau of Statistics of China

Migration data on p.263: "World Migration Report 2022" IOM Publications.

Data on p.263: © United Nations 2022 Department of Economic and Social Affairs (UN DESA)

Foreign-born population data on p.263: OECD (2023), Foreign-born population (indicator). doi: 10.1787/5a368e1b-en (Accessed on 02 February 2023)

Data on p.263: The World Factbook: Washington DC, Central Intelligence Agency, 2017

2021 Refugee and migrant data on p.264: UNHCR Data Portal licensed under a Creative Commons Attribution 3.0 International License. https://creativecommons.org/licenses/by/3.0/igo/

Image on p.264: Refugees on a boat crossing the Mediterranean sea, heading from Turkish coast to the northeastern Greek Island of Lesbos. Mstyslav Chernov/Unframe. https://upload.wikimedia.org/wikipedia/commons/c/c9/Refugees_on_a_boat_crossing_the_ Mediterranean_sea%2C_heading_from_Turkish_coast_to_the_northeastern_Greek_island_of_Lesbos%2C_29_January_2016.jpg. Licensed under CC BY-SA 4.0 via Wikimedia Commons. <https://creativecommons.org/licenses/by-sa/4.0>, via Wikimedia Commons

Map on p.265 adapted from ONS data www.ons.gov.uk, licensed under the Open Government Licence v3.0 & Ordnance Survey Data © Crown Copyright and database rights

Data on p.266, Henley Passport Index, 19 July 2022

Data on p.266: "Origins and destinations of European Union migrants within the EU". Pew Research Center, Washington, D.C. (19 June, 2017) https://www.pewresearch.org/global/interactives/origins-destinations-of-european-union-migrants-within-the-eu/

Data on p.266: "Europeans Credit EU With Promoting Peace and Prosperity, but Say Brussels Is Out of Touch With Its Citizens". Pew Research Center, Washington, D.C. (19 March, 2019) https://www.pewresearch.org/global/2019/03/19/ europeans-credit-eu-with-promoting-peace-and-prosperity-but-say-brussels-is-out-of-touch-with-its-citizens/

Data on p.267: "Key findings about U.S. immigrants" Pew Research Center, Washington D.C. (August 20, 2020). https://www.pewresearch.org/fact-tank/2020/08/20/key-findings-about-u-s-immigrants/

Data on p.267: Mexicans decline to less than half the U.S. unauthorized immigrant population for the first time". Pew Research Center, Washington, D.C. (12 June, 2019). https://www.pewresearch.org/fact-tank/2019/06/12/us-unauthorized-immigrant-population-2017/

Data on p.267: US Census Bureau

Page 270 contains Parliamentary information licensed under the Open Parliament Licence. https://www.parliament.uk/site-information/copyright-parliament/open-parliament-licence/.

South Sudan Human Development Index information on p.271: Human Development Index © United Nations Development Programme. Licensed under CC BY 3.0 IGO. https://creativecommons.org/licenses/by/3.0/igo/

Page 272 contains public sector information licensed under thee Open Government Licence v3.0. https://www.nationalarchives.gov.uk/doc/open-government-licence/version/3/

Financial Secrecy Index data on p.273: © Tax Justice Network

Gini Index data on p.274: World Bank: Gini Index — Ecuador. Source: World Bank, Poverty and Inequality Platform. Data are based on primary household survey data obtained from government statistical agencies and World Bank country departments. Data for high-income economies are mostly from the Luxembourg Income Study database. For more information and methodology, please see pip.worldbank.org. Licensed under CC BY-4.0. https:// creativecommons.org/licenses/by/4.0/

Monusco data on p.276: https://monusco.unmissions.org/en/background accessed Feb 2023 © 2023 United Nations.

Monusco data on p.276: https://peacekeeping.un.org/en/mission/monusco accessed Feb 2023 © 2023 United Nations.

Acknowledgements

Data on p.277: The WTO at 25. Assessing the economic value of the rules based global trading system

UK Loan data on p.277 © Crown Copyright licensed for re-use under the Open Government licence v3.0.
https://www.nationalarchives.gov.uk/doc/open-government-licence/version/3/

Data on p.277: International Monetary Fund. IMF Members' Quotas and Voting Power, and IMF Board of Governors

Registered nurses data on p.278: International Development Research Centre (IDRC).

Jamaica Primary school data on p.278: © United Nations Environment Programme

Montreal Protocol data on p.279: https://www.unenvironment.org/ozonaction/who-we-are/about-montreal-protocol accessed Feb 2023.
© United Nations Environment Programme

Image of Antarctica territorial claims on p.281 by Lokal_Profil. https://commons.wikimedia.org/wiki/File:Antarctica,_territorial_claims.svg.
Licensed under CC BY-SA 2.5 via Wikimedia Commons. https://creativecommons.org/licenses/by-sa/2.5/.

Page 282 contains ONS data www.ons.gov.uk licensed under the Open Government Licence v3.0
http://www.nationalarchives.gov.uk/doc/open-government-licence/version/3/

Image of Main Entrance into The Land Rover Works, Solihull, 2005 on p.284: https://www.geograph.org.uk/photo/85096.
© Peter lloyd and licensed for reuse under CC BY-SA 2.0. https://creativecommons.org/licenses/by-sa/2.0/

Offshore companies data on p.284: Transparancy International. Licensed for reuse under CC BY-ND 4.0.
https://creativecommons.org/licenses/by-nd/4.0/

Gini index data on p.287: The World Bank: Gini Index - Somalia. Source: World Bank, Poverty and Inequality Platform. Data are based on primary
household survey data obtained from government statistical agencies and World Bank country departments. Data for high-income economies are mostly
from the Luxembourg Income Study database. For more information and methodology, please see pip.worldbank.org. Licesned under CC BY 4.0. https://
creativecommons.org/licenses/by/4.0/.

Somalia FDI data on p.287: UNCTAD - World Investment Report 2021 © United Nations 2021

Map on p.292 uses data from the Department for Environment Food & Rural Affairs, © Crown copyright licensed under the Open Government Licence
v3.0. http://www.nationalarchives.gov.uk/doc/open-government-licence/version/3/

Denmark energy consumption data on p.293: Danish Energy Agency

Page 294 contains OS data © Crown copyright and database right 2017

Data on p.295 from World Population Prospects: The 2008 Revision Population Database, published by UN DESA © 2009 United Nations.
Used with the permission of the United Nations.

Page 297: © Crown copyright and database rights 2022 OS (100034841)

Figure 1 and Figure 2 on p.303: Marlene Kronenberg et al, Mass-balance reconstruction for Glacier No.354, Tien Shan, from 2003 to 2014, Annals of
Glaciology, Volume 57, Issue 71, pp92-102, reproduced with permission

Figures 3 and 4 on p.303: National Snow and Ice Data Center/T. Mote, University of Georgia

Figure 5 on p.303: National Snow and Ice Data Center/T. Mote, University of Georgia

Image on p.304: Rock Fall at Burton Cliffs © John Stephen https://www.geograph.org.uk/photo/3817946
Licensed under CC BY-SA 2.0. https://creativecommons.org/licenses/by-sa/2.0/.

Page 306: Statistics based on the Access to Public Open Space dataset created by Greenspace Information for Greater London, CIC (GiGL, 2013), using
Ordnance Survey data (2013, 2014), published on London DataStore (2015) and made available under UK Open Government Licence.
https://www.gigl.org.uk/our-data-holdings/keyfigures/

Image on p.306: The Palace of Westminster © Philip Halling https://www.geograph.org.uk/browser/#!/q=westminster+palace/page=13/image=5483649
Licensed under CC BY-SA 2.0. https://creativecommons.org/licenses/by-sa/2.0/.

Pages 306 & 307 contain ONS data www.ons.gov.uk licensed under the Open Government Licence v3.0
http://www.nationalarchives.gov.uk/doc/open-government-licence/version/3/

Page 307: © gal-dem

Figure 1 on p.308: Dr. Pieter Tans, NOAA/ESRL (www.esrl.noaa.gov/gmd/ccgg/trends/)

Figure 1 on p.308: Wiese, D. N., D.-N. Yuan, C. Boening, F. W. Landerer, and M. M. Watkins (2016) JPL GRACE
Mascon Ocean, Ice, and Hydrology Equivalent Water Height RL05M.1 CRI Filtered Version 2., Ver. 2., PO.DAAC,
CA, USA. Dataset accessed [2023-02-16] at http://dx.doi.org/10.5067/TEMSC-2LCR5.

Denmark windfarm graph on p.309 by Jesper Berggreen. Data source: Danish Energy Agency

Chinese FDI data on p.310: © Rhodium Group, used with permission.

Gini Index Data 2019 on p.311: Gini Index World Bank, Poverty and Inequality Platform. Data are based on primary household survey data obtained from
government statistical agencies and World Bank country departments. Data for high-income economies are mostly from the Luxembourg Income Study
database. For more information and methodology, please see pip.worldbank.org.Licensed under CC-BY 4.0 https://creativecommons.org/licenses/by/4.0/

Gender Inequality Index Data 2021 on p.311: Gender Inequality Index (GII), United Nations Development Programme.
Licensed under CC BY 3.0 IGO. https://creativecommons.org/licenses/by/3.0/igo/

GDP per capita Data on p.312: World Bank national accounts data, and OECD National Accounts data files..
Licensed under CC-BY 4.0 https://creativecommons.org/licenses/by/4.0/

Net migration rate 2019 data on p.312: United Nations, Department of Economic and Social Affairs, Population Division (2022). World Population
Prospects 2022, Online Edition. Licensed under CC BY 3.0 IGO https://creativecommons.org/licenses/by/3.0/igo/.

Page 313: © UNESCO

Mean CO2 concentrations data on p.314: Lan, X., Tans, P. and K.W. Thoning: Trends in globally-averaged CO2 determined from NOAA Global Monitoring
Laboratory measurements. Version 2023-02 NOAA/GML (gml.noaa.gov/ccgg/trends/)

Index

Index

Index

Index

Index